CALIBRATION OF FUNDAMENTAL STELLAR QUANTITIES

INTERNATIONAL ASTRONOMICAL UNION
UNION ASTRONOMIQUE INTERNATIONALE

CALIBRATION OF FUNDAMENTAL STELLAR QUANTITIES

PROCEEDINGS OF THE 111th SYMPOSIUM OF THE
INTERNATIONAL ASTRONOMICAL UNION
HELD AT VILLA OLMO, COMO, ITALY,
MAY 24–29, 1984

EDITED BY

D. S. HAYES

*Kitt Peak National Observatory, National Optical Astronomy Observatories,
Tucson, Arizona, U.S.A.*

L. E. PASINETTI

Department of Physics, University of Milan, Italy

and

A. G. DAVIS PHILIP

*Van Vleck Observatory, Wesleyan University, Middletown, Connecticut,
and Union College, Schenectady, New York, U.S.A.*

SPRINGER-SCIENCE+BUSINESS MEDIA, B.V.

Library of Congress Cataloging in Publication Data

International Astronomical Union. Symposium (111th: 1984: Como, Italy)
    Calibration of fundamental stellar quantities.

    Includes index.
    1.   Astrometry–Congresses.   2.   Astronomical spectroscopy–
Congresses.   3.   Photometry, Astronomical–Congresses.   I.   Hayes, D. S.
(Donald S.)   II.   Pasinetti, L. E. (Laura E.)   III.   Philip, A. G. Davis.
IV.   Title.   V.   Title: Stellar quantities: proceedings of the 111th Symposium
of the International Astronomical Union, held at Villa Olmo, Como, Italy,
May 24–29, 1984.
QB807.I57   1984   522   85–14501

ISBN 978-90-277-2110-5   ISBN 978-94-009-5456-4 (eBook)
DOI 10.1007/978-94-009-5456-4

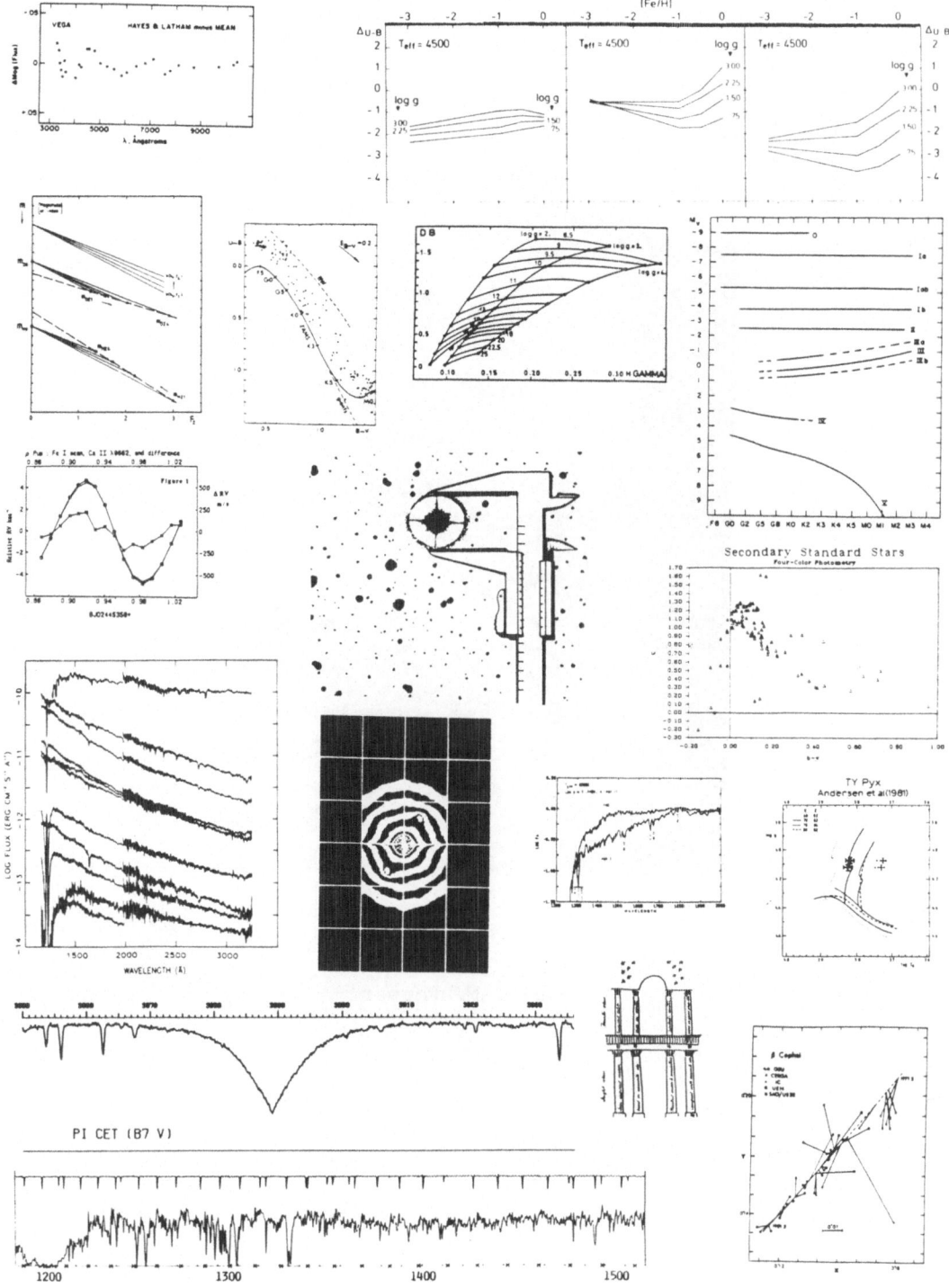

PI CET (B7 V)

WAVELENGTH (ANGSTROMS)

Daniel M. Popper

DEDICATION

Symposium No. 111 of the International Astronomical Union

and its proceedings recorded in this volume

are dedicated to

Daniel M. Popper

in grateful recognition of his contributions to

the Calibration of Fundamental Stellar Quantities

TABLE OF CONTENTS

2. CONTRIBUTED PAPERS

REMARKS BY THE CHAIRMAN OF THE SCIENTIFIC ORGANIZING COMMITTEE

The origin of IAU Symposium No. 111 was a conversation with L. W. Fredrick during an IAU Colloquium in Flagstaff, Arizona. Several speakers at that meeting had talked independently of their difficulties in calibration, and it seemed to us that a meeting organized to discuss such matters would bring together workers from many diverse fields to discuss problems fundamental to all our astronomical knowledge. The idea was ambitious and received much support, but many people also expressed reservations. The IAU Executive was sympathetic, but urged us to focus our ideas more clearly and to narrow the scope of the proposed meeting. Since most support for the original proposal came from stellar astronomers, the limitation to the "Calibration of Fundamental Stellar Quantities" came naturally. Commission 29 sponsored the new proposal and Commissions 5, 24, 25, 26, 30, 35, 37, 42 and 45 were co-sponsors. It is now clear that we could not have done justice to the original theme in one meeting: we hope others will attempt similar appraisals in other fields of astronomy.

An early decision of the Scientific Organizing Committee was to allow the full presentation of only the invited papers; contributed papers were all to be poster papers. This is reflected in the arrangement of the proceedings, in which the contributed papers are presented in somewhat abbreviated form, prominance being given to the invited review papers. In this way, we hoped to encourage full discussion of the invited papers. We also hoped to produce a volume that would serve, to some extent, as a handbook in its field. Invited speakers were asked to bear this aim in mind. It was fascinating, at the meeting, to see how each of them approached their task. How successful we have been in producing a handbook is for readers to judge.

The new proposal was submitted to the IAU Executive at about the time of Prof. Popper's seventieth birthday. We wished to recognize his many contributions to our knowledge of such fundamental stellar quantities as mass, radius and luminosity, and we are happy that he accepted our dedication and came to take an active part in the Symposium.

I would like to extend my thanks to all members of the Scientific and Local Organizing Committees for making all the necessary arrangements for the meeting. We are, of course, grateful to the IAU for its sponsorship and financial support of the Symposium. We also acknowledge with thanks the generosity of the following co-sponsors: Dipartimento di Fisica of the Università di Milano, the Ministero della Pubblica Istruzione, the Consiglio Nationale delle Richerche, the Assessorato alla Cultura e Informazione of the Regione Lombardia, the Centro di Cultura Scientifica "A. Volta" of Como, and the Banca Popolare di Milano. Particular thanks, amongst these, are due to the Director of the Dipartimento di Fisica, Prof. M. Pignanelli, the

Consiglio Nazionale delle Richerche, and the Assessorato alla Cultura e
Informazione. Finally, we would like to thank Dr. M. Bossi of the
Osservatorio Astronomico di Merate for his help to the LOC during the
Symposium, and the staff of the Villa Olmo.

                              A. H. Batten

PREFACE

    IAU Symposium No. 111, "Calibration of Fundamental Stellar Quanti-
ties", was held at Villa Olmo, Como, Italy, on May 24-29, 1984. Meetings held in the past ten years on related topics include: IAU Symposium
No. 109, "Astrometric Techniques", held at the University of Florida in
Jan., 1984, "The MK Process and Stellar Classification", held at the
University of Toronto in June, 1983, "Stellar Absolute Energy Distributions", an unpublished Joint Meeting (Commissions 25 and 45), held at
the General Assembly of the IAU in Patras, Greece in August, 1982, IAU
Colloquium No. 62, "Current Techniques in Double and Multiple Star Research", held at Northern Arizona University in May, 1981, the ESO Workshop: "Methods of Abundance Determination for Stars", held in Geneva in
March, 1980, "Problems of Calibration of Multicolor Photometric Systems",
held at Dudley Observatory in March, 1979, IAU Colloquium No. 48,
"Modern Astrometry", held at the University of Vienna in Sept., 1978,
IAU Colloquium No. 50, "High Angular Resolution Stellar Interferometry"
held at the University of Maryland in Aug., 1978, "Spectral Classification of the Future", held at the Vatican in July, 1978 and IAU Symposium No. 72, "Abundance Effects in Classification", held at the University of Lausanne in July, 1975. The present meeting was the first to
cover the broad range of the calibration of fundamental stellar quantities in one meeting.

    Nine commissions of the IAU co-sponsored the meeting. These were:
Commission 5 (Documentation and Astronomical Data), 24 (Photographic
Astrometry), 25 (Stellar Photometry and Polarimetry), 26 (Double and
Multiple Stars), 29 (Stellar Spectra), 30 (Radial Velocities), 35
(Stellar Constitution), 37 (Star Clusters and Associations), 42 (Close
Binary Stars) and 45 (Stellar Classification). The SOC consisted of A.
H. Batten (Chairman), M. Gerbaldi, B. Gustafsson, B. Hauck, D. S. Hayes,
C. Jaschek, G. Lyngå, H. A. McAlister, L. E. Pasinetti, A. G. D. Philip
and V. Straižys. The LOC consisted of L. E. Pasinetti (Chairman), E.
Antonello, G. Casati, M. Fracassini, L. Mantegazza and L. Pastori.

    Ninety-six astronomers from 19 countries attended the Symposium.
The chairmen of the sessions were:

I	D. M. Popper
II	M. G. Fracastoro
III	D. S. Evans
IV	J. Rountree
V	W. Gliese
VI	P. Couteau
VII	M. Gerbaldi
VIII	B. Nordström
IX	R. Hanbury Brown
X	M. Jaschek

            XI   K. Aa. Strand
            XII   I. N. Glushneva
            XIII   A. G. Davis Philip
            XIV   L. E. Pasinetti
            XV   R. F. Griffin
            XVI   G. A. H. Walker
            XVII   C. Jaschek
            XVIII   A. H. Batten

    In the four and one-half days of the Symposium 85 papers were given,
20 of which were spoken papers and 65 of which were poster papers.   The
proceedings of the Symposium have been edited from camera-ready copy
submitted by the authors.   The discussion was prepared from hand-written
sheets filled out by the participants in the discussion.   Those manuscripts which diverged too much from the correct style or from good
English were retyped.   We have made a great effort to increase the uni-
formity in format and language of the manuscripts.   Individual authors
have been made responsible for obtaining permission to copy material
from other publications.

    The host institution was the Università degli Studi di Milano, and
the Centro di Cultura Scientifica A. Volta hosted the meeting at Villa
Olmo, Como.   Financial support was received from:

            International Astronomical Union
            Department of Physics, University of Milan
            Italian Ministry of Education
            National Research Council (CNR, Italy)
            Regione Lombardia (Dept. of Culture and Information)
            Banca Popolare of Milan
            L. Davis Press

    The University of Milan published the abstract booklet.   The
Amministrazione Provinciale of Como and "Autunno Musicale" of Como
offered enjoyable concerts.   Chief Secretary Chiara de Santis of Villa
Olmo and her staff Manuela, Donatella and Barbara helped the LOC with
efficiency and enthusiasm.

    The editors wish to thank the members of the LOC, Dr. M. Bossi and
the staff of the Villa Olmo for handling the discussion sheets, and
especially Mary Bongiovanni, who typed up the discussion sheets and
passed them back to the participants to be checked.   The editors also
wish to thank the Director of Kitt Peak National Observatory for support
of the editorial process, and Monique Chapman, Pat Cochrane, Peggy
Stephens and Peggy Wiggins of the KPNO staff for retyping many of the
manuscripts.   Without this support the preparation of the proceedings
would have been much more difficult.

                                            D. S. Hayes
                                            L. E. Pasinetti
June, 1984   A. G. Davis Philip
Schenectady, New York

LIST OF PARTICIPANTS

Australia

M. S. Bessell   Mt. Stromlo and Siding Spring Observatory
J. Davis   Chatterton Astr. Dept., Univ. of Sydney
R. Hanbury Brown   Chatterton Astr. Dept., Univ. of Sydney

Austria

K. D. Rakos   Institute for Astronomy Vienna

Belgium

P. Magain   Institute for Astrophys., Univ. of Liège
J. Manfroid   Dept. of Astrophys., Univ. of Liège

Canada

A. H. Batten   Dominion Astrophysical Observatory
R. F. Garrison   David Dunlap Obs., Univ. of Toronto
C. G. Millward   Geophys. and Astr. Dept., U. of British Columbia
C. D. Scarfe   Dept. of Physics, Univ. of Victoria
G. A. H. Walker   Geophys. and Astr. Dept., U. of British Columbia

Czechoslovakia

D. Chochol   Astr. Inst. of the Slovak Academy of Sciences

Denmark

B. Nordstrüm   Copenhagen University Observatory, Brorfelde

France

G. Cayrel de Stobel   Paris-Meudon Observatory
P. Couteau   Nice Observatory
M. Froeschlé   C. E. R. G. A.
M. Gerbaldi   Inst. for Astrophys., Paris; Univ. of Paris
A. Heck   Strasbourg Observatory
G. Helmer   Nice Observatory
C. Jaschek   Stellar Data Center, Strasbourg
M. Jaschek   Strasbourg Observatory
M. O. Mennessier   Montpellier University
C. Meyer   C. E. R. G. A.
E. Oblak   Besancon Observatory
R. Papoular   Dept. of Astrophysics, CEN, Saclay
R. Peytureaux   Institute for Astrophysics, Paris
W. Tobin   Lab. for Space Astronomy, CNRS, Marseille

Germany, F. R.

H. Gass	Institute for Theoretical Astrophys., Heidelberg
W. Gliese	Astronomisches Rechen-Institut, Heidelberg
H. Jahreiss	Astronomisches Rechen-Institut, Heidelberg
D. Labs	Heidelberg Observatory
E. E. Lamla	Observatory, University of Bonn
H. Neckel	Hamburg Observatory
W. Seggewiss	Observatorium Hoher List, Bonn University

India

M. Parthasarathy	Indian Institute of Astrophysics, Bangalore

Italy

E. Antonello	Astronomical Observatory of Brera, Milano-Merate
G. Bertelli	Institute of Astronomy, Univ. of Padua
L. Bianchi	Astronomical Observatory of Turin
M. Bossi	Astronomical Observatory of Brera, Milano-Merate
A. Bressan	International School for Adv. Studies, Trieste
P. Broglia	Astronomical Observatory of Brera, Milano-Merate
P. Conconi	Astronomical Observatory of Brera, Milano-Merate
R. Faraggianna	Astronomical Observatory of Trieste
M. Fracassini	Dept. of Physics, University of Milan
M. G. Fracastoro	Astronomical Observatory of Turin
L. Galgani	Dept. of Mathematics, Univ. of Milan
G. Giuricin	Astronomical Observatory of Trieste
G. A. Guerrero	Astronomical Observatory of Brera, Milano-Merate
L. Mantegazza	Astronomical Observatory of Brera, Milano-Merate
M. Missana	Astronomical Observatory of Brera, Milano-Merate
C. Morossi	Astronomical Observatory of Trieste
E. Nasi	Astr. Obs. of Padua; Inst. of Astr., Bologna
L. E. Pasinetti	Dept. of Physics, University of Milan
L. Pastori	Astronomical Observatory of Brera, Milano-Merate
C. Rossi	Institute of Astronomy, Univ. of Rome
L. Rossi	Inst. for Space Astrophys., CNR, Frascati

The Netherlands

U. O. Frisk	Space Science Dept., ESA, Noordwijk
J. R. W. Heintze	"Sonnenborgh" Observatory, Utrecht
R. D. Wills	Space Science Dept., ESA, Noordwijk

Poland

J. Krełowski	Inst. of Astron., N. Copernicus Univ., Torun

Sweden

A. Ardeberg	Lund Observatory and ESO

B. Gustafsson   Stockholm Observatory
G. Lyngå   Lund Observatory
S. Wrandemark   Lund Observatory

Switzerland

R. Buser   Astronomical Institute, Univ. of Basel
J. Chmielewski   Geneva Observatory
B. Hauck   Institute of Astronomy, Univ. of Lausanne
L. Labhardt   Astronomical Institute, Univ. of Basel
G. Mathys   Institute of Astronomy, Zurich
F. Rufener   Geneva Observatory

Turkey

D. Kocer   Istanbul University Observatory

United Kingdom

Griffin, R. E. M.   Cambridge Observatories
Griffin, R. F.   Cambridge Observatories

United States

S. J. Adelman   Dept. of Physics, The Citadel, Charleston
R. C. Bohlin   Space Telescope Science Institute
A. D. Code   Washburn Obs., Univ. of Wisconsin
T. E. Corbin   U. S. Naval Obs., Washington, D. C.
R. Culver   Dept. of Physics, Colorado State Univ.
D. S. Evans   Dept. of Astronomy, Univ. of Texas
D. S. Hayes   Kitt Peak Nat. Obs., Nat. Optical Astr. Obs.
W. D. Heintz   Dept. of Astronomy, Swarthmore College
K. Janes   Dept. of Astronomy, Boston University
P. C. Keenan   Perkins Obs., Ohio State and Ohio Wesleyan Univ.
C. H. Lacy   Dept. of Physics, Univ. of Arkansas
H. A. McAlister   Dept. of Physics and Astr., Georgia State Univ.
G. J. Peters   Dept. of Astr., Univ. of Southern California
A. G. D. Philip   Van Vleck Obs. and Union College
R. S. Polidan   Lunar and Planetary Lab., Univ. of Arizona
D. M. Popper   Dept. of Astr., Univ. of Calif., Los Angeles
J. A. Rose   Institute of Astronomy, Univ. of Hawaii
J. Rountree   University of Denver
A. Slettebak   Perkins Obs., Ohio State and Ohio Wesleyan Univ.
K. Aa. Strand   Washington, D. C.
A. R. Upgren   Van Vleck Observatory

U. S. S. R.

I. N. Glushneva   Sternberg State Astr. Inst., Moscow

Vatican

R. Boyle   Vatican Observatory

LIST OF PICTURES

REVIEW PAPERS

THE CHOICE OF STANDARD STARS

Alan H. Batten

Dominion Astrophysical Observatory
Herzberg Institute of Astrophysics

ABSTRACT. Standard stars should normally be constant in the characteristic for which they have been chosen to be standards. Individually
they should be capable of testing the instrument used in their measurement, and collectively (usually together with a particular type of
instrument) they must define a system of measurement. A star should not
be adopted as a standard until several years' observation have demonstrated its constancy. The number of independent observations in that
interval is as important as its length. Many of these points are
illustrated by particular reference to standard-velocity stars. Even
100 observations of such a star may fail to reveal a detectable
variation that is present. A distinction is drawn between "primary"
standards, meeting the above criteria, and "reference" or "comparison"
stars which need not be so severely tested since their use is more
limited. Standard stars must be calibrated in some fundamental way.
Since random errors of observation can introduce systematic errors in
calibration, the use of the intermediate step of standard stars may
decline with increasing internal precision of observations.

1. INTRODUCTION

The theme of this symposium is the calibration of astronomical
measurements, especially of those that lead to knowledge of the fundamental stellar quantities of mass, radius, luminosity, temperature, etc.
We wish to take stock of the uncertainties, both accidental and
systematic, that may yet remain in our knowledge of these quantities;
not only because stars themselves are interesting, but because our knowledge of the whole universe depends on how accurately we know these
fundamental things about them. In one respect, even Copernicus could
not displace the Earth from the center of our Universe. Although we
have begun to observe from space, most of us are still bound to the
Earth and, both historically and logically, have had to explore the
Universe outwards from our home. There is a chain of inference
connecting our first measurements of the size and shape of the Earth,
through the determination of the dimensions of the Solar System and the

3

D. S. Hayes et al. (eds.), Calibration of Fundamental Stellar Quantities, 3–16.
© 1985 by the IAU.

parallaxes of stars, to our modern claim to know, at least approximately, the size and "age" of the Universe. Many of us would like to examine this whole chain, but we have been persuaded that that would be impracticable in one symposium. Our attention, therefore, is concentrated on a few links in the middle of the chain, and you are here because you have devoted much of your astronomical lives to forging those links. We hope, in the next few days, to discharge our responsibility to colleagues working on other links, by inspecting our own very carefully. In our necessarily close and even detailed inspection, however, we should remain aware of the whole chain of inference, and remember our colleagues who also advance our understanding of the Universe, whether their principal interest is closer at home or farther away than ours.

When we measure something about a star – velocity, luminosity, spectral type etc. – we often find it necessary or convenient to do so by comparison with one or more standard stars. Radial velocities are, indeed, measured by comparing the position of features in a stellar spectrum with those in a comparison spectrum that has been measured in the laboratory; but the rest wavelengths appropriate to the stellar features have been determined from the spectra of standard stars whose velocities have been determined (we hope reliably) by our predecessors. In classifying stellar spectra, we rely heavily on standard classifications, made by experts, to estimate – usually fairly easily and quickly – the type of our unknown spectrum. Photometers are usually calibrated by the observation of a set of standard stars rather than the measurement of the output of a standard lamp. The first kind of measurement is an example of a mixture of calibration by laboratory and stellar standards. The second is one of calibration only by standard stars, and it is hard to see how else one could proceed. Unless astronomers one day decide that spectral type is not a useful concept, and refer only to temperatures, surface gravities and abundances, classification is bound to be done by comparison with standards. In the third kind of measurement, there is indeed a clear choice of procedure, but the use of standard stars is overwhelmingly convenient.

2. WHAT IS A STANDARD STAR?

The last paragraph makes clear that standards are used and must be chosen for many different purposes. Because of this, we probably all use the term "standard" somewhat loosely at times. Not every star that is used as a reference or comparison is necessarily a standard. We do not always compare an unclassified spectrum directly with one of the standards defining the MK system, choosing instead spectrograms that we may have available of stars that have been reliably classified by comparison with the standards. The light variation of a variable star is usually measured with respect to a comparison star chosen – in order to minimize differential effects – to be close in the sky to the variable and, whenever possible, of similar magnitude and color. The star chosen is most unlikely to be one of the standards defining the photometric system in which the observer is working. The term "standard star" must be defined in such a way that we avoid confusion between genuine standards and mere reference or comparison stars. I propose that standards should meet the following requirements:

(i)   each individual star should be constant in the characteristic for which it has been chosen as a standard, within the smallest attainable errors of measurement.

(ii) together, the standards for a given characteristic, and the instruments used for their measurement, should define a system.

(iii) individually, the standards should be suitable for testing the performance of the instrument used for their measurement.

The first requirement seems almost too obvious to mention, but it should be the most important in the choice of a standard. Comparison stars for photometry are not always chosen with constancy of light as the sole – or even dominant – consideration. Thus it fairly often happens that a comparison star is itself found to be variable. If the variation is found to be periodic, we probably can salvage the observations. Sometimes we may have no choice but to refer observations of a star under study to a variable of known type. We should, nevertheless, hesitate to call a known spectroscopic binary a velocity standard, or a Cepheid variable a photometric or spectral-classification standard. If all members of a class of stars vary, it may be necessary to adopt some of them as reference stars, but we should not call them standards. Such a situation arises, for example, with M-type giants – all of which vary in light – some of which may be needed as reference stars in a photometric system (McClure 1976).

The second requirement emphasizes that usually we use a set of standards, not just a star. The number of stars in the set will depend, to some extent, on the characteristic being measured and the question will be discussed later. The whole set must be observed in a consistent fashion. Spectral-type standards must be observed at the same dispersion (even with the same spectrograph), with similar emulsions processed in a standard way (Morgan, Keenan and Kellman 1943; Keenan and McNeil 1976.) The UBV system is defined not only by the standard stars, but by the filter, photomultiplier and reflecting surfaces used and even the altitude of the observatory (Johnson and Morgan 1951, 1953; Johnson 1963). Thus a system is defined by a combination of selected stars and carefully specified instruments.

The third requirement underlines that a standard star is something more than a reference against which an unknown star is compared differentially. It can provide a guarantee that instruments are working properly. This is why standards should be observed frequently – preferably every night. If a radial-velocity standard consistently gives the "wrong" answer – it being assumed that the star is genuinely a standard in the sense defined and that wavelengths have been carefully selected – not only is at least an approximate "night correction" provided, but the observer is alerted to look for a possible maladjustment in his spectrograph. This is one reason why early-type stars should not be adopted as velocity standards. Even if new techniques of measurement (e.g. cross-correlation) improve the accuracy with which their velocities may be determined, there is no guarantee that any of them are

sufficiently constant in velocity to be used to detect instrumental
errors. Early-type "reference" stars may be very useful — precisely for
cross-correlation work, which requires the adoption of such objects for
several different sub-types — but, like the M-type giants, they should
not be called standards.

3. CRITERIA FOR THE CHOICE OF STANDARD STARS

Standard stars are usually chosen by individuals, even when they
are approved by a committee. The currently used radial-velocity
standards, for example, are largely the choice of Pearce, even though
they were formally adopted by I.A.U. Commission 30. The standards
defining the MK and UBV systems (Johnson and Morgan 1953) are obviously
the work of the originators of these systems and owe their wide acceptance more to their demonstrated usefulness than to any official
adoption. Endorsement by an appropriate I.A.U. Commission may be helpful, however, and may be obtained the more readily if definite criteria
for the selection of standards are adopted. The most important
criterion, reflecting the first requirement in the previous section, is
that a star should be constant with respect to the characteristic for
which it is proposed as a standard. This criterion can never be fully
met. There is always a possibility that an apparently constant star is
a long-period variable, or will be revealed as a variable when observations of higher internal precision become possible. Standards must
frequently be checked, therefore, and the "state of the art" of observation will limit the choice. In general, the higher the internal
precision of our observations, the fewer stars will be acceptable as
standards and the harder it will be to find them.
    Griffin (1975,1980) recently criticized I.A.U. Commission 30 for
what he regarded as too hasty acceptance of a new set of standard-
velocity stars. He offered improved values for the velocities of many
of the standards and unambiguously demonstrated the variable velocity of
one by determining orbital elements for it. He suggested that no star
should be adopted as a velocity standard unless its velocity had been
shown to be constant for at least ten years. This sounds like a counsel
of perfection, but Griffin's strictures were justified and could equally
well have been aimed at many of the stars adopted by the same Commission
in 1955. Recent studies (Batten et al. 1983, Andersen and Nordström
1983, Mayor unpublished) indicate that several of these standards may
have variable velocities, while McClure (1983) has shown that one
displays two spectra and is a spectroscopic binary. These discoveries
are partly the result of routine observation of the stars in question
with higher dispersions than were ordinarily available in the 1950s; but
they also reflect that many of the standards were, probably unavoidably,
assumed to have constant velocities on insufficient evidence.
    Suppose that a star being considered for adoption as a standard for
some quantity, x, is actually variable in x with a period P. Suppose
also that any deviation from x_0, the most usual value of x, that is
greater than or equal to δx can be detected by the observational technique employed. Suppose, finally, that x differs from x_0 by at least δx

through a fraction ϕ of the period; and that the rest of the time $|x-x_0| < \delta x$. These assumptions may seem artificial, but the velocity variation of a spectroscopic binary with an eccentric orbit and longitude of periastron near 0° or 180°, or the light-variation of an Algol-type eclipsing system with a very shallow secondary minimum, would nearly satisfy them. Under the assumptions, the probability that one isolated observation would yield a value of x detectably different from x_0 is simply ϕ. The probability that it would not do so is $(1-\phi)$. The probability that n independent observations would fail to reveal the variation would seem to be

$$(1-\phi)^n.$$

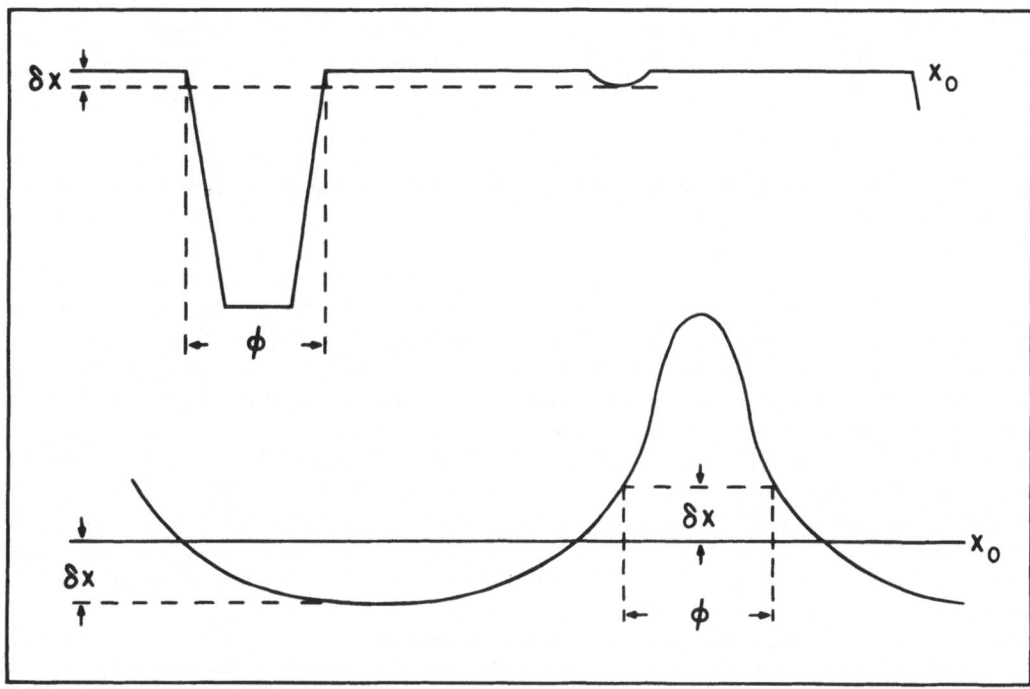

Fig. 1. Schematic light-curve and velocity-curve representing variations that can be discovered only by observations made during a specific interval ϕ (expressed as a fraction of the period).

This simple formula gives a lower limit to the probability that an existing variation will not be detected. If the period P is very long, the probability of detection will be diminished. Thus, if ϕ is 0.1 and P is 10 years, even an infinite number of observations will not reveal the variation unless some are made in the one year out of ten in which it would be detectable; while if $\phi = 0.1$ and P is ten months, we would expect to detect the variation after two or three years of serious observation. Particular values of the period (e.g. those close to one year) could also reduce the probability of detecting the variation, as could the finite length of time it takes to make an observation.

Frequent, regular observation throughout the observing season would
increase the probability of detecting a variation, perhaps by more than
the formula suggests. Nevertheless, the simple expression $(1-\phi)^n$ is a
useful first approximation and Table I shows the values it takes for
small values of ϕ and selected values of n:

Table I Probability that n Observations will not detect
a variation occuring during a Fraction ϕ of a Period P.

n \ ϕ	0.01	0.05	0.10
10	0.904	0.599	0.349
50	0.605	0.077	0.005
100	0.366	0.006	0.00003
1,000	0.00004	5×10^{-23}	2×10^{-46}

Thus, 1,000 observations will virtually always detect a variation that
exists, but 100 observations may quite often miss it and 50 are rarely
sufficient to demonstrate constancy.

It is instructive to compare this result with the records for the
I.A.U. standard stars listed by Pearce (1957). Of 35 stars fainter than
magnitude 4.3 for only two were there more than 30 observations available at the time of selection, and for only nine were there more than 20
observations. The situation was better for the 25 stars brighter than
4.3: only two were selected on the basis of fewer than 20 observations,
and for 12 (nearly half) there were 50 observations or more. Nevertheless, α Per - chosen after more than 250 observations - and α Car -
chosen after nearly 120 - had later to be deleted from the list. I
intend no criticism of Pearce's selections by these remarks; he probably
made the best choices possible at the time. It is not surprising, however, that new, more accurate observations lead to the rejection, or at
least the questioning of many of these standards.

A particularly interesting case history is that of HD 184467, now
known to be a binary and first proposed as a standard on the basis of 7
observations. It was about equally likely that these observations would
or would not show any variation in velocity, since $\phi \approx 0.1$. (The low-
dispersion observations of that time could not have resolved the
spectra, but might just have detected a velocity variation.) After
another 6 observations had been made, the star was adopted as a
standard. There is about one chance in four that 13 observations would
fail to reveal the variation. After Petrie and Pearce (1961) had
increased the number of observations to about 30, there was still a four
per-cent chance that the variation would escape detection. In fact,
they suspected that the adopted velocity needed correction. McClure
(1983) was the first to observe the star consistently at higher
dispersion and resolved the spectra with his third observation.
Although his value of δx was smaller, and his value of ϕ therefore
larger, than those of previous observers, he was rather lucky to resolve

the spectra so early in his own work, although - if all the observations
are regarded as one series - his discovery came at just about the stage
one might expect.

The frequency with which a prospective standard is observed is as
important as the number of years for which it is followed. In an ideal
world, I would not only agree with Griffin that prospective standards
should be observed for at least ten years, but I would add, they should
be observed at least ten times a year (spread evenly throughout the
observing season) from each of three or four observatories. In practice, such a policy would end in no standards ever being accepted, or,
at best, being adopted just when techniques of observation have improved
to the point at which we must begin the selection and testing all over
again. In discussing the frequency of observations, we have to think
carefully what we mean by "independent" observations. If our aim is to
eliminate the possibility of long-period variations, seven individual
observations on consecutive nights are not (apart from the increased
precision of their mean) of much more value than one single observation.
They cannot, for that purpose, be considered fully independent. Thus,
visiting astronomers at national or multi-national observatories can
make only a limited contribution to the selection of standards. They
should certainly observe standards, both for their own sakes, and to
help in the necessarily continual checking. As long as astronomers
require standard stars, however, they will need some observatories where
the resident staff has first claim on the observing time and where
unspectacular long-term programs are encouraged.

The second requirement for standard stars, that they define a
system, reminds us that there is no such thing as an isolated standard
(except, possibly, the Sun, regarded as the standard for abundance
determinations). The stars chosen must cover a range of the quantity
for whose measurement they have been selected as standards. Spectral-
type standards must cover the whole sequence of spectra. Photometric
standards must cover the range of colors and at least be capable of
extension to fainter magnitudes. Radial-velocity standards must cover
as wide a range of spectral types as is consistent with their being
standards. Pearce also wisely chose stars with velocities up to $\pm$ 100
km s^{-1}. This may not be so important for grating spectrographs as for
prism spectrographs, but it is a good principle to follow. There is no
unique answer to the question "how many?" standards. Moreover,
standards must be spread reasonably uniformly over the sky so that an
observer anywhere in the world can have at least some choice on any
night of the year. Thus the designation of as many as 60 radial-
velocity standards may not be unreasonable - but to check them all as
thoroughly as I have suggested needs a major effort of cooperation. A
photometric system may require hundreds of standards (McClure 1976).

Sometimes, however, the number of possible standards is strictly
limited by the "state of the art" and by the patience of the artist.
For example, there are excellent high-resolution atlases of the spectra
of three stars - the Sun (Minnaert et al. 1940, Delbouille, et al.
1973), Arcturus (Griffin 1968) and Procyon (Griffin and Griffin 1979).
The spectra of these stars are not standards in the full sense that I
have defined - that of Arcturus is known to be peculiar - but the very

existence of these atlases has resulted in these three spectra being used for reference and comparison. Even in these days, however, there are not many stars that can be observed at the requisite degree of resolution over a wide spectral range. This fact circumscribed the choice of stars to be observed in this way. Whatever theoretical principles we may lay down, the stars we select as standards are to some extent dictated by what our instruments can do, and, even more strictly, by what we ourselves are prepared to do. Not many of us are willing to take the time or the care to produce a high-resolution spectral atlas. Nevertheless, the selection of any kind of standard requires that kind of patient and not very glamorous work, if it is to be done properly.

The final requirement for standard stars, that they be suitable for testing instruments, also helps to determine their choice. As I have already suggested, this requirement rules out the adoption of early-type stars as velocity standards. Similarly one would hardly choose a Wolf-Rayet star – with its anomalous, strong, broad emission lines – as a photometric standard. When the inclusion, in a system, of some stars not suitable for testing is unavoidable (as with McClure's late-type giants) then those stars should not be regarded as, in the fullest sense, standard stars.

4. DIFFERENT SETS OF STANDARDS

The Universe (or even the Galaxy) has not, unfortunately, been so arranged that there are 20 or 30 stars, distributed around the sky, that can be adopted as standards for everything. Each set of standards must be adopted specifically for its purpose. The specific requirements of one purpose may contradict those of another. For example, although a rapidly rotating star may be unsuitable as a velocity standard, some of them must be included in any set of standards for the measurement of rotational velocities. Similarly, known binaries (of any kind) should be avoided for radial-velocity standards, but McAlister and Hartkopf (1983) have proposed a set of standards for speckle interferometry that are necessarily binaries. Each new method of observation is likely to create its own need for standards.

There may also be a need for hierarchies of standards. Each application in which standard stars are used will require certain primary or fundamental standards, that should be adopted only after the most careful investigation. Many people will adopt provisional standards of their own, hoping that eventually their system of measurement can be tied into a fundamental one. For example, Griffin (1969) found it convenient to adopt four personal velocity standards – which, by now, are probably better observed than many of the I.A.U. standards. Similarly, Glushneva (1983) has selected a number of standard stars for a spectrophotometric catalogue. Stars chosen in this way form a second level of standards. They usually have been chosen more or less carefully by someone who has a specific purpose in mind; they may or may not prove, later, to have a wider application. When fundamental standards have finally been chosen, these secondary standards can be related to them. At the lowest level are "comparison" stars – I am using the term

more widely than its narrow photometric sense. These are stars chosen
for specific differential measurements - often with convenience or
availability in mind, rather than the accuracy with which their relevant
properties are known. Sooner or later, these stars must themselves be
investigated.

The determination of radial-velocities by means of the objective
prism requires many standards, ideally one in every field, but certainly
some hundreds (Fehrenbach 1967). Obviously, it is impracticable to
choose and test so large a number of primary standards. Fortunately, it
is unnecessary to do so because of the relatively low precision of which
the method is capable. Secondary or tertiary standards would be sufficient and could eventually be chosen. Again, I emphasize that standards
- or comparison stars - are chosen with a particular instrument or
method in mind. Another example of radial-velocity work for which a
hierarchical ordering of standards is desirable has already been
mentioned - cross-correlation. For this method an appreciable number of
reference stars is required, so that fairly small intervals of spectral
type can each have at least one. There is not yet any general agreement
on which stars should be used.

5. FUNDAMENTAL CALIBRATION

The observant reader will have noticed that I have so far ignored
one important question: how do we measure, for the standard stars themselves, the quantities for which those stars were chosen to be
standards? There is an element of circular argument in the use of
standard stars. The question has been asked, in recent years: is the
Sun really of spectral type G2V? One answer (Garrison and Zimmerman
1983) might be summarized "the Sun is G2V because G2V is what the Sun
is". If we rephrase the question to ask if the Sun has an effective
temperature and luminosity similar to those of other stars classified as
G2V, the answer is not so trivial and will, I expect, be debated at this
symposium. I shall not join that debate in this review; I cite the
issue only to draw attention to a possible danger in the use of standard
stars. A similar problem is raised by standard-velocity stars: how do
we know their velocities? Historically, of course, spectrograms of the
sky or of solar-system objects were used. The radial velocities of
these objects can be calculated independently of any spectroscopic
evidence, from the kinematics of the solar system. Thus lines and wavelengths in the spectrum that give the "right" answer can be selected and
applied with some confidence to solar-type stars. The system can be
extended to other spectral types by observing stars in moving clusters
or in visual binaries of which one component is a solar-type star
(Petrie 1962). Extended objects such as the sky or planets, however,
illuminate the collimator differently from the way in which a stellar
point-source does. Moreover, if the spectrograph slit is not correctly
placed across the image of the disk of a planet, the planet's rotation
may affect the measured velocity. These objections are of more importance now that high-dispersion observations of high internal precision
are possible. Fortunately, the brighter asteroids are observable at

dispersions of 5 Å mm^{-1} to 10 Å mm^{-1} and have the same advantages as the major planets without, excepting possibly Ceres, the disadvantages. Preliminary results obtained from Victoria do suggest that the zero-point of the standard-star system may be slightly in error.

The circle of argument inherent in the use of standard stars can, then, be broken if a fundamental calibration is made. Not everyone need make it – after all, standard stars are used precisely because they are more convenient – but the fundamental calibration must sometimes be made. If we have all agreed on the same set of standards, at least an improved calibration is unlikely to do more than modify our zero-point. Blaauw (1963) has discussed – in connection with the Mount Wilson spectroscopic parallaxes – how a random error in the calibration of an observed quantity against some theoretical parameter can produce a systematic error in both the scale and the zero-point of the derived relation. Observational errors in the eye estimates of luminosity-sensitive line ratios combined with "cosmic scatter" amongst stars of similar spectral characteristics lead to the deduction of an erroneous relation between the estimates and the absolute magnitudes, and to an underestimate of the true scatter of individual points about the relation. Even if we could achieve infinite internal precision in our observations, cosmic scatter will still work in this way and, indeed, will become proprtionately more important as our precision increases. Lutz and Kelker (1973) have also drawn attention to systematic errors in luminosity calibrations based on trigonometrical parallaxes that can arise from the random errors in the parallaxes. Earlier treatments of both matters can be found in Trumpler and Weaver (1953).

A similar effect is found in radial-velocity standards, for rather different reasons. Dravins (1975) showed that observed velocities of solar-type stars, in particular, may differ by up to 0.5 km s^{-1} from the centers of mass of the stars. Granulation on the solar surface is evidence of (roughly) radial currents, rising and falling in the region where spectral lines are formed. At first sight, this might be supposed to increase only the random error in our knowledge of stellar veloci-ties. Dravins points out, however, that the hotter, brighter rising currents will make a greater contribution to the observed spectrum than will the darker, falling currents. Observed velocities, therefore, will tend to be systematically too negative. The amount of error will differ unpredictably from star to star, particularly since the temperature difference between dark and bright areas will, largely through blending and asymmetry, cause shifts in the rest positions of individual lines. Thus, regardless of the internal precision of the method of measurement, there is not only a possible systematic error, but a minimum random error in the determinations of radial velocities of solar-type (i.e. standard!) stars. Whether or not the possible zero-point difference between asteroids and stars, which we have found at Victoria with a dis-persion of 6.5 Å mm^{-1}, is related to this systematic error is not yet clear. The existence of a minimum random error, however, is of increasing importance as our potential internal precision improves. The minimum value was not large enough to be important for the old single-prism spectrographs – even in measures of solar-type stars. It is now comparable in size with the external scatter we find in our "conventional" high-dispersion observations. Methods are now being developed and used for which an internal precision of the order of 0.01 km s^{-1} is claimed (Campbell 1983, Serkowski 1976). Dravins has shown that no star is stable enough in observed velocity (since the effects of granulation will certainly change with time) to be used as a standard for measurements of such precision. Asteroids merely reflect the solar spectrum, so their spectra will contain the same sources of uncertainty.

The present century, in particular its last few decades, has been a time of significant increases in the internal precision of virtually all forms of measurement. In astronomy, the trend seems likely to continue for some time, and it may actually be accelerated by the increasing use of space instruments. An example is provided by the great increase in the precision of positional measurements expected from HIPPARCOS. I have pointed out that the difficulties of choosing standard stars increase with the level of precision attainable. I began the preparation of this paper convinced that the careful selection and testing of standard stars for many purposes would be of considerable importance for some time to come. Now, I am forced to ask myself if it is the most appropriate method of calibration at the highest levels of precision. Although it has its roots in the late nineteenth century, the use of standard stars is particularly characteristic of the twentieth. Is it necessarily to be of the same importance in the twenty-first? Calibration of some kind, of course, will be even more important. We must, as I have shown, calibrate our standard stars in some absolute or fundamental way. Difficult and time-consuming though this may be, we shall perhaps find that really high-precision observations should be directly calibrated, without the aid of the intermediate step of standard stars. As I finished this paper, I discovered - by chance - that, in the first year of the twentieth century, Belopolsky (1901) reported his attempts to measure, in the laboratory, Doppler shifts corresponding to velocities of the same order as those of stars. I believe the paper is largely forgotten and probably will not be referred to by any other participant in this symposium. Perhaps it was just 100 years ahead of its time.

6. ACKNOWLEDGMENTS

Many of the ideas presented in this review were developed after a conversation with Dr. J. Andersen of Copenhagen University Observatory. I am also grateful to my colleagues Drs. G. Hill, R.D. McClure and C.D. Scarfe for comments on an earlier draft of this paper.

REFERENCES

Andersen, J. and Nordström, B. 1983, Astron. Astrophys. Supp. 53, 287.
Batten, A.H., Harris, H.C., McClure, R.D., Scarfe, C.D. 1983,
    Pub. Dominion Astrophys. Obs. 16, 143.
Belopolsky, A. 1901, Astrophys. J. 13, 15.
Blaauw, A. 1963, in Basic Astronomical Data, ed. K.Aa. Strand,

Univ. of Chicago Press, p. 383.

Campbell, B. 1983, Pub. Astron. Soc. Pacific 95, 577.

Delbouille, L., Neven, L. and Roland, G. 1973, Photometric Atlas of the Solar Spectrum from λ3000 to λ10000, Liège.

Dravins, D. 1975, Astron. Astrophys. 43, 45.

Fehrenbach, Ch. 1967, in The Determination of Radial Velocities and their Applications, eds. A.H. Batten and J.F. Heard, Academic Press, London and New York, p. 149.

Garrison, R.F. and Zimmerman, L. 1983, J. Roy. Astr. Soc. Canada, 77, 78.

Glushneva, I.N. 1983, Soviet Astron. 27, 326.

Griffin, R.F. 1968, A Photometric Atlas of the Spectrum of Arcturus, Cambridge Philosophical Society.

_____. 1969, Mon. Not. Roy. Astron. Soc. 145, 163.

_____. 1975, Mon. Not. Roy. Astron. Soc. 171, 407.

_____. 1980, Mon. Not. Roy. Astron. Soc. 190, 711.

Griffin, R.F. and Griffin, R.E.M. 1979, A Photometric Atlas of the Spectrum of Procyon λλ3140-7470Å.

Johnson, H.L. 1963, in Basic Astronomical Data, ed. K.Aa. Strand, Univ. of Chicago Press, p. 204.

Johnson, H.L. and Morgan, W.W. 1951, Astrophys. J. 114, 522.

_____. 1953, Astrophys. J. 117, 313.

Keenan, P.C. and McNeil, R.C. 1976, An Atlas of Stellar Spectra of the Cooler Stars, Ohio State University Press.

Lutz, T.E. and Kelker, D.H. 1973, Pub. Astron. Soc. Pacific 85, 573.

McAlister, H.A. and Hartkopf, W.I. 1983, Pub. Astron. Soc. Pacific 95, 778.

McClure, R.D. 1976, Astron. J. 81, 182.

_____. 1983, Pub. Astron. Soc. Pacific 95, 201.

Minnaert, M.G.J., Mulders, G.F.W. and Houtgast, J. 1940, A Photometric Atlas of the Solar Spectrum, Amsterdam: Schnabel.

Morgan, W.W., Keenan, P.C. and Kellman, E. 1943, An Atlas of Stellar Spectra, Univ. of Chicago Press.

Pearce, J.A. 1957, Trans. Int. Astron. Un. 9, ed. P.Th. Oosterhof, Cambridge Univ. Press, p. 441.

Petrie, R.M. 1962, in Astronomical Techniques, ed. W.A. Hiltner, Univ. of Chicago Press, p. 63.

Petrie, R.M. and Pearce, J.A. 1961, Pub. Dominion Astrophys. Obs. 12, 1.

Serkowski, K. 1976, Icarus 27, 13.

Trumpler, R.J. and Weaver, H.F. 1953, Statistical Astronomy, Berkeley, Univ. of California Press, §1.51 and p. 369.

DISCUSSION

BESSELL: Now that many radial velocities are measured by cross correlation techniques in the visual and near infrared regions, there is a great need for non solar-like reference stars, such as M stars and carbon stars, where TiO and CNO or C_2 bands can be used. In particular, systematic velocity differences due to varying band saturation could be explored given a grid of reference velocity stars (based on metal lines) but with a range of TiO and/or CN strengths.

BATTEN: Your comment underlines my points about the need for different types and levels of standards for different purposes. The whole question of reference stars for cross-correlation needs careful consideration. Perhaps this should be done at the radial velocity colloquium next October.

BOHLIN: Could you please describe in more detail the problems especially associated with using B stars as radial velocity standards with a precision of 1 km/sec? What are the prospects of obtaining standards with this precision for the hotter stars that could be used in the UV down to 1150 Å?

BATTEN: The problems are the difficulty of measuring rotationally broadened spectral lines to that precision and the fact that no one knows if the velocities of B-type stars are constant to that degree.

KREŁOWSKI: One of the criteria that standard stars should fulfill is the lack of reddening effects in their spectra. This is the difficult problem in OB stars where unreddened objects are very scarce. Two of the early-type stars proposed as standards on a list distributed before the symposium, HD 3360 and 160762, were observed by the ANS satellite. These photometric data show clearly the presence of reddening effects.

JASCHEK: Dr. Batten mentioned that the first condition to be fulfilled by a standard star is that it be constant in the parameter concerned. I hope much more attention can be paid to stars which are chosen to be standards for one parameter, but which in fact are variable in another one.

GRIFFIN: In mentioning my four radial velocity reference stars Dr. Batten generously refrained from embarrassing me by reporting the discovery by one of his own colleagues at Victoria, Dr. McClure, that one of the four stars is itself a spectroscopic binary. This is an object lesson to me. The amplitude of the variation is about equal to the measuring error and although I have made more than a thousand observations of the star I have usually compared it with stars which are themselves variable so the variation of the reference star has not been apparent. The higher precision of the Victoria radial velocity spectrometer and the use of a laboratory reference instead of other standard stars has enabled the variation to be discovered and now that 't has been pointed out it is in fact easily traceable in the residuals of the stars with which I have compared it.

BATTEN: It is an object lesson to us all: even 1000 observations may not be enough.

THE USE AND ABUSE OF STANDARD STARS

R. F. Garrison

David Dunlap Observatory
University of Toronto

ABSTRACT. Most systems of classification or quantitative measurement
depend on standards. It is of the greatest importance for the user to
be aware of the mandate of a particular system. If the mandate is not
understood by casual users, a system can be either underutilized or
abused.

In the particular case of the MK system of spectral classification,
types are defined by the standard stars. They can be calibrated, and
the calibration may evolve with time, but the types are relatively stable
because they are defined by the standards. The autonomy of this powerful
system is crucial to its success, but some astronomers do not understand
the importance of this distinction. Recent suggestions to change the
spectral type of the Sun show an ignorance of the way the system works.

Precautions in the use of standard stars and the frequency of their
use depend on the particular system and on its mandate.

1. INTRODUCTION

The first step in any science is to classify the objects to be
studied. In astronomy, as in botany, this is a continuing, complex
process because of the number of objects. We can classify stars by
brightness, color, position, change of position, line spectrum,
variability, or other observable parameters. Once a particular classification scheme has been set up, it is possible to use it to segregate
peculiar objects, and thus to gain insight into the processes which
generate "normal" objects. Eventually, when there are enough objects in
a given "peculiar" class, the definition of "normal" can be extended to
include them. Such a system, carefully developed, can be used to infer
fundamental stellar quantities, which may be more or less directly
related to the quantity measured; sometimes the relationship is very
remote or very ambiguous. It is in the determination of this relationship between the measured and fundamental quantities that the calibration

D. S. Hayes et al. (eds.), Calibration of Fundamental Stellar Quantities, 17–29.
© *1985 by the IAU.*

enters. It is also where the greatest problems arise. This conference
has been called to delineate these problems, and possibly to throw some
light on them, though I must admit to a certain amount of cynicism about
the possibility of solving any of them definitively. Perhaps we can at
least clear up some misconceptions.

This introductory talk, by the nature of its title, must necessarily
be somewhat philosophical, though I will try to bring in some reality
through specific examples drawn mainly from the systems I know best, the
MK system of spectral classification and the UBV system of photometric
classification. However, most of the statements can be easily translated
to apply to other systems.

2. PHILOSOPHICAL CONSIDERATIONS

While it may be more difficult to discuss philosophical approaches
to science than to present data and discuss errors or interpretations,
it is very important to do so from time to time. The philosophical
basis of the MK system has been fully discussed by Morgan et al (1943),
Morgan and Keenan (1973), Morgan (1984), Mihalas (1984) and others (see
McCarthy, Philip and Coyne 1979 and Garrison 1984).

An important distinction that was made during the 1983 MK workshop
held in Toronto is that between the MK Process and the MK System. The
MK system of standards may not be applicable to other wavelength regions
or other resolutions, but the MK process can be used to set up any
classification scheme.

2.1 Definition of Terms

In this talk, the term classification will be used in its broadest
sense, including directly measured quantities. For example, the
measurement of parallax allows an ordering, or "classification", of
stars according to numerical parallax or distance.

Most, but not all, of the classification systems in astronomy use
standard stars to a greater or lesser degree. Others use more direct
measurements; for example, absolute fluxes may be measured with standard
lamps. The latter have their own problems and I will restrict my remarks
to the use of standard stars. Laura Pasinetti was going to discuss the
relative merits of the two approaches. I believe that they are not
competitive, but can both be used in a complementary way; perhaps she
would have come to a similar conclusion.

There is, however, some confusion about the meaning of the term
"standard star". In a letter to me on this subject, Johannes Andersen
made some useful distinctions and outlined several classes of standard
stars used for different purposes. In slightly modified form, these
are:
     a. Standard stars that define a system (e.g. the MK system).

  b. Standard stars that are used to transform observations from an
      instrumental system to a common system (e.g. the UBV system).
  c. Standard stars needed for occasional zero point checks (e.g. the
      radial velocity system).
  d. Standard stars that define internal zero point in an instrumental
      system (e.g. Griffin's system).
  e. Standard stars that define a unit of measurement (e.g. the Sun
      for masses, radii, luminosities, etc.).

   As Alan Batten (1985) has said, it is also important to distinguish
among primary standards, secondary standards, lists of carefully observed
stars, and comparison stars. Variable star comparisons are not the same
as systemic standards. Unfortunately, the data centers and the Standard
Star Newsletter have not been critical enough in their publication of
lists of standard stars. The lists seem to be indiscriminate, yet have
the appearance of IAU sanction. While those of us who work in the field
may understand, this is a very dangerous trend and it would be good to
have some discussion of the problem at this meeting.

   Finally, there are some areas of research where the concept of
"standard stars" has no meaning whatsoever. What is the meaning of a
list of "standard" radii, where the measurements are of necessity absolute
and vary only according to the measurement technique and errors? If
someone uses a better system of measurement, they may get different
values from the list of standards, yet there is no reason why they should
transform to the list. So, I suggest that the use of the concept of
"standard stars" should not apply in the case of fundamental stellar
quantities such as mass and radius.

2.2 The Mandate of a System

   Every system of classification or measurement in astronomy has been
created for a reason. In using the system, it is of the greatest importance to be aware of why it was created, how it was constructed, what its
useful limits are, how it has evolved, and what credibility it has
achieved in practice. This is what I call the "mandate" of the system.

   Astronomers who use a system casually are not always sufficiently
aware of the mandate of the system to use it intelligently, and this
leads to problems. To properly utilize the UBV system, for example, it
is essential to understand that it was the first of the modern systems
and was devised to provide a general reference frame. It remains useful
as a general reference frame and for observations of very faint stars,
but there are people who complain about the lack of precision. High
precision is not its present mandate and is probably not achievable
because of the broad bands. The standard stars are not defined better
than 0.01 or 0.02 magnitudes for all-sky coverage. If high precision is
required, some other system, such as the Strömgren system or the Geneva
system should be used. When the mandate is misunderstood, the system
will be abused by expecting too much of it or by overinterpreting the
data. It doesn't matter that Alpha Lyrae is slightly variable in UBV

because the variations are well below the limit of accuracy of the system.
It does matter that it is too bright for most people to observe, however,
and it is important to carefully establish secondary standards, but it
can still be used as the primary standard.

Similarly, systems are often underutilized because some astronomers
do not realize what can be done usefully and do not give the system
credit for its value for a particular task. Both of these conditions
are deplorable, but can be improved by a correct understanding of the
mandate of a system.

It follows from this, in my opinion, that it may not be possible to
define a set of standard stars which would be appropriate for all systems,
so I will not support such a move at this meeting. However, what we can
do is to obtain data in other systems for standard stars used in a
particular system; that I will support enthusiastically. For example,
MK standard stars may not be appropriate as standards in other systems,
but it is very useful to know their values in other systems.

2.3   The MK System Mandate

The mandate of the MK system is to describe the appearance of the
blue-violet spectrum of stars at moderate dispersion by reference to a
set of standard stars. It is not bound to match the color or the
effective temperature or anything else besides the spectrum itself.
Though, for convenience, the sequences are ordered roughly by effective
temperature, they are not exactly the same as effective temperature and
the ultimate test is whether the spectrum matches that of the standard
star using exactly the same observing technique for both the standard
and the unknown. Thus the MK system depends only on the standard stars
and nothing else. There may be small errors in the system of standards,
and those must be realized and corrected, but by and large it is a very
reliable tool, because it is based on standard stars independently of
the calibration.

For a given star, the range of effective temperatures found in the
literature is greater than the range of spectral types given by different
classifiers. What a mess we would have if the spectral type were changed
every time some theoretician came out with a new effective temperature.
It is best that the MK system remain autonomous. It has greater value
that way, to observers and theoreticians alike.

Similarly, there is not a one-to-one correlation between spectral
type and color anywhere in the HR diagram, though the general trends are
well correlated. Not all stars with the same color have the same
spectrum or vice versa, and it doesn't mean that the type is wrong any
more than it means that the color is wrong. They measure different parts
of the atmosphere. If they give different values, the confrontation
will teach us something. The interface between two autonomous systems
produces new information which is not available to either system alone.
By ignoring the difference or by assuming automatically that one or the

other is wrong, we are throwing away information. That is bad science, in my opinion, and we should be careful not to fall into that trap.

It irritates me to see superficial criticism of the MK system by others who use completely different systems to tell us what we should be seeing. We see what we see and we call it as we see it. It doesn't matter what the photometry says or what high dispersion shows, the comparison of the appearance of the spectrum with that of the standard stars is what determines the type. It is fine for various groups to try to define the colors of the Sun, but it is not okay for them to claim that therefore the MK classification of the Sun is wrong. The appearance of the spectrum of the Sun is halfway between that of Beta CVn at G0 V and Kappa Cet at G5 V, independently of abundance. The three stars form a consistent set of standards and if there are problems with other stars, it is the others that should be changed.

Stars which cannot be matched with the standards of the MK system can still be usefully described in terms of the system. Two examples come to mind. The Am stars, which are now given both a hydrogen type and a metallic-line type (e.g. A5mF2), can be described accurately by the new designations. The late-type population II stars are described in terms of the closest type and then an abundance parameter is given for out-standing anomalies (e.g. G8 III Fe-1,CN-2). In other words, if a star doesn't fit, describe how it doesn't fit. That is part of the power of the system.

3. PRACTICAL CONSIDERATIONS

Most of the problems are with users, not systems! It has been my experience that most of the differences among observers using a given system are not due to problems with the system itself, but are due to lack of proper attention to careful standardization, through use of standard techniques and standard stars. It is not surprising to me that an observer who uses high dispersion, unwidened, overexposed radial velocity plates for spectral classification or who uses a different filter set for photometry, might get a different result from others using the standard techniques. In some systems, the standard technique is more important than the standard stars. However, since this is a discussion of the use of standard stars, it will be assumed that the standard techniques recommended for the systems will be used.

I will not try to outline in detail all of the precautions that should be observed in all the systems of measurement being discussed at this meeting, since they depend on the mandate of the system. However, I would like to mention some key areas where I think there are problems and while I may mention particular systems, some of these difficulties apply to more than one system.

3.1 Precautions in the Use of Standard Stars

When using a system of standard stars, it should be obvious that the standards should be taken under the same conditions as the unknowns, but some astronomers still violate this basic rule. The use of large telescopes has actually increased the temptation, since telescope time is at such a premium and since the new detectors are too sensitive to be used on stars brighter than 10th or 12th magnitude. Thus there is a danger that the use of poor secondary standards, and fewer observations of them, will increase in the future.

As discussed in the 1983 MK workshop, it is important to carefully transfer the MK system to fainter stars using the traditional photographic techniques. Then, the carefully established faint secondary standards can be used with modern detectors. This is being done using the classification spectrographs on small University of Toronto telescopes in Chile and Canada, in conjunction with the prime focus MK spectrograph on the 3.6 meter Canada-France-Hawaii telescope. With the former, secondary standards will be established near 10th magnitude around the equatorial zone, and with the latter these will be extended to 15th magnitude. Thus, in a few years, there will be good standards for use with large telescopes and new detectors if time is made available. For the UBV system, this process is already underway (e.g. Landolt, 1983).

A good grid of MK standards is essential to good quality classifications, but how are these to be established for large telescopes? It is too expensive for individuals to each have their own set. Thus, it may be that libraries of standards taken with the new detectors will have to be set up so that observers can plug into the library by taking only a few standards as a check that the conditions are the same as those under which the more extensive grid was taken.

Another possibility is to use a smaller telescope of the same f-ratio with the same spectrograph. I have used the f/3.8 CFHT-MK spectrograph at the DDO 1.88 meter Newtonian focus, as well as at the DDO 60 cm Cassegrain focus with an f/15 to f/3.8 focal reducer. It is also possible to design a 15 or 25 cm telescope to mount on the spectrograph for use with a dedicated drive unit. With a minimum time for standards with the big telescope, it is thus possible to tie into a much larger grid of standards if proper precautions were taken. This is not the ideal way to do classification, but we have no choice with big telescopes, if we want to get time for classification work.

Taking data indiscriminately from the literature and treating it as standard is unfortunately an all too common practice. This is a problem for all systems. Most astronomers are becoming aware that, because of inadequate standardization on the part of some astronomers doing classification work, MK types given in the literature are not of uniformly high quality. Because photometric observations are given numerically, they are considered to be accurate, even though they may suffer from all sorts of errors. I call this the "deification of quantization" and most

scientists suffer from it. However, compilations of UBV data gathered by
the Strasbourg Data Center show that even for well-observed, non-variable
stars the values for B-V differ by more than 0.03 or 0.04 magnitudes and
for some it is as much as 0.10! Yet astronomers like Hardorp (e.g. 1982)
pretend to be able to distinguish between B-V=0.63 and 0.67 for the solar
analogs, taking UBV values (as well as MK types) indiscriminately from
the literature. For such critical work, it is essential to use the very
best values possible or to determine them on a uniform system.

   The "tyranny of the mean" has been criticized before, in many fields
but it needs mentioning here. It is not good enough to take photometry
or spectral types for stars or groups of stars randomly from the literature, average them, and use that value as "standard". Because of systematic effects, it is not enough to take the average of random radial
velocity data for a star from the literature and use that as "standard",
no matter how many thousands of measures are included. It is not enough
to use photometric measures of a random star, no matter how many there
are, as standard unless it is part of a well-defined system using standard
stars and standard techniques. These statements may seem obvious, but
systems are very often abused even today in these very ways.

   There are many other precautions which could be listed, but they all
reduce to one basic caution; think about the data being used and how it
has been obtained, and if you are doing the work, do it with an extreme
amount of care.

3.2 Frequency of Use of Standard Stars

   The frequency of observation of standard stars depends very much on
the system of measurement being used and its mandate. All too often,
shortcuts are taken when observing and in some cases (e.g. large telescopes) there is no alternative; however, if too many shortcuts are taken
the observations have less value and might as well not be made, so some
compromise must be reached.

   It has been said of photometry that the only way to adequately
determine the extinction and transformation coefficients is to observe
nothing but standards all night every night. Obviously that is not
practical and it has been found that well-determined mean coefficients
can safely be used with occasional checks during the night. This
procedure obviously favors the dedicated observer over the casual
observer because the former will have a more secure foundation over a
longer time period. It also means that large telescopes are almost
doomed to lower accuracy because nobody will be given time to adequately
determine mean coefficients. There may be no solution except to observe
as carefully as possible within the constraints. It is still necessary
to observe some standard stars each night. Fernie, at the David Dunlap
Observatory, has devised a promising technique using twin photometers on
two different telescopes, one of them dedicated to extinction determination.

The frequency of observations of standard stars for radial velocity is somewhat less critical, but it is nonetheless important and depends to a certain extent on the stability of the equipment.

For MK classification, it is not necessary to observe standards every night, because the equipment is usually sufficiently stable to allow comparison with a grid of previously observed standards. That is a big advantage. ·At the DDO in Toronto, there is a complete grid of MK standards taken with each spectrograph, and for each dispersion, devel-, oper, slit width, and slit length combination. Then, for each observing run, the most important standards are re-observed and at least one standard is observed each night, just as a check on the stability of the equipment and processing techniques. Because of very careful attention to standardization of techniques, differences have only rarely been detected. This is an example of how standard observing and reduction techniques are important in the use of standard stars. Classification is not a black art, as some have claimed, but can be successfully done by anyone who is willing to do the work carefully. Most of the differences in spectral type for a given star found in the literature are due to the use of non-standard techniques or inadequate attention to standard stars.

4. EXAMPLES

4.1 The Use of Standard Stars

The MK system, combined with photometry and absolute magnitude calibrations, has enabled astronomers to determine spectroscopic distances to a wide variety of objects, including black holes, Cepheids, and external galaxies, thus extending the system of parallaxes for nearby stars to the farthest reaches of space. Our knowledge of the structure of the Milky Way Galaxy was determined in 1951 by the careful use of standard stars, and it has since been extended by the complementary use of several different techniques. These advances were possible because the system was set up carefully, using standard stars, and because it is autonomous, yet can be calibrated in terms of fundamental quantities. One can even argue that the spectrum, color, position and motion are the real fundamental quantities, in the existentialist sense, because they are directly observable; however, not at this meeting.

The mandate of the MK system is to describe the appearance of the spectrum in terms of a set of standard stars, and not necessarily to give mass, radius and abundance. However, the accurate description of the spectrum allows astronomers to isolate interesting stars for more detailed observations to determine these fundamental quantities. In a similar way, photometric standard stars can be used to isolate stars with interesting colors and radial velocity standards can be used to isolate binaries or stars with extremely large motions. All of this can be used in a complementary way to learn more about the universe in which we live. The usefulness of the various systems depends partly on their autonomy and partly on their complementarity.

4.2 The Abuse of Standard Stars

Several references to the abuse of standards have been made in the comments above. Most problems can be avoided by a proper understanding of the mandate of a system and by careful observations and reductions. The controversy over the spectral type of the Sun is a case in point.

The Sun is an important standard for most astronomical systems because of its availability for detailed study. However, the very property which makes it so fundamental - its proximity - also renders it difficult to observe as a star. It is an extended source, is 30 magnitudes too bright, and cannot be observed at night; in most systems it is therefore impossible to observe with the same equipment at the same time as other standard stars unless an intermediate step in used.

Therefore Hardorp (1978), with good intentions, set out to find a solar analog among the stars. However, he used non-standard techniques and because of his misunderstanding of the mandates of the UBV and MK systems, he has muddied the waters considerably. Some of the problems were discussed at the Vatican Conference in 1978 (Garrison, 1979). Hardorp has also softened his approach recently.

In dealing with suggestions like this, it is useful to be aware of two dangerous effects, which I will call the "bandwagon effect" and the "don't rock the boat" effect. The first refers to the tendency for observers to ascribe differences in the observations to a fashionable interpretation, even when such leaps are not justified. The solution to this problem is to just "watch the parade" for awhile before carefully choosing a path. The second effect refers to the tendency of the "old guard" to try to keep the status quo. The solution is to rock the boat only when necessary and then VERY carefully, making sure that you understand the "mandate" of the system; otherwise you may be "all wet".

The idea of a search for solar analogs in all systems is a good one, but for the MK system it must be done using line spectra of comparable resolution to that used in MK classification because that is where the symbols G2 V originate. To use those symbols for another system is to abuse the MK system of standard stars. The biggest mistake made by Hardorp and others who have jumped on his bandwagon is to use non-standard techniques to infer results in the MK system. The Sun is G2 V by definition and other stars with the same classification have line spectra in the blue-violet region at moderate resolution that very closely resemble that of the Sun. It doesn't matter what B-V they have or what effective temperature someone thinks they have. It only matters that the spectra match. That is the test in MK terms. Hardorp has not made that test so he can say nothing about the MK type of the Sun except to suggest that we look at the spectra again. That is being done now.

One good result is that I have been stimulated to look for a solar analog using standard MK techniques. To this end, I have taken spectra on a homogeneous system of all the analogs suggested by Hardorp (1982),

Reitsema (1977) and others. The total comes to about 400 stars, which
includes other G stars brighter than 5th magnitude. We are not yet in a
position to publish definitive results, but already there are some
interesting classifications. The Cayrels suggested that HR 1318 B might
be a solar analog. Corbally and I find that it is G5 V, Fe-2 and does
not resemble the Sun at all, either at 120 Å/mm or at 67 Å/mm. To be
sure to avoid the "don't rock the boat" effect, I have tried to look at
the problem as carefully as I can and will be doing the classifications
blindly to make sure that I am at the mercy of the data. I hope that my
complaint is understood. I am not upset that there may be some mis-
classifications of early G stars in the literature, but that the MK types
have been criticized by people who have not looked at classification
spectra, and have in fact used much poorer data than was used for the
original types. That is a complete abuse of the MK system of standard
stars.

5. CONCLUSIONS

The conclusion of this paper is obvious. Classification of any
kind is a fragile process and great care must be taken to use standard
techniques and observe standard stars. New systems can and will be
presented and if they are good, they will eventually be acknowledged.
The MK Process, using an autonomous system of standards, which can then
be calibrated and recalibrated without changing the system, is very
powerful. It can be applied to other classification systems with great
value. The confrontation and complementary use of such autonomous
systems yields information which neither contains in isolation.

To paraphrase a famous American president, "Ask not what astronomy
can do for your system of standard stars, but what your system can do
for astronomy."

6. ACKNOWLEDGEMENTS

It is a great pleasure to acknowledge with warm gratitude the
hospitality of the Danish astronomical community at Brorfelde during a
month spent there just before this meeting. Their stimulating
discussions helped to focus my thoughts on this difficult subject. The
generous support of the IAU and the Natural Sciences and Engineering
Council of Canada is also gratefully acknowledged. Without their
support, such meetings could not be held so successfully.

REFERENCES

Batten, A. H. 1985, in IAU Symposium No. 111: Calibration of Fundamental
    Stellar Quantities, ed. D. S. Hayes, L. E. Pasinetti and A. G. D.
    Philip, (Reidel, Dordrecht), p. 3.

Garrison, R. F. 1979, in IAU Colloquium No. 47: Spectral Classification
    of the Future, eds. M. F. McCarthy, A. G. D. Philip and G. V. Coyne
    (Vatican Obs. Ric. Astron. Vol. 9), p. 23.
Garrison, R. F. 1984 (editor) The MK Process and Stellar Classification,
    (David Dunlap Obs., Toronto).
Hardorp, J. 1978, Astron. Astrophys., 63, 383.
Hardorp, J. 1982, Astron. Astrophys., 105, 120.
Landolt, A. 1983, Astron. J., 88, 439.
McCarthy, M. F., Philip, A. G. D. and Coyne, G. V. 1979 (editors), IAU
    Colloquium No. 47: Spectral Classification of the Future, (Vatican
    Obs. Ric. Astron. Vol. 9).
Mihalas, D. 1984, in The MK Process and Stellar Classification, ed. R.
    F. Garrison (David Dunlap Obs., Toronto), p. 4.
Morgan, W. W. 1984, in The MK Process and Stellar Classification, ed.
    R. F. Garrison (David Dunlap Obs., Toronto), p. 18.
Morgan, W. W. and Keenan, P. C. 1973, Ann. Rev. Astron. Astrophys., 11,
    29.
Morgan, W. W., Keenan, P. C. and Kellman, E. 1943, An Atlas of Stellar
    Spectra, (Univ. of Chicago Press, Chicago).
Reitsema, H. J. 1977, Bull. Amer. Astron. Soc., 9, 635.

DISCUSSION

HECK: Bob, I am very happy to hear your comments on the dangers of averaging photometric data and would like to call the attention of the audience to the papers on this matter by Manfroid, and Manfroid and myself which are displayed as poster papers.

LYNGÅ: What has been referred to earlier as peculiarities in the MK system, now often takes the shape of more specific comments, letters. etc. Is this a third dimension being introduced and does it change the mandate of the system as you see it?

GARRISON: There is no simple third dimension and most people do not realize this. The remarks, notes and additional symbols used by Keenan and Morgan are the best we can offer for now. It is not a simple problem. The mandate has changed in the sense that these symbols and notes are more accepted, but has not changed in the sense that our task is to describe the spectrum as consistently and completely as possible.

JASCHEK: Two comments. You referred to the "mandate" of a system. I guess this is an average of what the author said and what the colleagues thought of it ten years after. I hope that at this meeting we arrive at a better set of standards; the "Centre de Données Stellaires" has deliberately tried to get available lists published. They are scattered all over the literature and they should be published, assembled and critically discussed at the same place so that astronomers may use them more easily.

GARRISON: It is not the simple average, but as I mentioned, it does include the definition as well as the evolution of the system.

BIANCHI: When good standards are not available (e.g. in the case of faint stars of late spectral type in the Strömgren system) and brightness limits are set by the instrument, one is left with two possibilities; either to give up matching the colors of the star, or to observe a very large number of "non reliable" standards (i.e. for which very few measurements exist), hoping that the "bad" ones can be recognized and eliminated afterwards during the reduction. This is very dangerous and time consuming. Do you have any suggestion for what a wise criterion would be?

GARRISON: There is a third possibility which you did not mention; that is to establish carefully secondary standards yourself. People too often rely on others to do the really hard, unexciting work, but it is an essential step in doing the more exciting stuff like cosmology. Many theoreticians have become observers because of a similar problem. Why do you think they have done so? It is because they became frustrated with the data available and could not get anyone else to do what they needed in the way they wanted. You may just have to do the preliminary work on setting up standards before you can do the more interesting work on faint stars.

BESSELL: Mandates evolve with time. The UBVRI system in the E regions in the southern hemisphere has errors less than 0.01 magnitude. Also one should not ask what your system can do for astronomy but what system best provides the astrophysical parameter one requires.

GARRISON: Obviously, if your choose the best system for your astrophysical application, you will automatically do well for astronomy! That was implicit in my remarks. In my definition of "mandate" the evolution of the system was included, as well as credibility.

HECK: The problem you just raised, i.e. getting time on large telescopes for observing photometric standards is a very important one and also a more general one. We have also encountered difficulties in getting time on IUE for our low-dispersion reference atlas. Maybe we should think about some action in the course of this meeting for calling the attention of the various allocation committees to this problem. I definitely hope that this meeting will have an impact in this sense.

GARRISON: I agree. The same problem was discussed at length last summer at the MK Workshop in Toronto, as you will remember. Perhaps a letter to the directors of large observatories is in order.

SCARFE: You suggested that large-telescope time is too precious to use on standards. But earlier in your talk you pointed out that this condemns large telescope results to a lower accuracy than the best from smaller telescopes. Surely this means that we cannot afford not to use large telescope time for observing standards, to get the best use of these instruments.

GARRISON: Ah, that is precisely the problem. I do not know the solution except to put pressure on the large observatories.

POPPER: With respect to the search for a "solar analog" it is my impression that a purpose of the search is to a large extent to determine quantities difficult to observe for the Sun, such as (B–V), in order to be able to fit the Sun with its well established properties, into the stellar systems, rather than to re-evaluate the solar spectral type.

GARRISON: That is the reason for doing it, and (B–V) has been chosen as the interface, even though it is relatively insensitive and one could question the choice. The idea is a good one, but it must be done carefully.

CAYREL: I agree with Dr. Garrison that for an observer who is writing a program for a given research problem (in my case finding out the best solar analogs) it is much easier to observe a sample of stars already proposed for such a research by somebody else. Unhappily it may happen that the results are deceiving and that the stars of the chosen sample do not have the physical properties we wanted them to have.

Is 16 Cyg B similar to the Sun in type?

GARRISON: Not really. It is slightly later. Dr. Keenan and I agreed in November on G3 V for 16 Cyg B. I wanted it slightly later.

KEENAN: Dr. Garrison's type of G3 V is in agreement with mine, but I am not absolutely sure that Kap Cet remains constant at G5 V! That is the reason that I do not like to rely on individual stars, but prefer to use a cloud of standards.

GARRISON: You and Bill Morgan differ somewhat in your approach. He has been moving in the direction of single standards for "dagger" types and you prefer a cloud.

TRIGONOMETRIC PARALLAXES AND THEIR CALIBRATION

Arthur R. Upgren

Van Vleck Observatory

1. INTRODUCTION

The precision obtained in equating a very precise linear dimension to a very small and imprecise angular dimension sets a limit on the precision of the distance scale of the Universe. Within the Solar System, errors in distances are of the order of one part in one hundred million, but beyond its limits only a single star, Barnard's star, has a distance known to better than one part in one hundred, and distances known to one part in twenty from parallaxes are limited to only a few hundred nearby stars. Yet most other distance methods and results must ultimately be calibrated against distances to nearby stars derived from the heliocentric parallax method and its observations and uncertainties.

The early work on the calibration of parallaxes and their systematic and accidental errors was mostly done by Frank Schlesinger. At about the beginning of this century, Schlesinger introduced rigorous practices to the then new use of photography in astrometry in the course of his observations using the Yerkes refractor. Among them were the use of magnitude compensation devices, standardized photographic emulsions and filters, and constraints obtained by observing only near the meridian. He also introduced dependences which increased the efficiency in the reduction of the measures over Turner's earlier plate-constant reduction method, although with diminished rigor acceptable at that time.

Schlesinger concerned himself with parallax errors of all kinds. Perhaps he best described his concern in his George Darwin lecture (Schlesinger 1927) when he remarked, "the history of the measurement of stellar parallaxes has presented, more than any department of astronomy, a continual struggle between the necessities of the problem and the methods for attacking it, very similar to the conflict that has gone on between heavier and heavier artillery and stronger and stronger armour-plate. A source of error having once been revealed, it is seldom that much time has elapsed until methods for eliminating it or avoiding it have been devised. After such improvements have been applied, new but

31

D. S. Hayes et al. (eds.), Calibration of Fundamental Stellar Quantities, 31–51.

smaller sources of error come to light to challenge our patience and ingenuity". The calibration of parallaxes and of stellar luminosities and other properties from them, has continued to be a story of encountering and surmounting sources of error.

Schlesinger compiled and published the General Catalogue of Stellar Parallaxes (sometimes called the Yale Parallax Catalogue) in 1924 and a second edition in 1935. He made adjustments to the published errors in the parallaxes of each contributing observatory in order to more properly represent the true uncertainties in the data. Two independent methods were used for this purpose. In the first, the average difference without regard to sign was found from intercomparisons between the parallaxes of each observatory compared with every other observatory from which the external probable errors of each of the two series were estimated. The second method used differences between the trigonometric parallaxes of each observatory with spectroscopic parallaxes of the Mount Wilson Observatory.

Schlesinger looked into the constant differences between the various observatories and applied corrections based on the differences in order to remove them and thus place the parallaxes from all observatories onto a single system. These corrections are sometimes known as the Yale precepts and are published in the first two editions of the Yale parallax catalogue. Some differences occur between the two editions but these are mostly due to the addition of many more parallaxes in the later edition. The first edition lists trigonometric parallaxes for 1682 stars which were available in January 1924 whereas by January 1935, the closing date of the second edition, the stars with available trigonometric parallaxes had more than doubled, to a total of 3928 stars. The third edition of the catalogue was published in 1952 by Louise F. Jenkins and was renamed the General Catalogue of Trigonometric Stellar Parallaxes since it excluded spectroscopic and other distance determinations, unlike the two earlier editions. It contains 8832 parallax determinations for 5822 stars available by June of 1950. Later, Jenkins published a supplement to it now bound with it (Jenkins 1963) with parallaxes available at the end of 1962. It lists 730 new parallax determinations of 654 of these same stars and 632 new parallaxes of 577 additional stars, thus raising the total to 10194 parallaxes of 6399 stars with at least one parallax determination. These numbers may not be exact since observatories occasionally combine parallaxes made from two or more plate series or of two components of a binary star into a single published parallax. In both catalogue and supplement, Jenkins alters the Yale precepts somewhat from the earlier editions, adhering closely to Schlesinger's methods and conclusions. This third edition of the catalogue and its supplement together (hereafter abbreviated GCTSP) incorporate the corrections based on computed systematic differences between observatories in the final parallaxes adopted. The corrections assume a zero point for each observatory which is shifted into a zero point for the entire system of all observatories taken together. The zero point which Schlesinger and Jenkins adopted was that of the Allegheny Observatory since its parallaxes had been found to have the

lowest mean external error. A new fourth edition of the catalogue being
prepared by van Altena (1985) will be discussed later.

Schlesinger (1928) and later Hertzsprung (1952) determined that the
average external standard error of a parallax appearing in the GCTSP is
about 0".016 or 16 milliarcseconds. One milliarcsecond or 0".001 is the
unit of angular measure in most common use by astrometrists. A
milliarcsecond is frequently shortened to a "mas" and is equal to about
five (precisely $2 \pi /1.296$) nanoradians.

The years following the publication of the supplement in 1963 were
years of change in the determination of trigonometric parallaxes. The
changes have occurred in all aspects of parallax research including
observation, measurement and reduction as well as the evaluation and
analysis of published parallax errors. Major improvements have also
been made in the evaluation of systematic differences between observatories in the period and since, and in the application of corrections to
biased parallax samples which result in improved stellar luminosity
calibrations. The result has been the reduction of the size of the
external parallax error to less than half the error of the GCTSP stars.
The selection and use of parallax standard stars and regions for calibration between observatories has received much attention in the last
few years. Finally the parallax programs active in the last two decades
are based upon a vastly improved and realistic experimental design
wherein each parallax is likely to be of significance in deriving the
luminosity distributions of many kinds of stars.

Many reviews have been written of the parallaxes determined up to
the time of the completion of the supplement to the GCTSP. Among the
ones with extensive treatment of errors are the papers by Strand (1963),
Vasilevskis (1966), Gliese (1972), Upgren (1977) and Heck (1978). Some
of their conclusions are discussed later in this paper. Mention should
also be made of the recent review by van Altena (1983). He emphasizes
new techniques which promise still greater precision to parallaxes made .
in the future. But the results and developments achieved since the
publication of the GCTSP in 1963 have not received the attention given
to the earlier data. This review is intended to focus on and summarize
these recent improvements, beginning with the state of parallaxes in the
period just prior to 1963.

2. CONTEMPORARY TRIGONOMETRIC PARALLAXES

Observations, measurements and reductions for parallax changed little during the first sixty years of this century. The observational
requirements laid down by Schlesinger were described by him (Schlesinger
1924) and more recently by van de Kamp (1962) with some modifications.
With few exceptions the parallaxes listed in the GCTSP have been made
using long-focus refracting telescopes with apertures ranging from 50 to
100 centimeters and focal ratios of 10 to 20. The outstanding exception
has been at Mount Wilson where parallax observations were made with the

reflectors at that observatory by van Maanen. Photographic observations were made using standard emulsions and filters based on the properties of the visual or photographic refractors. The description by van de Kamp (1962) of the observational procedures developed at the Sproul Observatory is generally typical of the programs prior to that time. Measures were made on single-screw machines by hand, and typically between three and five comparison stars were measured along the x-axis (aligned in right ascension) only, since the parallactic stellar motion is mostly in right ascension. Reductions were made using the dependence method introduced by Schlesinger (1911,1924).

One of the most significant developments of recent years is the design and construction of an astrometric reflecting telescope by Strand resulting in the 1.5-meter reflector of the U.S. Naval Observatory at Flagstaff. At a conference on the cosmic distance scale, Strand (1958) called for the development of this telescope and pointed out that with minimized flexure and a comparatively coma free field, it would overcome the limitations of existing reflectors such as those at Mount Wilson. He described its features more extensively in the proceedings of a symposium on astrometry held at New Haven (Strand 1962). Although the parallaxes obtained with it are discussed later, it should be mentioned here that our knowledge of distances and luminosities of faint red dwarfs and white dwarfs is now almost completely based on the observations made with this telescope. Most other recent parallaxes continue to be obtained with conventional long-focus astrometric refractors.

The measurement of photographic plates has proceeded through two stages since the review by van de Kamp. About 1960, the single-screw measuring machines in use at the Van Vleck and U.S. Naval Observatories and others were replaced by two-screw machines with machine-readable output. The measures were still made by hand and hence no increase in measuring precision was realized for any one image. But other advances were made as a result of these second-generation machines. Among them were the elimination of accidental blunders since the transfer and recording of data were no longer made by hand. Furthermore a considerable reduction in measuring time per plate was realized along with a reduction of eyestrain and fatigue on the part of the measurer. The simultaneous introduction of computers further lowered the measuring time per plate through the elimination of the necessity for a careful and usually laborious orientation of each plate in direct and again in reversed mode made by hand by trial and error. The gain in measuring speed and computational ability allowed an increase in the number of reference stars to be measured as well as the simultaneous measurement of all star positions in both x and y coordinates. The second major achievement in measuring techniques of the last 15-20 years has been the replacement of the manual machines by automatic impersonal machines, beginning with those at the Lick and U.S. Naval Observatories and proceeding to the PDS microdensitometers now in use at Yale and elsewhere. Although they differ greatly in the details of engineering design and method of image centering, they share the advantages over

their predecessors of centering with a repeatability of less than one
micron (as opposed to about two microns for a typical hand measure) and
of the removal of the personal differences between different measurers.
From somewhat limited data, the parallaxes of the Van Vleck Observatory,
measured automatically at Yale and the U.S. Naval Observatory are found
to be more precise than their hand-measured counterparts. Stetson
(1974, see also Upgren 1977) found that the total variance in external
parallax error is reduced by about one-third for the automatically
measured data. There is no reason to suppose that this gain is not
realized at other observatories as well.

The use of computers has had perhaps the greatest single influence
in parallaxes since it has led to a thorough revision in the reductions
of the measures. As mentioned above, the time consumed in hand calculations placed several constraints upon the solutions. These included
measures in right ascension only, limitation of the reference frame to
three or four stars, and the use of dependences in the solution. The
dependence solutions incorporate several disadvantages in which some
precision is sacrificed; Eichhorn and Jefferys (1971) have given an
extensive description of these limitations. Chief among them is the
absence of residuals for the reference stars allowing errors in their
measurements or significant proper motion or change in position to
escape detection.

With time of computation no longer a constraint, dependence solutions have been replaced at most observatories by solutions employing
linear (and sometimes quadratic) plate constants. A rigorous and elegant approach to the plate-constant technique using a non-iterative
method was suggested by Eichhorn and Jefferys (1971) and applied to
examples by Eichhorn and Russell (1976). Both methods and their constraints are fully described and compared by Russell (1978) who also
describes the assumptions which reduce the non-iterative method to the
more conventional plate constant solution method. Few parallaxes have
been determined using the non-iterative technique, but a variant of it
sometimes referred to as the central-overlap method (Gatewood and
Eichhorn 1973) has been directly compared to the linear plate constant
approach by Upgren and Breakiron (1980) in two sets of parallax solutions for seven stars. Later Upgren and Breakiron (1981) made new
solutions for all 269 stars whose parallaxes had been determined at the
Van Vleck Observatory in the period from 1960 to 1980. For most
series, the parallaxes and proper motions of the two methods differed
only insignificantly.

3.   CALIBRATION OF INDIVIDUAL PARALLAXES

At the time of the publication of the supplement to the GCTSP in
1963, Schlesinger's analysis of parallax errors was still accepted. He
had established the precepts upon which it was based along with a zero
point for all parallaxes considered together adopted from those defined
by the Allegheny Observatory. Allegheny was one of four observatories

which together produced about 70% of the 10,194 parallaxes listed in that catalogue; the other three are the McCormick Observatory and the Cape and Yale Observatories in South Africa. Almost all of the remaining parallaxes had been determined at six other observatories. These are the Dearborn, Greenwich, Mount Wilson, Sproul, Van Vleck and Yerkes Observatories. Schlesinger's (1928) analysis showed external errors (here converted to standard errors) for five of these observatories ranging from 10 mas for Allegheny to 21 mas for Yale with intermediate values of 15, 18 and 20 mas for Mount Wilson, McCormick and Yerkes, respectively. Hertzsprung (1952) made a very simple but straightforward analysis of the combined parallaxes for the stars in the GCTSP and found the average external mean error for a parallax in it to be 16 mas, in good agreement with Schlesinger. Both of the conclusions have stood up under repeated analysis. Vasilevskis (1966) repeated and generally confirmed Schlesinger's results for individual observatories, and Upgren and Carpenter (1977) confirmed the Hertzsprung result.

The years since the GCTSP have seen a marked decline in the numbers of parallaxes published, due mainly to the decline in the number of observatories with fully active parallax programs. But in many ways, the limited results of recent years are of greater value. One of the three most productive of recent parallax programs is new; this is the U.S. Naval Observatory program using observations of the 1.5m astrometric reflector at Flagstaff which Strand had envisioned and developed. The initial program, its procedures and the selection of the stars included was described by Worley (1966) and the results for 485 stars were published in five lists between 1970 and 1978, and summarized by Harrington and Dahn (1980) and a sixth list with parallaxes of 97 additional stars has since been published by Dahn et al (1982).

The range in apparent visual magnitude of the 582 stars with published parallaxes is quite constrained, with 88 per cent of the total being nearly uniformly distributed between magnitudes +12 and +16. Only 42 stars are brighter than this interval and 29 are fainter. Almost all of the stars are found to lie between absolute visual magnitudes +10 and +15 and are fairly evenly distributed within this interval. About one-third of the stars are white dwarfs and form a distinct sequence in the M_v, B-V color-magnitude diagram. The remainder define a narrow lower main sequence whose only significant departure appears to be for the few stars brighter than +10 which lie below the main sequence. These stars may represent a high-parallax tail of a distribution of comparatively distant stars.

The other two most productive parallax programs are modifications of older ones which employ long focus refractors on the campuses of small liberal arts colleges in the Eastern United States. These are the Sproul and Van Vleck observatories with 0.6m and 0.5m refractors, respectively. The Sproul effort has not sought to determine parallaxes as its sole, or even its primary goal. It has instead concentrated on specific nearby stars using very long plate series in order to obtain masses of known astrometric binary stars or to detect the presence of unseen companions of single stars. Nevertheless, it has been one of the most productive and steadiest sources of new and accurate parallaxes. The program is being modernized under the supervision of W.D. Heintz, whose aims have been given in a recent review (Heintz 1978).

The Van Vleck program is a departure from earlier astrometric work done at that observatory. The earlier material included in the GCTSP was published by C.L. Stearns in eight papers appearing in the Astronomical Journal between 1930 and 1959, which together contain data for 259 stars. Final solutions and more extensive details for these same 259 stars are also given by Slocum, Stearns and Sitterly (1938) for the 130 stars appearing in the first three of Stearns' lists and by Stearns (1960) for the 129 stars comprising the last five of his lists. After 1960, the program was greatly modified by H.K. Eichhorn and again by A.R. Upgren. The changes along with every Van Vleck parallax published between 1960 and 1980 have been summarized by Upgren and Breakiron (1981). This compilation lists 342 parallax solutions for 269 stars along with their photometry. The summary includes the 13 lists of Van Vleck parallaxes appearing in the Astronomical Journal between 1968 and 1980 by Upgren and his collaborators. The parallaxes and proper motions given in this summary supersede the similar data of the earlier lists since the solutions were redetermined using the original measures in order to provide data to 0.1 mas (as opposed to one mas in the original solutions). The summary also includes a list of parallaxes by Eichhorn and several solutions for individual stars. Since its publication, three further lists have been published giving parallaxes for a total of 68 additional stars (Weis, Nations and Upgren 1983 and references cited therein). These raise the post-1960 total to 410 solutions for 337 different stars.

The current Van Vleck program has as its principal concern the parallaxes of stars on the middle and lower main and subdwarf sequences. Its program stars like those of the Naval Observatory are almost all too faint to qualify for Schlesinger's original program model which limits parallax observations to stars brighter than apparent visual magnitude 5.5 and of spectral class A0 and later, thus selecting stars which are mostly closer than 100 parsecs with true absolute parallaxes larger than 10 mas. It has concentrated on the lists of K and M dwarf stars of Vyssotsky and his colleagues at the McCormick Observatory which were detected and identified spectrophotometrically. These stars avoid the high-velocity bias characteristic of earlier lists of faint nearby stars which are based on proper motion. The only other observatory with a large number of parallaxes of these stars is McCormick itself. But most of its parallaxes were placed on the program before Vyssotsky's lists and they still incorporate a selection effect towards high transverse velocity. The Van Vleck program includes the Vyssotsky stars of low transverse velocity in their proper proportion and has greatly reduced the high velocity bias present among even the nearest of stars in the catalogue of nearby stars by Gliese (1969).

Other observatories have also been active during part or all of the
period since the GCTSP. Together they account for about half of all
parallaxes published in that interval. An approximate count for each
observatory through 1977 along with the breakdown of the GCTSP by observatory has been published previously (Upgren 1978). An exact updated
list including parallaxes published since 1977 must await the completion
of the new fourth edition of the Yale parallax catalogue but approximate
numbers for each contributing observatory can be given here. Since
1963, these numbers are as follows: U. S. Naval 582, Van Vleck 410,
Sproul 284, Yale 202, Allegheny 193, McCormick 179, Lick 128, Greenwich
78, Yerkes 71, and 20 are scattered among several other observatories.

The improvements in these modern parallaxes lead to a greater
precision than even the former Allegheny parallaxes, the most precise of
the data appearing in the GCTSP. Furthermore, and perhaps most important of all, they concentrate on the very nearby stars. Since the error
in the distance and therefore the intrinsic luminosity of a star is
determined by the error in parallax divided by the parallax itself, the
nearby stars are the ones with the most precisely known absolute magnitudes. This is illustrated by Gliese (1983) who finds 585 stars with
accurate photometry whose trigonometrically determined absolute magnitudes have errors that do not exceed 0.30 mag. His M_v, B-V color
magnitude diagram shows a very narrow main sequence with no stars bluer
than about A0, and only four of the nearest giant stars appear in it,
indicating a sample severely limited in distance.   Yet a glance at the
frequency distribution of the parallaxes in the GCTSP reveals that only
a few percent of them are sufficiently large to fulfill Gliese's conditions of accuracy. The frequency distribution is illustrated by Upgren
(1978) and also in an analytical study by Hanson (1980) along with those
of each of the four leading observatories contributing to it (Allegheny,
McCormick, Cape and Yale). They mostly reflect the basic program of
Schlesinger emphasizing the naked-eye stars. Hertzsprung's (1952) conclusion, mentioned above, that the standard error of a GCTSP parallax is
16 mas came from his realization that the lower part of the distribution
closely resembled a Gaussian distribution about the median value of +18
mas for all parallaxes. He and others have realized that this distribution appeared to reflect the uncertainty in the observational process.
The errors of the recent parallaxes are well known and are much smaller
than the GCTSP data.   The U.S. Naval Observatory claims 4 mas as the
average standard error and that of the Van Vleck Observatory has been
found to be 8 mas with only small variations from one star to another.
About half of the difference between the errors of the two programs
arises from the difference in focal lengths of the two instruments and
half is due to the fact that the Naval data has been measured on automatic machines with a smaller uncertainty in image-centering ability
whereas most of those of Van Vleck published to date have been measured
by hand. Recent unpublished Van Vleck parallaxes have been measured
using the PDS microdensitometer of the Yale Observatory. These as well
as a few others measured on the SAMM machine of the Naval Observatory,
suggest that automatic centering reduces the external error of Van Vleck
parallaxes from 8 mas to 6 mas, in line with other current parallaxes

and confirming Stetson's (1974) conclusion (see Section 2). Most of the recent data from the other observatories have individual errors falling between 4 and 8 mas.

Many plausible reasons have been advanced for the much greater precision of almost all of the recent parallaxes over the earlier ones but to date no thorough study has been made which evaluates the relative merits of each. Vasilevskis (1969) cites five possible reasons for the high precision of the first parallax results of the Lick Observatory program which was begun about that time. He states them as follows: (1) automatic guiding and measurement, (2) inclusion of magnitude and color terms into the plate reduction, (3) up to 24 properly selected reference stars for each parallax star, (4) parallax solution made simultaneously in both rectangular coordinates, and (5) careful analysis of plate constants and residuals.

We can extend his list to include several additional factors which are likely to influence the sizes of errors: (6) the number of plates upon which the solution is based, (7) the number of evening and morning epochs at which plates have been taken and (8) increased rigor in the parallax solutions arising mainly from the use of computers. Not all of these likely sources of improvement have been carried out at each observatory but collectively in some degree they are very likely to account for most, if not all of the error reduction.

The first of Vasilevskis' points, automatic guiding and measurement has not yet been available or implemented at all observatories due often to a lack of funding or support staff. The second and third are interrelated since an insufficient number of reference stars can lead to underdetermined solutions. As mentioned in Section 2 one of the limitations of the dependence method of solution is the inability to examine the residuals for the comparison stars. Since most of the recent parallaxes used one or another form of the plate constant method of solution in both coordinates, as opposed to dependences and measures in right ascension only for the earlier data, it is likely that large residuals have been eliminated only in the recent parallaxes. Although their cumulative effect cannot be evaluated, they would affect points 5 and 8 above.

The number of reference stars used was almost uniformly low in earlier measures but have more recently been variable among observatories. The parallaxes of Yerkes and Lick made since the new programs were started there by van Altena (1971) and by Vasilevskis (1975), respectively, average about 20 reference stars per field. For Van Vleck, the average number of reference stars is smaller, being about 10 with most series falling between 6 and 15. This smaller number is the result of a brighter limiting magnitude due to the much smaller aperture of the telescope (0.5m vs 1.0 and 0.9m for the Yerkes and Lick refractors, respectively). The Naval Observatory on the other hand, has included generally 4 to 6 reference stars per field similar to parallax programs of the past.

About one thousand or some ten percent of all published parallaxes now have simultaneous measures in both x and y coordinates and almost all of these are recent ones. The weight of the declination component is always much less than that in right ascension and is a function of the eccentricity and orientation of the parallactic ellipse, but it can be combined with the right ascension component into a single parallax of higher weight, providing no systematic differences are present. Recently Lutz and Upgren (1980) analyzed sets of data from four observatories, which at that time included 484 stars from the Naval, 248 from Van Vleck, 199 from Sproul and 75 from McCormick, the only sufficiently large samples of combined parallaxes from individual observatories. They concluded that in all four sets of data, the parallaxes in the two coordinates are measures of the same intrinsic quantity, but that all four observatories overestimated the precision of the y-parallaxes but not the x-parallaxes. No reason for the overestimation could be found, but the study showed the value of measures in both x and y, as well as the publication of all results to 0.1 mas rather than to 1 mas as had been the common practice in earlier work.

The increase in plates, epochs and time invervals (between first and last observation) is more uniform among modern programs. The number of plates per solution was frequently as low as 15 or 20 for GCTSP data, but few recent parallaxes have been made from less than 30 plates. The number of epochs has also risen from about 5 to 8 to about 8 to 12, and the time interval from 2 to 4 years up to 4 to 7 years for the modern data. Finally the program design has resulted in an increase in the number of useful parallaxes. The ratio of the external standard parallax error divided by the parallax, is a measure of the value of the parallax, and as is described in a later section, a value of 0.15 for this ratio is a useful threshold value to adopt when defining a good or high-weight parallax and we adopt this definition here. Since the standard error in absolute magnitude is related to the ratio by the quantity $5 \log_{10} e$ or about 2.17, this threshold corresponds closely to Gliese's limit of 0.3 magnitudes mentioned above. The GCTSP contains only 376 stars (or 4% of its total number) with good parallaxes, whereas the Naval and Van Vleck Observatories have produced 375 and 160, respectively, or 64% and 47% of their total output (see Upgren, 1983a for a more detailed distribution).

One other source of error should be mentioned here for completeness. Many new developments in technique and instrumentation promise a further decrease in parallax error in the future, to possibly as low as 1 to 2 mas. At that level of precision, the correction from relative to absolute parallax becomes of importance because its error is about of this same size. The presently active programs discussed here, however, have adopted the mean corrections to absolute given by van Altena (1974) based on the mean apparent magnitude and galactic latitude of the reference stars, which appear to be sufficient.

4. SYSTEMATIC ERRORS IN PARALLAXES AND THE ZERO POINT OF THE SYSTEM

At a conference on problems in astrometry held at Evanston in 1953, Schilt (1954) and Harris (1954) both devoted their attention to parallaxes, and to Schlesinger's and Jenkins' precepts in particular. Schilt found evidence to suggest that these systematic zero-point corrections were not sufficiently well established to warrant their use in the GCTSP but Harris made an independent evaluation of the external errors and concluded that the Yale precepts are generally reliable, at least for the longer parallax series. Schilt (1958) continued his analyses of discrepancies between parallaxes from different observatories and examined the possibility that the frequent absence of agreement between parallaxes for the same stars might be due to the parallaxes of the different sets of reference stars used. However, Vasilevskis (1966) suggested that this is not a serious problem. Strand (1958, 1963) also reviewed the sources of systematic and accidental error. He and Vasilevskis were critical of the Yale precepts as was Schilt earlier. In his review, Vasilevskis repeated Schlesinger's early (1928) analysis using much of the recent data available to him. Since these investigations, many more questions have been raised about the validity of the Yale precepts used to place all observatories on the common system in use in the GCTSP, as well as the size and nature of the precision of the parallaxes of each observatory.

In the GCTSP, Jenkins adopted a value of +3 mas for converting most Allegheny parallaxes into an absolute system. A few of the more recent ones were corrected by +2 mas. Comparisons to Allegheny determined the corrections for the other observatories. Schilt's (1954) analysis also compared others to Allegheny as the standard. He concluded that corrections amounting to only -1 mas were necessary for McCormick and Greenwich but the others required corrections between -4 and -6 mas. Most disturbing was the difference of -5 mas between the only two observatories located in the Southern Hemisphere, Cape and Yale, and the two largest northern contributors, Allegheny and McCormick. This implied a zero-point difference of that size between the parallaxes of stars at southerly declinations and the stars observed from north of the Equator with no satisfactory explanation for its existence.

The analysis of the precision of parallaxes at each observatory and the way that systematic differences have been dealt with in the GCTSP has made much progress since the mostly qualitative criticisms of Schilt (1954), Strand (1963) and Vasilevskis (1966) but their conclusions have generally been supported. In the last decade, since the reviews by Gliese (1972) and Upgren (1977) of the methods by which the information about both kinds of errors has been derived, much more rigor has been introduced into these problems. New studies of both systematic and accidental errors were begun in order to define the best possible system for the new fourth edition of the Yale Parallax Catalogue for which these data must be well evaluated. These studies have had the effect of quantitatively confirming the principal conclusions and concerns of the earlier work. The advances were made possible in part by the recent improvement in the ability of photometric systems to calibrate stellar luminosities and distances. Thus, Turon Lacarrieu and Creze (1977) and Norgaard-Nielsen (1977) used photometric parallaxes to calibrate the zero point of the parallax system and compared them to the trigonometric parallaxes of each observatory separately. Systematic differences were then obtained between the latter for each of the major contributors to the GCTSP data. Both conclude that sizeable systematic corrections of 3 mas are necessary, in contrast to Schlesinger, who believed that the precepts based on the Allegheny system were accurate to less than 1 mas. Later Hanson (1980) concluded that the absolute zero point of the GCTSP parallaxes without systematic observatory corrections can be confirmed to within 1 mas. Lutz (1978) presented a thorough review of most of these attempts along with the earlier ones. Although his review appeared only about a year after those of Upgren (1977) and Heck (1978), it addresses in detail for the first time the analyses of Turon Lacarrieu and Creze and of Norgaard-Nielsen and their use of photometric parallaxes, and covers both recent parallaxes and the upcoming revision of the Yale Parallax Catalogue. The subsequent series of papers by Hanson and Lutz and their colleagues (Hanson 1979,1980, Lutz 1979,1983, Lutz and Upgren 1980, Lutz, Hanson, Marcus and Nicholson 1981, Hanson and Lutz 1983) has re-examined the parallax system and its zero point, and the external parallax errors of individual observatories in an effort to determine the best precepts to use for a new parallax catalogue. They have also investigated the calibration of absolute magnitudes and the systematic effects involved. This last problem is as important as the others because it requires the evaluation of two corrections in order to remove two well-known biases which occur whenever mean absolute magnitudes are derived from trigonometric parallaxes. These are the Malmquist correction (Malmquist 1920) and the correction described by Lutz and Kelker (1973). They must be applied to samples of stars which are magnitude-limited and distance-limited, respectively.

The major conclusions of these papers have formed the bases for the new Yale Parallax Catalogue although in describing it, van Altena (1984) concludes that a unique system of zero points may not yet be possible to achieve. The premises upon which the new catalogue is based and the conclusions of van Altena and of the papers cited above represent much progress in the analysis of trigonometric parallaxes and their uncertainties. They also illuminate the difficulties in calibrating parallaxes individually or collectively and in deriving stellar distances and luminosities from them.

The recent papers which attempt to derive the external parallax errors of individual observatories include Hanson (1978), Schmidt-Kaler (1978), Upgren (1978), Hanson (1980), Lutz et al. (1981) and Hanson and Lutz (1983). The last three form a series dealing with the systematic effects in trigonometric parallaxes as well. They summarize most major previous conclusions and intercompare errors determined from comparisons between observatories, with photometric parallaxes and with parallaxes of member stars of nearby open clusters derived from the cluster modu-

lus. The agreement between all three methods is close and demonstrates
that for some observatories (e.g. McCormick and the recent Van Vleck
program) a single standard external error characterizes the data
satisfactorily. For others such as Allegheny, Yale and the new Lick
program, the error is a function of features peculiar to each individual
program. The standard errors of 10, 16 and 17 mas for Allegheny,
McCormick and Yale remain close to Schlesinger's original findings, for
example, while Lick's estimate of 6 mas and Van Vleck's of 8 mas are
substantiated.

On the second point dealing with systematic differences, much
recent work has also been done. The main problem encountered in
attempting to define a system, is the absence of standard stars on which
it could be based. Schlesinger and his contemporaries realized this and
addressed it more than once. He proposed a list of standard stars for
parallax observation (Schlesinger 1926) but its length of 171 stars was
unrealistically long and the stars included on it covered mostly only
brighter apparent magnitudes.

Later Strand (1958) made a more realistic appeal for the observation of a specified group of stars by all observatories. He recommended
as standards groups of subgiant K stars and high-velocity F stars, thus
recognizing that standards would have very precisely determined parallaxes. They would improve the calibration of luminosities of these
kinds of stars of high astrophysical interest in addition to their use
as parallax standards. Strand identified 16 dwarf stars of spectral
classes F,G and K with parallaxes listed in the GCTSP between 30 and 50
mas and with proper motions indicative of high velocity and possibly of
a subluminous nature as well. The GCTSP numbers of these 16 stars are:
422, 674, 757, 1395, 1857, 1890, 2697, 2810, 2863, 3044, 3425, 3496,
3552, 4852, 5092 and 5098. At his suggestion made in 1956, Allegheny,
Cape, McCormick, Sproul and Yerkes placed some or all of the stars on
their programs and he directed that Yerkes emphasize subdwarfs in their
program. Later, Van Vleck added all of the stars and Yale some of them
to their programs. Although no collaborative venture has yet been
undertaken to produce joint parallaxes from the participating observatories, the observational data is by now quite substantial, and some of
these stars will serve as calibrating standard stars in any future
collaborative investigation.

In 1978 the problem of parallax standards was raised anew (Lutz
1978, Upgren et al. 1978, Upgren and Lutz 1979) and a working group was
created for this purpose in 1979 by IAU Commission 24. This is not the
first such group; in 1955 the same commission designated a similar group
which resulted in Strand's initiative. But that was the period of the
lowest ebb in parallax work and the effort was not very successful
because so few programs remained active in the years immediately afterwards. The efforts of the present working group have been published
(Upgren 1982). The rationale for the stars chosen is most fully evident
in the paper featuring inter-observatory comparisons of parallaxes by
Lutz et al. (1981). Figure 1 of that paper illustrates the severe

shortcomings in basing the observatory corrections upon mean differences
between observatories as Schlesinger had done.   The chief problem lies
in the fact that the stars observed in common by two or more observatories are rarely representative of the stars constituting the majority of
the program of any one of them.   Lutz et al. display the distribution in
apparent magnitude and in declination for the five leading contributors
to the GCTSP; Allegheny, Cape, Greenwich, McCormick and Yale.   The
southern observatories overlap Allegheny and McCormick only near the
Celestial Equator, and Greenwich not at all.   Furthermore, Allegheny and
Yale adhere closely to the original Schlesinger program featuring bright
stars whereas the others do not.   The formulation of precepts from such
unrepresentative samples is further complicated by different distributions of the colors of the program stars, and by seasonal effects which
are closely correlated with right ascension.   The most severe of the
latter form are caused by the hemispheric difference from which arises
an out-of-phase temperature variation throughout the year.   Lutz et al.
conclude that the observatory corrections of the past are neither statistically nor physically justified.   For the new parallax catalogue, the
fact remains that no proper set of standards are available and the
heterogeneous collection of data compiled in it cannot be accorded
observatory corrections of much value.   But it is hoped that for any
future fifth edition of the catalogue, the success of the present program of standards will play a role.

     The report of the working group defines two closely related objec-
tives.   The first is the necessity for the ongoing monitoring of each
telescope in order to evaluate any changes which may affect the degree
to which it must by modeled; i.e. the plate constants beyond the customary linear and second-order ones, which may be necesary to transform
its projection characteristics into cartesian coordinates.   Most refractors are suitably represented by additional first and second-order terms
for magnitude, color or coma or their combinations.   Russell (1976,
1978) studied most active astrometric instruments from positions in the
field of Praesepe and the report recommends the extension of her treatment to include this cluster and two others, the Pleiades and IC 4756.
Regular evening and morning parallax observations of these clusters,
near the Equator and fairly evenly spaced in right ascension, will be
sufficent for this purpose.

     The second objective provides for parallax standard stars (or star
fields) scattered around the sky.   For this purpose, the report identi-
fies 72 stars or stellar systems in three lists.   The first two lists
differ only in priority; it is hoped and intended that the 20 stars
comprising the first list will be placed on all programs except in cases
where the declination is inappropriate.   The members of the working
group realized the hesitation on the part of some program directors to
commit more than a small portion of the telescope time to standards.
The second list of 26 additional stars is made in order to encourage
those whose time permits to extend their observations of standard regions.   The central stars on both lists are all fainter than apparent
visual magnitude 6.5.   Brighter stars make up a third separate list

since they require extensive magnitude reduction devices which may introduce magnitude effects such as the one that Hanson (1980) found in the Allegheny parallaxes. The program intended for the HIPPARCOS satellite, if successful, may reduce the necessity for the third list since all stars brighter than this limit will be included in its program. In selecting the standards, the group sought to optimize the following features: (1) widespread and uncorrelated distributions in right ascension, declination, apparent magnitude and color for the reasons found by Lutz et al., (2) absence of detectable orbital motion in the central star or stars, (3) availability of past parallax observations, (4) minimal disruption of presently active parallax programs, (5) availability of sufficient numbers of suitable reference stars and (6) astrophysical interest of the central star or stars. The success of this endeavor depends above all upon the willingness of parallax observers to devote a reasonable part of the available telescope time to calibration, in the same manner as do participants in photometric and spectroscopic programs.

The corrections to absolute magnitudes arising from errors in trigonometric parallaxes are not as easily applied as might be believed. Mention has been made of this difficult problem, which exists because it is not possible to define or obtain a sample of stars which is not limited in distance or apparent magnitude or both. Lutz (1983) has reviewed the Malmquist and Lutz-Kelker biases which result from samples with these two limitations. The Malmquist bias is the more easily evaluated of the two corrections, for samples drawn from a luminosity function whose form and space density are known or assumed. The Lutz-Kelker bias is not so easily handled because a true volume-limited sample cannot be obtained. Errors in parallaxes, unless they are extremely small with respect to the parallaxes themselves, produce a bias in any sample limited in observed parallax. Such a sample favors stars whose true parallaxes are smaller on average than the observed values. Lutz and Kelker (1973) sought to evaluate the correction to the absolute magnitudes computed from the parallaxes and their errors. They were successful for stars for which the error to parallax ratios are less than 0.175, a severe restriction. Our constraint of 0.15 and Gliese's equivalent limit of 0.3 in absolute magnitude are both based in part upon this limit. Since the errors are independent of the sizes of the parallaxes in most cases, the magnitude correction increases in size with increasing distance or decreasing observed parallax until this limit is reached. Lutz (1979) and others have tried to extend the range in distance over which the correction can be applied, but their efforts have not been entirely successful. The most significant consequence of this limit is the apparent inability of trigonometric parallaxes to determine the absolute magnitudes of distant luminous stars without introducing a bias affecting the luminosities by a large and unknown amount. Unless systematic effects of this kind become better understood, it remains doubtful that even large numbers of parallaxes of the very bright stars all of which are distant, will be of use in their luminosity calibration. For these stars, secondary distance methods of greater reliability will continue to be necessary and are usually available.

Perhaps the most significant of the new devices which may be used for parallax is HIPPARCOS, the new astrometric space satellite of the European Space Agency. This is one of many new and promising ways of surpassing the limit in precision of the long-focus refractor coupled with the photographic plate from which most of todays parallaxes have been determined. These methods are discussed by van Altena (1983). It should be noted here, however, that the calibration of stellar luminosities from parallaxes requires long and often tedious observations of a great number of stars of any one type, and that makers and users of new instruments and techniques are all too often mutually exclusive groups of people. The development of a method does not guarantee its extensive use. The advantage of HIPPARCOS, if it is successful at all, lies in the assurance of just this very necessary abundance of data in a reasonably short time. As a consequence, our discussion of parallax programs of the future will be mostly limited to this instrument.

Within a decade or two, the parallaxes of some 100,000 stars should be completed and reduced. This program is to include all of the 40,000 brightest stars in the sky; the faint limit in apparent magnitude of this group is about 8 visual or 9 photographic. Below this completeness limit stars would be included in decreasing numbers with increasing faintness to the detection limit of about 12 visual or 13 photographic. Although the parallax errors are not yet known, preliminary estimates are 2 mas for stars brighter than the completeness limit increasing to 5 mas at the detection limit. The effectiveness of this satellite in improving the absolute magnitudes of stars, using these errors, has been studied by Gliese (1979), Pagel (1979), Lutz (1983), Murray (1983) and Upgren (1983a,b). These investigators have described the constraints imposed by the Lutz-Kelker limit on magnitude calibration and the expected completeness and precision of the HIPPARCOS program upon nearby stars of all kinds. They conclude that the usefulness of the program in absolute magnitude calibration will be confined to the A, F and early G stars of the main and subdwarf sequences as well as the giant and subgiant sequences. For stars brighter than these, the space density is too low; too few of them lie within about 75 parsecs, the maximum distance at which the expected parallax error ratio is still smaller than the Lutz-Kelker limit. At the other extreme, they noted that too few of the stars intrinsically fainter than the early G dwarfs appear brighter than the completeness limit of the program, but this is no problem since the fainter stars are being adequately covered by the present ground based programs. By working in close collaboration with the ground based observers, and observing the same sets of standard stars, HIPPARCOS will make a great contribution to luminosity calibration by parallax.

The Hubble Space Telescope is a satellite also capable of high-precision astrometry. It is to be a 2.4-meter f/24 Cassegrain configuration of Ritchey-Chretien optical design. In the course of many other kinds of observations, it is expected to obtain parallaxes for a few hundred stars with standard errors of about 0.5 mas. If the calibration of the Malmquist and Lutz-Kelker biases can be improved and extended to parallax data of higher percentage error than 15 to 20 percent, then it may be possible to extend high-precision calibration of absolute magnitudes from parallaxes to greater distances than is now possible. In this event, some thinly populated luminous stars might be successfully observed and the bright limit in absolute magnitude of -1 to -2 expected from HIPPARCOS might be increased. Pagel (1979) has summarized the uses of small parallaxes and points out that parallaxes of at least the closest few stars as bright as absolute magnitude -5 may be useful, but only if no systematic errors are present. However, the number of parallaxes to be obtained with the Space Telescope is limited and it is likely that its major astrometric contribution will be made in two other areas. These are the nearby stars where the detection of small perturbations could lead to the discovery and study of brown dwarfs and large planets, and the parallax standard stars where its very high precision may establish the zero point of parallaxes from both the HIPPARCOS and the ground-based programs. Observations of the standards by the Space Telescope would increase the value of all recent and future parallaxes from other sources.

5. SUMMARY

It is perhaps fitting that a new edition of the Yale Parallax Catalogue be published at this time for it summarizes the period of the supremacy of the photographic plate and the long-focus refractor. In the last twenty years, the full potential of these conventional devices has been more fully realized. The field of positional astronomy is rife with new techniques and devices promising further gains in parallax precision. It is hoped that the best of these find dedicated astrometrists willing to make the lengthy parallax observations necessary for the calibration of distances and other properties of sufficient numbers of stars. But for another decade at least, our distance scale shall still be reliant upon the more conventional parallaxes, calibrated hopefully on standard stars and regions for the first time.

The author wishes to acknowledge the National Science Foundation for continuing support for the astrometric program of the Van Vleck Observatory, most recently under Grants AST-8121463 and AST-8318649.

REFERENCES

Dahn, C. C. et al. 1982, Astron. J. 87, 419.
Eichhorn, H. K. and Jefferys, W. H. 1971, Publ. Leander McCormick Obs.
    Vol. XVI, p. 267.
Eichhorn, H. K. and Russell, J. L. 1976, Mon. Not. R. Astr. Soc. 174,
    679.
Gatewood, G. D. and Eichhorn. H. K. 1973, Astron. J. 78, 769.
Gliese, W. 1969, Veroff. Astron. Inst. Heidelberg, No. 22.
Gliese, W. 1972, Q. J. R. Astr. Soc. 13, 138.
Gliese, W. 1979, European Satellite Astronomy, C. Barbieri and P. L.
    Bernacca, eds., University of Padua, Padua, Italy, p. 195.
Gliese, W. 1983, IAU Colloquium No. 76, The Nearby Stars and the Stellar
    Luminosity Function , A. G. D. Philip and A. R. Upgren eds.,
    L. Davis Press, Schenectady, N. Y., p. 5.
Hanson, R. B. 1978, IAU Colloquium No. 48, Modern Astrometry ,
    F. V. Prochazka and R. H. Tucker eds., Univ. Obs., Vienna, p. 31.
Hanson, R. B. 1979, Mon. Not. R. Astr. Soc. 186, 875.
Hanson, R. B. 1980, Mon. Not. R. Astr. Soc. 192, 347.
Hanson, R. B., and Lutz, T. E. 1983, Mon. Not. R. Astr. Soc. 202, 201.
Harrington, R. S. and Dahn, C. C. 1980, Astron. J. 85, 454.
Harris, D. 1954, Astron. J. 59, 59
Heck, A. 1978, Vistas in Astronomy 22, 221.
Heintz, W. D. 1978, IAU Colloquium No. 48, Modern Astrometry ,
    F.V. Prochazka and R. H. Tucker eds., Univ. Obs., Vienna, p. 82.
Hertzsprung, E. 1952, Observatory 72, 242.
Jenkins, L. F. 1963, General Catalogue of Trigonometric Stellar
    Parallaxes , Yale University Observatory, New Haven.
Lutz, T. E. 1978, IAU Colloquium No. 48, Modern Astrometry ,
    F. V. Prochazka and R. H. Tucker eds., Univ. Obs., Vienna, p. 7.
Lutz, T. E. 1979, Mon. Not. R. Astr. Soc. 189, 273.
Lutz, T. E. 1983, IAU Colloquium No. 76, The Nearby Stars and the
    Stellar Luminosity Function , A. G. D. Philip and A. R. Upgren eds.,
    L. Davis Press, Schenectady, N. Y., p. 41.
Lutz, T. E., Hanson, R. B., Marcus, A. H. and Nicholson, W. L. 1981,
    Mon. Not. R. Astr. Soc. 197, 393.
Lutz, T. E. and Kelker, D. H. 1973, Publ. Astron. Soc. Pacific 87, 617.
Lutz, T. E. and Upgren, A. R. 1980, Astron. J. 85, 1390.
Malmquist, K. G. 1920, Lund Astron. Obs. Medd., Ser. II, No. 22.
Murray, C. A. 1983, Proceedings of an International Colloquium on the
    Scientific Aspects of the Hipparcos Mission, Held at Strasbourg,
    France, p. 115.
Norgaard-Nielsen, H. U. 1977, Astron. Astrophys. 59, 203.
Pagel, B. E. J. 1979, European Satellite Astronomy, C. Barbieri and
    P. L. Bernacca, eds., University of Padua, Padua, Italy, p. 211.
Russell, J. L. 1976, Thesis, University of Pittsburgh.
Russell, J. L. 1978, IAU Colloquium No. 48, Modern Astrometry ,
    F.V. Prochazka and R. H. Tucker eds., Univ. Obs., Vienna, p. 355.
Schilt, J. 1954, Astron. J. 59, 55.
Schilt, J. 1958, Astron. J. 63, 173.
Schlesinger, F. 1911, Astrophys. J. 33, 161.

Schlesinger, F. 1924, Probleme der Astronomie (Seeliger Festschrift),
    Springer-Verlag, Berlin, p. 422.
Schlesinger, F. 1926, Astron. J. 36, 185.
Schlesinger, F. 1927, Mon. Not. R. Astr. Soc. 87, 506.
Schlesinger, F. 1928, Astron. J. 38, 189.
Schmidt-Kaler, Th. 1978, IAU Colloquium No. 48, Modern Astrometry,
    F.V. Prochazka and R. H. Tucker eds., Univ. Obs., Vienna, p. 55.
Slocum, F., Stearns, C. L., and Sitterly, B. W.(1938), Publ. Van Vleck
    Obs. Vol. I.
Stearns, C. L. 1960, Publ. Van Vleck Obs. Vol. III.
Stetson, P. C. 1974, Thesis, Wesleyan University.
Strand, K. Aa. 1958, Astron. J. 63, 152.
Strand, K. Aa. 1962, Astron. J. 67, 706.
Strand, K. Aa. 1963, Stars and Stellar Systems Vol. III, Basic Astro-
    nomical Data ,K. Aa. Strand ed., University of Chicago Press,
    Chicago, p. 55.
Turon Lacarrieu, C. and Creze, M. 1977, Astron. Astrophys. 56, 273.
Upgren, A. R. 1977, Vistas in Astronomy, 21, 241.
Upgren, A. R. 1978, IAU Colloquium No. 48, Modern Astrometry,
    F.V. Prochazka and R. H. Tucker eds., Univ. Obs., Vienna, p. 69.
Upgren, A. R. 1982, IAU Trans. Vol XVIIIB, p. 197.
Upgren, A. R. 1983a, IAU Colloquium No. 76, The Nearby Stars and the
    Stellar Luminosity Function, A. G. D. Philip and A. R. Upgren
    eds., L. Davis Press, Schenectady, N.Y., p. 57
Upgren, A. R. 1983b, Kinematics, Dynamics and Structure of the Milky Way,
    W. L. H. Shuter, ed., D. Reidel, Dordrecht, p. 15.
Upgren, A. R. and Breakiron, L. A. 1980, Astron. J. 85, 71.
Upgren, A. R. and Breakiron, L. A. 1981, Astron. J. 86, 776.
Upgren, A. R. and Carpenter, K. G. 1977, Astron. J. 82, 227.
Upgren, A. R., Gatewood, G. D., Lutz, T. E. and Ianna, P. A. 1978, Bull.
    A. A. S. 10, 649.
Upgren, A. R. and Lutz, T. E. 1979, Dudley Obs. Report No. 14, Problems
    of Calibration of Multicolor Photometric Systems, A. G. D.
    Philip, ed., L. Davis Press, Schenectady, N. Y., p. 235.
van Altena, W. F. 1971, Astron. J. 76, 932.
van Altena, W. F. 1974, Astron. J. 79, 826.
van Altena, W. F. 1983, Ann. Rev. Astron. Astrophys. 21, 131.
van Altena, W. F. 1985, IAU Symposium No. 109, Astrometric Techniques,
    H. K. Eichhorn ed., D. Reidel, Dordrecht, in press.
van de Kamp, P. 1962, Stars and Stellar Systems Vol. II, Astronomical
    Techniques, W. A. Hiltner ed., University of Chicago Press,
    Chicago, p. 487.
Vasilevskis, S. 1966, Ann. Rev. Astron. Astrophys. 4, 57.
Vasilevskis, S. 1969, Bull. A. A. S. 1, 209.
Vasilevskis, S. 1975, Publ. Lick Obs. Vol. XXII, Part IV.
Weis, E. W., Nations, H. L., and Upgren, A. R. 1983, Astron. J. 88, 1515.
Worley, C. E. 1966, Vistas in Astronomy 8, 33.

DISCUSSION

GLIESE: In your excellent paper you showed a color-luminosity diagram of 585 stars with accurate absolute magnitudes (standard error < 0.3). I presented that diagram last June at Middletown, having just received a preliminary version of van Altena's next parallax catalogue. Meanwhile some dubious cases have been cleared up and further photoelectric data became available. Today we have 681 stars with precise M_V data.

JASCHEK: I have seen claims that the U.S. Naval Observatory program provides parallaxes accurate to ±0.002 arcsec. Could you comment?

UPGREN: I think ±0.004 arcsec is the average external error of a USNO parallax.

STRAND: The USNO parallax program published so far had an external error of 0.004 arcsec. However the present use of fine grain plates has reduced this error to 0.002 arcsec. Another development is to extend the program beyond the 16th magnitude by use of a CCD camera and for bright stars experiments are being made with a metallic spot on the filter, reducing the parallax star by about 10 magnitudes relative to the comparison stars. Such filters have been used successfully by Dr. Pascu of USNO in observations of the satellites of Jupiter.

UPGREN: The Van Vleck Observatory and many others are also experimenting with fine-grain emulsions which promise a substantial reduction in external parallax error.

JASCHEK: There was a program at the US Naval Observatory to obtain colors and spectra of all parallax stars. Is that still going on?

STRAND: Photometric observations in the UBV system are made of all stars selected as candidates for parallax observations prior to being placed on the program to ascertain if they have measurable parallaxes.

KEENAN: May I ask, that for the benefit of those engaged in statistical calibrations, parallax observers should not limit themselves to stars expected to give "good" parallaxes, but should give us a complete magnitude-limited sample down to V = 5.0?

UPGREN: Hipparcos will produce parallaxes for all of the 40,000 stars brighter than about the eighth magnitude. However, ground-based telescopes may do well to observe some of the standards among these brighter stars if only to check on any zero-point errors in these data.

POPPER: How are the external errors of the newer parallax series evaluated? There has been no discussion of the parallaxes of visual binaries with good orbits. Must we wait for results from Hipparcos or will they be included in the bright star program at USNO mentioned by Dr. Strand?

UPGREN: Parallaxes of binaries have been done at the Sproul Observatory. Dr. Heintz may discuss this work later in this meeting. External errors of recent parallaxes are determined in a variety of ways including comparison with results from cluster members and spectroscopic parallaxes of distant stars among the reference stars.

BESSELL: If one wishes to use a parallax from a catalogue when should one use the Lutz-Kelker correction?

UPGREN: I'm not sure of the justification for applying that correction to individual stars.

MILLWARD: You showed us a table which indicates that Hipparcos will be able to obtain good parallaxes for 26 stars of magnitude -1.0 or brighter. How many of these stars are main sequence objects?

UPGREN: Most of these stars are main sequence objects, although the brightest star at M = -3 is a supergiant, Canopus.

JASCHEK: The Hipparcos limit of completeness is about 7.5 to 8.0 mag. After this limit and up to 12th mag. there exist a large number of programs of astrophysically important stars (Am, Ap, Miras, RR Lyr. etc.). There exist about 180 specific programs of this kind, so that eventually we shall get parallaxes for all these groups.

UPGREN: If the intrinsically bright groups among these kinds of stars are all very distant, it may be difficult to apply the Malmquist and Lutz-Kelker corrections in such a way to produce correct, unbiased luminosities. Secondary distance methods may still be better for some kinds of stars.

SEGGEWISS: How did you estimate the fraction of "good" parallaxes for the Hipparcos data?

UPGREN: By calculating the proportion of stars at each apparent magnitude which lie within the horizon determined by $\sigma/\pi < 0.15$ and the estimated precision of the Hipparcos data. Details are given in my paper in IAU Colloquium No. 76 held last June at Middletown.

THE DETERMINATION OF FUNDAMENTAL PROPER MOTIONS

Thomas E. Corbin

U. S. Naval Observatory

ABSTRACT Fundamental positions and proper motions are derived from absolute
catalogs. Taking recent Washington Six-Inch and Seven-Inch Transit
observing programs as examples, the observation and compilation of absolute
catalogs are discussed. The roles of quasi-absolute and differential
catalogs in the formation of fundamental catalogs are also examined.
Finally, the current effort to extend the fundamental system to fainter
magnitudes than are presently represented in the FK4 is described as well
as the expected composition of the forthcoming FK5 catalog.

INTRODUCTION

The stellar positions and proper motions of the Fourth Fundamental
Catalog (FK4), compiled by Fricke and Kopff (1963), form the reference
frame that has been the basis of celestial coordinates for the past twenty
years. At their respective epochs the mean positions are quite good
north of -30 degrees with mean errors of only 1 to 4 milliseconds of time
in right ascension and 2 to 5 hundredths of an arcsec in declination.
However, south of -30 degrees the errors are generally two to three times
as great. Even so, it is not these errors that are the main sources of
difficulty in using the FK4 at current epochs, but rather the 60 to 75
years of proper motion that must be applied to the majority of the stars.
For example, using the formal errors listed in the FK4, an error of 0.18
arcsec is to be expected at 1985 for star 452. Of this amount 0.17 arcsec
is due to proper motion. In addition to this, modern observations have
shown that the systematic and individual errors of the FK4 proper motions
published in the FK4 are considerably larger than had been estimated,
especially in the Southern Hemisphere (Schwan, 1985). This situation,
combined with the early epochs of the FK4 positions, led Fricke and Gliese
(1968) to propose the preparation of an FK5.

D. S. Hayes et al. (eds.), Calibration of Fundamental Stellar Quantities, 53–70.
© 1985 by the IAU.

A new fundamental catalog would not only take advantage of recent observations to reduce both random and systematic errors, but would also extend the fundamental catalog in its range of magnitudes and improve its distributions of stars over the celestial sphere. The improvement of the FK4 positions and motions involves three stages: 1.) The zero points of the system are redefined to more accurately represent the dynamically defined fundamental reference frame. To achieve this, absolute catalogs of observed stellar positions are required. These are also often referred to as fundamentally observed catalogs. 2.) Systematic errors of the FK4 positions and motions that cause distortions of the reference frame must be corrected. This step requires absolute and quasi-absolute catalogs. 3.) The individual random errors of the FK4 positions and motions are reduced through the application of the observed positions in absolute, quasi-absolute and differential catalogs. In addition, an extension to fainter magnitudes and an improved distribution of fundamental stars over the sky was proposed by Fricke (1973) and will involve the addition of stars from the "Catalog of 1987 Supplementary Stars to FK4" (FK4 Sup), Fricke (1963) and from the list of International Reference Stars (IRS), Scott (1962). It is the purpose of this review to discuss how absolute observations are currently made, how these are combined to produce absolute star positions and how these positions may be used to yield fundamental positions and proper motions. In addition, the role that the FK4 Sup and IRS stars can play in the FK5 will be examined.

1. OBSERVATIONS

It is appropriate to discuss the observations first because not only are they the first step in the process of determining fundamental positions and motions but also they are the most important. The ultimate quality of the fundamental catalog depends on observed absolute positions that are as free as possible of instrumental error and that are part of a program that has been conducted in such a manner as to allow the determination of a reference frame that is independent of existing catalog positions. Since transit circle observations have determined the reference frame in the past and will do so again in the FK5, they will be discussed here.

In order to make absolute observations the orientation of the telescope must be accurately known. In the Washington programs, the collimation, level, nadir point with respect to the local gravity and azimuth are measured every two to three hours, as is shown in Table I. In fact, during the W L-50 program at El Leoncito, Argentina (Hughes, Smith and Branham 1984), observers were encouraged to take the first two sets of instrumental constants after sunset only one hour apart because of the rapid temperature changes occurring then. Very small mechanical shifts in the mounting of the instrument or its optics can have serious effects on the observations if not corrected. A shift of

TABLE I

INSTRUMENTAL CALIBRATION
FOR ABSOLUTE PROGRAMS

QUANTITY	FREQUENCY	ACCURACY
AZIMUTH OF MARKS	MONTHLY	+0.08"
MARKS − 1 TOUR	2 − 3 HOURS	− .30
COLLIMATION	2 − 3 HOURS	.25
LEVEL	2 − 3 HOURS	.15
NADIR	2 − 3 HOURS	.35
CIRCLE −DIAM CORR	4 TIMES/PROGRAM	.07
− ONE READING	EACH STAR	.06
CLOCK CORRECTION	EACH TOUR	.10
REFRACTION	PUBLISHED TABLES	.03
TEMPERATURE	EACH STAR	.10 °C
DEW POINT	EACH STAR	1.0
BAROMETER	EACH STAR	.1 mm
VAR OF LATITUDE	BIH & SOLUTION	0.03"
FLEXURE	PROGRAM AVERAGE	.02
SCREW ERROR − RA	1 TIME/YEAR	.02
SCREW ERROR − DEC	1 TIME/YEAR	.02
PIVOT ERRORS	2 TIMES/PROGRAM	<.005
CLAMP DIFFERENCES	REVERSE MONTHLY	.02
EQUINOX CORRECTION	PROGRAM AVERAGE	.14
EQUATOR CORRECTION	PROGRAM AVERAGE	.02

only 10 microns in one of the mounting cages of the Washington Six-Inch Transit Circle can produce a shift of 2.3 arcsec in the azimuth. In addition, there are characteristics of a transit circle that need to be calibrated less frequently but with high accuracy. For example, the corrections to the divisions on the graduated circle are now measured four times during a program. In addition the circle is monitored every six months. This allows the corrections to be applied with an accuracy of 0.24 microns, corresponding to 0.07 arcsec in the pointing of the instrument in declination. (Rafferty and Klock, 1982) The instrument's pivots, micrometer screws and tube flexure must also be evaluated. Because a transit circle must operate under a wide range of environmental conditions a large number of calibrations will not necessarily lead to an accurate determination if they are not also taken over a similar range.

An example is the determination of flexure in the Washington transit
circles. For many years, measures of the coefficient of the flexure were
made on cloudy nights when the instruments were not otherwise occupied.
However, during the recently completed W 6-50 program a flexure measure
was made with each collimation. This practice has permitted an extensive
examination of the behavior of the flexure covering the same range of
conditions as the observations, including the important daytime observatons.
Miller (1984) has found that in previous programs the flexure has been
treated too simply and that if the temperature and even the rate of change
of the temperature are not taken into account then errors in computing
the flexure can amount to as much as $\pm$0.55 arcsec.

In order for an observing program to produce absolute results,
several requirements should be met. To obtain independent determinations
of the azimuth of the instrument, observations of circumpolar stars must
be made twelve hours apart. Generally, these can be made only between
the autumnal and vernal equinoxes. During the other six months the
meridian marks must be relied upon to monitor the azimuth. Since the
marks are usually viewed through several hundred feet of air near the
surface of the ground, it is important that there be one north and another
south of the instrument. It has recently been determined at the Washington
site that the images move several arcseconds with periods of around one
or two minutes. Fundamental stars within 20 to 30 degrees of the equator
should be observed at a rate of one every 30 to 40 minutes. Each night's
observing should have a good distribution of these stars in declination.
If the program is to contribute to the determination of the equator and
equinox of a fundamental reference frame, then either the Sun, Mercury
and Venus or the minor planets must be observed. It is, of course,
preferable to observe both groups. Mars and Jupiter can contribute to
the equator correction (Branham, 1984), but observations of Saturn, Uranus
and Neptune mainly contribute to improved orbits for these objects. The
Moon should also be observed. Branham concludes that, theoretically,
the Moon can contribute strongly to the equinox determination, but that
the difficulty of observing the lunar limb and applying the appropriate
corrections reduces the lunar contribution. Finally, if the goal of a
program is to provide new absolute positions for the improvement of a
fundamental catalog, then each observing tour should contain stars from
the catalog well distributed over declination as well as right ascension.
This allows the $\Delta\alpha_\delta$ and $\Delta\delta_\delta$ terms to be corrected. In the forthcoming
Washington Six-Inch/Seven-Inch program, each observing tour will contain
at least two FK5 stars for every 15 degrees of declination.

2. FORMATION OF AN ABSOLUTE CATALOG

In order for a program to give absolute positions the observing
tours must be reduced in such a way that the results are independent of
any existing catalog. Two recent programs will serve to illustrate this
process. In the W 5-50 program (Hughes and Scott, 1982) FK4, IRS from
-30° to $+5^\circ$, the stars of IAU Resolution No. 17 and several small lists

were observed. The observations of the azimuth stars in the September
through March period served two purposes. First the upper and lower
culmination observations served to establish the azimuth of the instrument
independent of the FK4, and second, these same observations served to
correct the stars' positions for determining the azimuth of the instrument
during the summer. Averaging this value on a monthly basis, the average
value of the azimuth of the marks was then determined for each similar
time period. This, combined with the mark measures made during the tours
allowed an azimuth independent of the FK4 to be applied to each object
observed in each tour, the marks now serving as interpolating devices.
At this point the right ascension system of the observations is independent
of the FK4 as far as the azimuthal orientation of instrument is concerned,
but is tied to the FK4 with respect to origin and systematic errors of
the type $\Delta\alpha_\alpha$. This is because each tour has been reduced using a clock
correction that is derived from the FK4 positions of the selected clock
stars between $+30^\circ$ to -30°. However, due to the high quality of modern
clocks and the accurately calibrated orientation of the instrument it is
possible to pair tours either during the same night or morning-evening
of consecutive nights to derive periodic corrections to the clock star
system. In the Six-Inch Programs these corrections have been given the
form:

$$\Delta\alpha = A \sin(\alpha) + B \cos(\alpha) + C \sin(2*\alpha) + D \cos(2*\alpha).$$

Note that this solves for periodic errors within the FK4 right
ascension system but cannot correct the origin. The periodic corrections
were added to each star's position and then new clock corrections were
computed for each tour. Additional individual corrections to the clock
stars were determined by differencing each star in a tour with the improved
mean clock correction for that tour. Thus final corrections of periodic
plus individual corrections could be determined for each clock star and
applied. At this point final clock corrections were computed and applied
to all of the observations in each tour. Since the instrument can operate
with its pivots first in one position and then reversed, every object
was observed in both positions. (These positions are called clamp east
and clamp west.) Reversals were made monthly. Comparing the results made
on the two clamp positions and averaging the results gave the final
positions on the instrumental system with the FK4 equinox as the origin.

In the second case, the W L-50 observations, it was possible to
determine the corrections to the positions of the azimuth stars in a much
more rigorous manner. In the W 5-50 program the large number of partly
cloudy nights made it necessary to combine upper and lower culmination
observations of the azimuth stars on a monthly basis. At El Leoncito
the many clear nights made it possible for most tours observed during
the fundamental azimuth period to be paired with at least one other.
This was important for two reasons: the positions of the southern azimuth
stars are much poorer than those in the north and Leoncito is a seismically
active area. Thus it was possible to confirm the stability of the marks
over periods as short as 24 hours. Again, analysis of the paired upper
and lower culmination observations resulted in corrections to the catalog

positions of the azimuth stars that were linked only to the FK4 equinox.
These corrections were then used to compute improved azimuths of both
summer and winter tours leading to improved clock corrections for the
tours. In turn, the improved clock correction was used to further refine
the azimuth. In all cases three iterations of the azimuth were sufficient
to reduce the change between the last two iterations to $\pm$0.0005 second
of time. Finally, the observations of the clock stars were analyzed by
pairing tours only from the same night. No functional fit was assumed,
but rather new corrections to the individual stars were computed relative
to the nightly mean clock correction. Included in the analysis were
personal equations of the observers. Surprisingly, several showed
significant values for the program, ranging from $+0^s.017$ to $-0^s.012$. New
clock star residuals were computed for each night by applying the individual
corrections to each star, personal equations, if any, and change in the
equation in the equinox. Only three iterations of this process were
necessary. From this, a final value of the clock correction for each
tour was derived and applied to each object observed in the tour. Again,
the point is reached where the results represent a system that is
independent of the FK4 except for the zero point in right ascension.
The adherence to the FK4 equinox was checked by summing the final observed
minus computed residuals of the FK4 stars that were not clock stars.
The result was $+0^s.0015$.

In reductions of the declinations, the starting points are the
assumed latitude of the instrument and the readings of the nadir taken
each tour. The observed positions in declination are related to the
zenith point through the circle of the instrument and the individual
corrections to its inscribed divisions. In addition, a model of the
atmospheric refraction must be applied to the observed zenith distances.
In the W 5-50 and W L-50 the Pulkovo Refraction Tables were used, the
third and fourth editions, respectively. The variation of latitude used
in the W 5-50 was taken from the results of the Washington PZT. Errors
in the assumed latitude, refraction and flexure are determined by an
analysis of the circumpolar FK4 stars observed at upper and lower
culmination. The success of such an analysis depends strongly on the
latitude of the observing site. At the Washington site $(+38^{\circ}\ 55')$ it
was possible to separate all three variables, and so the flexure measured
from the collimators was not used. The El Leoncito site $(-31^{\circ}\ 48')$ is
even closer to the Equator than Washington and the flexure term (sin z)
becomes inseparable from the refraction (tan z), where z is the zenith
distance. Thus a large number of flexure measures were made and the
coefficient so determined was used without further correction. Otherwise
the the same procedure was used as in the W 5-50 to compute preliminary
declinations. In both programs these procedures yielded declinations
independent of the FK4 system.

At this stage the results are not yet absolute. In right ascension
the zero points are tied to the FK4 equinox. In declination it will be
seen that residual errors in the solutions for the corrections to latitude
and refraction cause the positions to be on a system that is in each case
free of the FK4 but that should not be used to correct the FK4 declinations.

To make the results in each coordinate absolute, observations of solar system objects are required. In the W 5-50 the results of observing the Sun, Mercury, Venus, Mars and Jupiter were used to correct the equinox and equator of the catalog. The Sun, Mercury and Venus were observed in conjunction with FK4 stars bright enough to be seen in the daytime. These stars were used to connect the daytime observations with those made at night. The observations of Mars and Jupiter made in each tour were reduced along with the stars in that tour. Differences between the observed and ephemeris positions of the Sun and planets were then used to compute corrections to the catalog's equator, equinox and to the orbital elements of the planets and the Earth. This resulted in corrections of $+0\overset{..}{.}431 \pm 0\overset{..}{.}033$ in right ascension and $-0\overset{..}{.}347 \pm 0\overset{..}{.}032$ in declination. The right ascension correction also represents the correction to the FK4 equinox at the mean epoch of the catalog. The declination correction was used to adjust all of the declinations in the catalog by defining the final correction to the latitude, refraction and flexure. The provisional system was established by using a measured coefficient of the flexure of $+0\overset{..}{.}0123$ and initial adjustments of the latitude and variation of latitude determined from the upper and lower culminations of the circumpolar stars. The provisional correction to declination is

$$\Delta\delta = +0\overset{..}{.}353 - 0\overset{..}{.}057 \tan z + 0\overset{..}{.}012 \sin z.$$

This gives the adjustment determined at the pole. The adjustment determined at the equator, from the solar system objects, is

$$\Delta\delta = -0\overset{..}{.}193 + 0\overset{..}{.}015 \tan z + 0\overset{..}{.}225 \sin z.$$

The sum of these,

$$\Delta\delta = +0\overset{..}{.}193 + 0\overset{..}{.}015 \tan z + 0\overset{..}{.}225 \sin z,$$

is then the total adjustment to the declination system. This includes the correction to the preliminary value of the coefficient of the flexure, and the three values are the final corrections to the latitude, refraction and flexure, respectively. It is seen from this that the final correction to the declinations retains the location of the pole that was defined by the upper and lower culmination observations and yet brings the system into agreement with the dynamically defined equator. The W L-50 reductions generally followed the above procedure but differed in a few important details. First, and most important, the Sun and planets were not observed in this program. The minor planets Ceres, Pallas, Juno and Vesta were observed, however, and the large number of clear nights at El Leoncito permitted over a thousand minor planet observations to be taken during the six year course of the program. This strong body of data allowed Branham (1979) to conclude that equinox and equator corrections could be derived from these observations alone. Another difference was that after the provisional corrections to the refraction and latitude were applied to the declinations a catalog mean <O-C> was formed for each FK4 star and the FK4 <O-C>'s of each tour were compared to this mean. This comparison showed no zenith distance variations, but a number of tours

had large zero point differences in declination relative to the mean.
The average difference for each tour was applied to all of the observations.
The final equinox correction derived from the minor planets is +0''713
+0''138, and the equator correction is -0''056 +0''016. Again, the equator
correction was used to correct the latitude and refraction in such a way
that the fundmentally determined pole remains unchanged and the correction
at the equator is -0''056.

3. FUNDAMENTAL POSITIONS AND PROPER MOTIONS

The intention of the somewhat lengthy discussion above, illustrating
how absolute catalogs are observed and reduced, has been to stress the
necessity of freeing the results from systematic error as much as possible
and rigorously relating them to a dynamically defined inertial system.
If these processes are not successfully carried out, no amount of analysis
or modeling can prevent a catalog's results from adversely affecting a
fundamental catalog if they are included in the solutions for the
fundamental system. This has become especially true in the past twenty
years when there have been so few observatories producing absolute results.

3.1 Determination of the Equinox and Equator

The system of FK4 positions and proper motions, combined with
Newcomb's precession, define a reference frame at any given dates of
epoch and equinox. Using absolute catalogs it is possible to compare
the FK4 at the date of each catalog with the FK4 stars in that catalog
in order to evaluate the corrections needed to the FK4 equinox and equator.
Relying primarily on observations of the Sun made at Cape, Greenwich,
Washington, Ottawa, Breslau and Pulkovo in combination with results of
lunar occultations and minor planet results, Fricke (1982) compared the
FK4 equator and equinox with absolute determinations over a range of
epochs from 1906 to 1971. Figure 1 shows this comparison. Fricke found
that not only does the right ascension zero point represented by the FK4
at 1950.0 need correction, but that the the correction is also epoch-dependent.
Thus the error in the FK4 equinox is found to be

$$E(T) = +0^{s}.035 \pm^{s}.003 + (0^{s}.085 \pm^{s}.010) (T - 19.50)$$

where T is in centuries from 1950. A similar study of the FK4 equator
showed no need for correction. Thus the absolute catalogs are used to
correct the zero points of the fundamental system. It should be noted
that the correction derived to the FK4 equinox is really defined only
over the range of declinations covered by the Sun and Moon, but is assumed
to apply at all declinations. Deviations from this assumption are
corrected at a later step. (See section 3-b.)

Correcting the FK4 equinox in order to define that of the FK5 has
the great advantage of making the transition to an improved fundamental
system with as little discontinuity as possible. However, Smith (1985)

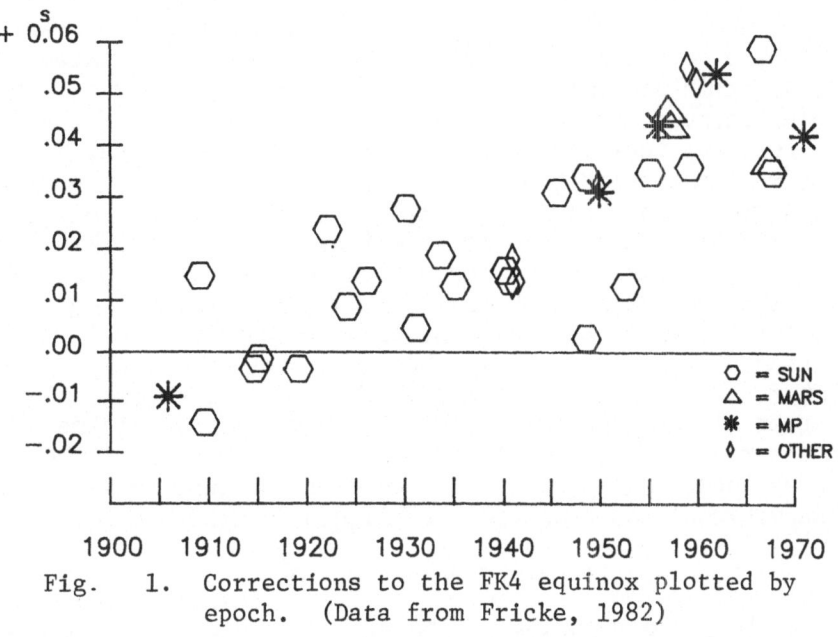

Fig. 1. Corrections to the FK4 equinox plotted by
          epoch. (Data from Fricke, 1982)

points out that new observing programs will make an independent fundamental
catalog possible.   Currently the U. S. Naval Observatory is making
preparations to observe the FK5 and the IRS in a simultaneous program
involving the Six-Inch Transit Circle in the Northern Hemisphere and the
Seven-Inch Transit Circle in the Southern Hemisphere (Hughes, 1982).
The Sun Moon, planets and minor planets will also be observed, and one
of the intended results of the program will be absolute positions of the
FK5 and IRS stars.   With estimated mean epochs around 1989 these positions
could be combined with the other absolute catalogs observed after 1900
to produce an independent fundamental catalog.   In this case the zero
points would be determined by first removing the effect of the error in
Newcomb's value of the lunisolar precession (Fricke, 1977) from each of
the absolute catalogs. This is done by precessing the observations of
the Sun, planets and minor planets to the date of observation with the
precession that was used to compile the catalog and then precessing to
J2000. Using a modern ephemeris, new solutions for the equinox and
equator corrections are made. Also, other time-dependent corrections
can be made at this point, eg. elliptic aberration and inaccuracies in
the variation of latitude that was initially applied to the catalog
observations. The mean observed positions of the stars are brought to
J2000 by the same route and the equator and equinox corrections applied.
The zero points of the fundamental reference frame are then defined at
any given epoch by combining all of the corrected catalogs and solving
for the mean positions and proper motions.

3.2 The Fundamental Mean Positions and Proper Motions

Thus far the discussion has centered about the absolute catalogs for only they can define the zero points of the fundamental frame. Once these points are established, however, the quasi-absolute catalogs can make a valuable contribution to the fundamental system. The quasi-absolute catalogs are observed and reduced in much the same manner as the absolute catalogs. However, they lack the observations of solar system objects required to make the solutions for the corrections to the equinox and equator and so cannot contribute to defining the zero points of the fundamental system. These catalogs are independent of the FK4 otherwise and as such can help define the corrections to the FK4 system once they are brought into agreement with the improved equinox and equator. A combination of the absolute and quasi-absolute catalogs will result in the definition of the fundamental system over the entire sky. Now the zero point of the right ascension system is defined not only at the equator but at all declinations. This is only possible if the catalogs have been observed and reduced in such a way that errors of the form $\Delta\alpha_\delta$ have been avoided. Again, the dependence of the fundamental system on proper observational and reduction techniques is seen.

3.3 The Final Mean Positions and Proper Motions

Unfortunately the majority of observed catalogs of stellar positions are neither absolute nor quasi-absolute but are differential. A good example is the AGK3R (Scott, 1963), which contains the northern half of the IRS. A differential catalog is one in which a list of stars is observed in conjunction with fundamental stars. The orientation of the instrument is determined each observing tour, but this is to assist in bringing the observations into the system of the fundamental catalog. When reductions are made the purpose is to duplicate the fundamental system as closely as possible in the observed positions of both the fundamental and list stars. This is not to say that the positions of the fundamental stars are reproduced, but rather the goal is to adhere to the fundamental system in an average way in each part of the sky. Within this average there are differences between the observed positions of the individual fundamental stars and their catalog positions. Thus while differential catalogs cannot contribute to improving the fundamental system, they can help improve the positions and motions of the individual stars within the system. In order to do this a differential catalog must be first brought into the improved fundamental system that has been defined by the absolute and quasi- absolute catalogs. To accomplish this the positions of the improved system are brought to the equinox and individual epochs of the fundamental stars' observed positions in the differential catalog. Individual differences are then computed and these are averaged to give systematic differences over regions of the sky. Right ascension and declination differences are treated separately. Bien et al. (1978) have developed a very elegant treatment of this problem which not only functionally maps the differences with respect to right ascension and declination but also with respect to magnitude. The systematic differences are then applied to the observed positions in the differential catalog to bring it into the improved fundamental system. At this stage the catalog can assist in determining the mean positions and proper motions of the individual stars within the improved system. The data from the absolute, quasi-absolute and differential catalogs are then combined to give the final mean positions and proper motions of the improved fundamental catalog. Table II summarizes the attributes of these three types of catalogs.

TABLE II

CHARACTERISTICS OF OBSERVED CATALOGS

TYPE OF CATALOG	OBSERVATIONS REQUIRED	CONTRIBUTION TO THE FUNDAMENTAL SYSTEM
ABSOLUTE	CONSTANTS 2 − 3 HOURS MERIDIAN MARKS FUNDAMENTAL AZIMUTH PAIRED CLOCK CORRECTIONS SOLAR SYSTEM OBJECTS FLEXURE MEASURES CIRCUMPOLAR SOLUTION FOR LATITUDE AND REFRACTION	EQUINOX & EQUATOR CORRECTION OF SYSTEMATIC ERRORS IN THE EXISTING SYSTEM IMPROVEMENT OF MEAN POSITIONSITIONS AND PROPER MOTIONS
QUASI− ABSOLUTE	SAME AS ABSOLUTE EXCEPT SOLAR SYSTEM OBJECTS ARE NOT OBSERVED	SAME EXCEPT DO NOT CONTRIBUTE TO EQUATOR & EQUINOX
DIFFERENTIAL	INSTRUMENTAL CONSTANTS AZIMUTH AND CLOCK STARS FUNDAMENTAL STARS AS PART OF THE OBSERVING LIST	IMPROVEMENT OF MEAN POSITIONS AND PROPER MOTIONS

4. THE FAINT FUNDAMENTAL EXTENSION

The stars in the FK4 as a group have better observing histories than any others that could be used to define a reference frame. They also are bright with apparent magnitudes that average 4.85. In fact, only 13% of the FK4 is fainter than apparent magnitude 6.0. This situation has often made it difficult to use the FK4 because the objects to be referred to the FK4 system are usually somewhat fainter than this. Another difficulty is that the distribution of FK4 stars over the celestial sphere is not even. These conditions exist because the growth of the fundamental catalog from Auwers' FC (1879) with 539 stars to the FK4 with 1535 stars has generally been controlled by the lists of stars that the observatories happened to select for their programs. Until the appearance of the FK4 Sup there has never really been a list to guide observing efforts toward the requirements of future fundamental catalogs. Fricke (1973) realized that the FK5 would afford the best opportunity to both extend the magnitude range of the fundamental system and improve the distribution over the sky. An extension would also help focus observing efforts. The extension stars are to be selected from two lists: the FK4 Sup and the IRS. The FK4 Sup contains almost 1100 stars in the range 5.5 > m > 6.5 out of a total of 1987 stars. The IRS on the other hand has 32,000 stars in the range 7.0 > m > 9.0, or 84% of the total. (See Table III.)

TABLE III

CHARACTERISTICS OF THE IRS

MAGNITUDE		SPECTRAL TYPE	
< 6.5	0.9 %	B & O	2.6%
6.5 − 7.0	4.1	A	12.8
7.0 − 7.5	9.2	F	14.3
7.5 − 8.0	16.4	G	17.3
8.0 − 8.5	26.4	K	49.7
8.5 − 9.0	31.6	M & OTHER	3.2
9.0 − 9.5	11.0	LATE TYPES	
9.5 −10.0	0.5		

FAINT FUNDAMENTAL CANDIDATES
SOUTH OF −30 DEGREES DECLINATION

MAGNITUDE		SPECTRAL TYPE	
< 6.5	0.3 %	B & O	0.6 %
6.5 − 7.0	1.7	A	5.0
7.0 − 7.5	17.6	F	10.7
7.5 − 8.0	33.2	G	18.6
8.0 − 8.5	30.3	K	60.2
8.5 − 9.0	12.6	M & OTHER	4.8
> 9.0	4.4	LATE TYPES	

In addition each of these catalogs is generally composed of the best-observed stars in its respective magnitude range. During the past year the U. S. Naval Observatory has selected a list of 2160 Faint Fundamentals in collaboration with our colleagues at the Astronomisches Rechen-Institut. These are IRS stars in the magnitude range 6.5 to 9.5. To select the stars the sky was divided into blocks of 22 square degrees. The IRS star with the best observational history was usually chosen, although considerable effort was given to balancing the distributions in apparent magnitude and spectral type as well. In the Southern Hemisphere the selection was rather difficult to make. The primary reason for this can be seen in Figure 2. South of −30 degrees the observational histories of the IRS are poor. Whereas in the north only 4% of the stars selected

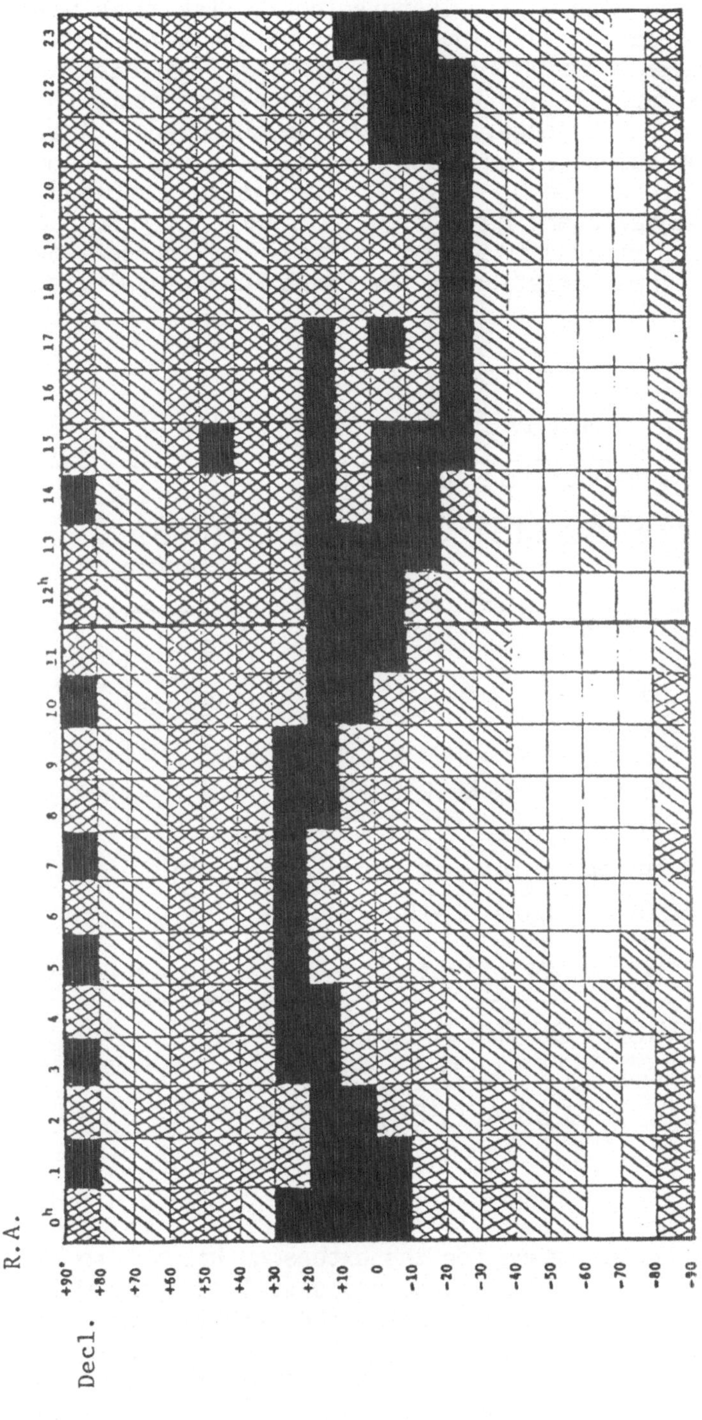

Fig. 2. The observational histories of the International Reference Stars. The averages are for blocks of 1^h in right ascension by 10° in declination.

have observational histories of 5 catalog positions, in the south 30% of th stars chosen have histories of 4 or 5 catalogs, mostly from the zone south of -30 degrees. It was also difficult to select a list that was balanced in magnitude and spectral type in this part of the sky. A look at Table III shows the reason; the IRS south of -30 deg., selected at the Cape Observatory, is heavily populated with late-type stars and concentrated in the $7.5 < m < 8.5$ range. Stars within one degree of an FK4 star were generally not selected, and double stars with separations under 50" and a magnitude difference $m_2 - m_1 < 4.0$ were also excluded. In the polar regions additional stars have been selected in order to provide sufficient stars for evaluating $\Delta \alpha_o$ terms in making catalog comparisons. The general characteristics of the Faint Fundamental list are shown in Table IV.

TABLE IV

CHARACTERISTICS OF THE
FAINT FUNDAMENTAL LIST OF 2160 STARS

MAGNITUDE		SPECTRAL TYPE	
< 6.5	0.8 %	B & O	6.3 %
6.5 — 7.0	9.0	A	19.7
7.0 — 7.5	20.1	F	16.8
7.5 — 8.0	21.4	G	19.8
8.0 — 8.5	22.0	K	32.5
8.5 — 9.0	19.9	M & OTHER	4.9
9.0 — 9.5	6.9	LATE TYPES	

INTERVALS OF MEAN ERROR OF PROPER MOTION ("/CENT)

	0.0/0.1	0.1/0.2	0.2/0.3	0.3/0.4	0.4/0.5	0.5/0.6	>0.6
RA	2 %	14 %	37 %	29 %	13 %	5 %	1 %
DEC	1	11	34	32	15	6	1

AVERAGE MEAN ERRORS OF THE PROPER MOTIONS
RIGHT ASCENSION: 0.31 ARCSEC/CENTURY
DECLINATION: 0.33 ARCSEC/CENTURY

The selection of the FK4 Sup stars for the extension has yet to be made. It is intended that about 1200 of these stars will be included in the FK5. The anticipated result of combining the FK4, 1200 FK4 Sup stars and the selected 2160 Faint Fundamentals is shown in Figure 3. Such a list should greatly improve the capability of the fundamental system to provide a reference frame at least to the ninth magnitude with a good distribution of stars in all parts of the sky.

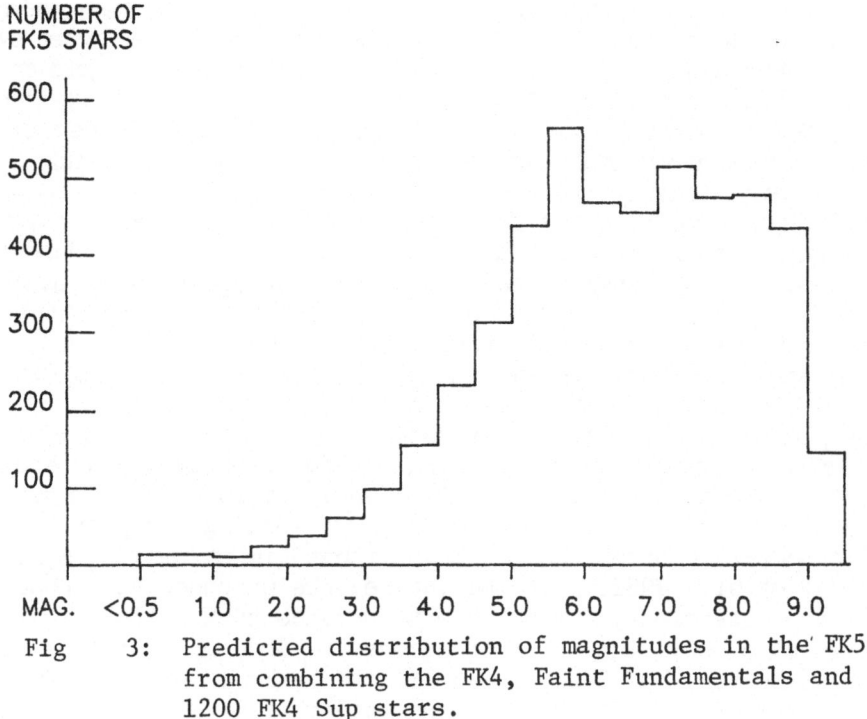

Fig   3:   Predicted distribution of magnitudes in the FK5
from combining the FK4, Faint Fundamentals and
1200 FK4 Sup stars.

Finally, it is important to point out that the extension stars cannot
be used to define the system of the FK5. They do not have sufficiently
extensive observational histories to have their positions and motions
determined from only the absolute and quasi-absolute catalogs. Although
the FK4 Sup stars are better represented in these catalogs than the Faint
Fundamentals, they have nevertheless received special attention in the
observing programs for only about the past 20 years. Thus the extension
stars will have their FK5 positions and motions computed by first reducing
their observed positions in catalogs to the FK4 system. These positions
will then be used to compute positions and motions on the FK4 system for
the extension stars. The improved positions and motions (FK5 system) of
the FK4 stars will be compared to the FK4 in order to define the conversion
from FK4 to FK5, and this conversion will then give the positions and
motions of the extension stars on the FK5 system.

5. THE RADIO FRAME

In recent years it has become clear that radio interferometry has
the potential to make a substantial contribution to the fundamental

reference frame. There is no doubt that a selected list of radio sources with optical counterparts at great distances that are not affected by 1. noticeable proper motion, 2. time variable structure such as jets or hot spots which develop, radiate intensely and then disappear, 3. wavelength dependent differences in the location of the radio-image center or 4. other astrometrically unacceptable characteristics such as the eccentricity of the radio image relative to the optical image, could provide an order of magnitude improvement in the precision of the fundamental reference frame defined in terms of those individual sources. Such a frame, however, must be linked with the stellar and planetary systems of observations. At present there is a program underway to do just this. First, transit circle observations are made of reference stars that surround selected radio sources (Dick and Holdenried 1982). Next, these fields are photographed with the USNO twin 20-cm astrograph. The star positions from these plates then provide a reference for the USNO 1.55-m astrometric reflector, which is capable of photographing the faint optical counterparts of the radio sources (Harrington et al., 1983). This effort is guided by the lists provided by the IAU-Commission 24 Working Group on Optical / Radio Astrometric Sources for the Establishment of an Inertial Reference Frame (Witzel et al., 1982). In the future, the proposed Hipparcos and Space Telescope astrometry may be able to provide such a link with the radio system at a very high level of accuracy. Indeed, optical positions at accuracies better than 0.01 arcsec will be needed if the full potential of the radio system to contribute to the fundamental reference system is to be realized.

6. ACKNOWLEDGEMENTS

I would like to thank our colleagues at the Astronomisches Rechen-Institut, especially W. Fricke and H. Schwan, for many useful discussions and exchanges of information concerning the FK5 and the Faint Fundamental stars. I would also particularly like to thank the following U. S. Naval Observatory personnel: C. Smith, J. Hughes and S. Gauss for their many helpful comments and discussions during the preparation of this paper, and E. Holdenried, R. Miller and T. Rafferty for their help in the preparation of Table I.

REFERENCES

Bien, R., Fricke, W., Lederle, T. and Schwan, H. 1978, Veroff. Astron. Rechen Inst. Heidelberg, 29.
Branham, R. 1979, Astron. J., 84, 1399.
Branham, R. 1984, "Equator and Equinox Determinations from Simulated Planetary Observations", in press.
Dick, S. and Holdenried, E. 1982, Astron. J., 87, 1374.
Fricke, W. 1963, "Preliminary Supplement to the FK4 (FK4 Sup)," Veroff. Astron. Rechen Inst. Heidelberg, 11.

Fricke, W. 1973, _Mitteilungen des Astron. Rechen Inst. Heidelberg_, Serie A, No. 68.

Fricke, W. 1977, _Veroff. Astron. Rechen Inst. Heidelberg_, **28**.

Fricke, W. 1982, _Astron. Astrophys._, 107, **L13**.

Fricke, W. and Gliese W. 1968, _Highlights of Astronomy_, ed. L. Perek (Reidel: Dordrecht), p. 301.

Fricke, W. and Kopff, A. 1963, "Fourth Fundamental Catalog (FK4)," _Veroff. Astron. Rechen Inst. Heidelberg_, **10**.

Harrington, R., Douglass, G., Kallarakal, V., Smith, C. and Guetter, H. 1983, _Astron. J._, **88**, 1376.

Hughes, J. 1982, _Southern Stars_, **29**, 245.

Hughes, J. and Scott, D. 1982, "Results of Observations Made with the Six-Inch Transit Circle 1963-1971," _Publ. U.S. Naval Obs._, Ser. 2, part III.

Hughes, J., Smith, C. and Branham, R. 1984, "Results of Observations Made with the Seven-Inch Transit Circle at El Leoncito, Argentina 1967-1973," _Publ. U.S. Naval Obs._, in preparation.

Miller, R. 1984, private communication.

Rafferty, T. and Klock, B. 1982, _Astron. Astrophys._, **114**, 95.

Schwan, H. 1985, in _IAU Symposium 109, Astrometric Techniques_, ed. H.K. Eichhorn (Reidel, Dordrecht), in press.

Scott, F.P. 1962, _Astron. J._, **67**, 690.

Scott, F.P. 1963, "The System of Fundamental Proper Motions," in _Basic Astronomical Data_, ed. K. Strand, (Univ. of Chicago Press, Chicago), p. 11.

Smith, C. 1985, _IAU Symposium 109, Astrometric Techniques_, ed. H.K. Eichhorn (Reidel, Dordrecht), in press.

Witzel, A. and Johnston, K. 1982, _Abhandlungen aus der Hamburger Sternwarte_, Band X, Heft 3.

DISCUSSION

GLIESE: T. Corbin mentioned the weakness of the FK4 in southern declinations (declinations < -50°). I will not discuss here the possible sources of such errors. Only one point: we have fairly reliable catalogues of absolute positions observed since 1960 but probably not only is the fundamental proper motion system south of -50° in the FK4 erroneous but also the positions at the mean epoch of the FK4 system (at about 1930). We need these positions again for deriving the proper motion system of the FK5. Therefore, I have some doubt whether we may succeed in producing an "ideal" proper motion system for the FK5 even if the system of positions of the FK5 should be correct at its mean epoch.

CORBIN: The effect that Dr. Gliese refers to is found mostly in the early catalogues, in which the stars were observed as they passed by fixed wires. Since these observations were not usually made using screens to dim the light of the bright stars, the observer's judgement as to when a star crossed each wire was often affected by the magnitude of the star - whence the term "magnitude equation" in these catalogues.

WILLS: Hipparcos will measure about 100000 stars of magnitude down to about B = 12 and after 2.5 years of observation will provide for stars of B = 9 positions and parallaxes accurate to 0.002 seconds and proper motions accurate to 0.002 seconds/year.

POPPER: Why is it necessary to observe solar system objects for defining the equator in addition to the definition of the pole using circumpolar stars. Diffraction? Too large an arc to measure?

CORBIN: The upper and lower culmination observations define the pole, but for purposes of contributing to the equator solution of the fundamental system it is best to use the dynamically defined value derived from observations of solar system objects because: 1) Flexure north and south of the zenith is not necessarily the same. 2) The same may be true of refraction. 3) The equator may not be 90° from the pole as measured by the circle of the instrument. Thus using the solar system objects as well as circumpolar observations defines the declination in a fundamental way at two points on two sides of the zenith and is a much stronger solution.

POPPER: Are recent techniques, such as photoelectric timing of transits and an Atkinson type instrument being introduced or do they represent too great a departure from the techniques used previously?

CORBIN: The instrument that observed the Perth 70 Catalogue had a photoelectric micrometer that employed a series of slits. This instrument was quite successful - especially in right ascension. The U.S. Naval observatory is about to begin a program with the 7" transit circle in New Zealand in which an image disector will replace the observer. The El Leoncito program has shown that it is very desirable to replace the observer with a detector because the observer generates heat and the observers show personal equations, especially in right ascension in spite of the "impersonal" micrometer.

FRACASTORO: I agree that photoelectric multislit measures with transit instruments are much more precise than the so called impersonal measurements.

BESSELL: Is there any published measurement of the relative reliability of the various southern catalogues, i.e. G. C., Cape, New Yale, Sydney?

CORBIN: Not really. This has been one of the great problems in the southern hemisphere. A comparison using Perth 70 would give differences around epoch 1970, but for a definitive comparison, especially for the proper motions in these catalogues, one should wait for the southern IRS Catalogue. I hope this will be available within a year.

POSITIONAL ASTRONOMY AND STELLAR MASSES

W. D. Heintz

Department of Astronomy, Swarthmore College

ABSTRACT. In this paper are surveyed the combinations of data needed for mass determinations, the impact of radial-velocity and speckle observations, the 1983 catalog of visual-binary orbits, the current material for main-sequence and red-dwarf masses and the controversial methods and results on low mass objects.

Masses are a delicate link between stellar properties, since they follow from the cube of observed quantities, and in turn enter evolutionary models with a substantial power. As an illustrating although rather unrealistic thought: If the age of a star could be found directly with a precision akin to that of orbits, and we could reason the opposite way, orbit dimensions would be found as a tenth root, with so small an error that there would be no further need for nocturnal gymnastics with micrometers, spectrographs, and the like.

The recent Fourth Catalog of Orbits of Visual Binaries (Worley 1983) contains 847 objects, including 80 for which second solutions are listed. Less than 40% of the entries come from the 1970 catalog; the others are revised or new. We have based selection and grading on computer-generated lists of residuals, and have examined some hundreds of other orbits (including a substantial number before 1970). We have also reobserved almost all the listed pairs during the last decade. Datacenter files were searched for recent photometric information, and published radial velocities were examined.

The total of 847 pairs are only 1% of the Washington Double Star Index (Worley 1984). On the other hand, only 5% of the orbits qualify for good mass determinations, i.e., less than 100 binary components. (This number does not include eclipsing pairs). The evident reason is that several kinds of data and conditions need to be combined, and if just one of them is not pinned down to a few percent, the resulting masses will not be reliable. Even some data of ostensibly good quality will probably be inaccurate owing to the presence of

D. S. Hayes et al. (eds.), Calibration of Fundamental Stellar Quantities, 71–80.
© *1985 by the IAU.*

undetected subsystems.

The classical case is the combination of an orbit with a trigonometric parallax and a photographic mass ratio; the latter may also give some indication of a possible third mass. Heintz and Borgman (1984) give a list of new mass ratios recently measured at Swarthmore. As always, the parallaxes are generally the weakest data. Masses would be within much better reach if the typical standard error of a parallax went from 5 mas (milli-arcsec) to 3 or 2. However, this will not be easy to accomplish, as it requires - among other things - an adequate time interval as well as good seasonal distribution of measures, lest some random error in the proper-motion term carry over perilously into the smaller parallax. This linkage is probably a major cause for the often large differences between nominal and actual parallax errors.

How reliable are visually measured separations, on which most visual binary orbits and masses are based? The currently active micrometer observers have no significant systematic differences between them, which is reassuring but not strict evidence. Direct comparison with photography reaches only down to 2" or 1.5. Orbits with large changes of angular separation afford a sensitive criterion, since systematic errors at the small separations should show as deviations from the Law of Areas - yet they are rarely noticed. Recently another control has been added by the speckle measures. Lists of standard stars for visual separations have been proposed, for close pairs in particular by P. Couteau (1969), and it is suggested that measures with new techniques include some of these pairs.

Now for the bad news: We have virtually no masses from the southern sky. Most parallaxes are of substandard precision, and no photographic series (except α Cen) is long enough for a mass ratio. Secondly, the visual pairs are very strongly selected in favor of near-equal luminosities, i.e., of fractional masses between 0.4 and 0.5. Statistical incompleteness increases rapidly when the magnitude difference Δm exceeds 2 mags. Speckle is less impeded by large Δm values. There is no reason to believe that lower-mass components in non-interacting binaries behave abnormally, but there is not much evidence in support of their normality either. Finally, the photometry of components is a familiar headache. Earlier work was done visually by wedges and polarization, and much of these results seem to be good and solid; but the speckle CCD will be able to contribute also in this respect. Incidentally, the magnitude estimates by visual observers - as I believe to have seen from numerous comparisons - are mostly better than their reputation. Weed out a few of the estimates, and derive systematic corrections for some others, and then a large body of potentially valuable data reposes in the card file on double stars. The catalog by S. Wierzbinski (1969) was quite good but it could be redone now with much increased material, estimates as well as photometric comparisons.

Radial velocities contribute to the study of visual binaries particularly for many objects outside the range for good parallaxes, and especially in recent years when spectroscopically found pairs can be speckled or vice versa. There have not been too many objects of combined analysis, simply because they are in the spectroscopic long-period range. The observational effort is lengthy, the RV amplitude usually small, and the lines often annoyingly blended. The problem that systematic errors and changes in some earlier spectrographs are not well known compounds the difficulty of combining data from various epochs.

A blessing in the disguise of disappointment lies in the frequent discordance between observed radial velocities and those predicted from visual orbits. The two kinds of observations are complementary in that they strengthen each others weak points; what the positional measures leave poorly determined, may make a large effect in the RV curves. The bearing on the accuracy of masses is obvious. In particular, when part of an orbit cannot be reached visually, or only with difficulty, the uncovered arc could cause errors in the elements e and ω, which transfer via the inclination substantially into the semiaxis major and the mass. An uncertainty of i enters the computed orbit dimension in opposite directions (through cos i and sin i, respectively), and may be revealed by a disagreement of the spectral type with the computed parallax. In brief, good mass results will come from the positional/spectroscopic combination, provided we are aware where potentially harmful uncertainties are. Another outcome of speckle observations is to be mentioned in this context, viz., the negative observations, which reduce the upper limits for small separations more than visual observers can achieve.

It may be time for a new compilation of masses from visual and interferometric pairs, but I have delayed that for several reasons. Some systems are under current investigation; the orbit catalog was not completed, and the Yale parallax catalog (the final version of which is not yet available) may shed light on the question of systematic errors of parallaxes. It was noticed long ago that the inclusion of marginally reliable binary-star parallaxes (0."04 to 0."02) tended to shift the mass-luminosity line slightly toward lower masses; but I am not sure if this is real, and is caused by a systematic error, such as derived in the catalog by Jenkins (1952).

Speaking of calibration, we have to mention the Hyades. Of course, we expect them to match the field stars in that almost every one should be double. Ten orbits are recorded, three of them with spectroscopic subsystems. The objects β 552 and 70 Tau are still under study; but the estimated masses indicate already a difference of their distance moduli by 0.8 or 0.9 mags. Compared with the tangential diameter of the Hyades, it seems that they are really a "globular" cluster.

Generally, however, we cannot afford to be choosy when it comes to mass calibration; we have to grab results where we can get them. Only for the middle main sequence (types A to K) has the mass-luminosity relation not changed in 25 years and seems to be safely pinned down. Exploration of the overluminous, expanding masses is being aided by photometric measuring systems designed for evolutionary classification. The simplest indicator of such stars is the dynamical parallax in comparison with the spectral type. I will not take up the subject of dynamical (or orbital) parallaxes here; they have been overused and over-discussed in the past. They can serve as luminosity indicators, and may show when an object is in some way peculiar. Note that dynamical parallaxes are sensitive to the apparent magnitude - a quantity often not well known for faint binaries.

The mass array for the Population II main sequence seems not within easy reach. Catalogs of proper-motion stars contain numerous high-velocity objects; but it will be a long way to find out which of these are resolvable binaries and are also metal-poor.

The mass-luminosity graph for the red-dwarf segment of the main sequence is reproduced in Fig. 1 (from Heintz 1983). A few revisions and additions in the latest years have not changed the feature of a surprisingly wide spread, which cannot be explained by parallax errors.

Moving further down the mass scale, the orbit catalog lists 23 unresolved pairs, mostly from U.S. Naval Observatory results. Some of them are still shaky as the series of their measurements are short or weak. Most of the unseen companions are probably in the stellar range, but one or another black-dwarf mass should be among them. This subject leads to another part of high-accuracy positional astronomy, and is characterized by two features (not necessarily related): It is still executed by photography, and it has a controversial history.

Unseen companions reported from apparently variable proper or orbital motions have had a very high mortality rate - even more than stars recorded as having variable radial velocities earlier, and found constant upon reinvestigation. Evidently some stronger evidence is required than merely a few consecutive plus or minus residuals of positions, namely a significant correlation between at least two telescopes and/or a significant recurrence of an orbital pattern. With an observed amplitude of 50 mas, one can feel as reasonably sure about the case as one could with a spectroscopic range over 10 km/s; but effects at the levels of 20 mas or of 4 km/s need to be checked more carefully. Unfortunately, the black dwarfs in the expected mass range of 10 to 50 mS (millisuns) will be, with few exceptions, below the 20 mas level, and hypothetical planetary objects (under 5 mS) are well below 10 mas. Much larger instrumental effects being known, it should be standard practice to ascertain beyond reasonable doubt that an alleged feature is not caused by the equipment. But planets are such a crowdpleaser that one can get by with almost anything.

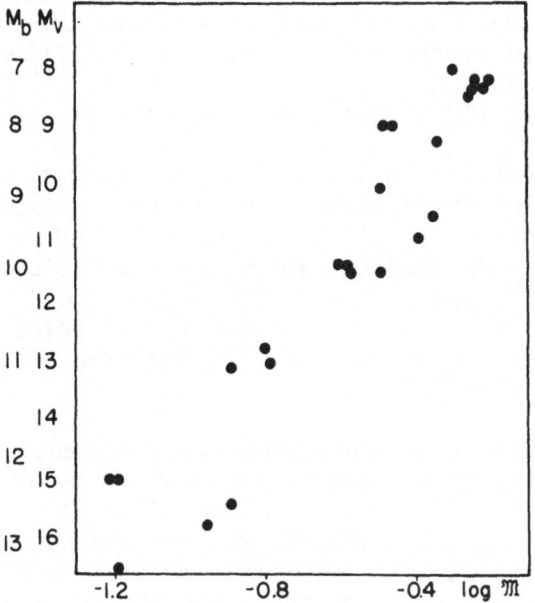

Fig. 1. The lower part of the mass-luminosity diagram.

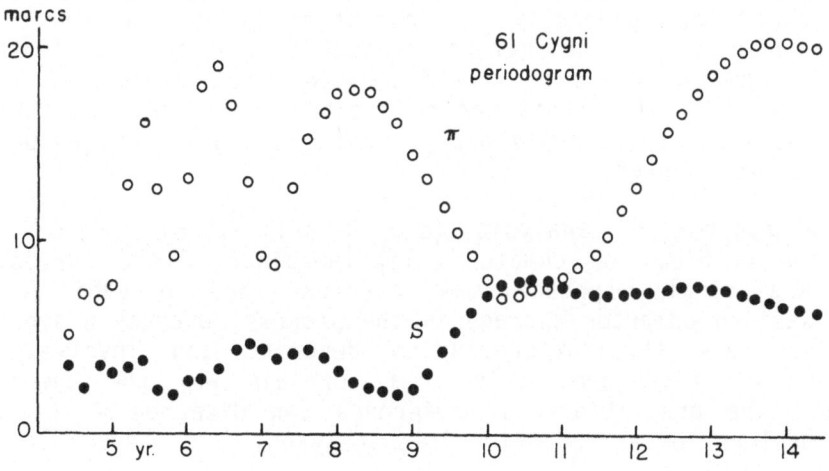

Fig. 2. Periodogram of the relative positions of 61
Cygni measured with the Poulkovo (Π) and
Swarthmore (S) refractors.

Criticism of suspect results need not even necessarily invoke systematic errors. The binomial theorem suffices to show that even a considerable number of random results, say, at the ±30 mas level, need not cancel out to below ±10 mas, but that there is a 10% or 20% chance of a pseudo-systematic effect remaining that level.

In Fig. 2 (presented at a conference in Tübingen, 1978) is shown a periodogram of the often discussed and probably best-known case of suspected planetary objects, 61 Cygni, as a straightforward instance of inadequate study. The Poulkovo refractor had two periods with very serious residuals of over 50 mas, separated by 25 years. The overtones of that time interval show plainly as peaks in the period spectrum. At the Swarthmore refractor (S) the same relative positions of the pair show no period.

It has been argued that photography is excellent, and also, that it is poor and inefficient, needing replacement by other techniques the sooner the better. In this fray one alternative seems to have been largely overlooked, i.e., whether photography has ever been given a fair try by coupling it with adequately general reduction methods - and this over almost a century during which the technique has been in use!

A data-processing method may be useful for parallax reductions but for nothing much else; the study of parallax errors - restricted to a narrow bandpass of 1 yr^{-1} - is not too significant for the analysis of long-term variations that may mimic periods. For instance, the often-heard statement that more than four or five reference stars on the plate do not help the formal precision of the parallax could be generally underwritten (since geometric and magnitude balances of the reference star frame are at least as important), provided it is clearly understood that it applies to the parallax only, that all errors can be treated as random, and that the conclusion does not necessarily apply outside the one-year bandpass in which it has been tested.

The approaches of analysis (cf. Russell 1978) in some ways resemble the problems of complex eclipsing-binary light curves. A rigorous analytical approach may overload and destabilize the solution, wasting computer storage in the process, whereas a synthetic approach is less straightforward to describe and involves more "educated-guess" technique. This may explain why the theorist's delight and the practitioner's preference can disagree - for light curves as well as for astrometric field analyses.

I have not published much on the small-amplitude, low-mass suspects in recent years, but that does not mean satisfaction with the status quo; it had rather extraneous causes related to past history. At the IAU General Assembly in Brighton in 1970 we discussed the alleged discoveries of low-mass objects in the 1960's, the apparent field distortions at the Sproul refractor (qualitatively studied in

1961), and the question whether the neglect of correcting for the latter was related to the former. At that time I did not believe in serious errors, but consented to look into the matter. Reaching an appalling conclusion in July 1972, I indicated the intention of thoroughly reinvestigating the "planets". Within one week administrative decisions occured which had the effect of forestalling the project forthwith. More interesting still, in December 1976 a befuddled administration restored my access to research support, but under the explicit condition that my work not parallel that of the local observatory. Which goes to show that the proof of extrasolar planets requires the use of subtle measures. Since this muzzle was dropped in late 1982, tests by remeasurement of selected material have begun, in order to check on the telescope errors and on the hypothesis I had published in 1976 from limited data. The decade since has elapsed without progress toward better material at this telescope, and some other things have happened: several cases whose reality had been questioned were studied at other instruments, and invariably resulted in an absence of confirming correlations. And the alleged planets have gotten into the propaganda machinery of the Extra-terrestrial Intelligence campaign. To be sure, SETI is a legitimate scientific endeavor, but it can also be misused as an ideology or a pseudo-religion. The way some of the literature distorts or altogether suppresses astronomical evidence smacks of pseudo-scientific UFO fanatism and leaves some room for concern. However, the improvement in data processing gives hope that the reiteration of past truths will not remain an embarrassment to astrometrists much longer. The course of events - slipshod analyses followed by equally inadequate cover-ups - has cost us at least 20 years in learning what actually goes on in high accuracy imaging. More regrettable still, a field analysis is quite laborious, with more than ten times the work and costs of an ordinary parallax; thus I have to maintain a moderate pace, lest the output of other data be impaired. Some effort along this line is needed, however, since issues other than ETI are riding thereon. In particular, there is the question of a continuous mass frequency relation into the range of black dwarfs, and that pertains to the problem whether binary star and planet formation are due to similar processes (there seems to be evidence that they are not). Currently debated among cosmologists is also the issue whether the missing mass can or cannot be explained in terms of very large numbers of stars in the 10 to 100 mS mass range.

The current status of the data on the cases of suspected lowest mass companions (under 25 mS) is thus:

BD + 4°3561 (Barnard's star):
    Four series from different telescopes, and no significant
    correlation between any two of them. Periods had been
    calculated from limited time spans, but over the whole
    interval the effect, if any, in the residuals seems
    aperiodic. Under reinvestigation as a standard test field.

BD + 36°2147 and AD Leonis:
    Found spurious on reinvestigation, and dropped from lists of
    binaries.
ε Eri:
    Disallowed in two reinvestigations, and rated spurious.
EV Lac:
    Original orbit (calculated by the author in 1972) erroneous,
    caused by uncorrected observations. Unclear residuals since,
    but the reference background on the Swarthmore plates is very
    poor; reanalysis therefore of doubtful value as only a small
    part of the material is useful. The star would be of
    particular interest because of UV flares.
Σ 2398:
    New measures at a noise level below the formerly suspected
    small amplitude, and clearly no evidence for a submotion.
61 Cygni:
    Original results from weak data now disallowed.
Stein 2051:
    Presumed period not yet covered; reinvestigation due in five
    or ten years. (The last three cases are distinguished from
    the former by a better reference frame on our astrometric
    plates).

Other unresolved objects presumably belong to higher mass ranges.

Perhaps it is still premature to compare output with effort.
Father Stein had to measure 2050 rather uninteresting doubles before
he hit on a remarkable, red triple of large parallax, which possibly
contains the first-known black dwarf. A large number of binaries is
still under observation, with more being discovered, but only a
fraction of them will be rewarding. Which pair will have an eclipsing
or astrometric subsystem, a variable component, or something else
worth noting, cannot be foreseen. Many spectroscopic pairs sit in the
catalogs with what appear to be garden-variety orbits - until some of
them become exciting on reinvestigation fifty years later because of
element or line changes.

The progress in mass determination has been slow in quantity, but
probably better in quality, considering the smaller error ranges in
quite a few instances. The results always have come from large data
bases: survey coverage as well as enhanced attention to specific
objects. Micrometers and astrometric and spectrographic plates may be
phased out by the end of this century; new techniques may be able to
further reduce error limits and - as I hope - may improve the
tractability of systematic effects. They are likely to increase the
demands on large telescope time and the unit cost of observation.
Ways will doubtlessly be found to reconcile wide range surveys with
the concentration of individual, promising objects. It can be stated
as a cosmological principle that exciting results are apt to come from
objects which nobody before had time to put on the observing
program.

This presentation is dedicated to the memory of our late colleagues and friends F. Zagar and J. Fleckenstein of the Osservatorio di Milano. It was at the Schiaparelli Symposium in 1960, which they had marvelously organized, that I first addressed the topic how to unscramble orbital and proper motions.

REFERENCES

Couteau, P. 1969, Astron. Astrophys., 2, 126.
Heintz, W. D. 1983, Lowell Obs. Bull, 167, 11.
Heintz, W. D. and Borgman, E. R. 1984, Astron. J.,,89, 1068.
Jenkins, L. F. 1962, General Catalogue of Trigonometric Stellar
    Parallaxes. (Publ. Yale Obs.).
Russell, J. L. 1978, in IAU Colloquium No. 48, Modern Astrometry, ed.
    F. V. Prochazka and R. H. Tucker (Institute of Astron.: Vienna)
    p.355.
Wierzbinski, S. 1969, Contr. Obs. Wroclaw, 16.
Worley, C. E. (ed.) 1984, USNO/USSDC tape.
Worley, C. E. and Heintz, W. D. 1983, Publ. U. S. Naval Obs., 24, part
    7.

DISCUSSION

STRAND: I agree with the speaker that new technology will bring greater accuracy which will be of importance, but it will not necessarily be useful in certain cases. An example is the determination of the mass-ratio in visual binaries, where observations over half a century, perhaps more, are necessary. For such a problem the archival value of the photographic plates is important.

HEINTZ: Agreed; there is no comparable means of storage. We must be aware, however, of a possible psychological bias that "what is new must be better". Older methods may lose some credence (and support) although they are demonstrably not inferior.

POPPER: In the discussion of parallaxes this morning, we heard of the old story of systematic differences in parallaxes between different institutions. Even though the formal errors in the parallaxes may be small, one might be concerned about their reliability.

    In your reference to your new catalogue of visual binaries you mentioned that there were 40 - 50 pairs suitable for mass determination. On what basis was the acceptability of the parallax decided?

HEINTZ: The formal error was under 10%. In borderline cases, consideration was taken of the agreement of multiple parallax determinations, or of confirmation by spectral type.

STRAND: A chapter in "Basic Astronomical Data" lists the visual binaries for which the masses of the components have been determined with a precision of 30% or better, which require the parallaxes to be known to better than 10%.

EVANS: Dr. Bjorn Petersen has done a series of high dispersion high signal-to-noise ratio spectra of AD Leo and EV Lac as flare stars, which may resolve some of their duplicity problems.

MASSES, RADII AND LUMINOSITIES FROM ANALYSIS OF ECLIPSING BINARIES

Daniel M. Popper

Department of Astronomy
University of California, Los Angeles

ABSTRACT. The reliability of values of the fundamental properties of stars derived from eclipsing binary analysis is discussed in terms of general concepts. Since the principles involved in the determination of masses and radii are simple, the heart of the matter is the care and judgment with which the relevant spectrographic and photometric observations are obtained and analyzed. Problems in the evaluation of the radiative properties require special attention.

1. INTRODUCTION

My task is to give you my views on the requirements eclipsing binary systems must satisfy in order to qualify for consideration as standards of reliability of masses, radii, and luminosities. Those views can be stated very simply. The systems must be favorable ones, both spectroscopically and photometrically, the observations must be of high quality, and their analysis carried out with care, understanding, and good judgment. For details, read my papers. That is really the essence of what I have to say, but I'm afraid more is expected of me. Since the concepts I have just stated are elementary, their elaboration is also pretty elementary, for which I apologize.

The topic may be discussed under four headings: Principles, observations and their analysis, results, and choice of stars as standards or calibrators.

2. PRINCIPLES

The basic principles involved in the direct determination of the masses and radii of the components of double-lined eclipsing binaries have been well established for a very long time, namely the dynamics of two-body motion and the geometry of eclipses of limb-darkened stellar discs. The most important complications are gravity brightening and mutual irradiation in the light curves, for which the principles are somewhat less well established. In the most favorable systems - well-separated, nearly spherical stars - these effects are unimportant,

81

D. S. Hayes et al. (eds.), Calibration of Fundamental Stellar Quantities, 81–96.
© *1985 by the IAU.*

but can become of great importance in some semi-detached systems and
all contact systems. Insofar as luminosities are concerned, the
principle is that the surface flux correlates with other radiative
properties of a star (e.g., color index) directly obtainable from
observations. Radius and surface flux give the luminosity.

The only really new approach introduced into the field in recent
decades has been the use of computers for light curve analysis. I
mention it here because of its importance, although no new principles
are involved, and discussion belongs under analysis of observations.

In making the comment that there have been no new approaches, I
am limiting the discussion to the classical problem. Close binary
systems with degenerate components - X-ray binaries, binary pulsars,
cataclysmic variables - do require and have brought forth new
concepts and assumptions (Bahcall 1978; Robinson 1976), as well as
their associated sources of uncertainty.

3. ANALYSIS OF OBSERVATIONS

In discussing problems of analyzing spectrographic and
photometric data, I shall limit myself primarily to the simplest
cases of well-detached systems. It is these that provide the best
results, of the highest accuracy, freest of assumptions, and
presumably most suitable as standards or calibrators. I use the word
"presumably," since, before coming to this meeting, at any rate, I
have not had a clear understanding of the concept of calibration of
quantities more than one or two steps removed from observation. I'll
return to this matter briefly under the choice of stars as
standards.

It is, of course, essential in any evaluations based on
observational data that careful attention be paid to the determinacy
of the results. This is particularly the case for results under
consideration as standards. The matter of determinacy or reliability
is at the heart of the matter. I presume that I am standing before
you because, in my work, I have tried to pay particular attention to
those aspects of the observations and their analysis that can, if
sufficient care is not taken, lead to untrustworthy results.

I'm afraid much of what I have to say may appear critical or
even negative. But the very essence of decision making with respect
to standards is critical examination of details. If all details of
the observations and their analysis are completely satisfactory,
there is little to be said. The more they are unsatisfactory, the
more comments are required to show what it is that needs to be
improved. It's like refereeing a paper - the better the paper, the
shorter the referee's report. Another way of looking at the matter
is to point out that there are many more unsatisfactory methods of
observation and analysis than satisfactory ones.

3.1. Spectrographic Considerations

One of my first motivations in looking into problems of close binaries was to try to understand some outstanding discrepancies. In particular, the dissertation of A. B. Wyse (1934) at the Lick Observatory had emphasized differences between the surface flux ratios for a number of binaries evaluated, on the one hand, from the published spectral types and, on the other, from the depths of the two eclipses. As the opportunity became available to me to accumulate photometric and, primarily, spectrographic material to try to understand the problem, it gradually became apparent that the difficulty lay principally in the assignment of the spectral types of the fainter components. It is difficult, at best, to estimate a trustworthy type from weak lines. It is exceptionally difficult if the lines of the seconary are, in fact, too weak to be visible. After all, if a star is known to be a spectroscopic or eclipsing binary, it has *ipso* *facto* two components. Moreover, if the period and epoch are known, one knows about where to look for the lines of the second component. It is certain that if one doesn't look for the lines of the secondary, he'll not see them. By "looking" I mean inspection of photographic spectra; study of microdensitometer scans of a plate; examination of the traces on an oscilloscope screen; observation of the output of a Griffin-type or Coravel radial velocity device; analysis of the results of cross-correlating the digital output of a reticon, digicon, CCD, or whatever with that of a standard star, etc., etc. The principles are the same.

There are several possible outcomes of the looking process. 1) The lines of the secondary are not said to be detected. 2) The lines of the secondary are not detected, but are mistakenly thought to be. 3) The lines of the secondary are really detected, but they give systematically incorrect velocities because they are blended with lines of the primary. 4) The lines, or at least some of them, of the secondary are detected and are not blended, so that usable orbits can result, and perhaps spectral types as well. The outcome depends, of course, not only on the true nature of the system and the care of the observer, but also on the resolution and signal/noise of the observations, and the region of the spectrum examined. Only in the fourth case are the results useful. The dividing line between conditions 3) and 4) may not always be clear. Special problems arise in O-B systems where, as demonstrated by Andersen (1975), even apparently well-resolved lines can have overlapping damping wings, giving rise to systematic effects.

On studying both old material and new material of my own, it became evident that the source of the discrepancies in surface flux ratios noted by Wyse was the assignment of spectral types to components falling in one of the categories 2) or 3). Once this situation was realized, it naturally caused one to wonder how securely the masses and radii were established. I must add that the problem of ascribing orbits - and masses and radii - on the basis of inadequately resolved or of nonexistent lines has not completely disappeared from the literature. As a consequence of such

considerations, I have taken the position that I must be personally
satisfied, from examining my own material or that of others, as to
the reality and resolvability of component lines before I consider a
system to be satisfactory for the purposes we are discussing.

In that earlier era (before about 1950), the quality of the
spectrographic material available was inadequate except for a
relatively small number of systems. A fundamental advance was the
introduction at some institutions of efficient grating spectrographs
with greater spectral resolution and broader spectral coverage, along
with more sensitive photographic emulsions. The broader spectral
coverage allowed one, for example, to observe at longer wavelengths,
where the lines of a cooler secondary component would be enhanced.
It was these advances that led to most of the results now considered
to be of high quality. The use of oscilloscopic scanning devices has
resulted in more efficient measurement of spectrograms.

We are currently in another period of improving spectroscopic
technology - digicons, reticons, CCD's, micro-channel plates,
radial-velocity meters, cross-correlation techniques, and presumably
more to come. The most effective ways of applying these devices to
binary star problems is a matter of concern. Perhaps the most
striking contribution to date is their ability to obtain results for
systems with resolved lines of greater magnitude difference between
components than heretofore. But use of the new generation of
detectors is no guarantee that the pitfalls referred to earlier will
automatically be avoided. And it is not outside the realm of
possibility that new and as yet unexpected effects may arise in these
advanced techniques that will require special care if they are to be
overcome. One new kind of effect that I wonder about is in data
handling and reduction. In the old days one <u>looked</u> at his plates,
<u>measured</u> them, with his eyes as an essential tool, and perhaps even
<u>plotted the results by hand</u> to see what they looked like. Now, with
most output in digital form, the tendency is increasingly to feed the
observations into a computer and, without having to examine any
intermediate steps, to accept the results. I have seen just enough
of this procedure in my home environment to realize that the fact
that a computer program has been used by many people does not
guarantee that it is free of errors that can degrade results. Let me
give an example of what I consider to be inadequate data handling.
Spectrographic orbits are often published in which the velocities of
the two components are combined into a single solution, without
solving the two separately. A bad example of what this generally
unwise practice can lead to may be seen in a paper soon to appear on
V624 Her, a bright, double-lined Am binary (Popper 1984). The old
prismatic velocities have been analyzed over the years by three
astronomers: the spectroscopic observer, a photometric observer, and
a third person. All used the same questionable procedure, and none
of them noted that the residuals for the two components differed
systematically by 7 km s^{-1}! The reason for the difference is not
important in the context of this discussion. Its having been missed
is a consequence of poor procedures and of not looking at the
results. I have given (1974) a long list of reasons why it is unwise

to assume equal systemic velocities for the two components except in special circumstances, and will not repeat it here. To this list may be added the consequences of the cross-correlation technique when the two components differ sufficiently in type that different standards of velocity are used for the two components. The difference in the adopted velocities for the two standards should not be assumed to be precisely equal to the true difference.

3.2. Photometric Considerations

These comments have thus far had to do primarily with the determination of spectroscopic orbits. In some respects the situation is parallel for photometric observations. The great observational advance came, of course, with the replacement of the photographic plate and visual photometer by the photomultiplier, at about the same time as the introduction of efficient grating spectrographs.

A second great advance in photometric studies of eclipsing binaries came with the availability of high-speed, high-capacity digital computers. They are a great help in reducing the observations, though here also one should not underestimate the advantages of <u>looking at</u> the results at various stages of the reduction procedures. But it is in analyzing light curves that the computer's influence is vastly greater than in the spectrographic case. The spectrographic orbit has a simple analytic form, and it makes no difference in the results whether one makes use of a hand-crank desk calculator or a state-of-the-art computer. But the computer has revolutionized the analysis of photometric observations of eclipsing binaries. There being essentially no analytic relations between the observed quantity and the results to be extracted, the only effective procedure is model fitting, involving many numerical integrations. In the simplest cases of spherical or nearly spherical stars, and with 75% or more of each star covered at mid-eclipse, which are just those cases that can give the most determinate masses and radii, the details of the computational procedure may not be critical, but use of a satisfactory computer program leads, in such cases, to objective results and to rational evaluation of their uncertainties. This matter of the non-critical nature of the method of analysis in simple cases is not to say that there may not be pitfalls for the unwary in applying someone's computer program uncritically.

In more distorted systems, both the increase in the number of parameters defining the system and some uncertainties in the model ("reflection," gravity brightening) can cause a solution of the light curve to be less secure. Similarly, systems in which smaller fractions of the stars are covered will give less determinate solutions. Systems in these categories should probably be avoided as standards or calibrators of fundamental stellar properties, although the results may be important for studying specific problems. With respect to reflection and gravity effects, I find a tendency among users of computer programs to assume that the computer model is without uncertainties, so that derived quantities, including mass ratios, are subject only to observational uncertainties. It is my opinion that the modeling of some of these variations is not completely secure, and that one often needs to consider potential effects on the derived results of uncertainties in the models.

Perhaps the greatest potential pitfall is to have too much confidence in a formal photometric solution that is derived from inadequate data or for a system that is intrinsically indeterminate. Use of a computer cannot, despite its great powers and despite heroic efforts of investigators, make determinate that which is not. As is well known, determination of the light ratio of the components from the line spectrum can provide additional information that can rescue an otherwise indeterminate photometric analysis. Use of modern spectral detectors, with linear response and large signal/noise, should improve greatly our ability to determine ratios of line strengths. If the two components differ in temperature and/or surface gravity, the differential use of model atmospheres may be capable of converting ratios of line strengths to a light ratio. I would feel more comfortable about the use of model atmospheres in this problem if there were a body of empirical equivalent widths for a wide range of temperatures and gravities that were found to be in agreement with predictions from model atmospheres. The use of stars in a cluster for such an empirical test would eliminate the uncertainties due to differences in composition, and the differences in gravity and temperature might be placed on a reasonably sound basis.

I have referred to the use of computers in both spectrographic and photometric analyses. Programs have also been developed and employed for combining the two kinds of analysis into one grand program (e.g., Wilson 1979). It is my opinion that only in special circumstance is the use of such a program desirable. The natures of the two kinds of data and of the two kinds of analysis are so different, each with its own special problems and idiosyncracies, that one should always look at each separately to see if everything is consistent. Let me illustrate my point by a couple of examples. A preprint I saw recently made use of a combined program. One assumption made was that the phasing (e.g., epoch of conjunction) was the same for both spectrographic and photometric observations. That is not necessarily a safe assumption if the two kinds of observation were obtained at different epochs because of uncertainties or changes in the period. In the same preprint, the photometric and spectrographic observations of a well-detached system with spherical stars were combined to derive the mass ratio. The authors were surprised to find that the inclusion of photometric data caused a slight change in the mass ratio from the value obtained from the radial velocities alone. They concluded that there was some subtle information hidden in the photometry contributing to the evaluation of the mass ratio. A glance at the photometric observations showed what this "information" was. One deviant point in the light outside eclipses, probably a poor observation, forced the computer to conclude that there was the equivalent of a non-zero coefficient of the cos 2θ term in the light variation outside eclipses - ergo, a mass ratio is evaluated. Such inappropriate mixing of data can readily cause a degradation of the derived results. This is also an example of how necessary it is to examine observations carefully. There are situations where a combined program may be appropriate, for example, in eccentric orbits where significant information on the eccentricity may be contained in both kinds of observation, although separate solutions should always be carried out as well to test for consistency. Another useful case is the combination of photometric observations in two or more wavelength bands, with the assumption of common geometry, although here also separate solutions can reveal unexpected effects.

In both spectrographic and photometric analysis it is essential, if the results are to be taken seriously, that realistic uncertainties of the important quantities be evaluated. This is perhaps the most personal aspect of all, depending on the investigator's judgment as well as on his powers of analysis, so that an objective comparison of the results by different investigators may be difficult to carry out. If the spectrum lines of the components are well resolved in the available material, the formal mean errors of the spectrographic elements may be accepted as realistic. The situation is not so clear in the case of a photometric orbit, where the interplay between the elements that produce the light variation may be complex, subtle, and not simple to understand. Differences between the results by different investigators of the same system show that formal mean errors based on internal agreement tend to be overly optimistic. One needs to be particularly cautious in accepting the formal mean errors that are produced when certain parameters are kept fixed and others allowed to vary. In my own work I have attempted not to underestimate the various sources of error, particularly in photometric solutions, but it is difficult to form objective judgments.

3.3. Radiative Properties

Thus far in our discussion, the focus has been on solutions of spectrographic and photometric orbits, primarily for the purpose of obtaining masses and radii. The radiative properties of the stars are also required. Without specifying them, we don't have a picture of the kind of star we are talking about. They are also required, given the radius of a star, to calculate its luminosity. In evaluating the fundamental properties of stars in binary systems of all kinds, eclipsing, visual, and resolved spectroscopic, I have found the specification of radiative properties the most consistently vexing and difficult of all problems.

The various interrelated radiative properties are color indices, surface fluxes, and effective temperatures. One might also include spectral types.

The flux and temperature scales are discussed by others at this symposium, and I may have comments on their presentations. The flux and temperature scales are usually calibrated in terms of color

indices, so one observational problem is to obtain the color indices
of the two components. It is <u>essential</u> for this purpose that the
photometric observations be carried out on one of the best calibrated
photometric systems in two or more wavelength bands unless the
components are very similar; in that case, the photometric indices
need not be evaluated throughout the light curve. In systems where
the difference in radiative properties of the components is not great
(often the case in double-lined systems), the difference in color
index between the components may not be best determined in the usual
solutions of the light curves. Two additional, although not
completely independent, methods are available for evaluating the
individual color indices. If an eclipse is deep enough and is well
observed in two (or more) wavelength bands, the color index of the
star being eclipsed is just the color index of the light lost
(account being taken of the difference in limb darkening in the two
bands). The second method employs the relation between color index
and surface flux in a standard band (usually V). The flux <u>ratio</u> is
usually well determined from the depths of the two minima (for
circular orbits) even if the light ratio is not well determined in
the solutions. With the color index of the combined light, the flux
ratio, and the flux-color index calibration, the individual color
indices and fluxes may be obtained. It is a particular complaint of
mine that many, if not most, of the computer-generated solutions of
light curves that are published do not treat this matter of
individual color indices and flux ratios satisfactorily, so that one
cannot evaluate the radiative properties well. Beware of results
that purport to give temperatures or bolometric light ratios without
showing those radiative properties that come most directly from the
observations - the flux ratios and light ratios in the observed,
carefully calibrated wavelength bands. Then one does not have the
available information he needs to obtain absolute fluxes and
luminosities. In most computer programs I am aware of, one assigns a
temperature to one component and derives the temperature of the other
by means of some often unstated assumptions about the relation of
surface fluxes to temperatures via black-body curves or model
atmospheres. Such temperatures, although perhaps required in some
intermediate steps in the program, should never be listed as
<u>evaluations</u>. The only satisfactory approach is to apply the relation
between a well-calibrated color index and absolute surface flux or
temperature. Such relations are a major topic of this symposium.

The situation is more complicated for distorted than for
spherical stars. For distorted components, the radiative quantities
vary over the surface, and models must be used. But even in these
cases, the investigator should give some kind of mean flux ratio and
color index, derived from the observations, along with information
about the assumed variation of flux over the surface. Distorted
stars are, for this as well as other reasons, generally less suitable
as standards than nearly spherical stars.

Let me give two examples of the kind of problems one may
encounter with respect to radiative properties. I recently received
a preprint of a study of a detached eclipsing binary. Both

radial-velocity and photometric observations were of high quality.
The photometry had been carried out in what were presumably standard
B, V, and R bands. The two components differ considerably, with a
mass ratio, q = 0.55, and a ratio of the radii, k = 0.83. Careful
analysis using several indices and MK classification gave the
spectral type of the primary A1.5V, or B-V=+0.04. From the
photometric solutions in B and V, the difference in color index
between the components is Δ(B-V)=0.51, giving for the secondary
B-V=+0.55. From the V and R solutions, Δ(V-R)=0.56, giving V-R for
the secondary +0.62, which corresponds to B-V=+0.80. And the flux
ratio in V, 6.68, corresponds to Δ(B-V)=0.66, giving B-V=+0.70.
0.55, 0.70, 0.80 -which is correct, and why the differences? Perhaps
the photometry was a poor match for the standard system. Direct
comparisons with standard stars, which should always be carried out,
had not been undertaken. Until this matter is resolved, the very
good work on the system is not particularly useful, at least insofar
as the secondary component is concerned. And it is just such
systems, with considerable difference between components, that
provide the best tests for evolutionary models.

My second example is taken from recent literature. The spectrum
of the primary component of a well-known bright detached eclipsing
binary has been classified in several independent studies in the
range A2 to A4, and color indices are in agreement. The new study
concludes that the true type is B9.5 by the following reasoning. The
masses and radii of the components being known, one evaluates the
temperature of the secondary (why the secondary I'm not quite sure),
an F star, by the strange method of entering "standard" tables
relating mass, radius, and temperature. Photometry gives the flux
ratio, and from this the temperature difference. The resulting value
for the primary is 10600K, and therefore it must be B9.5! We have
here a beautiful example of the use of two questionable practices
together. The first demonstrates how not to evaluate spectral types
or temperatures. The second demonstrates that "standard" tables can
be badly misused.

This problem of evaluating the radiative properties is most
difficult of all when there is a large magnitude difference between
the components. The greater the difference in properties, the better
the system is as a test of evolutionary models, since the models must
satisfy, with a single age and composition, stars of quite different
properties and stages of evolution. While the radiative properties
(luminosity, effective temperature) of the primary component may be
soundly based, those of the secondary may be only poorly known. As
our techniques for observing weak lines of secondaries and of
measuring faint components in resolved binaries improve, the demands
on the photometric differences between the components increase.
Unless the technology for obtaining properly calibrated photometric
differences keeps pace with the technology for orbit determinations,
the value of the latter will not be fully realized.

3.4. Additional Considerations

There is a fundamental observational attribute of a star, in addition to its mass, radius, and luminosity (or other radiative quantity). That is its chemical composition. Calibration of composition is also the subject of a presentation here. Such calibrations have not, by and large, been particularly successfully applied to binary stars, nearly all of which are disk or arm objects. The most striking departure among well-analyzed stars is the Am characteristic, and that is generally thought not to reflect interior composition. The tendency has been, rather, to evaluate composition parameters by comparing mass, radius, luminosity to models and come out with age and composition. Direct verification of the results so derived is not easy to obtain.

Throughout this discussion I have presented general, qualitative concepts. I hope no one has been waiting for my views on how spectrograms should be measured, how lines should be chosen, what orbit program is best, whose program for light curve analysis ought to be used, and so on. While I do have ideas on these matters, I consider them less controversial than what has been discussed. If anyone is interested in my views on these details, I recommend again that he read my papers.

Let me conclude this section on analysis of observations by relating a nightmare I occasionally have. I am visiting one of my European friends at his home institution. He investigates eclipsing binaries. "Let's do V5555 Centauri," he says. He sits at his console and starts interacting:

```
        ID?       ESO-JA-524
        TYPE?     ECLIPSING BINARY
        NAME?     V5555 CEN
        DATES?    MAY 14-MAY 20  (It is now May 1)
        PHOTOM?   STRÖMGREN 4-COLOR
        SPECTRO?  CORAVEL
        PROGRAM?  W-D COMBINED (Wilson-Devinney photon + spectr)
        START?    YES
```

On May 25 my friend returns to his console.

```
        ID?       ESO-JA-524
        NAME?     V5555 CEN
        RESULTS?  YES
```

Immediately there appears on the screen:

```
     M1 2.734 ± 0.022      M2 1.945 ± 0.018
     R1 2.577 ± 0.011      R2 1.850 ± 0.011
 LOG TE1 4.032 ± 0.013  LOG TE2 3.940 ± 0.009
  LOG L1 1.900 ± 0.057   LOG L2 1.246 ± 0.045
                LOG AGE 9.111 ± 0.004
                Y 0.273 ± 0.013 Z 0.027 ± 0.004
                VROT    SYNCHRON
```

"But....but....but..." I stammer. "But....but....but. How good
were the nights? How many times was each minimum covered? What
standards were used? What does the spectrum look like? How well
resolved are the lines? What is the period? On what basis were the
temperatures obtained?..?..?..?..?" "Why do you ask all these
irrelevant questions? Everything is taken care of in the program and
the ± values tell you all you need to know about how good the
observations are." "But...but...but... What models were used? What
mixing length? What opacity tables?" "Stop asking irrelevant
questions. Everything is in the program. We use only the best
models. You can trust our results."

To me this is a horrible nightmare. But perhaps to some of
you it would be paradise.

4. RESULTS

Values of the masses, radii, and luminosities of the components
of eclipsing binaries appear in the astronomical literature from time
to time, both in original investigations and in compilations. It is
for each scientist who wishes to make use of the results to evaluate
the quality of published work. In the preceding sections of this
presentation, I have discussed some of the matters that I take into
consideration in my own evaluations. My Annual Review article of
1980 (Popper 1980) gives results for those systems that appeared to
me at that time to be the most definitive in each of several
categories of binaries (including visual and resolved spectroscopic
binaries). Improved results are now available for several of those
systems, and a number of new ones can, of course, be added. I am
discussing with Johannes Andersen the possibility of preparing a
supplement to the Annual Reviews compilation. I might say once more
that the most vexing single problem, where I consider the results
most subject to uncertainty, is in the evaluation of surface fluxes
and temperatures, which carry over into the evaluation of the
luminosities.

It might be useful at this point to survey the HR diagram
briefly from the standpoint of reliable data on stellar masses and
radii in particular. Only components in detached systems are
relevant. The main sequence band from about B8 to F8 is the most
heavily populated with good data. There is a small number of earlier
B stars equally well known. But most B-type binaries suffer, for a
variety of reasons, from a lack of well-resolved lines. The problem
of treating properly lines that are somewhat blended has not been
adequately addressed. There are no detached O-type binaries with
first-rate results, the best being a couple of what appear to be
contact systems. For the main sequence of types G to M, eclipsing
binaries have thus far yielded only the two M-type systems, YY Gem
and CM Dra, the latter a high-velocity system, although the G8
system, HS Aur, under investigation, should lie in this gap if there
are no difficulties with it. It is here that the visual binaries are
much more numerous. Whether they can ever be expected to provide
masses of comparable accuracy (5%?) to that attainable for eclipsing

systems is highly problematical. It may be possible to find, among
known eclipsing systems of types late F and G, some with fainter,
cooler companions with lines measurable with sensitive detectors of
high signal/noise. UV Psc, FL Lyr, and RT And are possible
candidates. The matter of fluxes and temperatures of the fainter
components is a particularly difficult one here for two reasons.
First, and already discussed, is the problem of obtaining a good
color index because of the domination of the hotter star. Second,
and more basic at this time, is the lack of an absolute flux scale
for main sequence stars between the Sun and YY Gem (M1). There is no
objective evidence that the (V-R)-flux relation ("Barnes-Evans"
relation) for giant stars is valid for cool main sequence stars.

As main-sequence stars expand, their possibility of eclipsing
increases, and there are a number of detached eclipsing binaries with
components in the 1 to 2 $m_\odot$ range in which the more massive component
appears to have evolved well across the Hertzsprung gap, but not so
far as to fall prey to mass exchange. These systems show RS CVn
characteristics, and it is not completely clear that their properties
have not been affected by mass loss through stellar winds. Radius
determination in most of these systems is subject to uncertainties
because of their unstable light curves.

Eclipsing binaries have yet to produce good masses and radii for
typical cool giants, although the Copenhagen discovery and work on
TZ For should give results for a system apparently similar to
Capella, the cooler component of which may be a typical giant. The
selection against large stars is that if they are close enough to
eclipse and to give adequate radial-velocity changes, they become
semi-detached as the more massive components expand. The number of
supergiant eclipsing systems amenable to analysis has not increased
beyond ζ Aur, 31 Cyg and a questionable VV Cep in recent decades.
Mass determinations of the most luminous stars of all spectral
classes are non-existent. Direct information on the masses of stars
of chemical composition clearly different from that of the local
population (e.g., halo stars) is also non-existent.

If we wish to extend our discussion to results for semi-detached
systems, we find that our supply of velocity curves of the faint
Roche-lobe filling secondaries has been significantly augmented in
recent years by application of the new generation of detectors. But
the masses and radii of the components suffer in their determinacy in
most cases because of absorption in the spectrum produced by
non-photospheric material, affecting the velocities of the primaries,
which are often of small amplitude to begin with. They also suffer
because of difficulties with the light curves arising from the large
amount of "reflected" light as well as from possible distortions by
gas streams.

For stars in contact configurations, the approach to both radial
velocities and light curves requires principles and concepts of
analysis that are beyond those I have discussed here and that I have
not pursued. A critical review of results in this field would take
us outside the limits of this discussion. The same comments apply to
the analysis of X-ray and cataclysmic binaries, both of which make

use, in some degree, of radial velocity and photometric observations.

5. STANDARDS

As the organizers of this Symposium are aware, I have not been completely clear about the choice or even the concept of standards or calibrators among the components of close binaries. One can think of several approaches. First might be a list of the best-determined masses, etc. As noted above, this is basically the approach adopted for the most part in my Annual Review article (Popper 1980). The criterion for inclusion in such lists could be the accuracy with which the properties are thought to be known, e.g., all stars with masses and radii known to within ± 5%. There should also be some criterion for the surface fluxes or luminosities, since a mass is not of a great deal of interest unless one knows what kind of star is being considered.

A second approach could be to tabulate the best systems for each box in the HR diagram. The quality of the data would differ greatly over the diagram, some regions being essentially blank. Tables of this kind could serve as handy references for someone who just wished a general idea. It is possible, of course, to misuse "standard" tables of this kind. I have already given you an example of what I consider to be such misuse from the recent literature. This example illustrates a general problem with "standard" tables. Their application to a particular case assumes a conformity among stars, while it is just their diversity, even among well-behaved objects, that makes them fascinating. Nevertheless, the use of "standard" tables derived from observations is preferable to basing properties on models alone, as must be done for stars of a kind for which no masses are known directly from observations.

A third approach to the selection of standards might be to provide help in understanding some problem of particular interest. For example, the most informative binaries for testing predictions of evolutionary models should be those with the components having quite different masses and other properties. Differences in composition and age are eliminated, and one has the simplest evolutionary problem. One difficulty here is that, from the observational standpoint, the greater the difference in properties of the components, the more difficult the observations are likely to be. As another example, one might be interested in observations to test his prediction of what happens to mass-exchange systems. He might look for results that appear most nearly to match models he has found amenable to theoretical treatment, rather than those with the best-determined properties. This is likely to be a theorist's approach. Other kinds of questions, for which there are no direct answers, might be: what is the mass of a star at the turn-off point in a globular cluster, or what is the mass of the most luminous star in a galaxy, and so on. Thus, the very concept of a "standard" or calibrator requires a determination of the use to which the standards are to be put, a problem addressed in an earlier presentation here.

In all fairness, it must be pointed out the direct determination of stellar masses, in particular, the most fundamental of all the properties of a star, has failed to provide information for many categories of stars, so that the seekers for answers to many questions have, of necessity, had to rely on models rather than on observational material.

As for the future, I have already referred to improvements in instrumental and computational techniques that have played and will continue to play crucial roles in improving and extending our knowledge of the fundamental properties of the components of binary stars. Of equal importance in the realization of these goals will be the existence of astronomers deeply interested in obtaining fundamental data of high quality, as much as a service for our science as for solving specific problems of interest to them; the availability of observing time at major facilities for such general programs with long-range objectives; and finally, the willingness and patience to accumulate enough data, spectroscopic and photometric, and to subject it to the detailed, painstaking analysis required if the results are to be worthy of consideration as standards.

REFERENCES

Andersen, J.   1975, Astron. Astrophys. 44, 355.
Bahcall, J.N.   1978, Ann. Rev. Astron. Astrophys. 16, 241.
Popper, D.M.   1980, Ann. Rev. Astron. Astrophys. 18, 115.
        This reference contains citations of many spectrographic
        and photometric investigations of numerous authors.
Popper, D.M.   1984, Astron. J., in press.
Robinson, E.L.   1976, Ann. Rev. Astron. Astrophys. 14, 119.
Wilson, R.E.   1979, Astrophys. J. 234, 1054.
Wyse, A.B.   1934, Lick Obs. Bull. 17, 37.

Papers on spectrographic problems, not cited in the text:
Popper, D.M.   1966, Trans. IAU Vol. X11B, ed. J.-C. Pecker,
        (Academic Press: New York) p. 485.
Popper, D.M.   1967, Ann. Rev. Astron. Astrophys. 5, 85.
Popper, D.M.   1970, in Spectroscopic Astrophysics,
        ed. G.H. Herbig, (University of Calif., Berkeley), p. 441.
Popper, D.M.   1970, in IAU Colloquium No. 6, Mass Loss
        and Evolution in Close Binaries, eds. K. Gyldenkerne
        and R.M. West (Copenhagen U. Publ. Fund: Copenhagen), p. 13.
Popper, D.M.   1981, Astrophys. J. Suppl. 47, 339.

Papers on light curve analysis, not cited in the text:
Popper, D.M., and Etzel, P.B.   1981, Astron. J. 86, 102.
Popper, D.M.   1981, Rev. Mex. Astron. y Astrofiz. 6, 99.
Popper, D.M.   1984, Astron. J. 89, 132.

DISCUSSION

JASCHEK: Two comments on your talk, in which you allude to quite general problems. It was true in the past that most photoelectric observations were made to an accuracy of ±0.01 in an undefined system, so the observations were totally useless. This applies specifically to observations prior to 1955. I thought this had improved, but from your talk I see that this still goes on.

The other situation that you alluded to is the disappearance of the observational material the observations are based on. Instead of plates, data are put on tapes and soon erased. We are on the way of becoming an observational science with just a very thin observational record base!

POPPER: That is a most unfortunate state of affairs.

LACY: I would like to point out that there are two eclipsing binaries (YY Gem and CM Dra) composed of main sequence stars which are used in the determination of the red part of the Barnes-Evans relation and they do not depart significantly from the mean relationship which is determined mainly by red giants. Also, on a different topic, I recognize your example of the misuse of "standard tables" as being that of deLandtsheer's paper on YZ Cas which appeared recently in Astron. Astrophys. I was asked to referee that paper and subsequently recommended strongly on two occasions that the paper should not be published until the data were reanalyzed in a more reliable manner. Nevertheless, the editor chose to ignore my extensive criticisms of the analysis and conclusions, and published the paper essentially unchanged from its initial form. As a result of this editorial lassitude we now have in the literature a bad analysis whose flawed results can now lead unsuspecting workers to erroneous conclusions.

POPPER: The absolute flux scale for main sequence stars from type F to M remains very poorly defined.

FRACASTORO: I wonder whether it would not be useful to add in your list another category of people, namely those who do identify spectral lines in a spectroscopic binary, and then deduce radial velocities, assigning them to pure Keplerian motion of the star. It is rather frequent, instead, that these velocities are altered by the contribution of circumstellar matter. From Batten's Catalogue of Spectroscopic Binaries, a Barr effect results even when orbits having eccentricities e > 0.6 are selected. Therefore, several radial velocities must be spurious.

I have a second point. You have shown some discrepancies resulting when $(B-V)_c$ or temperatures are deduced from observations made in different groups of colors. In my opinion, this might depend on the fact that in a Planckian mode, two colors would be sufficient, whereas the spectral bands U, B ,V etc. have been selected redundantly with the aim of getting the maximum of astrophysical information. This is just the opposite of the Planckian model and may explain the inconsistency of some results, as a consequence of the different viewpoint for laboratory and astronomical photometry.

HEINTZE: Lacy stated that the YZ Cas results Popper showed in his introductory talk may not be used in my poster paper to be presented next Tuesday. However this temperature has been checked carefully. From a high dispersion IUE short wavelength spectrum the metal abundance turns out to be about 10X the solar abundance. From a low dispersion IUE long wavelength spectrum (calibrated) and the Kurucz (1979) models it turns out to be impossible to get a T_{eff} lower than 10000 K.

POPPER: My reason for not referring to YZ Cas by name is that I merely wished to point to what I considered examples of poor practices, rather than to discuss the UV observations.

HEINTZE: I agree with Popper, that the study of detached eclipsing binaries, which at the same time are double line spectroscopic binaries does not give T_{eff} directly but flux ratios as a function of wavelength next to the very precise masses and gravities. However, we desperately need the effective temperatures of these components in order to compare with evolutionary tracks. Modern analysis methods of light curves in any case give a model-effective temperature of one component if that of the other component is assumed or measured. According to me the best way to find T_{eff} for components that obscure one another totally is to observe the energy distribution in a long as possible wavelength region, as has been done by Plavec for SX Cas, U Cep and RW Tau.

POPPER: If wide ranging spectrophotometry is not available, and it usually is not, the T_{eff} is best evaluated through well calibrated indices, such as (b-y) or (B-V). Flux ratios of the components in the V band, for example, are also of importance. My objection was to the custom, by means of programs for light curve analysis, of giving values of T_{eff} rather than of those quantities derived most directly from the light curve, such as flux ratios, light ratios in the observed bands, etc. T_{eff} is <u>not</u> a suitable parameter in light curve analysis.

CHOCHOL: It is not easy to distinguish if a system is detached or semi-detached, especially in the case of early-type binaries. Are you sure that all the binaries used in your work are detached systems?

POPPER: Until the analysis of a light curve has been carried out, one cannot be certain. With one possible exception, I think all the B-type systems, for which I have considered the properties well determined, are detached.

THE CALIBRATION OF INTERFEROMETRICALLY DETERMINED PROPERTIES OF
BINARY STARS

Harold A. McAlister

Department of Physics and Astronomy
Georgia State University

ABSTRACT. With the advent of speckle interferometry, high angular
resolution has begun to play a routine role in the study of binary
stars. Speckle and other interferometric techniques not only bring
enhanced resolution to this classic and fundamental field but provide
an equally important gain in observational accuracy. These methods
also offer the potential for performing accurate differential
photometry for binary stars of very small angular separation. This
paper reviews the achievements of modern interferometric techniques in
measuring stellar masses and luminosities and discusses the special
calibration problems encountered in binary star interferometry. The
future possibilities for very high angular resolution studies of close
binaries are also described.

1. INTRODUCTION

    Interferometry has been applied intermittently to the examination
and measurement of close visual binaries (where the word "visual" is
used here in the context of direct resolution by any means) since
Anderson and Merrill used Michelson interferometry at Mt. Wilson during
1919-21 primarily for resolving Capella (Merrill 1922). It was not
until speckle interferometry was first proposed by A. Labeyrie (1970)
and subsequently demonstrated by Labeyrie and his collaborators (Gezari
et al. 1972) that an interferometric technique of exceptional accuracy
was routinely applied to a wide sample of binary stars. Visual
interferometry, particularly as developed and pursued by W. S. Finsen
at Johannesburg (Finsen 1971), provided a gain in resolution while
maintaining the same level of accuracy as micrometry. A photoelectric
version of Michelson interferometry provided promising initial results
(Wickes and Dicke 1973), but no sustained effort using that approach
followed upon the initial success. For a complete listing of modern
interferometric approaches to binary star astrometry, the reader is
directed to the catalog of McAlister and Hartkopf (1984), hereafter
referred to as MH.

D. S. Hayes et al. (eds.), Calibration of Fundamental Stellar Quantities, 97–119.
© 1985 by the IAU.

Speckle interferometry employing photon detecting cameras at 4-meter class telescopes is capable of resolving 15th magnitude binaries with separations down to 0."025. Most of the speckle results have so far been for objects brighter than 8th magnitude, and there is much useful work to be done in this regime, but it is likely that further speckle programs will push resolution toward a significantly fainter limit and give greater emphasis on employing analysis procedures capable of extracting differential photometric information from speckle pictures. The potential in this latter area has already been demonstrated, and a great deal of very useful information is likely to follow in the coming few years as close visual pairs are resolved photometrically as well as spatially.

Relatively few in number but important results have come from long baseline optical interferometry to date. As has been pointed out many times, the principal reason for the short supply of accurately known stellar masses and luminosities is not due to any paucity of binary stars. It is instead the result of the inability of any one technique to determine all the necesary parameters and the shortage of systems amenable to study by more than one technique. Where speckle interferometry using single telescopes has increased the overlap between visual and spectroscopic binaries, long baseline interferometry will subject virtually every spectroscopic binary to direct resolution.

Interferometric measurements of binary stars can be expected to make substantial contributions to the calibration of stellar masses and luminosities. When long baseline instruments are fully functional, we can look forward to direct measurements of the diameters of the individual components of binaries and hence to major contributions in calibrating emergent fluxes and effective temperatures of stars of known mass. Obtaining such complete and fundamentally important data for stars should serve as strong motivation to the few groups around the world who are working in the extremely challenging area of long baseline optical interferometry.

2.   INTERFEROMETRICALLY DETERMINED MASSES AND LUMINOSITIES

The contributions to the empirical mass-luminosity relation made from combining interferometric and spectroscopic observations are shown in Figure 1 in which are also plotted the points considered to be reliable by Popper (1980). The values for α Virginis were determined with the intensity interferometer at Narrabri (Herbison-Evans et al 1971) which concurrently determined the angular diameter of α Vir A. The orbital solution of α Aurigae by Merrill (1922), which has been only slightly modified by extensive modern measurements (McAlister 1981), and its combination with spectroscopic data is discussed by Batten et al (1978). Speckle interferometry has been responsible for the remaining additions to the mass-luminosity relation in Figure 1 with specific references to the individual studies available from MH. Popper (1980) considered only the analyses of α Aur, α Vir and 12 Per

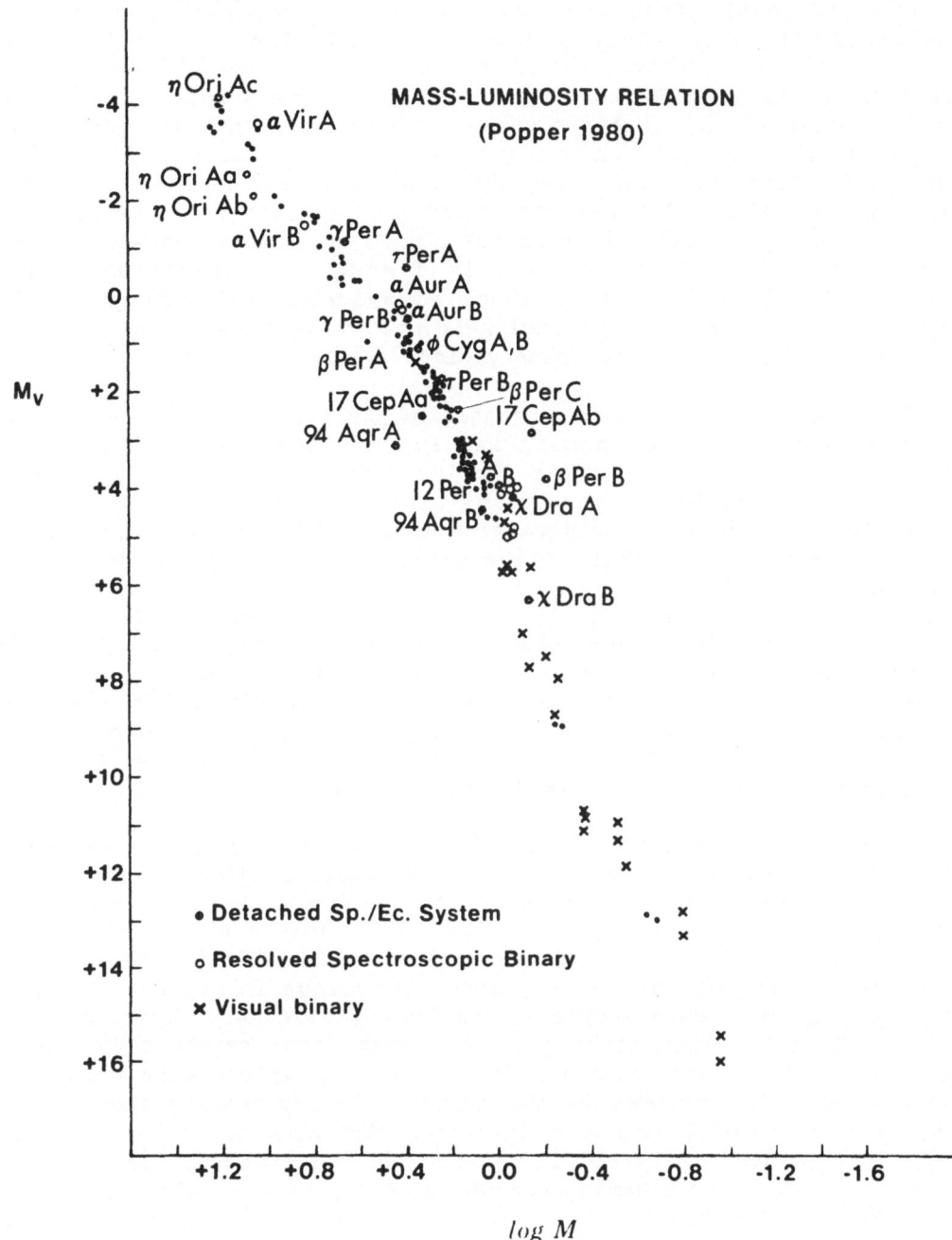

FIG. 1. Points added to the empirical mass-luminosity relation by interferometry are shown plotted against reliable other determinations as given by Popper (1980). Complete references to the individual analyses can be found in the catalog of McAlister and Hartkopf (1984).

to be sufficiently free from significant future revisions to be considered reliable. Analyses too recent to be included in Popper's review for χ Dra and φ Cyg yielded formal errors in the masses of less than 10%. It is possible that the spectroscopic mass ratio for χ Dra may be revised (F. C. Fekel, private communication) thus affecting the conclusion that the components are excessively luminous for their masses. The mass ratio for φ Cyg should be less of a problem since the components are of similar luminosity and the velocities are relatively free of blending effects. In this case, however, the system always has a separation less than 0".04 and is frequently unresolved. The residuals to the six existing speckle measurements of φCyg are no greater than 0".002 and it seems unlikely that the orbital elements will change appreciably as further observations are accumulated.

Interferometric techniques capable of 0".025 resolution are likely to continue to yield new masses and luminosities as spectroscopic binaries with long period orbits receive sufficient coverage of their relative visual orbits. Such systems often need modern spectroscopic observations, particularly to determine mass ratios, and small velocity amplitudes make this a challenging task. Of the 118 binaries newly resolved by interferometry which are listed in Table I (see MH for a complete list of references), 39 are spectroscopic systems of which 11 are double-lined. It is probable and certainly highly desirable that the fraction of spectroscopic binaries in this sample could be increased or that at least the mass ratios determined for systems exhibiting high eccentricity as they undergo periastron passage. Parallaxes forthcoming from the HIPPARCOS mission would then fill in the missing information for mass determinations.

As of early 1984, there were 3363 measurements of 824 systems derived from accurate interferometric observations with an additional 1863 negative examinations for duplicity (see MH). The mean measured separation is 0".32 while the median is 0".21. Thirty eight percent of the measurements are of separations no larger than 0".16 and thus could not be generally provided by any other technique in standard usage. The remaining measurements have accuracies significantly improved over those obtained by other techniques for binary star astrometry. More than 90 percent of these data are from speckle interferometry with the majority being for systems in the northern hemisphere. Although the existing programs of binary star interferometry seem to be healthy and likely to continue, the field is by no means overcrowded. A program resident in the southern hemisphere would be especially welcome.

TABLE I.

Binary Stars First Directly Resolved by Interferometry

Name		$\langle\rho\rangle$	MK Type	Name		$\langle\rho\rangle$	MK Type
HR 132	51 Psc	0."22	B9V	HR 5953	δ Sco	0.16	B0IV
HR 178	–	0.17	A7m	HR 5985	β Sco	0.11	B2IV-V
HR 233	–	0.05	B9V + G0III-IV	HD 144641	–	0.12	G5
HR 439	–	0.13	K0Ib + B9V	HR 6084	σ Sco	0.37	B8Vp
HR 483	–	0.55	G1.5V	HR 6148	β Her	0.09	G8III
HR 539	ζ Cet	0.06	K0III	HR 6168	σ Her	0.04	B9V
HD 12483	–	0.23	G5	HR 6237	–	0.05	F2V
HR 640	55 Cas	0.10	B9V + G0II-III	HD 155095	–	0.13	B8
HR 645	φ Per	0.04	G8III	HR 6388	–	0.04	K3III
HR 649	ξ¹ Cet	0.06	G6II-III	HR 6396	ζ Dra	0.05	B6III
HR 763	31 Ari	0.08	F7V	HR 6410	δ Her	0.10	A3IV
HR 788	12 Per	0.05	F9V	HR 6469	–	0.06	F9V
HR 793	μ Ari	0.05	A0V	HR 6485	ρ Her	0.29	B9III
HR 825	–	0.19	A5Ia	HR 6560	–	0.16	A5V + G5III
HR 838	41 Ari	0.30	B8V	HR 6588	ι Her	0.16	B3IV
HR 854	τ Per	0.05	G4III + A4V	HD 163640	–	0.09	A2
HR 915	γ Per	0.22	G8III + A2V	HR 6697	–	0.10	G2V
HR 936	β Per	0.07	B8V	HR 6742	W Sgr	0.12	F8Ib
HR 1010	ζ² Ret	0.05	G2V	HR 6779	ο Her	0.06	B9V
HR 1084	ε Eri	0.05	K2V	HD 167570	17 Sgr	0.27	A3 + G5
HR 1129	–	0.06	G0III + A3V	HR 6927	χ Dra	0.06	F7V
HR 1252	36 Tau	0.04	G0III + A4V	HD 171347	–	0.16	A2
HR 1331	51 Tau	0.09	F0V	HR 7059	5 Aql	0.13	A2V
HR 1346	γ Tau	0.40	K0III	HD 178452	–	0.12	A2 + G5
HD 283571	RY Tau	0.04	–	HR 7262	ι Lyr	0.08	B6IV
HR 1411	θ¹ Tau	0.15	K0III	HR 7377	δ Aql	0.13	F3IV
HR 1497	τ Tau	0.17	B3V	HR 7417	β¹ Cyg	0.42	K0III + B9V
HR 1569	6 Ori	0.33	A3V	HD 184467	–	0.08	K0
HR 1708	α Aur	0.05	G5III + G0III	HR 7441	9 Cyg	0.04	A0 + F5
HR 1788	η Ori	0.05	B1V	HR 7478	φ Cyg	0.04	G8III + G8III
HR 1808	115 Tau	0.09	B5V	HR 7536	δ Sge	0.05	M2II + A0V
HR 1876	37 Ori	0.05	B0III	HD 187321	–	0.40	A + G0
HD 37614-5	–	0.14	A + G	HD 190429	–	0.12	Oe
HR 2001	–	0.16	A4V	HR 7735	31 Cyg	0.03	K2II + B3V
HR 2002	132 Tau	0.04	G8III	HR 7744	23 Vul	0.24	K3III
HR 2130	64 Ori	0.05	B8III	HR 7776	β Cap	0.06	K0II-III + B8V
HD 41600	–	0.10	B9	HD 196088	–	0.05	A0 + G
HR 2304	–	0.04	A2V	HR 7906	α Del	0.14	B9IV
HR 2425	53 Aur	0.06	B9	HR 7921	49 Cyg	0.24	G8IIb
HR 2605	40 Gem	0.08	B8III	HR 7922	–	0.11	B6III
HD 52822-3	–	0.16	F5 + A	HR 7963	λ Cyg	0.03	B6IV
HR 2846	63 Gem	0.04	F5V + F5V	HR 7990	μ Aqr	0.04	A3m
HR 2861	65 Gem	0.04	K2III + K5III	HR 8047	59 Cyg	0.21	B1e
HR 2886	68 Gem	0.18	A1V	HR 8059	12 Aqr	0.05	G4III
HR 3109	53 Cam	0.04	A2p	HR 8119	I Cep	0.05	B0II
HR 3485	δ Vel	0.62	A1V	HR 8164	–	0.10	M1Ib + B2V
HR 3880	19 Leo	0.13	A7V	HR 8238	β Cep	0.18	B1III
+20°2465	Gl 388	0.08	M4Ve	HR 8264	ξ Aqr	0.03	A7V
HR 4365	73 Leo	0.04	K3III	HR 8417	ξ Cep	0.05	A3m
HR 4544	–	0.17	K0III	HR 8485	–	0.52	K3III
HR 4689	η Vir	0.12	A2IV	HR 8558	ζ¹ Aqr	0.06	F6IV
HR 4785	β CVn	0.11	G0V	HR 8572	5 Lac	0.11	K7Ib + A8V
HR 4905	ε UMa	0.05	A0p	HR 8650	η Peg	0.04	GII-III + F
HR 4963	θ Vir	0.49	A1IV + Am	HD 215318	–	0.15	F8 + A5
HD 126269	–	0.05	F5 + A0	HR 8704	74 Aqr	0.07	B9III
HR 5435	γ Boo	0.07	A7III	HR 8762	ο And	0.33	B5III + A2p
HR 5472	–	0.04	F3V	HR 8866	94 Aqr	0.13	G5IV
HD 136406	–	0.36	K0	HR 9003	ψ And	0.28	G5Ib + A0V
HR 5747	β CrB	0.05	A8III	HR 9064	ψ Peg	0.19	M3III

3.   THE ACCURACY OF INTERFEROMETRY

The effects of systematic errors in angular separation measurements on mass and distance determinations for spectroscopic binaries resolved by interferometry depend upon the specific manner in which the data from the two complementary techinques are combined. The propagation of a relative scale error into the errors for the masses with a factor of three increase from Kepler's Third Law does not usually occur since the interferometry is most often used to solve for the inclination factor in the spectroscopically determined values for $M\sin^3 i$. The error in the distance determination would be linearly related to a scale error, and thus the luminosities would have relative errors sensitive to scale errors by a factor of two.

Interferometric measurements are of potentially very high accuracy and hence deserve high weight in the solution of a binary star orbit. With the current sample of several thousand measurements, it is possible to judge the level of accuracy of the data within and among the various groups of observers employing interferometric techniques. An earlier discussion of this topic (McAlister 1978) concluded that there is little evidence for significant systematic errors in speckle observations when compared to orbits of high quality. This has also been indicated more recently by Worley (1983) who found average residuals for 170 speckle observations from orbits of high quality of

$$\langle \rho\Delta\theta \rangle = + 0\overset{..}{.}001 \pm 0\overset{..}{.}007$$

$$\langle \Delta\rho \rangle = -0\overset{..}{.}009 \pm 0\overset{..}{.}015$$

It is interesting to note that the dispersion is twice as large in the separations as it is in the position angles. In both cases the dispersions are significantly larger than the mean differences. As will be seen later, this unfortunately does not imply that such orbits should be used for calibration purposes. Furthermore, comparison of speckle and micrometer measurements obtained nearly simultaneously shows no significant systematic difference at the limit of accuracy set by the micrometry. For 31 pairs of observations in the sense (visual-speckle) Worley (1983) found

$$\langle \Delta\theta \rangle = + 0\overset{.}{.}34 \pm 3\overset{.}{.}06$$
$$\langle \Delta\rho \rangle = - 0\overset{..}{.}021 \pm 0\overset{..}{.}042$$
for $\qquad\langle \rho \rangle = \quad 0\overset{..}{.}38 \pm 0\overset{..}{.}19$

The dispersion in the diffences represents about 11 percent of the average separation and is probably inherent in the micrometer measures of such close pairs. Speckle interferometry should be routinely capable of measuring such separations to better than 1 percent accuracy, as speckle results are essentially derived from measuring the separation of two Airy disks having diameters on the order of $0\overset{..}{.}03$. A level of 1 percent accuracy for binaries with separations of $0\overset{..}{.}4$ essentially requires the centroiding on an Airy disk to an accuracy of about 10 percent. There is really little excuse for doing much poorer than this.

To assess the internal accuracy of interferometric measurements it is most profitable to examine the results for systems which have been extensively observed by various groups. Unfortunately such systems are rare. One binary star which has received a great deal of attention by interferometrists is β Cephei for which 42 measurements have been made between 1971 and 1981 by five different groups using at least six different telescopes. During that time span, β Cep closed in separation from about 0.''25 to 0.''17 while remaining nearly fixed in position angle. Figure 2 is a representation of these measurements in Cartesian coordinates. The solid line in Figure 2 is a linear fit to the GSU/KPNO 4-meter observations (filled circles) while the dashed line is an extrapolation ignoring elliptical motion back to the early French speckle results. In this straightforward attempt to compensate for orbital motion, the 18 GSU/KPNO 4-meter measurements show dispersions in angular separation of $\pm$0.''0026 and in position angle of $\pm$0.°55. The 7 GSU/KPNO 2.1-meter measurements show a degradation in accuracy with dispersions of $\pm$0.''0094 and $\pm$2.°2. While the 4-meter speckle observations approach a 1 percent level of accuracy, the 2.1-meter results emphasize the advantage of using the largest possible telescopes for speckle interferometry, not just to achieve the highest resolution but also to obtain the greatest accuracy.

The system comprising the star θ Virginis has been extensively observed in the course of the GSU speckle program but very little by other observers. A linear fit to 19 GSU/KPNO 4-meter measurements shows dispersions of $\pm$0.''0048 and $\pm$0.°53 and are again at the 1 percent level for this system which has an average separation of 0.''495.

It is seen that speckle interferometry can (and indeed should) achieve an internal accuracy of 1 percent. Although a perusal of the data in MH shows scattered indications of systematic differences among the observers, there does not yet exist sufficient overlap in the data to bring all interferometric results on a "system" that appears to be free of systematic error. The effects of these errors on the calibration of stellar masses and luminosities are certainly no worse (and perhaps considerably better if the consumer scrutinizes the data carefully) than the complementary spectroscopic, micrometer and astrometric material that joins with the interferometric results to yield fundamental astrophysical parameters.

4. CALIBRATION OF SPECKLE INTERFEROMETRIC OBSERVATIONS

4.1 Background

This section will concentrate primarily on observations of binary stars by speckle interferometry although discussions of standard stars and standard orbits pertain to any high angular resolution technique.

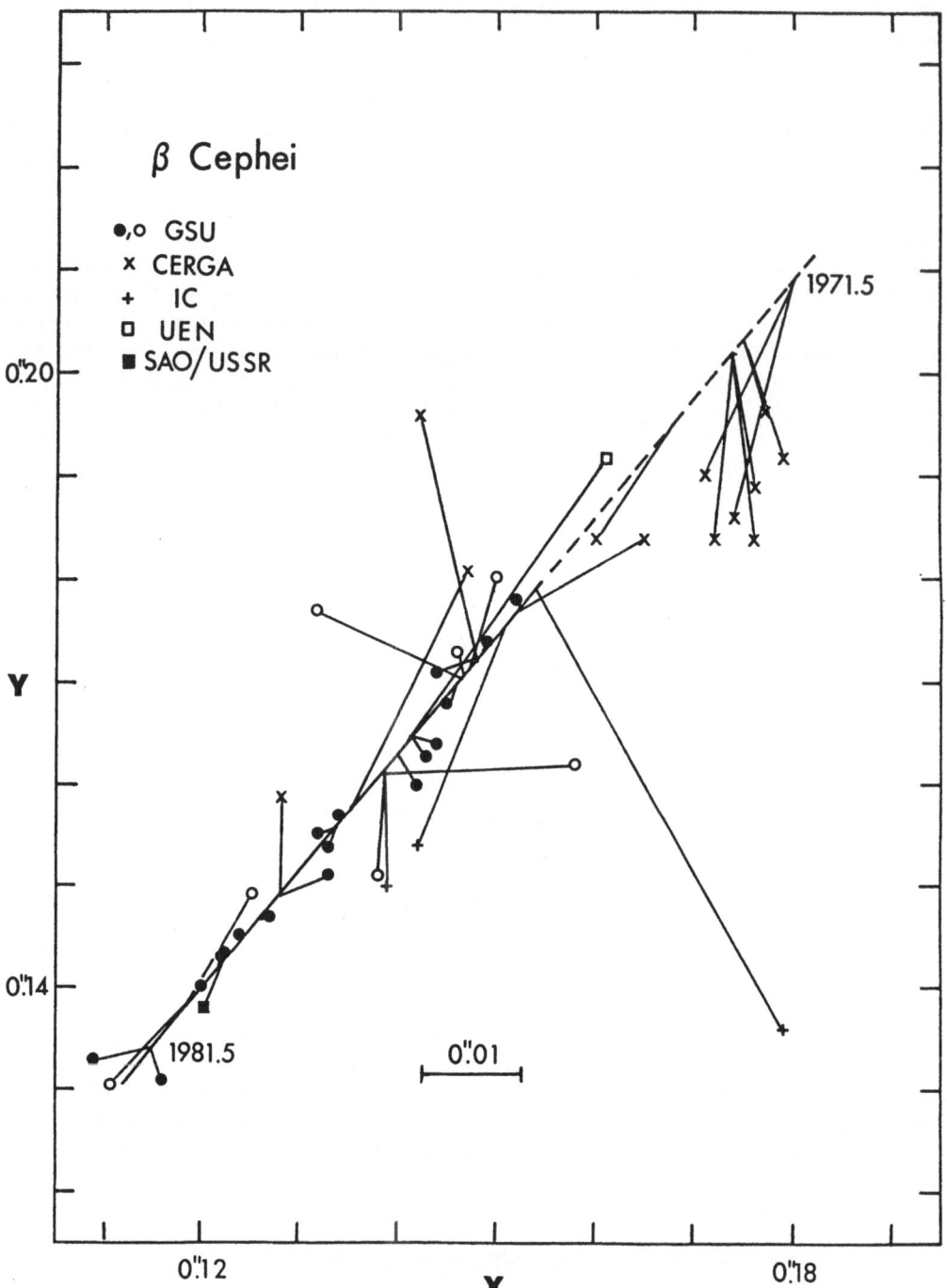

FIG. 2. Speckle interferometric measurements of the binary star
β Cep are shown for the interval 1971-82 during which time the separation closed from 0.''26 to 0.''17 with little change in position angle.

Because of its straightforward applicability speckle interferometry is more likely to be widely applied to high angular resolution measurements than other techniques, particularly long baseline optical interferometry, and there is a significant body of data from speckle work which can be used as a solid basis for this discussion. Reviews of speckle interferometry are given by Bates (1982), Dainty (1975), Worden (1977), Labeyrie (1978) and McAlister (1983).

Speckle pictures (or interferograms as some prefer to call them) are obtained during short time exposures which freeze the instantaneous effects of atmospheric turbulence. Speckle interferometry is essentially an examination of the spatial frequencies present in a speckle picture with the frequency cutoff determined by the aperture of the telescope. This permits the method to make diffraction limited measurements of spatial structure. If in one dimension $O(x)$ is the instantaneous object intensity, then the instantaneous image intensity will be given by

$$I(x') = \int O(x)P(x'-x)dx$$

where $P(x')$ is the instantaneous point spread function induced by atmospheric and telescopic effects. Taking the Fourier transform of this equation utilizes the convolution theorem to deconvolve the atmospheric and instrumental effects to obtain

$$i(u) = o(u)T(u)$$

where $i(u)$ and $o(u)$ are Fourier transforms of the image and object intensities and $T(u)$ is the instantaneous transfer function and is the Fourier transform of the point spread function. The squared modulus of $T(u)$ is the modulation transfer function (MTF) and it is the average value of the MTF which is normally utilized in speckle interferometry where the power spectrum

$$W(u) = \langle |i(u)|^2 \rangle = |o(u)|^2 \langle |T(u)|^2 \rangle$$

is calculated or determined in an analog fashion from a series of speckle pictures of an object.

An alternate and entirely equivalent processing procedure operating in the spatial rather than spatial frequency domain utilizes the average autocorrelation of the instantaneous image intensities

$$C(x) = \langle \int I(x') I(x'-x)dx' \rangle .$$

Dainty (1974) describes the effects of the transfer function on $C(x)$ and shows that atmospheric seeing completely dominates over telescope aberrations under typical conditions. Figure 3 summarizes the analysis approaches of speckle interferometry to the special problem of binary star astrometry and Figure 4 shows a single speckle picture of a Aur with a composite autocorrelation of many such

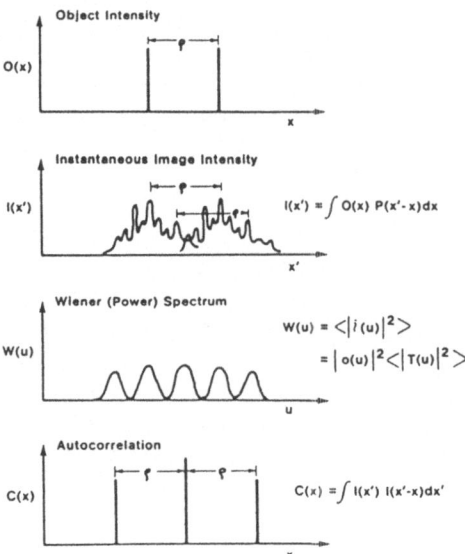

FIG. 3. A summary of the origin of speckle patterns and the analysis of speckle data for binary stars using either power spectrum or autocorrelation analysis is shown.

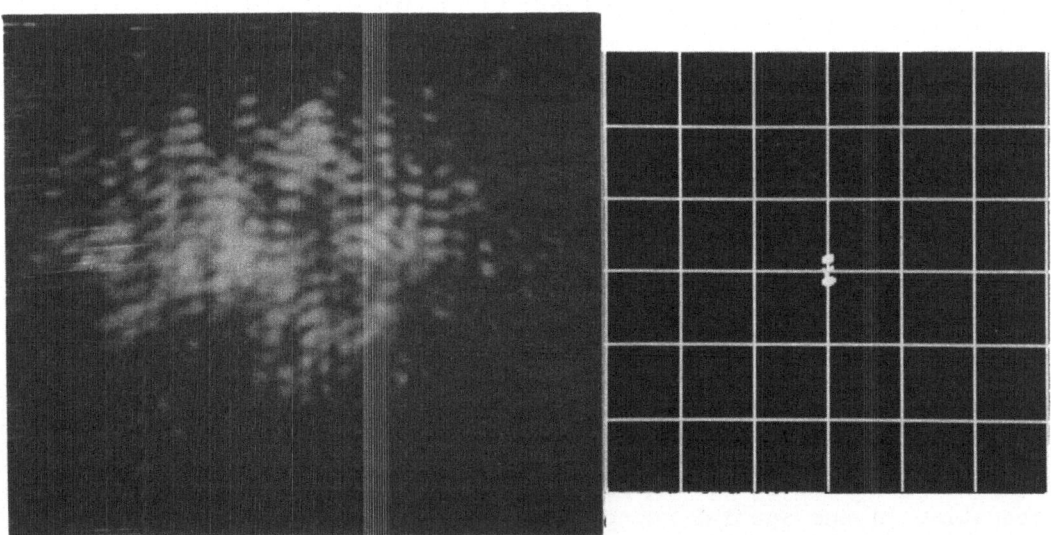

FIG. 4. A speckle picture of α Aur obtained with the GSU ICCD speckle camera at the KPNO 4-meter telescope in January 1984 is shown on the left where the picture is approximately 1" across. Individual pairs of speckles clearly show the 0.''05 binary nature of the star oriented in the north-south direction. The picture on the right is the output from the hardwired digital vector-autocorrelator in use at GSU to reduce speckle data.

pictures produced by a hardwired vector autocorrelator at GSU.

4.2 Atmospheric Effects

The atmosphere limits the applicability and accuracy of binary star speckle interferometry by the effect of non-isoplanicity and by introducing a major seeing component into W(u) or C(x). The limits of isoplanicity are under normal circumstances sufficient to permit separation measurements of up to several arcseconds - well into the realm of standard micrometer and photographic techniques - although the detailed influence of the transition from isoplanicity to non-isoplancity (which is equivalent to considering the transition from very high correlation to very low correlation between the speckle images of the components of a binary) has by no means been thoroughly studied. It is expected that loss of complete isoplanicity decreases the precision with which separations can be measured and may introduce systematic errors into differential photometry, but these effects enter at separations larger than those normally measured by speckle interferometry.

Atmospheric seeing effects are definitely of concern, however, and must be considered in the measurement of binary star separations, stellar angular diameters and differential photometric properties. Figure 5 is a representation of the analysis of a sequence of digital speckle pictures of the binary star ADS 4241 taken with the GSU ICCD speckle camera (McAlister et al 1982) at the KPNO 4-meter telescope in January 1984. ADS 4241 has an approximate angular separation at that epoch of $0\overset{''}{.}25$. In Figure 5, the practiced eye can perhaps pick out the double speckled structure in the picture which is about $1\overset{''}{.}5$ x $1\overset{''}{.}3$ in size. Figure 5(b) is an integrated vector autocorrelogram of about 10 seconds of a stream of pictures of ADS 4241 recorded at 30 pictures per second. Contours show the overall structure of the seeing component which dominates the autocorrelogram as well as the two spikes indicative of the double star geometry at position angles of about 150° and 330°. A first approach at compensating for the seeing component is shown in Figure 5(c) which is the result of rotating the autocorrelogram 90° in position angle and subtracting it from the unrotated original version. This procedure would adequately subtract out the seeing component if the autocorrelogram were perfectly axially symmetric about the zero spatial component. Unfortunately, this is essentially never the case due to telescope aberrations, atmospheric dispersion, detector effects, etc., but this "rotation" algorithm is useful in locating the binary star spikes which are often small effects on the overall seeing slope. The radial profiles in Figure 5 (d) represent cuts through the autocorrelogram of 5(b) where the central profile passes through an estimate of the coordinates of the double star spike and the upper and lower profiles are cut in counterclockwise and clockwise directions at 30° intervals from the spike. The central profile clearly shows the asymetry of the spike resting on the sloping background of the seeing component. In order to compensate for seeing, third order polynomial fits to the top and

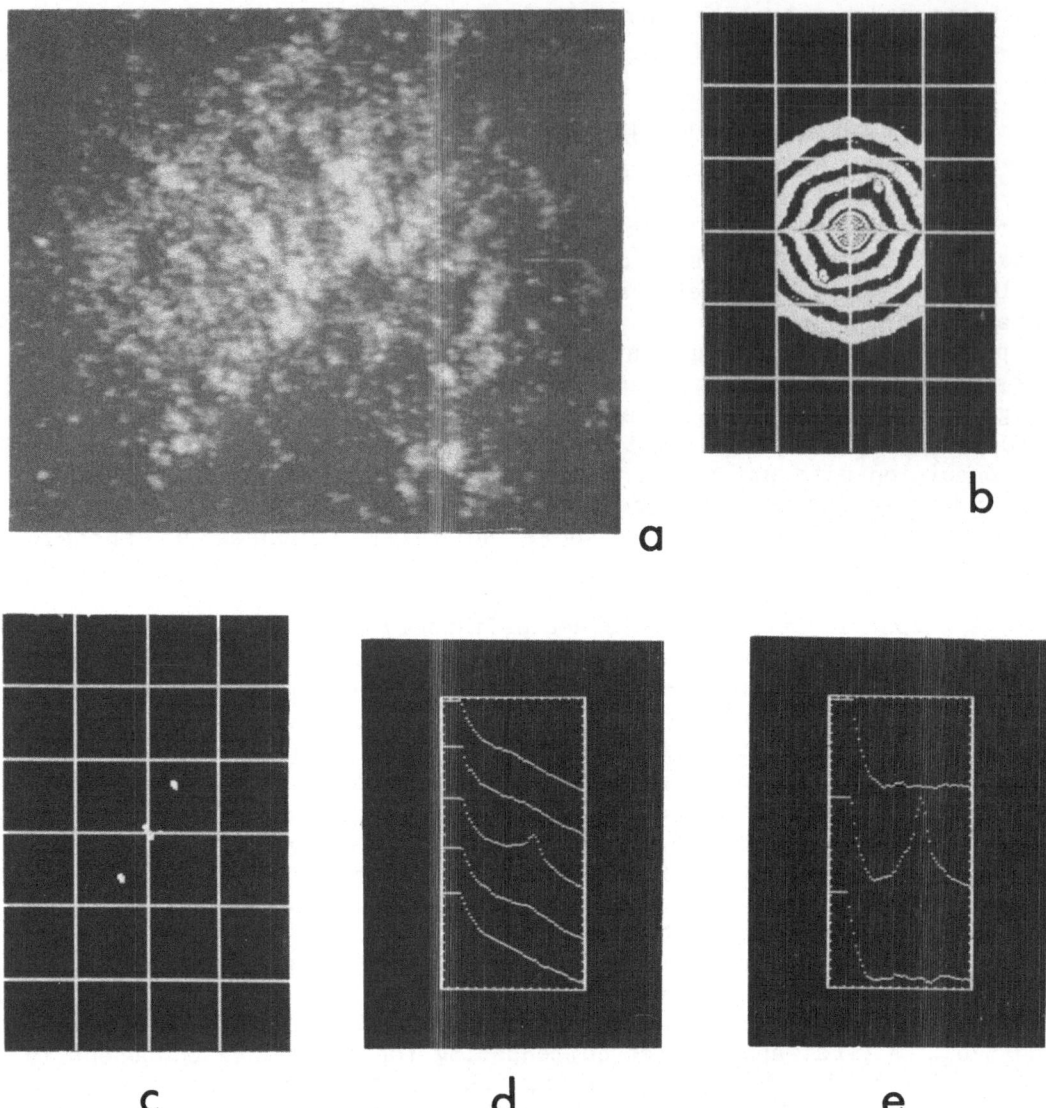

FIG. 5. A single speckle picture is shown in (a) of the 0."25 binary
ADS 4241 taken with the GSU speckle camera at the KPNO 4-meter telescope
in January 1984. The integrated autocorrelogram is shown in (b) and
indicates the broad seeing induced component. The result of approximately correcting for seeing by rotating the autocorrelogram by 90°
and subtracting it from the original is shown in (c). Radial profiles
through the double star spike in the autocorrelogram and through the
adjacent regions are shown in (d) and (e) before and after a more
precise seeing correction.

bottom profiles over a radial range encompassing the radial location of the spike in 5(d) were used to subtract out the sloping background by interpolating the coefficients of the polynomials to every point in the autocorrelogram between the two outer radial cuts. Figure 5(e) shows the resulting seeing corrected profiles which have been rescaled in intensity. In this case, the r.m.s. fluctuations in the background are at a level of 50 units in the relative intensity scale while the intensity of the binary star spike is approximately 2800 units. Centroiding the tops of the spikes in 5(d) and 5(e) shows a 5 percent increase in the separations deduced from the seeing corrected spike in comparison to the uncorrected spike with little effect on the position angle. This is precisely as would be expected and shows that failure to correct for atmospheric seeing effects in speckle interferometry can easily lead to systematic errors in separation measurements that could propagate to 15 percent errors in mass determinations.

One should ideally compensate for seeing effects by observing a nearby single star as close in time as possible to the program star observation. In practice this tends rarely to produce satisfactory compensation because of non-stationary seeing conditions and variations in the instrumental response and in the analysis of the separate data sets. This is more of a problem where the goal is to measure angular diameters or extended structures where the precise shape of the seeing corrected autocorrelogram is needed than in the case of binary star measurement where the background is only to be flattened. Seeing similarly affects measurements of fringe spacings made from power spectra. The systematic effect on the final separation measurement is, however, significantly reduced because the fringe spacing is a differential measure from a set of fringes which all tend to be shifted toward the central fringe by similar amounts, at least to first order. (McAlister 1978).

4.3 Angular Scale Determination

In transforming the linear measurement from an autocorrelogram or power spectrum of speckle data to an angular separation on the sky with an accuracy of 1 percent or better is somewhat of a challenge. Few large telescopes have had their focal lengths determined with this accuracy and so a laboratory measurement of the magnification of a speckle camera system is usually insufficient.

A very effective way of measuring the scale in the focal plane of the speckle camera is to turn the telescope into a Michelson interferometer by placing a double slit mask over the entrance aperture. This may be cumbersome for telescopes of large aperture, and it is possible to select an intermediate pupil at which such a mask can be placed. The projection effects of extrapolating this mask onto the entrance pupil require a knowledge of the focal length of the telescope and the location of the mask in the beam. This procedure was followed for the 4-meter telescope on Kitt Peak (McAlister 1977) and allowed a scale and orientation calibration with limiting accuracies of ±0.6

percent in angular separation and $\pm0^{\circ}.2$ in position angle. The double slit calibration mask has been used during more than 25 observing runs on that telescope with two speckle camera systems and has permitted us to establish what we believe to be a very uniform geometric calibration for our series of binary star speckle measures.

4.4 Calibration Standard Stars

A purely internal scale calibration procedure is ideally desirable but is occasionally not practicable. This is particularly the case where itinerant instruments are used at remote telescopes on a visiting basis, a circumstance which is now almost standard practice. This brings up the question of standard stars or standard orbits. It seems a pity when one uses techniques with potential high accuracy of angular measure to flip through a catalog, pick out a convenient binary star and adopt its published separation as a scale calibration. This can result (and has!) in scale errors of 50 percent or worse.

Choosing binary star orbits, even those which can be judged as definitive on the basis of orbital coverage, as standards for scale determination can also be treacherous. Comparison of the "Fourth Catalog of Orbits of Visual Binary Stars" (Worley and Heintz 1983) with the catalog of interferometric measurements (MH) shows 21 systems having definitive orbits which have received modern attention from interferometry. Table II contains a summary of the residuals to seven of these orbits for which there are more than a handful of interferometric measurements. If we consider only the GSU/KPNO measurements, which probably have the most uniform internal calibration, then it is seen that three of the orbits in Table 2 (ADS 1598, 9617, 11060) yield average residuals insignificantly differing from their dispersions and within 2 percent of the average separations of the systems. The remaining four orbits (ADS 1123, 6650, 8804, 14073) show statistically significant systematic residuals ranging from -6.7 to $+8.8$ percent of the average separations observed. In every case there is no indication of any significant systematic problems with position angles. This result does not contradict the conclusion of Worley (1983) quoted in Section 3 since the average percentage residual in the separations for the orbits in Table II is only $+0.8$ percent. Although there is no overall systematic trend between interferometric measurements and orbital ephemerides, there are strong differences when one considers orbits on an individual basis. Unfortunately, the latter procedure is what one follows in selecting orbits for calibration purposes. To further complicate matters, it can be expected that the degree of agreement with a particular orbit may vary with mean anomaly.

One orbit which perhaps can be used for calibration purposes is that of α Aur as shown in Figure 6 (McAlister 1981). Residuals to that orbit for 56 observations using four different interferometric techniques by seven different observing teams are

TABLE II. Average Residuals of Speckle Observations
to Definitive Orbits

ADS	$<\Delta\rho>$ All (N) GSU (N)	$<\rho>$ $\Delta\rho/\rho$	$<\Delta\theta>$ All (N) GSU (N)
1123	+0".025±0".013(14) +0.026±0.009(9)	0".296 +8.8%	+0°.5±2°.1 −0.2±1.5
1598	−0.005±0.008(11) −0.003±0.004(10)	0.665 −0.5%	−0.5±1.3 −0.2±0.9
6650	−0.044±0.015(14) −0.046±0.007(8)	0.852 −5.4%	−0.9±1.6 −0.5±0.7
8804	+0.032±0.019(25) +0.033±0.006(19)	0.389 +8.5%	+1.0±1.6 +1.2±0.5
9617	−0.005±0.010(12) −0.008±0.006(10)	0.453 −1.8%	+1.0±1.5 +0.8±0.9
11060	+0.004±0.005(20) +0.003±0.003(16)	0.343 +0.9%	+0.1±0.8 +0.1±0.7
14073	−0.036±0.012(27) −0".038±0".005(18)	0.569 −6.7%	−1.0±2.9 −0°.1±0°.9

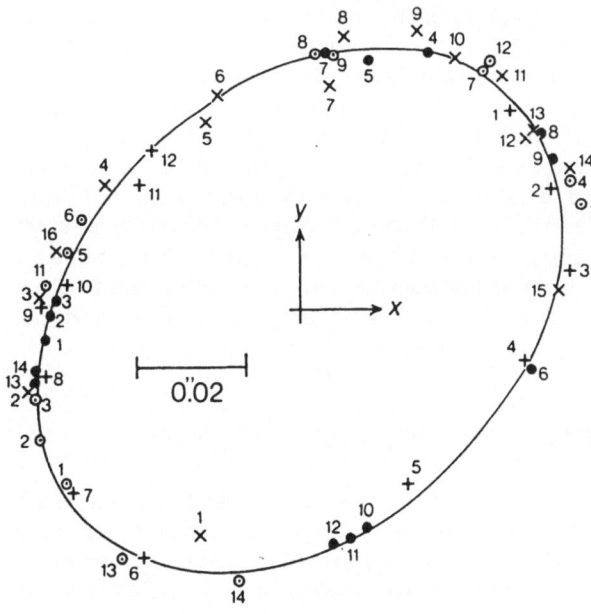

FIG. 6. Measurements from four interferometric techniques of the resolved spectroscopic binary α Aur are shown along with the apparent relative orbit (from McAlister 1981)

$$\langle \Delta \theta \rangle = - 0.06 \pm 2.14$$
$$\langle \Delta \rho \rangle = + 0.0001 \pm 0.0021$$

and the internal error of the calculated semimajor axis is +0.0001. Further thought shows that even this orbit is not well suited to calibration since only the largest telescopes can resolve it in the first place and then measure it with an accuracy of ± 0.002 which is already about 4 percent of the average separation of 0.055. An accuracy of 1 percent is beyond reach if Capella is the only source of calibration. It must be concluded, therefore, that there are presently no binary star orbits which can be used unequivocally for calibrating angular scale to an accuracy of 1 percent.

As an alternative to standard orbits, it has been suggested (McAlister and Hartkopf 1983) that all interferometric observers adopt a set of standard stars for binary star interferometry. This list is reproduced here in Table III. Extensive observation of these objects will eventually lead to standard orbits in some cases. It will more generally define a set of slowly moving systems, such as θ Virginis, whose geometry can be frequently measured by (hopefully) several groups employing independent scale calibration. These objects, which are distributed all over the sky, can thus serve as tie-in stars to a system of accurate absolute angular calibration.

4.5 Photometric Calibration

Interferometric techniques offer the potential for performing photometry of the individual components of close visual binaries. This is extremely important since no other technique, except for the highly restricted method of lunar occultations, can obtain this information over the separation regime accessible to interferometry. Several approaches have been tried for speckle interferometry using actual data (Hege et al 1983, Cocke et al 1983, Bagnuolo 1983, Weigelt and Wirnitzer 1983, and Baguolo and McAlister 1983) and all rely upon the ability to perform accurate intensity calibration of speckle pictures. With the increasing use of solid state detectors in speckle cameras, such calibration is certainly feasible. Although "speckle photometry" is still in a developmental stage, there is every reason to believe that accurate photometry will be forthcoming for the components of binary systems once generally considered to be beyond the reach of photometric resolution.

5.   POTENTIAL FROM VERY HIGH ANGULAR RESOLUTION INTERFEROMETRY

While interferometry using single telescope apertures has definitely enhanced the potential of binary star astrometry, the application of long baseline interferometry will quite literally revolutionize the field. Simple inspection of Kepler's Third Law shows that, for a given distance, an increase in limiting resolution from 0.025 as in the case of speckle interferometery at a 4-meter telescope

TABLE III

Standard Stars for Binary Star Interferometry

HR	ADS/Name	SAO	α 1950	δ	m	Δm	N_{obs}	t_f t_l 1900+		<P.A.>	<Sep.>	comment
132	51 Psc	109262	00^{h}29^{m}48.^{s}8	+06°40'47"	5.7	0.7	6	77.7	81.5	95.°4±0.°6	0."236±0."005	no orbit
404	1123	129277	01 21 49.6	−07 10 30	5.9	0.2	14	75.6	80.9	217.2 0.8	0.346 0.009	P = 16.1 yr
719	Kui 8	110542	02 25 24.9	+01 44 16	6.5	0.4	12	76.9	81.7	32.5 0.2	0.510 0.001	no orbit
1199	Kui 15	111469	03 49 20.1	+06 23 10	5.7	0.1	8	76.9	80.9	208.4 0.2	0.645 0.003	no orbit
−	Stt 97	76954	05 02 36.5	+22 59 38	6.7	1.5	8	76.9	80.9	153.4 0.3	0.362 0.001	no orbit
1946	4265	94759	05 38 24.2	+16 30 35	4.9	0.4	10	76.9	80.9	237.9 0.2	0.347 0.001	P = 95.4 yr
2678	5795 AB	152394	07 04 19.8	−11 12 57	5.4	1.3	4	76.9	80.9	116.5 0.2	0.557 0.002	no orbit
2982	6313 AB,C	115839	07 40 31.4	+00 18 33	6.2	1.8	5	76.3	80.2	228.5 0.4	0.818 0.008	no orbit
3269	Fin 346	116630	08 17 12.2	+04 06 23	6.1	0.0	9	76.9	80.9	74.9 0.4	0.274 0.003	no orbit
3744	7382 AB	136861	09 24 47.6	−09 00 21	6.5	0.0	4	76.9	80.2	196.2 0.5	0.349 0.001	no orbit
4347	8086	156528	11 10 01.4	−18 13 39	6.1	0.4	10	76.0	81.4	333.3 0.4	0.250 0.004	P = 233 yr
4789	Wrh	82390	12 32 21.6	+22 54 15	4.8	2.4	10	76.3	81.5	12.4 0.3	0.356 0.001	no orbit
4963	θ Vir Aa	139189	13 07 21.5	−05 16 21	4.4	1.5	27	76.0	81.5	325.3 0.2	0.495 0.001	P = 228 yr
−	9392	120673	14 46 24.2	+06 09 46	6.7	0.0	8	78.3	81.5	296.5 0.8	0.363 0.010	P = 228 yr
5654	Cou 189	101429	15 09 47.7	+19 09 47	5.9	1.9	8	78.1	81.5	145.5 0.1	0.461 0.003	no orbit
5850	9758	101699	15 40 50.3	+13 49 33	6.4	1.0	9	76.4	81.5	2.5 0.1	0.682 0.003	no orbit
−	9932	140981	16 05 42.9	−09 58 09	6.9	0.3	6	78.6	81.5	192.5 0.3	0.384 0.007	P = 55.0 yr
6627	10795	103106	17 44 55.5	+17 42 51	5.7	1.3	12	78.3	81.7	267.5 0.3	0.574 0.002	no orbit
6795	11111 AB	123187	18 07 04.6	+03 59 00	5.7	1.2	12	78.3	81.7	333.3 1.0	0.339 0.002	P = 270 yr
7285	12160 AB	104602	19 10 19.8	+16 45 40	6.7	1.3	4	76.6	81.7	137.9 0.1	0.681 0.001	no orbit
7497	12808 AB	105168	19 40 12.8	+11 42 27	5.3	1.2	8	76.5	81.7	76.3 0.1	0.456 0.001	no orbit
7882	14073 AB	106316	20 35 12.2	+14 25 12	3.6	1.0	24	75.6	81.7	182.5 1.5	0.538 0.008	P = 26.6 yr
−	14648 Aa	145118	21 04 45.8	−08 26 13	8.1	0.3	2	78.6	81.7	90.7 0.1	0.256 0.001	no orbit
8566	15988	127551	22 27 26.2	+04 10 38	5.5	1.4	10	76.6	81.7	117.1 0.1	0.818 0.005	P = 140 yr
8739	16428	108307	22 56 41.6	+11 27 40	5.8	1.3	8	78.6	81.7	297.3 0.5	0.600 0.002	P = 270 yr
8890	16708	165658	23^{h}20^{m}02.^{s}0	−15°18'50"	5.2	0.7	6	76.5	81.7	102.°0±0.°2	0."412±0."003	P = 63.2 yr

to the 0".0002 limit as an example for an interferometer with a 300-meter baseline results in a gain in sensitivity to shorter periods by a factor of $(0".025/0".0002)^{3/2} \cong 1400$. Where speckle interferometry now resolves systems having periods of years, a long baseline interferometer will resolve binaries having periods of hours! A two solar mass binary at a distance of 100 pc could be resolved as long as the period exceeded 18 hours corresponding to a semi-major axis of only 0.020 A.U. A system of the same total mass could be resolved from a distance of 1000 pc with a period as short as 23 days.

Observations of double-lined spectroscopic binaries offer the most promising returns from very-high angular resolution interferometry of binary stars. Distances determined for these objects are independent of the effects of interstellar extinction (and can in fact be used as a probe of the interstellar medium) and can potentially penetrate to distances well beyond the effective limit for trigonometric parallaxes while preserving the directness and uniqueness of a simple geometric technique. Single-lined spectroscopic binaries do not so easily yield a complete set of orbital and physical parameters since the linear scale of the relative orbit is not obtainable from the radial velocities of just one component. If the trigonometric parallax is known and the apparent orbit is determined interferometrically then the resulting mass sum can be combined with the spectroscopically determined mass function to give the individual masses. The HIPPARCOS astrometry satellite should eventually provide parallaxes of great value in this application.

By determining spectroscopic parallaxes for the objects included in the "Seventh Catalogue of the Orbital Elements of Spectroscopic Binary Systems" of Batten et al.(1978), Halbwachs (1981) calculated the expected angular separations at nodal passages. Of the 978 member systems in the catalogue, 683 binaries or 70% of the entire sample are predicted to have nodal separations exceeding 0".0002 and would thus be good candidates for resolution by an interferometer having a baseline of 300 meters. Conservatively considering only double-lined systems as prime candidates due to their anticipated small magnitude differences and limiting the program to objects north of declination -20°, we find a sample of 180 double-lined spectroscopic binaries likely to be resolved by a 300 meter baseline interferometer. Of these, 102 systems are predicted to have separations greater than 0".0010 and are almost certain to be resolved. Figure 7 presents a histogram of this last sample as a function of the MK spectral types available for the objects compared to the distribution currently available for the mass-luminosity relation. Important aspects of Figure 7 are the new large numbers of early type main sequence components as well as the substantial collection of evolved stars for which masses, luminosities and in many cases effective temperatures could become available.

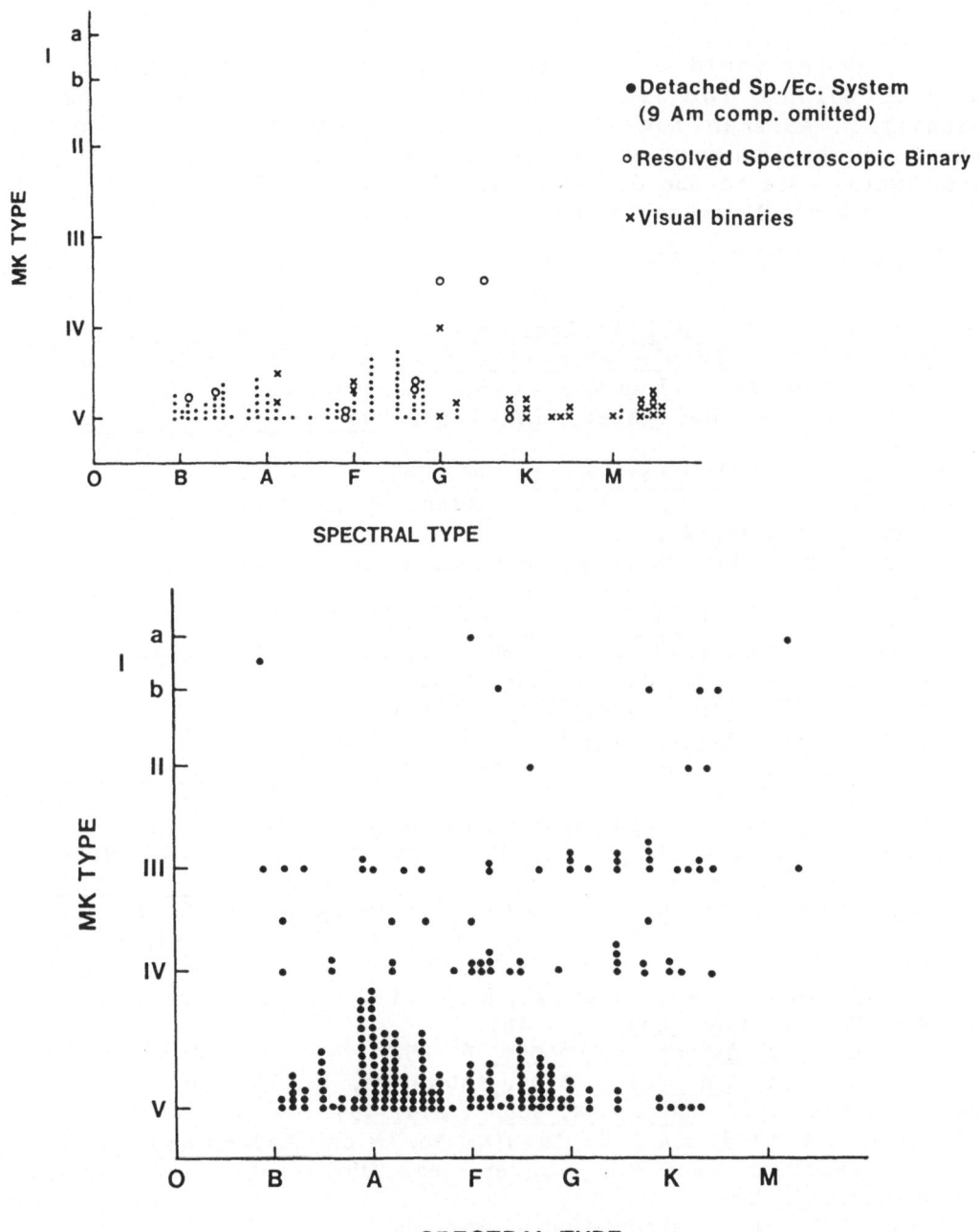

FIG. 7. The distribution of MK spectral types among the components of binaries considered by Popper (1980) as having reliably determined masses is shown in the upper histogram. The lower histogram shows the distribution of the components of double-lined spectroscopic binaries from the catalog of Batten et al (1978) which can be expected to have angular separations exceeding 0".0010 and thus be excellent candidates for long baseline optical interferometry.

ACKNOWLEDGEMENTS

The author would like to express grateful appreciation to the National Science Foundation and the U. S. Air Force Office of Scientific Research for their continued support of the speckle interferometry program at Georgia State University. The valuable contribution made to the GSU speckle effort by Dr. William I. Hartkopf are also gratefully acknowledged.

REFERENCES

Bagnuolo, W.G. 1983, I.A.U. Colloquium No. 62 on Current Techniques in Double and Multiple Star Research, R.S. Harrington and O.G. Franz, eds.,(Lowell Obs. Bull. No. 167, Vol. IX, No. 1), p. 180.

Bagnuolo, W.G. and McAlister, H.A. 1983, Pub. Astron. Soc. Pac. 95, 992.

Bates, R.H.T. 1982, Physics Reports, 90, 203.

Batten, A.H., Fletcher, J.M., and Mann, P.J. 1978, Pub. Dominion Astrop. Obs. 15, No. 5.

Cocke, W.J., Hege, E.K., Hubbard, E.N., Strittmatter, P.A., and Worden, S.P. 1983, I.A.U. Colloquium No. 62 on Current Techniques in Double and Multiple Star Research, R.S. Harrington and O.G. Franz, eds,(Lowell Obs. Bull. No. 167, Vol. IX, No. 1), p. 159.

Dainty, J.C. 1974, Mon. Not. R. Astr. Soc. 169, 631.

Dainty, J.C. 1975, Topics App. Phys. 9, 255.

Finsen, W.S. 1971, Ap. Space Sci, 11, 13.

Gezari,D.Y., Labeyrie, A., and Stachnik, R.V. 1972, Astrophys. Journ. 173, L1.

Halbwachs, J.L. 1981, Astron. Astrophys. Supp. 44, 47.

Hege, E.K., Hubbard, E.N., Cocke, W.J., Strittmatter, P.A., Worden, S.P., and Radick, R.R. 1983, I.A.U. Colloquium No. 62 on Current Techniques in Double and Multiple Star Research, R. S. Harrington and O.G. Franz, eds,(Lowell Obs. Bull. No. 167, Vol. IX, No. 1), p. 185.

Herbison-Evans, D., Hanbury Brown, R., Davis, J., and Allen, L.R. 1971, Mon. Not. R. Astr. Soc, 151, 161.

Labeyrie, A. 1970, Astron. Astrophys. 6, 85.

Labeyrie, A. 1978, Ann. Rev. Astron. Astrophys. 16, 77.

McAlister, H.A. 1977, Astrophys. Journ. 215, 159.

McAlister, H.A. 1978, I.A.U. Colloquium No. 48 on Modern Astrometry, F.V. Prochazka and R.H. Tucker, eds (University Observatory: Vienna), p. 325.

McAlister, H.A. 1981, Astron. Journ. 86, 795.

McAlister, H.A. 1983, I.A.U. Colloquium No. 62 on Current Techniques in Double and Multiple Star Research, R.S. Harrington and O.G. Franz, eds,(Lowell Obs. Bull. No. 167, Vol. IX, No. 1), p. 125.

McAlister, H.A., Robinson, W.G., and Marcus, S.L. 1982, Proc. S.P.I.E. 331, 113.

McAlister, H.A., and Hartkopf, W.I. 1983, Pub. Astron. Soc. Pac. 95, 778.

McAlister, H.A., and Hartkopf, W.I. 1984, "Catalog of Interferometric
    Measurements of Binary Stars," GSU CHARA Cont. No. 1.
Merrill, P.W. 1922, Astrophys. Journ. 56, 43.
Popper, D.M. 1980, Ann. Rev. Astron. Astrophys. 18, 115.
Weigelt, G., and Wirnitzer, B. 1983, Optics Lett. 8, 389.
Wickes, W.C., and Dicke, R.H. 1973, Astron. Journ. 78, 757.
Worden, S.P. 1977, Vistas in Astron. 20, 301.
Worley, C.E. 1983, I.A.U. Colloquium No. 62 on Current Techniques in
    Double and Multiple Star Research, R.S. Harrington and O.G.
    Franz, eds, (Lowell Obs. Bull. No. 167, Vol. IX, No. 1), p. 1.
Worley, C.E., and Heintz, W.D. 1983, Pub. U.S. Naval Obs. Vol. XXIV,
    Part VII.

DISCUSSION

EVANS: Dr. Davis has emphasized the limitations of the occultation
method and while, I agree with him, I feel I must emphasize what has
been achieved so far. It has produced the major proportion of measured
angular diameters. In some favorable cases, such as Alpha Tauri, where
the repeatability of occulation observations has produced some twenty
results; the mean value is uncertain by less than half a percent. The
observed values range down, for later types, well below those values
accessible to the speckle interferometrists. The observations do not
require large telescopes, complicated equipment or large amounts of time
so that even amateurs with relatively small telescopes can contribute.
The possibility of multicolor observations on the same occasion is of
value in the detection of possible variations with wavelength. As to
errors: these are proportionally greater the smaller the star observed,
but one needs to approach published values with reserve since methods of
reduction vary quite widely. The quotation of the formal errors of a
multiparameter fit usually give an unduly optimistic view of the
situation. One should try a variety of trial diameters near the result
of a multiparameter fit and assess critically the range over which
diameter values might be acceptable. In certain cases there is no well
defined best fit.

    The occultation program has a future in the detection of close
binaries, which in favorable cases are much closer than those accessible
by speckle interferometry. Magnitude and color differences can be
obtained from single observations and in some cases the ambiguity of
quadrant of the speckle results can be removed. If numerous
observations of the same system on the same or different occasions are
available, conventional data for the separation and position angle can
be derived and orbital elements improved or even derived. It is notable
that the histogram of separations for A-type binaries differs from that
given Dr. McAlister. For example, for the brighter A-type stars where
observational selection is less severe, the numbers increase quite
sharply for separations starting at the point where his begin to fall
off.

FRACASSINI: As regards the discovery of double stars by means of interferometry, I will ask something about the problem of Alpha Lyrae. In one of the first publications of the Narrabri researchers, Prof. Hanbury-Brown mentioned the problem of Alpha Lyrae, whose effective temperature, derived from angular diameter measures, is lower than that of Alpha Canis Majoris in spite of its earlier spectral type. In this connection Prof. Hanbury Brown mentioned the hypothesis by Petrie of the duplicity of Alpha Lyrae. That is not a trivial problem for the researchers of the Department of Physics at the University of Milan and their colleagues of Brera-Merate Observatory who have proposed Alpha Lyrae as a standard for the calibration of angular diameter determinations and, as I. N. Glushneva will say in her paper, this star is carefully studied by the researchers at the Sternberg State observatory in Moscow. Is there any news about this problem?

McALISTER: We have observed Alpha Lyrae on several occasions with our speckle camera and find no evidence of a companion. Our detectibility would be restricted to separations in excess of about 0.03 arcseconds and a magnitude difference less than about two magnitudes. Within these constraints, then Alpha Lyrae appears to be single to us.

BATTEN: High-dispersion observations of Alpha Lyrae at Victoria reveal no sign of variation in the radial velocity.

POPPER: I was disappointed that you did not discuss the application of your technique to the determination of color indices and magnitude differences (as described in your abstract) in binaries with good orbits. This information is often lacking so that fundamental properties of the components are not well known (but see K. D. Rakos, Astron. Astrophys. Suppl. 47, 221). At this time, this use of speckle observations could be of equal or greater importance than the astrometric results.

McALISTER: I did not say much about photometric determination from speckle interferometry because at the present time there are very few solid results available for demonstration. There are several very promising algorithms for extracting differential photometry from speckle data, as mentioned in my paper, and I believe that this potential may even be more important than the accurate astrometry speckle is providing. One member of our speckle group is devoting a major part of his time to implementing these algorithms and we aim, ultimately, at providing a catalogue of magnitude differences and color indices for binaries with separations of a few arcseconds down to about 0.03 arcseconds.

POPPER: Perhaps the most interesting result thus far from your work is for Chi Dra. The revised orbit, combined with published radial velocities, leads to a mass for the F7 V primary considerably less than one solar mass. No other late F star with well-determined properties has such a low mass. A critical observation is the radial velocity separation of the components. A very preliminary result by Tomkin and

Fekel at McDonald Observatory shows that the spectral lines of the components are not clearly resolved except possibly at maximum separation. An underestimate of the separation by 1 or 2 km/s could remove much of the discrepancy in the mass of the primary, which has been interpreted as a consequence of a non-standard chemical composition. We await the results of further observations by Tomkin and Fekel.

McALISTER: I am glad to hear that a revised mass ratio for Chi Dra may be forthcoming. That, coupled with the more recent speckle results, will certainly warrant a re-analysis of this now puzzling system and may indeed show that it is not anomalous as it now appears.

STRAND: What is the limiting magnitude with the CCD system?

McALISTER: 15th mag.

STRAND: I believe that you will be able to obtain measures of G 107-70, the close binary with white dwarf components, for which we now have a very precise orbit except for the value of the semi-major axis, which is of the order of 0.6" from 61-inch plates. With a parallax known to 1%, combined with a speckle interferometric measurement of separation, the system will give masses with a precision of a few percent.

McALISTER: We have not yet made a deliberate attempt to measure very faint binaries, but I do recall that we have very nice results on a 15th magnitude system, which I believe is Ross 29. I do not remember if GL 107-70 is on our program, but I will certainly see that it is added if it is not already there.

GARRISON: I would like to second what was said by Dr. Popper about providing photometric information and to encourage you to extend your work to overlap with some the area-scanner work in the region of separations of 1 - 5 seconds of arc.

    Chris Corbally at Toronto (now at the Vatican Observatory) finished a thesis last year in which he studied MK types for close visual binaries with separations of 1 - 5 seconds. The only photometric data available for comparison with his types are from the area scanner work of Hurly and of Rakos. Unfortunately there is little overlap between these two sources and they disagree with each other in those few stars. We can get good MK types for stars as close as 1" in good seeing, so this is important for studies of stellar evolution theories.

HEINTZ: The inclination usually is (along with the eccentricity) the element least reliably defined in astrometric (unresolved) orbits. I have a photographic series on Chi Draconis which I expect to complete shortly.

THE MK CLASSIFICATION AND ITS CALIBRATION

Philip C. Keenan

Perkins Observatory
Ohio State and Ohio Wesleyan Universities

1. THE STANDARD STARS OF THE SYSTEM.

I have been asked to describe the system of spectral classification as it has developed from the original Yerkes Atlas (Morgan, Keenan, Kellman, 1943) until to-day. I use the word "developed" because any system that is to remain useful must be flexible enough to adapt not only to improved techniques of measurement but also to new theoretical insights into the variables that actually determine the energy spectrum of a star in all its fascinating but sometimes frustrating detail.

The observed criteria on which the classification of normal stars rests are illustrated in detail in the published atlases, of Morgan, Abt and Tapscott (1978) for the stars of early type, and of Keenan and McNeil (1976) for types later than F8. In addition, there are beautiful reproductions of most of the standard stars of solar composition in the atlas prepared in Japan by Yamashita, Nariai, and Norimoto (1977). I shall not consider here the criteria of classification, but confine the discussion to the resulting set of temperature types and luminosity classes, and then look more carefully at a third variable: chemical composition.

Let us start with the first dimension of classification: the temperature types. When the MK system was developed it was decided to retain the basic notation of the Henry Draper Catalogue because of the great usefulness of that work. This is admittedly a somewhat awkward notation, for not all of the decimal subdivisions of the main HD types (F, G, etc.) are equally meaningful. That is not too serious a difficulty, however, if we are consistent in defining our subtypes to represent approximately equal differences in the spectra. Thus, as you know, it was necessary to drop several decimal subdivisions in type G, where the subtypes are G0, G2, G5 and G8. Similarly, there are no full subtypes between K5 and M0; K7, for example, is counted as a half subtype later than K5 and earlier than M0. On the other hand, near B0 and A0 the spectra are changing so rapidly with temperature that we count O9.5 and B9.5 as full subtypes.

D. S. Hayes et al. (eds.), Calibration of Fundamental Stellar Quantities, 121–136.
© *1985 by the IAU.*

The subtypes in the MK system are not arbitrary, but were developed gradually by experience with standard classification spectrograms with scales of 70 to 120 Å/mm (or resolutions of about 2Å). The ones that we are using now are listed in Table I. Of course, if better resolutions or greater spectral ranges are available, one introduces fractions of the subclasses. Conversely in low-resolution spectral surveys, it is obviously not meaningful to use all of the subtypes. Such surveys thus do not attain the accuracy of the standard MK types, but can be said to be approximately on the same system. This holds true, naturally, for all the dimensions of classification.

It follows, I think, that the standard stars that establish the scale and zero points of the system should have their classification indices defined more precisely than the average observer needs. If we could rely upon the constancy of the stars, we might be satisfied with a small number of basic standards, but in the real universe we cannot count upon any star later than G0 (or upon supergiants of any type) to maintain its spectral features unchanged even over short human time intervals. This is just as true of brightness and color also. Consequently, we need a grid of as many standard stars as possible, well distributed over the sky.

The MK spectroscopic boxes, bounded by full temperature subtypes and luminosity classes, are shown in Table I. Representative standard stars are included, and for types O, B, A and F are taken from the Atlas of Morgan, Abt and Tapscott. For the later types the list of standards furnished to Dr. Pasinetti's IAU Working Group on Standards, and published by Dr. Jaschek in the Bulletin of the Strasbourg Centre de Donnees Stellaires (Keenan 1983) allowed more of the boxes to be filled. For these types the many stars that fall between the boxes (e.g. 46 LMi, K0+ III-IV) have been omitted. I hope that the list of standards can soon be revised and extended, particularly to more stars in the southern hemisphere, if we do not exhaust the patience and funds of the publisher.

Relatively little work has been done on providing comprehensive lists of accurate MK types for faint stars. Some recent catalogs that do reach stars fainter than V = 8 are listed in Table II. The table does not include low-resolution surveys, but only the results of programs designed to give types as accurate as the MK standards that exist for bright stars. Also omitted from Table II are lists of special groups of stars, such as the catalog of types of barium stars by Yamashita and Norimato (1981) and that of S and SC stars by Keenan and Boeshaar (1980).

2.   CALIBRATION IN TERMS OF T_e AND M_v.

It will not be necessary to consider the problem of the reduction of MK types to effective temperatures here, for that topic is discussed in the review papers of Hayes (1985), and of Gustafsson (Gustafsson and Graae-Jørgensen 1985). Pending the publication of a really extensive catalog of effective temperatures, on some generally accepted scale, reference can be made to the

TABLE 1

MK TEMPERATURE SUBCLASSES

TYPE	V	IV	III	II	Ib	Iab	Ia
O4	HD 46223						
O5	HD 46150						
O6	HD 199579						
O7	15 Mon						
O8	HD 46149						
O9	HD 46202		ι Ori		19 Cep		
O9.5						BS 7589	
B0	τ Sco				69 Cyg		ε Ori
B0.5			ε Per				
B1	ω^1 Sco			β CMa	ζ Per		κ Cas
B2	β Sco(ft)	γ Peg	π^4 Ori		9 Cep		χ^2 Ori
B3	29 Per	ι Her				o^2 CMa	55 Cyg
B5	HD 36936		τ Ori		67 Oph		η CMa
B7	BS 1029		β Tau				
B8					13 Cep		β Ori
B9							
B9.5	HD 19803						
A0	α Lyr		α Dra		η Leo		BS 1040
A1							
A2	HD 23948	λ UMa					α Cyg
A3	+48°944					BS 641	
A5	HD 23886						

TABLE I (continued)

Type	V	IV	III	II	Ib	Iab	Ia
A7			θ^2 Tau				
A8							
F0	HD 23585	HD 27397	ζ Leo		α Lep		ϕ Cas
F2	78 UMa	β Cas			89 Her		ι^1 Sco
F3	BS 1279		20 CVn				
F5	HD 27534	δ Cas			α Per		HD 10494
F7			θ^2 Tau				
F8	HD 27808	BS 8732	υ Peg		γ Cyg		δ CMa
F9	BS 7162						
G0	η Cas A β CVn	ζ Her η Boo			ζ Mon α Aqr	HD 91629	
G2	Sun	ϕ Vir η Boo			ζ Mon α Aur		
G5	κ Cet	μ Her		β Lep β Cnv	9 Peg	BS 2513	
G8	61 UMa	β Aql	ϵ Vir 71 Oph	BS 1270	ϵ Gem		
K0	σ Dra BS 7368	η Cep	κ Per η Cyg	HD 179870			AX Sgr
K2	ϵ Eri HD 109011		ϵ CrB κ Oph	π^6 Ori σ Oph	ϵ Peg		
K3	BS 8832 BS 9038		BS 3751 α Tuc	ι Aur π Her	41 Gem 145 CMa		
K4	HD 160964 HD 216803		17 UMa κ Pyx	V418 Cen	BS 7475		
K5	61 Cyg A		γ Dra N Vel		BS 8726		

TABLE I (continued)

Type	V	IV	III	II	Ib	Iab	Ia
M0	HD 232979 HD 19305		τ Aq.r				
M1			ν Vir BS 5995	BS 3126	BS 6693		
M2	HD 1326		BS 8989 HD 107003		HD 10465		
M3			μ Gem 104 Gem	τ Aur			
M4	γ Leo C		BS 3577	δ² Lyr ρ Per	BO Car	HD 91095	
M5			BS 9047			HD 94599	
M6							
M7				BK Vir			
M8				R Dor RX Boo			

diagrams of Keenan (1982) showing how the spectral types vary with photometric indices that are sensitive to temperature.

The calibration of luminosity classes in terms of absolute magnitudes is likewise a continuing problem. For the early-type stars the calibration tables given by Blaauw (1963) remained in use for many years. They were based on a careful evaluation of the data available at that time, and owed much to the work of Schmidt-Kaler on the more luminous stars. It is known, however, that many of the results (and not only the earliest ones) were affected by serious systematic errors. It is not always clear whether the samples averaged were limited by magnitude or by distance (or by both). This hinders the comparison of luminosities for field stars with those for members of clusters. As our surveys go deep enough to reach completeness for at least the stars of higher luminosity classes in the open clusters with best determined distances, their sample space is the volume occupied by the cluster. Consequently, I hope that all future statistical determinations of mean absolute magnitudes for field stars will likewise be reduced to constant volumes of space.

Later improvements included calibration of the four-color photometry and H^β intensities (cf. Crawford [1978, 1979]), but the results have not all been referred to current MK types. For the O-stars Lesh

(1979) combined the reclassification and calibration by Walborn (1972) with her own results. Their values for class III are plotted as crosses in Figure 1. More recently a new calibration of Hγ intensities in O- and B-stars has been made by Millward and Walker (1985), and for class V their luminosities are shown by the continuous line in the figure. For types B, A and F, Grenier et al. (1984) have derived new absolute magnitudes from statistical parallaxes. Their calibration curves for classes III and V are shown by the open circles connected by dashed lines. The revised calibration of Schmidt-Kaler (1982) for supergiants of classes Iab and Ib is given by the dotted curves.

The differences between these calibrations, and between them and the old one of Blaauw, rarely exceed a half magnitude except for the F-type stars of class III. The luminosity classification of A- and F-stars is made difficult by the frequent presence of rotation, large magnetic fields, local abundance effects on their surfaces, and the small separation of luminosity classes V, IV, III and II in this part of the HR diagram. The current program of re-classification of A-type stars by Abt (1984) should permit better curves connecting luminosity class and M_v to be constructed.

TABLE II

LISTS OF MK TYPES OF FAINTER STARS

Range of Types	Range of V	Reference
B – G	6.5 – 9.0	Roman, N.G., 1978, A.J., 83, 172.
O9 – A3 (In Orion)	To 9.0	Abt, H.A., Levato, H., 1977, Pub. A.S.P., 89, 797.
A – K (In UMa stream)	To 9.0	Levato, H., Abt, H.A., 1978, Pub. A.S.P., 90, 429.
O – A (Near gal. anticenter)		Chromey, F.R., 1979, Astr. Jour., 84, 534.
A3 – M5		Krug, P.A. et al, 1980, Monthly Monthly Notices R.A.S., 190, 237.
167 HD Stars	To 8.5:	Jensen, K.S., 1981, Astr. Astrophys. Suppl., 45, 455.

Turning to later types, we find more published types, but many problems remain. Not even for the cooler dwarfs are there enough nearby stars with individual trigonometric parallaxes good enough to settle all questions about the width of the main sequence.

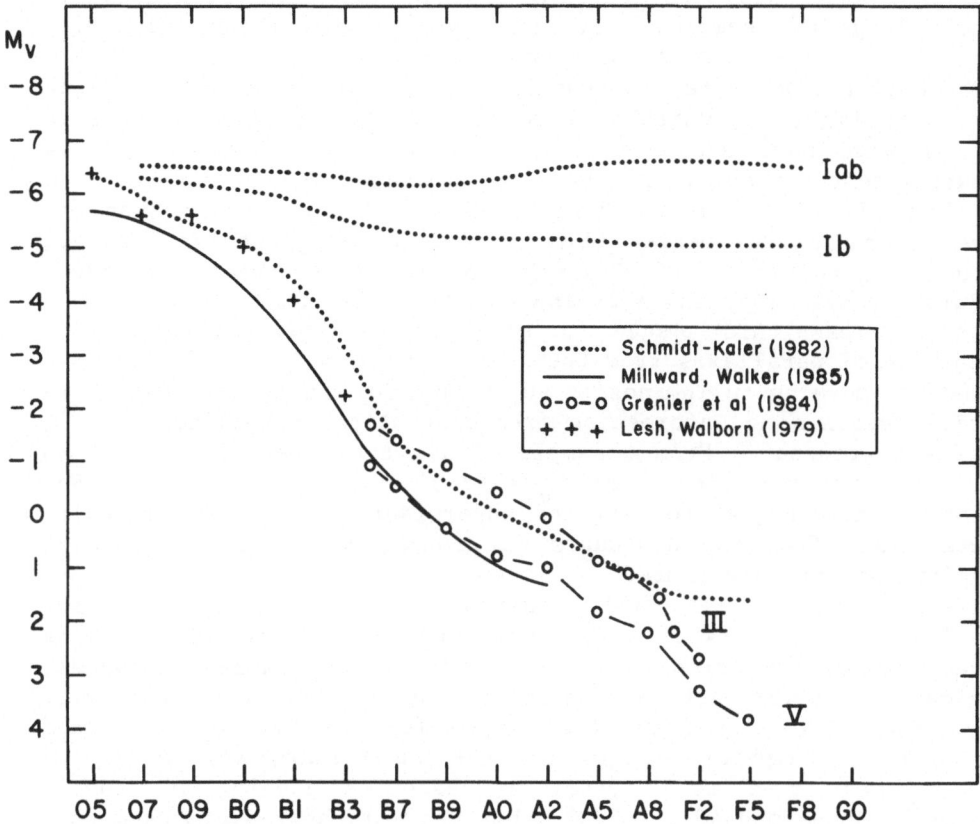

Fig. 1. Published calibrations of luminosity classes for early-
         type stars.

Mean trigonometrical parallaxes are useful for the yellow and red
giants. For class III stars of types G, K and M, and solar composition, meaningful averages have been found if the apparent magnitude is
limited to V ≤5.0 mag. (Egret, Keenan, Heck, 1982). Some earlier
solutions have included stars down to V = 6.5, but when that is done
the statistical corrections become large and uncertain. Pending any
large improvement in the accuracy of parallaxes that will be coming
from space telescopes, the most immediate gains that can be hoped for
will come from better parallaxes for southern stars.

For these same giants statistical parallaxes have given equal
precision when the samples were extended to stars about two magnitudes
fainter – to about V = 7.0. The solutions carried out by Egret at the
Centre de Donnees Stellaires at Strasbourg have employed the maximum
likelihood formulae developed by Heck (1975), which in turn were based
on the earlier work of Rigal (1958) and Jung (1970).

In the immediate future the potential gain in accuracy is greater
for the method of statistical parallaxes, since there are many as yet

unclassified stars brighter than the seventh magnitude which could be used to enlarge the sample. For this purpose approximate classification will not do - we must use a homogeneous set of types.

The calibration curves for Population I giants that we are using are shown in Figure 2, which is based on our 1982 paper. We hope to repeat it with more adequate samples in the near future. It is interesting that our solution gives a distance modulus of 3.42 for the three class III giants in the Hyades, about 0.1 mag. larger than the modulus recommended by Hansen (1980). Four G8-G9 giants in Praesepe also suggest a correction of +0.1 mag. to our calibration if we adopt a modulus of 6.20 (Upgren, Weis and DeLuca 1979) for the cluster. A shift of at least this amount for bright stars will probably result from the use of parallaxes from the revised catalog.

When we move up in luminosity to the bright giants of class II we can still obtain some information from statistical parallaxes, but it is of lower accuracy. From a sample of all types between G0 and K3 the Strasbourg computations gave $M_v = -2.05 \pm 0.60$ for class II. For these stars, however, as for all the supergiants, our best luminosity estimates come from the distances of groups, mainly open clusters, with which they are associated.

Recently Ronald Pitts and I reviewed all the evidence for individual late-type stars of high luminosity that seemed to us to be reasonably good. We tried to use the most recent cluster distances, but in many cases were not able to improve on the values adopted by G. Hagen (1970, 1974). Her spectral types for the bright giants and those of R. M. Humphreys (1970) for the supergiants are on the MK system. There is a more recent and extended catalog of cluster distances by Janes and Adler (1982), but many of their distance moduli are old ones, and the catalog must be used with caution.

We plotted the absolute visual magnitudes for each star (Keenan and Pitts 1984, Fig. 1), and found that the lines of constant luminosity were as nearly horizontal between G0 and about M2 as one can tell in view of the rather large scatter of the points.

The resulting calibration curves for these luminous stars also are shown in Figure 2. The uncertainty of the calibration curves is about $0\overset{m}{.}2$ for the giants but probably approaches one magnitude for the brighter supergiants. It should be emphasized that this calibration applies to Population I stars; for Population II a separate calibration needs to be made.

3. INDICES OF CHEMICAL COMPOSITION

From the early days of the Harvard classification it has been recognized that such objects as the carbon stars display spectra differing so widely from those of stars of solar composition that special indices must be devised for their classification. (For the carbon stars, in fact, the problem has never been satisfactorily solved.)

Especially among the hotter stars the complications mentioned in Section II exist, and in the Ap stars lead to composition differences between the parts of the stellar atmosphere viewed at different times.

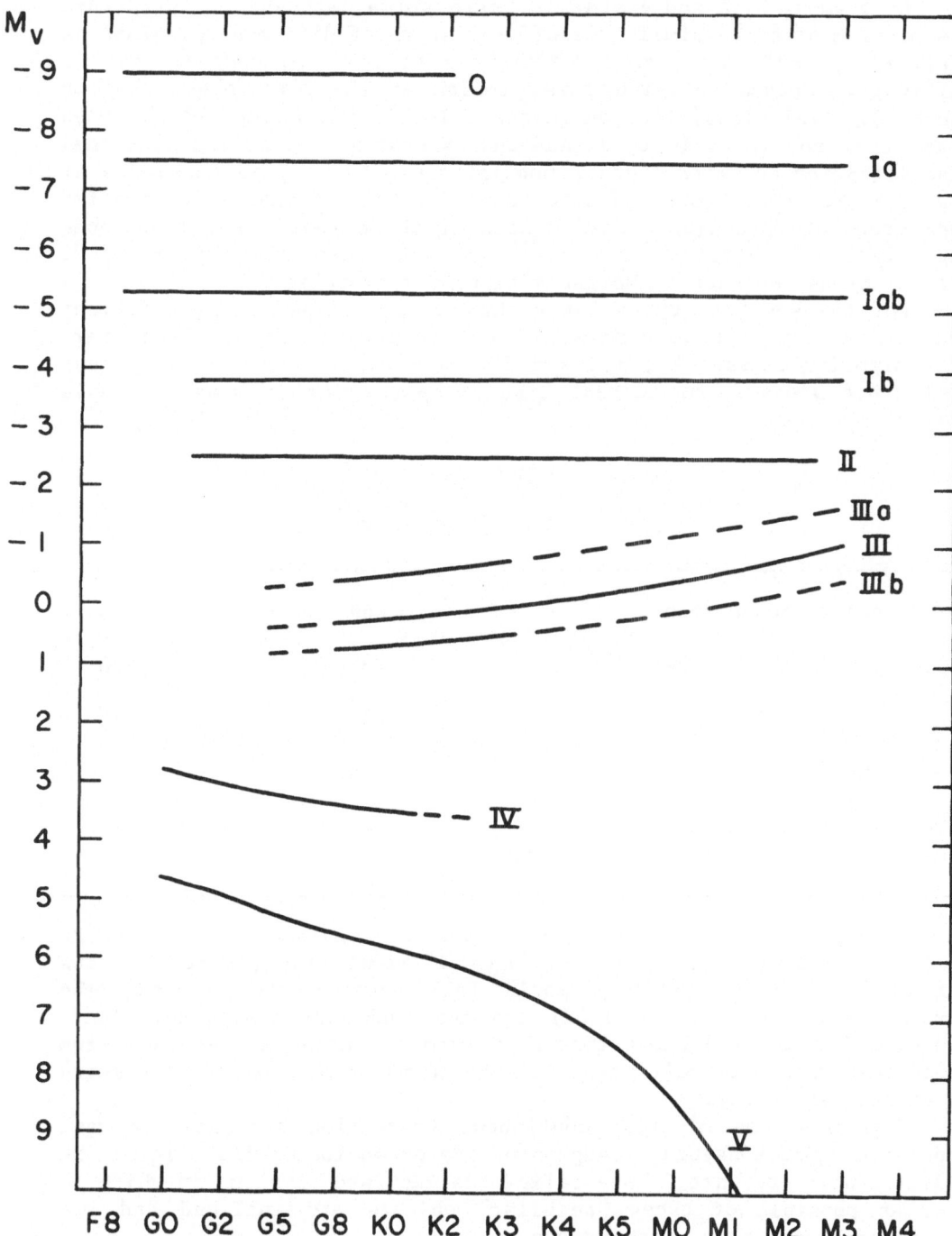

Fig. 2. Calibration of MK luminosity classes for types later than F8. For supergiants the luminosities are based on clusters and binary companions. For classes IIIa, III, and IIIb the mean curves derived by Egret, Keenan and Heck (1982) are shown. For classes IV and V the individual trigonometric parallaxes were plotted and smooth curves drawn through them.

Even among B, A and F stars of more normal atmospheric structure the problem of successfully classifying stars of different populations (defined by their systematic differences in observed composition) is difficult. Morgan reviewed this problem at the Toronto Workshop on the MK Spectral Classification (Morgan, 1984), and expressed the view that it is not feasible to extend the MK system to a multidimensional classification to embrace other populations. Rather, he considers it necessary to set up separate autonomous, systems of classification for each group of "peculiar" stars. Each of these would be set up analogously to the original MK system. This he terms the "MK process". Reference must be made to Morgan's paper for examples.

For the spectral types later than G0 the situation is different and in some ways not so complex. Consider the sample of G5-K5 stars of luminosity classes II, III and IV. The distribution among groups of the 426 stars in my current list of best types is shown in Table III.

TABLE III

SPECTRAL GROUPS, G5 TO K5

Normal solar composition	269 stars	63%
Slightly weak metal lines	33	8
Weak lines (Population II)	53	12
Strong metal lines (SMR)	28	7
Other peculiarities (Ba, strong CH, weak CH, etc.)	44	10

Although the sample is not unbiased, it is evident that if the MK system were limited to stars of "normal" composition, more than one out of every three examined would fall outside the system. The largest fraction of these (20%) are the weak-line (metal-deficient) stars, and it is well known that they form a continuous sequence from halo stars with extremely weak lines to normal solar-composition stars to strong-line stars.

The existence of such continuous transition sequences between population groups naturally suggested the extension of the original MK classification to what I have called the "Revised" system, in which as many as possible of these "peculiar" objects are included and are characterized by abundance indices.

For the spectra with moderate line weakening it has long been known that several of the usual MK criteria can be employed if sufficient care is used (Keenan and Keller, 1953). This conclusion is supported from the theoretical side by the calculations of Foy (1979), based on model atmospheres. He emphasized, however, that the

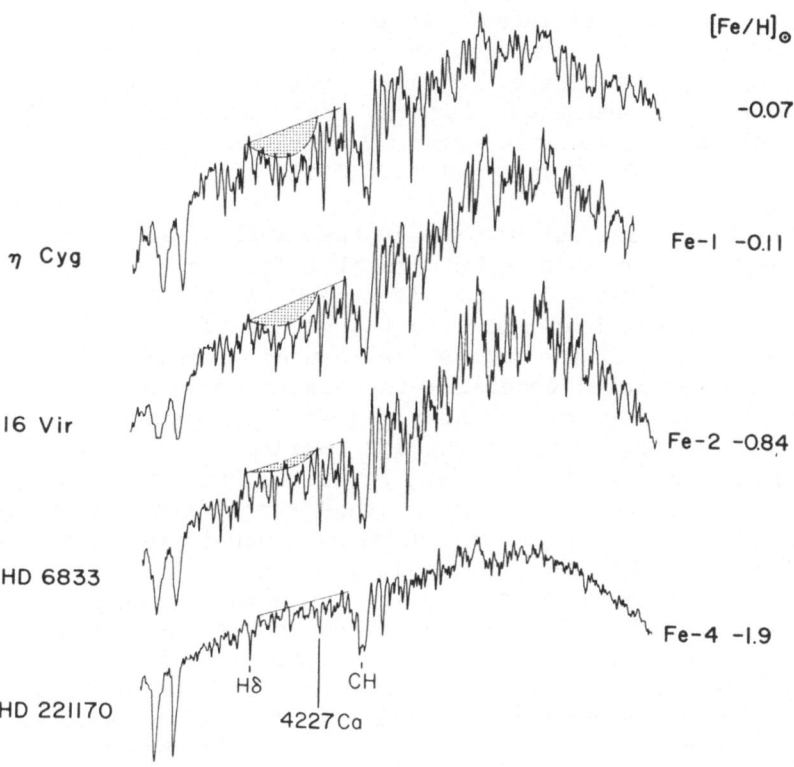

Fig. 3. Effects of metal deficiencies on spectra of K0 III stars. The metallicity index is shown at the right, followed by published values of [Fe/H]$_\odot$. The dotted areas indicate approximately the absorption by the blue CN band.

calibration of temperature types and luminosity indices in terms of physical variables must be done separately for stars with markedly different chemical compositions.

In assigning composition indices the classifier must face a problem. He is caught between Scylla and Charybdis. If he uses a separate index for each of the more conspicuous elements or compounds, he can be accused of making the system too complicated. If he tries to simplify things by using only one index he runs the danger of throwing quite different stars into the same spectroscopic box.

In designing the system it seemed safer at the beginning to use several indices whenever there was a suspicion that the abundances were not all closely correlated. As an example, a sequence of linear intensity plots of KO giants with progressively greater metal deficiences is shown in Figure 3. The logarithmic metal deficiencies shown at the right are taken for the most part from the catalog of Morel, Bentolila, Cayrel and Hauck (1981). For η Cyg the determinations published by L. Gratton et al. (1981) and R. G. Gratton (1983) were averaged.

The tracings bring out strikingly the familiar sensitivity of the blue CN depression to even slight metal deficiencies. In contrast, only the strong lines of the iron-peak metals are visible at classification dispersions and do not show clearly visible weakening until the abundances of the metals are reduced several fold.

This behavior holds generally for giants between G8 and K2, and led to the use of the two abundance indices, CN and Fe, for these stars. Now, however, enough types are available to allow us to say that a single metal abundance index, Fe, is sufficient to characterize these stars. Our classification is thus simplified. I should add that Morgan's work on stars of different populations in the main sequence led him to the same conclusion.

This leaves us with another problem. When the lines are weaker than in stars like HD 221170 we can no longer see them well enough to classify the stars on the usual small scale spectrograms. Figure 4 reproduces the photographic spectrum of such a star, BS 5270 = HD 122563. In the upper spectrum all the absorption features except H, K, Hδ, Hγ and the G-band have practically disappeared. When the scale is increased to about 20 Å/mm, however, in the lower spectrum, the lines are visible, even though much weaker than in the comparison star ϵ Vir. The usual line ratios can be estimated and the star classified, though with somewhat lowered accuracy, on the Revised MK system. This illustrates the flexibility of the classification, for there are few stellar spectra as featureless as that of BS 5270, for which various abundance analyses give $[Fe/H]_o \approx -2.6$. Our metal index for it is Fe -5.

Another small group of chemically peculiar stars is distinguished by spectra with the G-band of CH abnormally weak. The lines of the common metals and of hydrogen, however, are nearly normal for the temperature types of these class III giants. The CN bands are generally weakened in their spectra, but not so much as the CH band. This behavior has been shown by atmospheric analyses of such typical CH-weak stars as 37 Com, BS 6766, and BS 6791 to be due to extreme

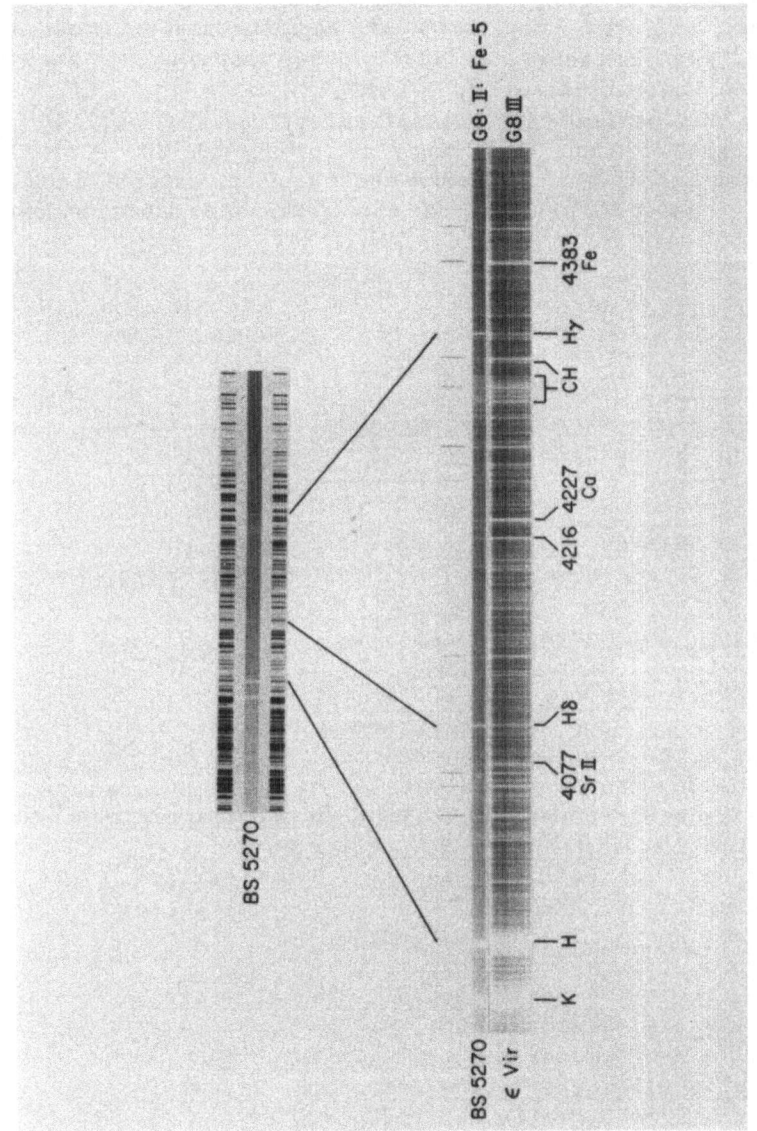

Fig. 4. The extreme metal-deficient star BS 5270 = HD 122563 (G8: II; Fe −5). The upper spectrum shows an ordinary classification plate at 76 Å/mm; the lower one an exposure at 20 Å/mm, taken by Walter Mitchell. The companion star, ε Vir, has essentially solar composition.

deficiency of carbon, coupled with a moderate excess of nitrogen, in their atmospheres (extensive references will be found in the papers of Cottrell and Norris 1978, and of Sneden and Pilachowski 1984).

We have generally characterized the CH-weak stars by negative CH and CN indices, but the appearance of the spectra of most of them is sufficiently consistent to justify the dropping of the CN index in classifying these carbon-poor stars.

In summary, the Revised MK classification can be applied to all but a few percent of the stars later in type than G0. For the two-thirds of these that have approximately solar composition no abundance index is needed; for most of the remainder one abundance index suffices.

REFERENCES

Abt, H.A. 1984, Personal communication.
Blaauw, A. 1963, in Basic Astronomical Data, ed. K.Aa. Strand, (Univ. Chicago Press, Chicago), p. 383.
Cottrell, P. L. and Norris, J. 1978, Astrophys. J., 221, 893.
Crawford, D.L. 1978, Astron. J., 83, 48.
Crawford, D.L. 1979, Astron. J., 84, 1858.
Egret, D., Keenan, P.C. and Heck, A. 1982, Astron. Astrophys., 106, 115.
Foy, R. 1979, Astron. Astrophys., 78, 25.
Gratton, L., Gaudenzi, S., Rossi, C. and Gratton, R.G. 1981, Mon. Not. R. Astron. Soc., 201, 807.
Gratton, R.G. 1983, Mon. Not. R. Astron. Soc., 202, 231.
Grenier, S., Gomez, A.,E., Jaschek, C., Jaschek, M. and Heck, A. 1984, unpublished.
Gustafsson, D. and Graae-Jørgensen, U. 1985, in IAU Symposium No. 111:Calibration of Fundamental Stellar Quantities, eds. D.S. Hayes, L.E. Pasinetti and A.G. Davis Philip, (Reidel:Dordrecht), p. 303.
Hagen, G. 1970, Publ. David Dunlap Obs., 4.
Hagen, G. 1974, Thesis, Univ. of Toronto.
Hanson, R.B. 1980, in IAU Symposium No. 85, Star Clusters, ed. J.D. Hesser (Reidel:Dordrecht), p. 71.
Hayes, D.S. 1985, in IAU Symposium No. 111:Calibration of Fundamental Stellar Quantities, eds. D.S. Hayes, L.E. Pasinetti and A.G. Davis Philip, (Reidel:Dordrecht), p. 225.
Heck, A. 1975, Thesis, Univ. Liege.
Humphreys, R.B. 1970, Astron. J., 75, 602.
Janes, K. and Adler, D. 1982, Astrophys. J. Suppl., 49, 425.
Jung, J. 1970, Astron. Astrophys., 4, 53.
Keenan, P.C. 1982, Publ. Astron. Soc. Pacific, 94, 299.
Keenan, P.C. 1983, Inform. Bull. CDS, No. 24, 19.
Keenan, P.C. and Boeshaar, P.C. 1980, Astrophys. J. Suppl., 43, 379.

Keenan, P.C. and Keller, G. 1953, Astrophys. J., 117, 241.
Keenan, P.C. and McNeil, R.C. 1976, An Atlas of the Spectra of
    the Cooler Stars:Types G,K,M,S and C. (Ohio State Univ.
    Press, Columbus).
Keenan, P.C. and Pitts, R.G. 1984, in preparation.
Lesh, J.R. 1979, in IAU Colloquium No. 47:Spectral
    Classification of the Future, eds. M.F. McCarthy,
    A.G.D. Philip and G.V. Coyne,
    (Vatican Obs. Ric. Astron. Vol. 9), p. 81.
Millward, C.G. and Walker, C.A.H. 1985, in IAU Symposium No. 111
    Calibration of Fundamental Stellar Quantities, eds.
    D.S. Hayes, L.E. Pasinetti and A.G. Davis Philip,
    (Reidel:Dordrecht), p. 377.
Morel, M., Bentolila, C., Cayrel, G. and Hauck, B. 1981,
    A Catalogue of [Fe/H] Determinations, 2nd version,
    C.D.S., Strasbourg.
Morgan, W. W. 1984, in The MK Process and Stellar Classification,
    ed. R. F. Garrison (David Dunlap Obs., Toronto), p. 18.
Morgan, W.W., Abt, H.A. and Tapscott, J.W. 1978, Revised MK
    Spectral Atlas for Stars Earlier then the Sun, (Yerkes Obs.
    and Kitt Peak Nat. Obs.).
Morgan, W.W., Keenan, P.C. and Kellman, E. 1943, An Atlas of
    Stellar Spectra, (Univ. of Chicago Press, Chicago).
Rigal, J.L. 1958, Bull. Astron. Paris, 22, 171.
Schmidt-Kaler, T. 1982, Landolt-Börnstein Neue Ser., Gruppe
    VI, 2b, 14.
Sneden, C. and Pilachowski, C.A. 1984, Publ. Astron. Soc.
    Pacific, 96, 38.
Upgren, A.R., Weis, E.W. and DeLuca, E.E. 1979, Astron. J., 84,
    1586.
Walborn, N.R. 1972, Astron. J., 77, 312.
Yamashita, Y., Nariai, K. and Norimoto, Y. 1977, An Atlas of
    Representative Stellar Spectra, (Univ. of Tokyo Press,
    Tokyo).
Yamashita, Y. and Norimoto, Y. 1981, Ann. Tokyo Obs., 18, 125.

DISCUSSION

GARRISON: I am glad to hear your remarks about the applicability of the MK system to other populations. I have always believed that unusual spectra could be described in terms of the MK system of standards and I have said so in my paper here as well as in other places.

KEENAN: Absolutely.

GARRISON: I am a bit concerned about the suggestion to reduce the peculiar line strength indices to one dimension because of the very few really peculiar objects. Can we agree on a convention that when a type such as Fe-2, or something like that, is given, we can assume that CN is similarly weak; and if it is not, we will give a separate CN index? There may be some confusion, however, for types given previously. How will people know whether an Fe-2 alone (pre 1984) means Fe-2, CN-2 or Fe-2, CN 0? I guess it probably will be so rare that you can republish the few in that category.

KEENAN: Yes, there will be just a few stars so peculiar that the use of several indices is justified.

CAYREL: I am delighted with your classification of the extreme Population II stars.

KEENAN: Thank you!

JASCHEK: Who carried out the calibration for the high-luminosity stars?

KEENAN: Pitts and I are doing that. Schmidt-Kaler did it originally. We are updating his work.

HOW PRECISE ARE SPECTROSCOPIC ABUNDANCE DETERMINATIONS TODAY?

Giusa Cayrel de Strobel

Paris-Meudon Observatory

ABSTRACT.   It is shown that a great breakthrough has occurred in the accuracy of spectroscopic abundance analyses with the introduction of solid state light detectors, such as Reticons and CCDs. Because of uncontrolled systematic errors in photographic photometry, abundances derived from high dispersion photographic spectra can hardly be known with an accuracy better than 0.3 dex. This is well exemplified by the recent finding that the observational scatter is large even in the equivalent widths of the Utrecht Solar Atlas. A fortiori these uncertainties are present in the [Fe/H] stellar abundance Catalogue, chiefly based in its present form on photographic material. For the future the calibration of the [Fe/H] Catalogue with spectra taken with Reticon detectors is recommended. A signal/noise ratio of 300 to 500 is more important than an improvement in spectral resolution with a low signal/noise ratio. Then, the remaining uncertainties in the abundances will mostly reflect inaccuracies in atmospheric parameter determinations of the models and in the assumptions underlying model computations.

1. INTRODUCTION

At present the impact of chemical abundance results on almost all the other branches of astrophysics is very important. You can hardly attend a meeting without assisting in an animated "coffee break" discussion between partisans of a high helium abundance ($Y = 0.28$) and partisans of a low helium abundance ($Y = 0.23$) in the present interstellar medium. Another hot discussion can be heard between believers that all globular clusters have the same helium abundance and those who do not. This latter group can again be divided in two subgroups: subgroup I believes that in globular clusters the abundance of helium is correlated with the abundance of metals; subgroup II believes that in some globular clusters the abundance of helium is anticorrelated with that of metals. Coming to abundance problems in stars nobody can testify better than myself about the great quantity of ink which has been spilled in writing papers and counter-papers on the

137

D. S. Hayes et al. (eds.), Calibration of Fundamental Stellar Quantities, 137–162.

existence or not of a super-metal-rich (SMR) population in our own and
in other galaxies.

Abundance determinations are also the cornerstone of the study of
atomic or turbulent diffusion in stellar envelopes. They provide
important clues about stellar interiors, such as the study of the
depletion of lithium along the main sequence of open clusters undertaken
by Duncan and Jones (1982) and by ourselves (Cayrel et al. 1984). Even
cosmology is linked to abundance determinations with the well known
argument about the primordial elements left behind by the Big Bang. F.
and M. Spite (1982) have recently discovered that lithium is still
present in the atmospheres of the halo dwarfs hotter then 5600° K. The
exact determination of the abundance of lithium in these extreme halo
stars has a strong cosmological implication supporting a low density
Universe ($\rho_B = 1.5 \times 10^{-31}$ g cm^{-2}). This density is in good agreement
with the density deduced from the abundance of deuterium and favors an
open Universe. So, those people who do not like the idea of living in a
closed Universe may be very happy. These are a few examples of the
present interest in abundance determinations but above all, research on
abundances in stars and in the interstellar medium is important in the
determination of the chemical evolution of our Galaxy and to a larger
extent in that of the Universe.

Only very recently in the history of astronomy have abundance
studies become fashionable. Table I contains an exhaustive list of
meetings on abundance problems set up by the international body of
astronomers. The first IAU Symposium dedicated exclusively to abundance
determinations in stellar atmospheres was held in 1964. Between the
first IAU Symposium in 1953 and the 26th Symposium nobody claimed that

TABLE I

CONFERENCES ON STELLAR ABUNDANCES

1953	Co-ordination of Galactic Research (IAU Symposium No. 1)
1964	Abundance Determinations in Stellar Spectra (IAU Symposium No. 26)
1970	Symposium on the Nuclear History of the Galaxy to honor the 60th Birthday of J. Greenstein (unpublished)
1975	Abundance Effects in Classification (IAU Symposium No. 72)
1980	ESO Workshop on Methods of Abundance Determination for Stars
1981	Cambridge (UK) Workshop on Arcturus (summarized by Trimble and Bell 1981)
1982	Systematic Effects in Abundance Determinations for Metal Poor Stars (held during the XVIII IAU General Assembly and published in the Publ. Astron. Soc. Pacific; see Bonsack 1983)

it was necessary to have chemical abundances discussed during an IAU meeting. Part of this lack of interest during the first half of our century in spectroscopic abundance studies was due to the fact that no difference was found in chemical abundance between the Sun and the first stars analyzed in detail. Indeed, the first stars analysed in detail by Unsöld and coworkers, by K. O. Wright and R. Cayrel were mostly B and A dwarfs and giants, which undoubtedly represent a nearly perfect specimen of solar composition Population I stars. Therefore a very strong belief in the uniform chemical composition of the Universe arose among astrophysicists at the end of the forties. We had to await the discovery by Greenstein in 1957-59 of a metal deficiency by factors between 100 and 200 in globular cluster stars to become aware of large abundance differences between old and young stellar populations. As we can see on Table I, chemical abundances have since then been discussed far and wide, and because our Symposium deals with calibrations we may contribute to it by presenting some very good abundance data in a sample of well chosen stars.

I shall not speak during my talk about specific calibration problems so much as the comparison between sets of equivalent widths of standard stars taken with different spectrographs. Several years ago I inherited from K. O. Wright the chairmanship of a subcommission of Commission 29 dealing with line intensity standards. I have to confess that I let it die out because I could not impose a discipline upon the users of standard stars.

Even though the calibration of photometric abundances with spectroscopic results is of fundamental importance I shall not discuss it, either. I dedicated a paper to it during the ESO Workshop on Abundances in 1980 (Cayrel de Strobel 1980). What I shall now discuss are abundance results derived from high resolution spectroscopy. In Section 2, I shall compare Reticon spectra with photographic spectra and in Section 3, I shall discuss the relationship between spectrophotometric accuracy and the accuracy of abundances based on Reticon observations. In Section 4 I shall present the new edition of the [Fe/H] Catalogue and in Section 5 I shall give the conclusions.

2. RETICON SPECTRA VERSUS PHOTOGRAPHIC SPECTRA

One of the greatest breakthroughs in high resolution spectrography has been the development of efficient, wide-dynamic-range solid state imaging devices such as Reticon photodiode arrays. They enable the accurate measurement of lines on the part of the curve of growth which is still nearly linear. Indeed, high dispersion spectroscopic analyses remain the only primary method for deriving heavy element abundances in stars.

The visibility of a weak spectral line, i.e., its emerging from the noise of the spectrum, does not depend only upon the signal to noise ratio, S/N, of the spectrum on which we want to measure the line. We shall now discuss the coupling between resolution and S/N with the help of the spectra of standard stars taken with different spectrographs and different detectors. Table II shows S/N ratios and resolutions measured on photographic, electronographic and Reticon spectra. This Table is divided into six columns. Column 1 gives the name of the telescope, spectrograph, and detector with which the spectrum has been taken; Column 2 gives the name of the object, the identification number of the spectrum, when available, or the year in which the spectrum was taken, the apparent magnitude and spectral type of the object. Column 3 indicates the S/N ratio measured in three very clean windows: A, B, C in the spectrum of each object. Please note that the windows of the photographic, photoelectric and electronographic spectra are not the same as the windows of the Reticon spectra which are taken in a redder region of the spectrum. From this Table we can follow the evolution of the S/N ratio achieved on the same object at about the same spectral resolution i.e., ~0.22 Å. Around the years 1969-70 we wanted to begin a detailed analysis of a solar type dwarf in the Hyades. We choose for this purpose the 8.1 mag. dwarf VB64. As we can see we get a spectrum having a S/N ratio of about 3/1 with the OHP 152 Coudé on IIa 0 plates in 6 hours and 30 minutes. Needless to say, we did not begin a detailed analysis of VB64. We did better in 1975 with the Lallemand electronographic Camera, obtaining a spectrum of VB64 in 4 hours having a signal/noise ratio of about 30/1 on the same 152 Coudé telescope. We observed VB64 again with the Reticon on the Canada-France-Hawaii (CFH) Telescope in 1980 and 1981 and we obtained on the average a S/N ratio of 250/1 to 300/1 in about 2 hours exposure time. This was the beginning of our era of Reticon spectroscopic observations. Since then, we also have obtained excellent Reticon spectra from the Coudé Auxilliary Telescope (CAT) at ESO.

TABLE II

COMPARISON OF SIGNAL/NOISE RATIOS OBTAINED WITH DIFFERENT DETECTORS

TELESCOPE, SPECTROGRAPH, DETECTOR	OBJECT	INDENT.	V	SP.	S/N			DISP. (Å/mm)	RES. Å	EXP.	
					A	B	C				
152 Coudé OHP	Sun(Moon)	4390	–	G2 V	17	21	15	12.4	0.27	0h	07m
	Sun(Moon)	4391	–	G2 V	14	24	16	12.4	0.27	0	08
IIa0 plates	Sun(Moon)	4398	–	G2 V	12	15	18	12.4	0.27	0	05
	HD76151	4369	6.0	G3 V	27	19	21	12.4	0.27	0	50
	HD76151	4377	6.0	G3 V	17	13	18	12.4	0.27	1	08
	HD76151	4395	6.0	G3 V	20	19	23	12.4	0.27	1	30
	VB64	–	8.12	G2 V	~3	~2	~4	12.4	0.27	6	30
152 Coudé OHP	Sun(Moon)	CE 682	–	G2 V	36	24	21	7.0	0.15	0	01
Electr. Camera	HD76151	CE 684	6.0	G3 V	49	28	17	7.0	0.15	1	00
Definix Plt.	VB64	CE 680	8.12	G2 V	29	41	23	7.0	0.15	4	00
Mt. Wilson photograph	Sun (Utrecht)		-26.7	G2 V	218	70	155	0.33	0.06	–	
Sac. Peak photoelect.	Sun (Beckers)		-26.7	G2 V	304	84	172	0.20	0.02	–	
Mt. Wilson photograph	Procyon (Griffin)[2]		0.38	F5IV-V	124	132	154	1.5	0.03	–	
Mt. Wilson photograph	Arcturus (Griffin)		-0.04	K1 III	116	95	99	1.5	0.03	–	
CFHT	Sun(Moon)	1980	–	G2 V	460	280	630	4.8	0.23	0	05
	Sun(Moon)	1981	–	G2 V	380	400	400	4.8	0.23	0	05
Coudé + image slicer	HD1835	1980	6.39	G2 V	–	160	270	4.8	0.23	0	30
+ 1872 pix. Reticon	ε Vir	1981	2.83	G8 III	480	360	420	4.8	0.23	0	05
	VB64	1980	8.12	G2 V	280	270	240	4.8	0.23	2	20

Another achievement was the installation of a holographic grating at Canada-France-Hawaii Telescope. The first Reticon spectra we have obtained with this grating are excellent and they combine high S/N ratio with high resolution (0.1 Å). As an example, CFH Reticon tracings of five stellar spectra are presented in Fig. 1. These tracings belong (from the bottom) to two **Hyades** dwarfs, to two Halo dwarfs and to the Moon. The first four stars are all fainter than 8th mag. The tracings of the spectra are very compact because they show the whole 133 Å large Reticon spectrum region. The spectra were obtained with the cooled Reticon array of 1872 photodiodes with 15μm wide pixels. With the 830 groove mm^{-1} mosaic grating the effective dispersion is 71 mÅ per pixel and the resolution (FWHM of the instrumental profile) is between 0.20 and 0.25 Å.

The five stars of Fig. 1 are all early G type stars, but whereas the spectra of the Hyades dwarfs very much resemble that of the Sun, the spectra of the Halo dwarfs are typically very weak-lined. The most interesting feature in these spectra is the presence, absence or weakening of the lithium line at 6707.8 Å. The reason Li is stronger in VB73 is that this star is somewhat hotter than VB64 and therefore its convective zone is shallower and lithium burning in this star is less important (Cayrel et al. 1984). In an analogous way we can explain the presence of Lithium in one Halo star, HD 194598, and its absence in the other (Spite and Spite 1982).

We usually took Reticon spectra centered at H_α, 6750 and 8550 Å for each program star. The H_α line was our criterion of effective temperature and the Ca II triplet was our criterion of chromospheric activity. The spectral region centered at 6750 is very interesting because besides the 6707 Li I doublet it also contains many weak and very weak Fe I lines. The spectral regions cited above are shown in Fig. 2 for the visual binary ξ UMa. In the Bright Star Catalogue both components of ξ UMa have the same spectral type: G0 V (V=4.944 mag.). But in comparing the tracings of each spectral region we see that the spectra of the two components are not exactly the same: ξ UMa A is slightly hotter than ξ UMa B (the H_α profile is more developed and neutral FeI lines are weaker in ξ UMa A). From the two H_α profiles and, more importantly from the two visible components of the Ca II triplet we also see that the chromospheric activity of ξ UMa A is less pronounced than that of ξ UMa B. Very puzzling is the fact already noted by Duncan (1983), that contrary to what is expected the more active component does not show the lithium line at 6707 Å. We are carrying out a detailed analysis of ξ UMa and we shall try to understand the whys and the wherefores of this absence.

Bruce Campbell, Roger Cayrel and myself have already taken Reticon spectra (principally at CFHT but also at ESO) of main-sequence solar type and later stars of the Hyades, Pleiades and of the Ursa Major stream. The magnitudes of the Hyades stars were 8 to 9 mag., those of the Pleiades stars 10 to 11 mag., and the Ursa Major stars were 6 to 8 mag. stars. We worked on two other projects: bright field solar analogs

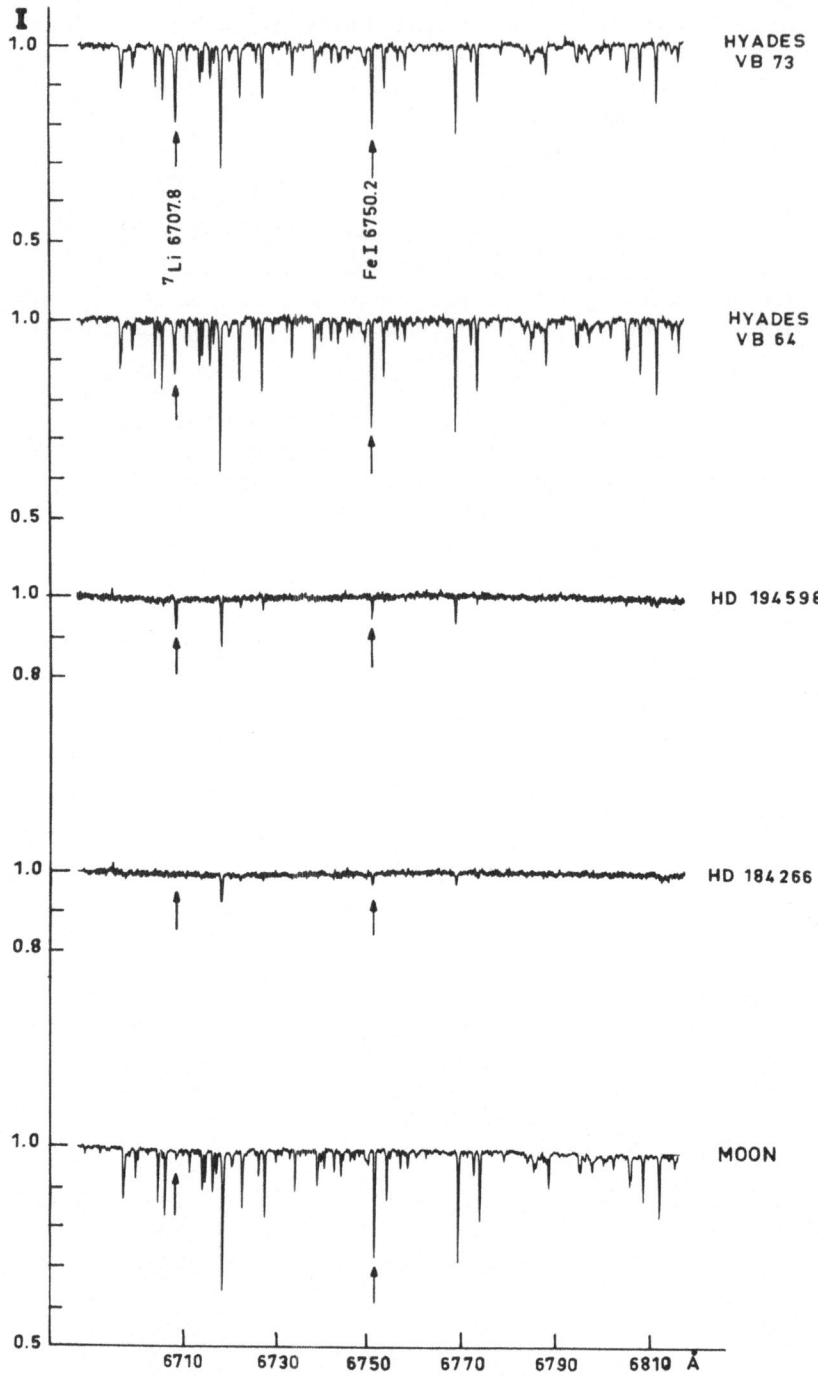

Fig. 1. Comparison of Reticon CFHT spectra of two Hyades and two Halo
solar dwarfs. The Moon is taken as comparison object. Please
note the presence of the resonance ^{7}Li line at λ6708 Å in the
two Hyades dwarfs and in the Halo dwarf HD 194598.

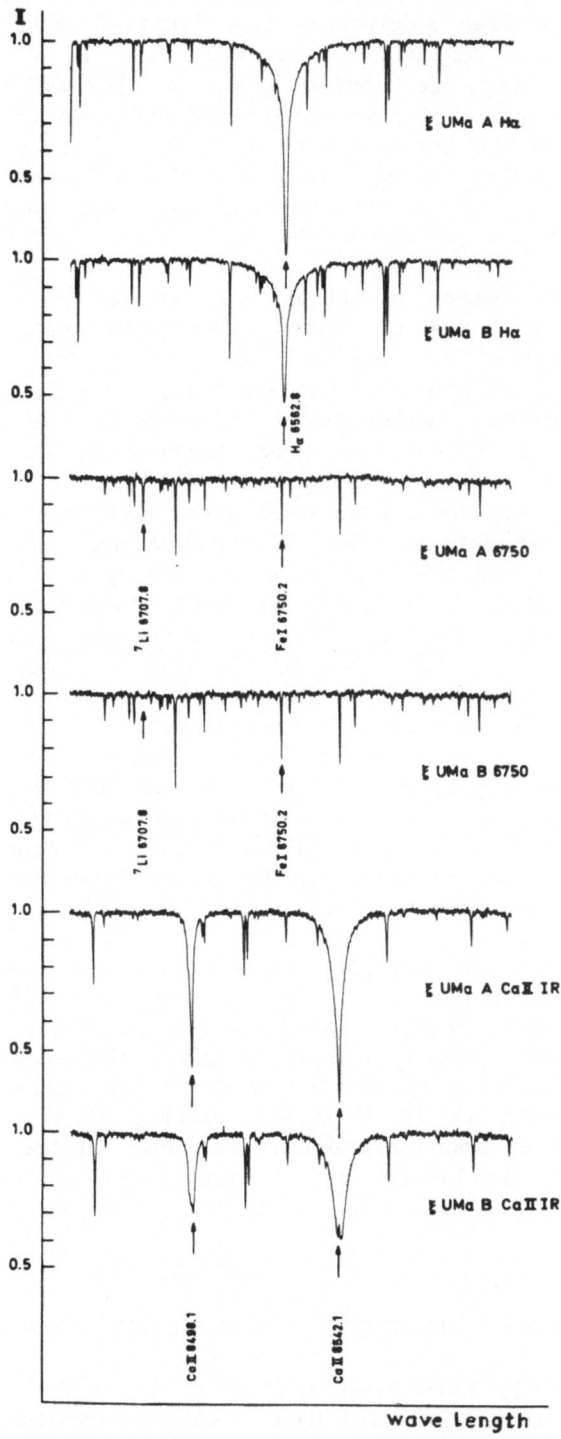

Fig. 2 CFHT Reticon spectra centered at H_α at $\lambda 6750$ Å and at $\lambda 8550$ Å
of the visual binary ξ UMa.

and bright SMR candidates. We also observed at ESO high parallax stars which have not yet been submitted to a detailed analysis. Up to now only one project has really been carried out: 12 Hyades solar dwarfs (11 cluster stars and one moving group star) have been analyzed in detail. For these stars we have determined differentially the abundance of iron relative to the Sun and with a lesser accuracy the abundance of other metals. Spectra of the Moon and Ceres have been obtained as comparison objects. This differential technique avoids the use of oscillator strengths.

Accurate measurements of lines weak enough to be on the nearly linear part of the curve of growth have been obtained. Equivalent widths were determined by detailed line profile fitting with special attention to the placement of the continuum and to contamination by blends of weak lines. Temperatures relative to the Sun were derived from the profile of H_α for the seven hottest stars. This procedure avoids the uncertainty of the solar color. For the four coolest stars (V-K) and (V-I) colors have been used after calibration of the indices through the H_α observations. The 12 program stars being all unevolved dwarfs, the gravity was taken to be equal to log g = 4.5. We have used a grid of flux constant line blanketed model atmospheres kindly provided by Bengt Gustafsson (1978). Stark broadening and self-resonance broadening for the theoretical H_α profile have been calculated following Vidal, Cooper and Smith (1971) for Stark broadening and Cayrel and Traving (1960) for self resonance broadening.

The details of the spectroscopic analysis of the Hyades dwarfs are given in the paper we have just finished writing on this subject (Cayrel, Cayrel de Strobel and Campbell 1984). From this paper are taken a few tables and figures concerning photometric data (Table III), equivalent width comparisons (Fig. 3 and 4), comparisons between observed and theoretical H_α profiles (Fig. 5,6,7, and 8), a few examples of curves of growth (Fig. 9,10,11, and 12), a table (Table 4) containing the atmosphere parameters: effective temperature, gravity, micro-turbulence and iron abundance [Fe/H], and a diagram of [Fe/H] vs. T_{eff} (Fig. 13). The last column of Table 4 gives the error in the mean of the iron abundance. This error is very small particularly for the first five stars, in which it is 10 times smaller than the error usually attributed to a metal abundance determination. In the next section, we shall discuss the influence of a high signal/noise ratio on the abundance determination of stars.

3. FROM SPECTROPHOTOMETRIC ACCURACY TO ABUNDANCE ACCURACY

This paper could have been called: "How precise are equivalent widths today?" This title would have been less ambitious but the goals are equally as important. Without precise equivalent widths of the weak lines we cannot pretend to know accurately the abundance of the chemical elements. Therefore we first have to discuss the spectrophotometric

TABLE III
Photometric data of observed Hyades dwarfs

HD number	VB	m_v	Sp. type	$B-V^{(1)}$	$B_2-V_1^{(2)}$	$(G-I)_6^{(3)}$	$(V-I)_{10}^{(4)}$	$(V-K)_{10}^{(4)}$
28344	73	7.85	G1V	0.60	0.350	−0.16	0.79	1.42
28992	97	7.94	G1V	0.63	0.359	−0.14	0.82	1.46
27859	52	7.80	G1V	0.60	0.343	−0.16	0.86	1.48
1835		6.39	G2V	0.66	0.398	−	−	−
28099	64	8.12	G6V	0.66	0.396	−0.12	0.89	1.54
27685	39	7.86	−	0.68	0.415	−0.09	0.93	−
26756	17	8.46	G5V	0.70	0.408	−0.04	0.93	1.64
28805	92	8.66	G8V	0.74	0.447	−0.02	0.96	1.70
27732	42	8.86	G9V	0.76	0.453	−0.01	1.01	−
+21°612	21	9.15	K0V	0.82	0.499	+0.06	1.05	−
+17°734	79	8.96	K0V	0.83	0.508	+0.11	1.06	1.88
27771	46	9.11	K1V	0.87	0.533	+0.12	1.13	1.93

(1) from Nicolet (1978)   (3) from Sears and Whitford (1969)
(2) from Rufener (1980)   (4) from Carney (1982)

accuracy of the equivalent widths we have obtained and then the error of the abundances we have derived.   If one observes with a signal/noise ratio, S/N, i.e. a photometric accuracy $\varepsilon = N/S$, with a detector having a pixel size $\delta\lambda$, then what is the photometric accuracy achievable in the measurements of the equivalent width, W, of a weak line? A good order of magnitude estimate can be obtained by considering that the equivalent width,

$$W = \int \frac{F_c - F_\lambda}{F_c} \, d\lambda,$$

is obtained by taking the difference of the flux integrated over a band width $n\delta\lambda$ in one small spectral interval without lines (continuum measurement) and in one containing only the line to be measured:

$$W = \frac{1}{F_c} \left(\int F_c d\lambda - \int F_\lambda \, d\lambda \right) \simeq \frac{1}{F_c} \left(\sum_i^n F_i - \sum_j^n F_j \right).$$

The F_i and F_j are the individual fluxes for each pixel, the index i is taken to cover the continuum interval and the index j is taken to cover the line interval. Because the errors on all F_i and F_j are independent, the expected error on the parenthesis is $\sqrt{2n} \cdot \delta F$ if each flux is measured with the accuracy, $\delta F \simeq \varepsilon F_c \delta\lambda$.   The relative error of F_c is much smaller because there is no destructive effect by difference on this factor, so one has for a weak line:

$$\delta W \simeq \sqrt{2n} \left(\frac{\delta F}{F_c} \right) \simeq \sqrt{2n} \, \varepsilon\delta\lambda \simeq \frac{\sqrt{2n} \, \delta\lambda}{S/N}.$$

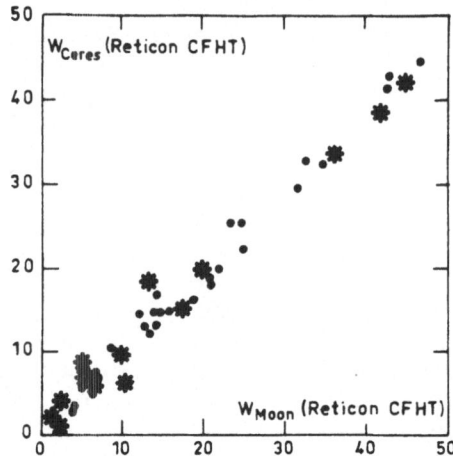

Fig. 3. Comparison of solar equivalent widths from CERES CFHT Reticon
spectra and solar equivalent widths from MOON CFHT Reticon
spectra. Dots: Fe-lines, asterisks: other elements. The
equation of the regression line is: $W_{Ceres}= 0.94\ W_{Moon} + 0.86$.
The correlation coefficient is 0.991.

In practice if n ≃ 4 to 6 the absolute error on W is about 3 times the
pixel size divided by the S/N ratio. With n = 5, S/N = 250 and δλ =
0.072 Å the formula gives:

$$\delta W \simeq \frac{\sqrt{10}}{250} \times 0.072\ \text{Å} \simeq 0.001\ \text{Å or 1 mÅ.}$$

We have actually checked this order of magnitude with our spectra of
Hyades dwarfs, estimating δW independently by taking the r.m.s. of the
difference of the W´s of several spectra of the same object. We found

$$\sigma = \delta W_{observed} = 1.7\ \text{mÅ.}$$

The agreement is not bad if one considers that in addition to the purely
photometric error, which amounts to 5% for a 20m Å line at a S/N = 250,
one should add an error which is much more difficult to estimate. This
error comes from the fact that it is not true that the band used for
measuring the continuum is free of lines, nor is it true that the band
used for measuring the weak line contains only that weak line. A simple
inspection of Delbouille (1973) and Rowland (1966) solar atlases (or the
atlas for the integrated solar spectrum of Beckers et al. (1978)) shows
that there are very weak unidentified lines everywhere making the
position of the continuum uncertain at the level of a few parts per
thousand. A 0.25% difference in the location of the continuum for the
two bands produces an additional:

$$0.0025 \times n \times \delta\lambda = 0.0009\ \text{or 0.9 mÅ}$$

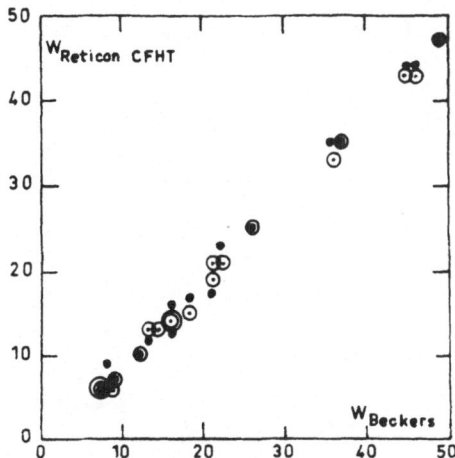

Fig. 4. Comparison of solar equivalent widths from CFHT Reticon spectra
and solar equivalent widths from Beckers et al. (1976). Dotted
circles: MOON, dots: CERES. The equation of the regression
line is: $W_{CFHT} = 0.95 \ W_B - 0.18$, and the correlation
coefficient is 0.993.

error, which is of course included in our empirical determination of
δW. This makes the total random error of the equivalent width more like
the experimental value: 1.7 mÅ than the theoretical value 1 mÅ.

The next step is to translate the error in the equivalent width
into the error in the calculated abundance. As the absolute error δW
is independent of W the relative accuracy first increases with W until
the effect of the slope of the curve of growth and the uncertainty in
which curve of growth to use (uncertainty in microturbulence and
damping) produces an opposite effect. In our case this accuracy is best
for a 20 mÅ line for which the slope of the curve of growth is about 0.8
and the effect of the uncertainty in microturbulence and damping is
still small. In practice one sees from Table IV that the average
standard deviation is about 0.07 dex (17%) on the abundance derived from
a single line at S/N = 250 (see column σ_1). One should note that with a
spectrum having a signal/noise ratio of 50 instead of 250 the error of
1.7 mÅ is going to become something like 6 mÅ moving the best equivalent
widths further up and degrading the standard error to 0.2 dex or so.

A second source of error limiting our knowledge of abundances is in
the inadequacy of the model atmospheres used in computing the lines for
disentangling the effects of variations of effective temperature,
gravity and general metallicity from genuine abundance effects. A
golden rule to observe is to use an homogeneous grid of stellar
atmospheres to do this, and not to use models of different origin for
the two stars to be compared (the analyzed and the comparison star).

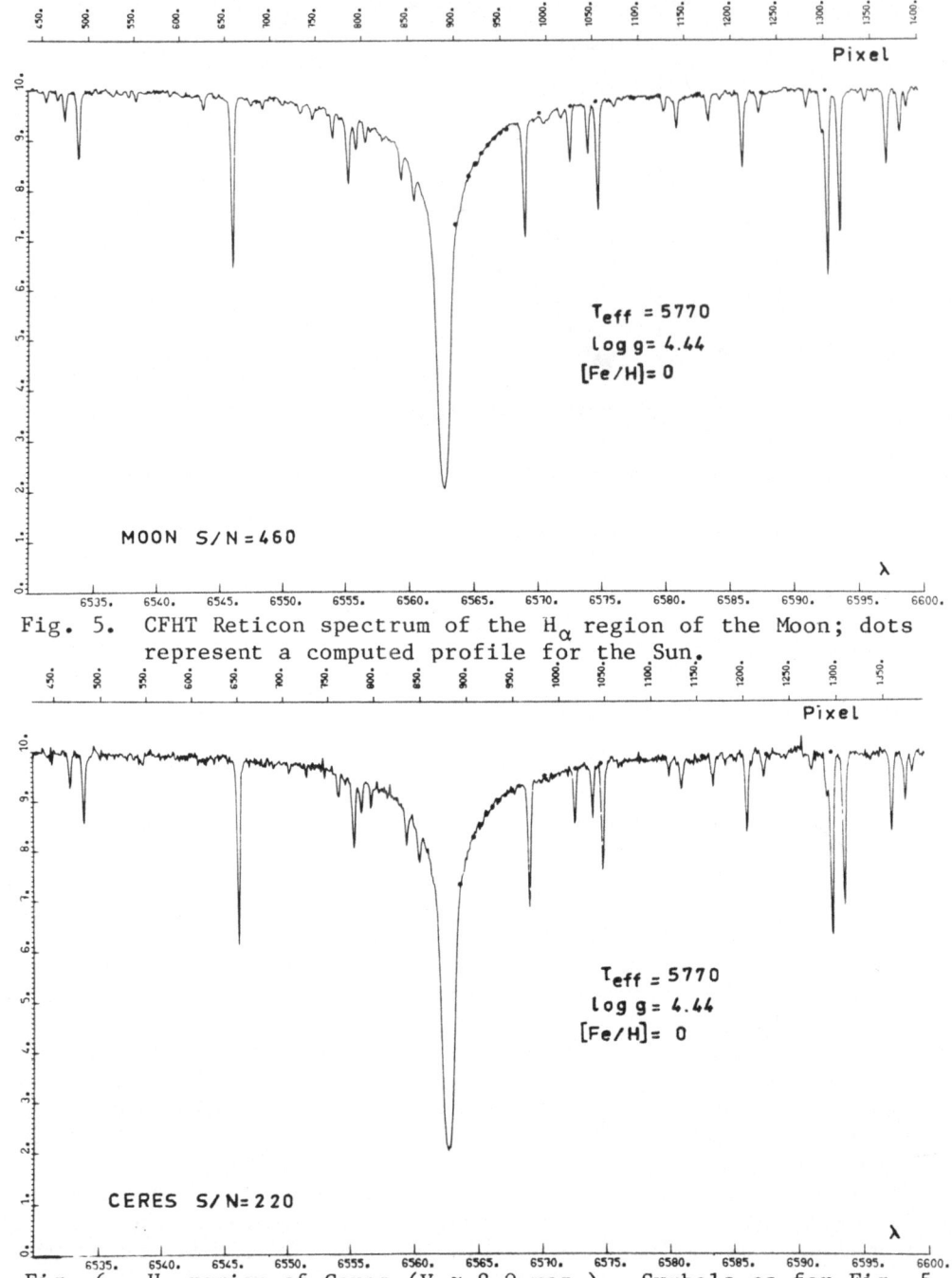

Fig. 5.   CFHT Reticon spectrum of the H_α region of the Moon; dots
           represent a computed profile for the Sun.

Fig. 6. H_α region of Ceres (V $\simeq$ 8.0 mag.).   Symbols as for Fig. 5.

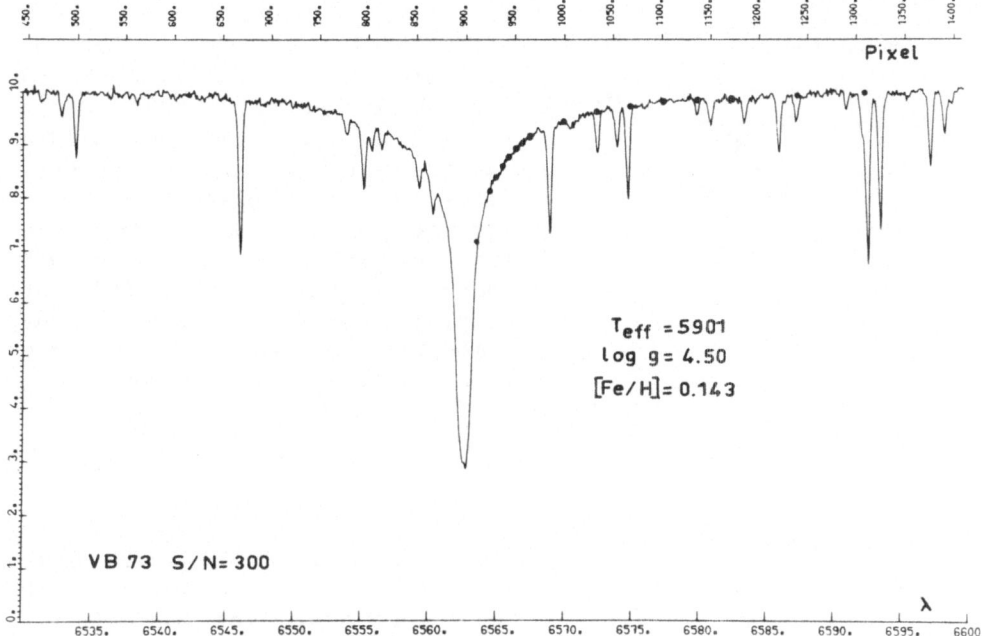

Fig. 7.   CFHT Reticon spectrum of the H_α region of the Hyades dwarf
VB73: dots represent a computed profile.

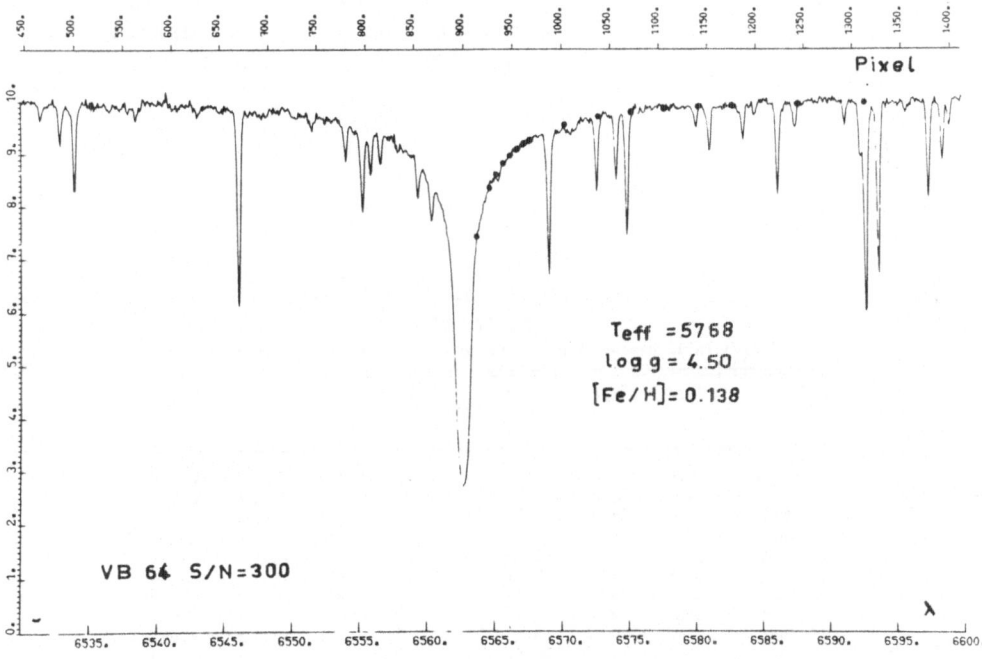

Fig. 8. H_α region for VB64.   Symbols as for Fig. 7.

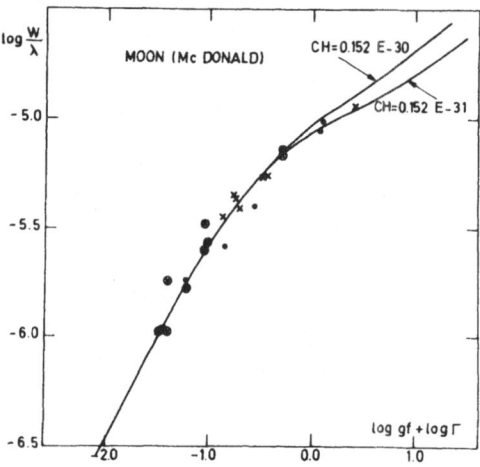

Fig. 9. Curve of growth constructed with Reticon equivalent widths of Sun (Moon) from Branch et al. (1980) Dots: H_α region lines, crosses: 6750 Å region lines, circled points: lines having solar oscillator strengths only. Solid lines are two theoretical curves of growth computed with two different damping constants, and with the atmospheric parameters contained in Table IV for the Sun (Ceres).

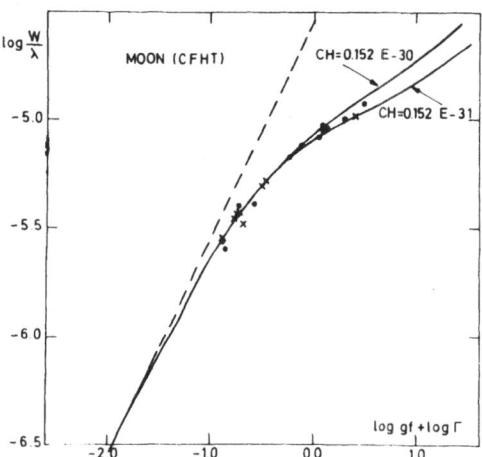

Fig. 10. Curve of growth constructed with Reticon CFHT equivalent widths of Sun (Moon). Here only lines having abso-lute oscillator strengths have been plotted. The abscissae are the same in Figs. 9 and 10. Note that the two observational curves of growth do not intersect the abscissa on the same point.

TABLE IV

[Fe/H] in 12 Hyades Dwarfs derived from about 35 Weak Iron Lines

Star	T_{eff}	θ_{eff}	log g	n number of lines	[Fe/H]	σ_1	σ_2
Ceres	5770	0.8735	4.44	45	+0.010	±0.069	±.035
VB73	5901	0.8541	4.50	43	+0.143	±0.066	±.035
VB97	5859	0.8602	4.50	36	+0.063	±0.081	±.035
VB52	5837	0.8635	4.50	41	+0.028	±0.066	±.035
HD1835	5774	0.8729	4.50	41	+0.165	±0.069	±.035
VB64	5768	0.8738	4.50	42	+0.138	±0.035	±.035
VB39	5622	0.8965	4.50	40	+0.028	±0.090	±.045
VB17	5568	0.9052	4.50	32	+0.095	±0.118	±.06
VB92	5540	0.9097	4.5	44	+0.137	±0.076	±.05
VB42	5398	0.9337	4.5	26	+0.101	±0.070	±.07
VB21	5293	0.9522	4.5	25	+0.093	±0.068	±.08
VB79	5235	0.9628	4.5	25	+0.136	±0.067	±.09
VB46	5169	0.9750	4.5	26	+0.070	±0.084	±.10

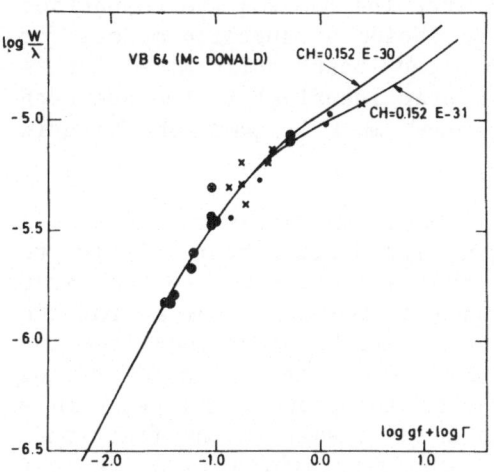

Fig. 11. Curve of growth constructed with Reticon equivalent widths of VB64 from Branch et al. (1980). Circled dots are lines with solar oscillator strengths.

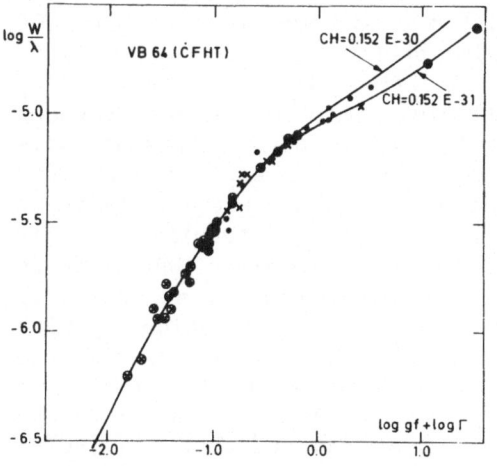

Fig. 12. Curve of growth constructed with CFHT equivalent widths of VB64. Same symbols as in the former Figs. Note the small dispersion of the observational curve of growth.

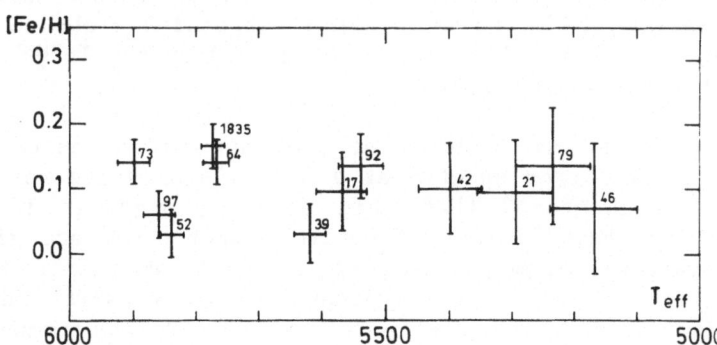

Fig. 13. [Fe/H] values for the 12 Hyades dwarfs in terms of effective temperature.

This mistake has been made several times when the Sun was the comparison star, because of the existence of a large choice of specific models for the Sun. The point here is not to choose the "best" solar model but is to find [Element/H] = 0 when comparing a star identical to the Sun with the Sun. This implies that the solar model must be part of the grid used for representing the program stars.

We shall limit our estimate of the "modelling" error to the simple case in which one looks for the abundances of a dwarf G star relative to the Sun. In a dwarf the neutral lines are so insensitive to the exact value of the gravity that the lion's share of the error comes from the uncertainty in the temperature. This error can be determined from the H_α wing strength. A gain is realized here, too, with a high S/N ratio, which allows one to discriminate a smaller temperature change. In a solar type star a 100 K change produces a 10% change in the fractional depth of H_α at 4.0 Å from the center of the line, which is about four times what can be detected with a signal/noise ratio of 250. For lines with an ionization potential of 7 eV and an excitation potential of 3 to 4 eV, an error of 25 K produces an error in the Fe abundance of about 0.03 dex, which has to be compounded with the photometric error of 0.06 dex already quoted. Of course the random photometric error of 0.06 is smaller if several lines are available. Optimistically it is reduced by a factor of $\sqrt{n}$ if n good lines are available. This is the case for our spectroscopic material of the Hyades. For 7 out of 12 stars we have determined their [Fe/H] abundance and the other atmmosphere parameters T_{eff}, g, and ξ_t, with more than 40 good weak lines. For the faintest stars and for VB17 we choose about 25 good lines to determine the same parameters. Table IV gives the values of these parameters. Note that we have chosen the same microturbulence for all of the stars i.e., $\xi_t = 1.0$ km s^{-1}. The last two columns contain the standard deviation σ_1 with respect to the mean of [Fe/H] derived from one line only, and the estimated error σ_2 in the mean including effects of the error in temperature. From Table IV and Fig. 13, which represents the abundance versus T_{eff} relation for the program stars, we can see that the [Fe/H] value has a non-zero dispersion within the Hyades, the individual values ranging from 0.03 to 0.165 dex. This dispersion might be explained by spurious effects on the line strengths caused by the large chromospheric activity of these young stars.

In conclusion, an accuracy of 0.05 dex is now quite possible for abundances derived from spectra at a high signal/noise ratio. But yet, it should be remembered that this accuracy degrades if the objects compared do not have similar effective temperatures and gravities, the role of departures from LTE and other model weaknesses becoming more relevant. In particular, comparisons between a dwarf and a giant are probably dominated by such effects and not by the computable errors given in this paper.

A third source of error in an absolute abundance determination is the lack of reliable oscillator strengths for weak lines. We have therefore worked strictly differentially with respect to the Sun in carrying out the abundance determination of the Hyades dwarfs. But this is not always possible, especially for hot stars, and oscillator strengths can then become the cornerstone of abundance determinations.

4. THE [Fe/H] CATALOGUE

The aim of our Reticon observations of field stars and nearby open cluster stars of solar type and later is to obtain for these stars very reliable abundance determinations. The aim of the publication of the[Fe/H] Catalogue is to see which stars have been given a detailed anaylsis, possessing therefore detailed metal abundance determinations. The two aims are not at all the same. The Catalogue has been built up from very heterogeneous spectroscopic material: the [Fe/H] abundances contained in it come from many telescopes, many spectrographs, many detectors (chiefly photographic plates) and many authors, who in their analyses used all kinds of model atmospheres. The abundance analyses we have begun at CFHT and ESO are based upon excellent observing material and a homogeneous set of model atmospheres. The features that the "Reticon" and the "Catalogue" abundances have in common is that they are all "spectroscopic" abundances determined by coarse or detailed analyses based on reasonably well resolved spectrographic observations. The first list of metal abundance determinations was compiled by Cayrel and Cayrel de Strobel (1966). The authors took as the metal/hydrogen parameter the logarithmic difference between the relative abundance of iron in a star and the relative abundance of iron in a standard star. This difference is written in the form:

$$[Fe/H]_{std}^{star} = \log (Fe/H)_{star} - \log (Fe/H)_{std}$$

In Table I of the paper by Cayrel and Cayrel de Strobel (1966) only five columns are given: column 1 contains the designation of the star, 2, its HD number, if any, 3, spectral type, 4, the value of [Fe/H], and 5, the bibliographic source. The number of stars contained in this table is 154.

It was Bernard Hauck, at the beginning of the seventies, who had the idea of publishing a "Metal Abundance" Catalogue. When the first Catalogue was compiled (Morel et al. 1976) it could have been called as well "A Stellar Atmospheric Parameters Catalogue" because together with distance and photometric parameters it contains other atmospheric parameters such as: effective temperature, gravity and micro- turbulence. Note that the chemical abundance, effective temperature, gravity and microturbulence are true physical parameters. On the other hand the photometric index (V-K) is an effective temperature indicator, but it does not give the effective temperature directly, the index needing to be calibrated. The [Fe/H] Catalogue could be very useful to astronomers interested in stellar atmospheres and stellar structure, but

the values of the atmospheric parameters given in it for each star have
to be reliable. How reliable they are, we do not know. The Catalogue
is and has been very useful to photometrists interested in calibration
problems. It has also been very useful in studies such as those on the
fine-structure of the HR diagram for the solar neighborhood stars by
Perrin et al. (1977) and on the status of evolution of F, G and K field
stars by Cayrel de Strobel and Bentolila (1983). But as a matter of
fact even if the Catalogue contains the true physical parameters of 1035
stars (see the 1984 edition), the values do not have the accuracy of the
determinations presented in the first part of this talk. Since the
first list of [Fe/H] determinations by Cayrel and Cayrel de Strobel
(1966) the number of new stars being submitted to a detailed analysis is
increasing continuously: this increase seems to be remarkably constant
(about 100 new stars and 200 new analyses per year). Table V, taken
from the 1984 edition of the Catalogue, presents the growth of data.

We thought that it would be useful to include two appendices in
this abundance review. In Appendix 1 we list the individual values of
effective temperature gravity and [Fe/H] for six spectroscopic standard
stars. Appendix 2 gives a list of spectroscopic standards, compiled by
the author and kindly revised by Mercedes Jaschek. These tables are
self explanatory. However, I want to call the attention to the [Fe/H]
abundance of α Lyr in Appendix 1.

Castelli and Faraggiana (1979) have analyzed the UV spectrum of α
Lyr by means of IUE observations and found [Fe/H] = −1.36. This very
low value of [Fe/H] from IUE spectra was confirmed although not quite so
drastically by other authors. What has happened to give these
results? It is not a suitable analysis? Is it an error in the
reduction of the observations? Before we can understand this discrepancy I suggest not including this UV value in the mean. This is an
example of how careful we have to be in using the [Fe/H] Catalogue,
because α Lyr is not the only star for which we have found iron-
abundance discrepancies between different authors.

TABLE V

GROWTH OF DATA OF THE [Fe/H] CATALOGUE

year of publication	Number of stars	Number of [Fe/H] determinations
1966	154	204
1976	515	973
1980	628	1109
1981	707	1298
1984	1035	1921

5. CONCLUSION

We have seen that the photometric accuracy of the equivalent widths measured on Reticon spectra has improved by almost an order of magnitude over those obtained by older conventional techniques. Under such conditions we have been able to measure lines really located on the linear part of the curve of growth, even in the case of stars of 8th to 9th apparent magnitude such as the Hyades dwarfs we have observed. Unfortunately very few of the weak 4-5 eV lines have measured oscillator strengths. This is not a handicap if we work strictly differentially with respect to the Sun, as for solar-type dwarfs. But for G and K giants and for O, B, A and M stars the absence of oscillator strengths can become a very great handicap in obtaining reliable abundances. We have seen that if a homogeneous grid of models is used both for representing the standard star and the analyzed star (such as that of Gustafsson 1978), the abundance, effective temperature, gravity and microturbulence of the star can be determined without bias. But for hot stars and cool giants the models one uses have to be computed with a non-LTE assumption if we want to improve the abundance results.

The tremendous breakthrough of high signal/noise, high resolution spectrographic observations has made possible the observing of faint stars which high dispersion spectroscopists never would have dreamed of observing even a few years ago. It is happily no longer true what Alan Batten said at the beginning of this conference: at present there are available more stars than Procyon and Arcturus observed at high resolution and high signal/noise ratio, suitable for accurate detailed spectroscopic analyses. Between Procyon and Arcturus and the Hyades dwarfs we have observed at the Canada-France-Hawaii Telescope, the flux ratio is about 3×10^3. I hope that the CCD and Reticon techniques will continually improve so the high dispersion spectroscopists can penetrate deeper and deeper into our Galaxy.

ACKNOWLEDGEMENTS

I cannot finish this talk without acknowledging that the abundance results of the 12 solar type Hyades could not have been realized without Bengt Gustafsson who provided us with very suitable models, Bruce Campbell who installed the Reticon at CFHT and worked with Roger Cayrel and myself on this project, and last but not least Gordon Walker who constructed the CHFT Reticon. Special thanks go to Claire Bentolila for her precious help in the preparation of the data for this paper. The paper has been revised by the editors of this Symposium. Sylvie Boucherie has typed the first version, and Pat Cochrane of KPNO has typed the final version. Many thanks to them also.

REFERENCES

Beckers, J.M., Bridges, C.A., Gilliam, L.B., 1976, A High Resolution Spectral Atlas of the Solar Irradiance from 380 to 700 Nanometers, Air Force Geophysics Laboratory, Tr 76-0126.

Bonsack, W. K. 1983, Publ. Astron. Soc. Pacific, 95, 93.

Branch, D., Lambert, D. L. and Tomkin, J., 1980, Astrophys. J., 241, L83.

Cayrel, R. and Traving, G., 1960, Z. für Astrophys., 50, 239.

Cayrel, R. and Cayrel de Strobel, G., 1966, Ann. Rev. Astron. Astrophys., 4, 1.

Cayrel, R., Cayrel de Strobel, G., Campbell, B., Mein, N., Mein, P. and Dumont, S., 1983, Astron. Astrophys. 123, 89.

Cayrel, R., Cayrel de Strobel, G. and Campbell, B. 1984, in preparation.

Cayrel de Strobel, G. 1980, ESO Workshop on Methods of Abundance Determinations for Stars. P. E. Nissen and Ed. K. Kjär, (ESO Report Sept. 1980).

Cayrel de Strobel, G. and Bentolila, C. 1983, Astron. Astrophys, 119, 1.

Cayrel de Strobel, G., Bentolila, C., Hauck, B. and Duquennoy, A., A Catalogue of [Fe/H] Determinations, 1984 Edition, Astron. Astrophys. Supp. Series, in press.

Duncan, D. K. 1981, Astrophys. J., 248, 651.

Griffin, R. F. 1968, A Photometric Atlas of the Spectrum of Arcturus $\lambda\lambda 3600-8825$ Å, (Cambridge, The Philosophical Society).

Griffin, R. and Griffin, R., 1979, A Photometric Atlas of the Spectrum of Procyon $\lambda\lambda 3140-7470$. (Institute of Astronomy, Cambridge).

Gustafsson, B. 1978, unpublished.

May, M., Richter, J. and Wichelmann, J. 1974, Astron. Astrophys. Suppl., 18, 405.

Moore, C. E., Minnaert, M.G.J. and Houtgast, J. 1966, Second Revision of Rowland's Preliminary Table of Solar Spectrum Wavelengths. (Utrecht Observatory).

Morel, M., Bentolila, C., Cayrel, G. and Hauck, B., 1976, in IAU Symposium 72, Abundance Effects in Classification, Eds. B. Hauck and P. C. Keenan, (Reidel: Dordrecht) p.223.

Perrin, M. N., Hejlesen, P. M., Cayrel de Strobel, G. and Cayrel, R. 1977, Astron. Astrophys. 54, 779.

Spite, F. and Spite, M. 1982, Astron. Astrophys., 115, 357.

Trimble, V. and Bell, R. A. 1981, Quart. J. Roy. Astr. Soc., 22, 361.

Vidal, C. R., Cooper, J. and Smith, E. W. 1971, J. Quant. Spectrosc. Radiat. Transfer 11, 263.

APPENDIX 1

INDIVIDUAL VALUES OF [Fe/H] ABUNDANCES OF SPECTROSCOPIC STANDARD
STARS CONTAINED IN THE [Fe/H] CATALOGUE

α LYR = HD 172 167; A0V

AUTHORS	T_{eff}	log g	$[Fe/H]_o^*$
HUNGER (1960)	-	-	-0.3
STROM and STROM (1966)	-	-	+0.05
STROM et al. (1966)	9000	3.8	+0.06
ALLER and ROSS (1967)	10080	3.5	+0.2
CONTI and STROM (1968)	9509	4.0	-0.1
PRZYBYLSKI (1968)	8692	-	-0.1
STROM et al. (1968)	9509	3.7	-0.1
GEHLICH (1969)	9164	4.0	-0.25
SMITH (1974)	9692	4.0	+0.02
BOYARCHUK and SNOW (1978)	9692	-	-0.9
BOYARCHUK and SNOW (1978)	9692	-	-0.5
CASTELLI and FARAGGIANA (1979)	9692	4.1	-1.36
DREILING and BELL (1980)	9692	3.9	0.0
SADAKANE and NISHIMURA (1981)	9692	3.94	-0.58

PROCYON = HD 61421; F5 IV-V

AUTHORS	T_{eff}	log g	$[Fe/H]_o^*$
GREENSTEIN (1948)	6222	-	+0.22
WRIGHT (1951)	6720	-	-0.40
EDMONDS (1965)	6450	4.0	-0.29
MERCHANT (1966)	6300	-	+0.03
POWELL (1970)	6630	-	+0.07
GRIFFIN (1971)	6540	-	0.00
HASEGAWA (1975)	6720	-	+0.74
TOMKIN and LAMBERT (1978)	6630	4.0	-0.15
KATO and SADAKANE (1982)	6630	4.0	+0.02

HD 219 134 K3 V

AUTHORS	T_{eff}	log g	$[Fe/H]_o^*$
CAYREL de STROBEL (1964)	4582	-	0.00
CAYREL de STROBEL (1966)	4710	4.50	+0.10
CAYREL de STROBEL et al. (1970)	4710	4.50	-0.01
STROMBACH (1970)	4710	4.50	0.00
OINAS (1974)	4667	4.40	-0.21
PERRIN et al. (1975)	4667	4.50	0.00
OINAS (1977)	4710	4.50	+0.20

ε VIR = HD 113226; G8 III

AUTHORS	T_{eff}	log g	$[Fe/H]_o^*$	REFERENCE STAR
GREENSTEIN and KEENAN (1958)	4421	-	+0.04	SUN
CAYREL and CAYREL (1963)	4941	2.7	+0.01	SUN
HELFER and WALLERSTEIN (1964)	5305	-	-0.15	γ, δ, ε TAU
CAYREL de STROBEL (1966)	4941	2.7	-0.03	SUN
HELFER and WALLERSTEIN (1968)	4421	2.45	-0.15	γ TAU
CAYREL de STROBEL et al (1970)	4941	2.70	-0.06	SUN
STROM et al. (1971)	4990	3.00	-0.1	SUN
VAN PARADIJS (1973)	4990	2.85	+0.04	SUN
BLANC VAZIAGA et al. (1973)	4941	2.70	+0.02	SUN
CAYREL et al. (1977)	4990	2.60	0.00	SUN
SNEDEN et al. (1978)	4990	2.75	-0.02	SUN
BRANCH et al. (1978)	4990	3.00	+0.17	SUN
HEARNSHAW and NEWBURGH (1979)	5040	2.70	-0.05	SUN
LAMBERT and RIES (1981)	5305	3.22	+0.21	SUN

μ LEO = HD 85503; SMR K GIANT

AUTHORS	T_{eff}	log g	$[Fe/H]_o^*$	REFERENCE STAR
STROM et al. (1971)	4755	2.7	+0.1	γ, δ, ε TAU, εVIR
BLANC-VAZIAGA et al. (1973)	4460	2.20	-0.08	εVIR
OINAS (1974)	4460	2.4	-0.01	SUN
PETERSON (1976)	4420	2.3	+0.03	SUN
PETERSON (1976)	4421	2.3	-0.11	SUN
BRANCH et al. (1978)	4541	2.35	+0.48	SUN
LAMBERT and RIES (1981)	4710	2.82	+0.11	SUN

HD 122 563; HALO K-GIANT

AUTHORS	T_{eff}	log g	$[Fe/H]_o^*$
WALLERSTEIN et al. (1963)	4065	-	-2.9
PAGEL et al. (1965)	4271	-	-2.65
BELL and PAGEL (1967)	4200	-	-2.6
WOLFFRAM (1972)	4582	1.2	-2.72
SNEDEN (1973)	4624	1.2	-2.7
SNEDEN (1974)	4582	1.2	-2.75
SPITE and SPITE (1978)	4582	0.9	-2.6
SPITE and SPITE (1979)	4582	0.9	-2.5
SPITE and SPITE (1980)	4582	0.9	-2.5
LUCK and BOND (1981)	4582	0.80	-2.59
BESSELL and NORRIS (1981)	4667	0.7	-2.7
STEENBOOK (1983)	4582	1.2	-2.61
LUCK and BOND (1983)	-	-	-2.35

APPENDIX 2

SOME STANDARD STARS FOR SPECTROSCOPIC ANALYSES

STAR	HD	V	B-V	Sp. Type	T_{eff} K	log g	$[Fe/H]_o$*	Remarks
10 Lac	214680	4.88	-0.20	O9 V	37450	4.45		no Fe lines in the spectrum
τ Sco	149438	2.82	-0.25	BO V	35000	4.35		
γ Cas	5394	2.47	-0.15	BO.5 IVe	-			"
ζ Per	24398	2.85	+0.12	B1 Ib	27000			"
γ Peg	886	2.83	-0.23	B2 IV	21910	3.7	+0.04	
105 Tau	32991	5.89	+0.19	B2 Ve	-	-	-	
55 Cyg	198478	4.84	+0.41	B3 Ie	14260			
134 Tau	38899	4.91	-0.07	B9. 5 V	11455	-	-	very narrow lines
α Lyr	172167	0.03	0.00	AO V	9692	3.94	-0.06	
Hertzsprung II 2507	23964	6.74	+0.06	B9.5 Vp	-	-	-	Pleiades cluster star
β UMa	95418	2.37	-0.02	A1 V	10286	4.30	+0.78	UMa nucleus star
θ Vir	114330	4.38	-0.01	A1 V	9510	4.0	0.00	
63 Tau	27749	5.64	+0.30	A1 m	7640	4.4	+0.57	Hyades cluster star
29 Cyg	192640	4.97	+0.14	A2 V	8000	3.9	-0.80	
α Cyg	197345	1.25	+0.09	A2 Iae	9160	1.13	+1.00	
-	161817	6.97	+0.16	A2 V pec	7695	3.0	-1.30	Halo horiz. branch
15 Vul	189849	5.64	+0.30	A4 III	8000	3.5	0.00	
θ Cep	195725	4.22	+0.20	A7 III	7640	4.0	+0.13	
30 L Mi	90277	4.74	+0.25	FO V	6720	-	+0.20	
	106516	6.11	+0.46	F6 V bvw	5930	4.3	-0.50	
α C Mi	61421	0.38	+0.42	F5 IV-V	6630	4.0	-0.15	
41 Cyg	195295	4.01	+0.40	F5 II	6720	2.50	-0.05	
γ Lep A	38393	3.60	+0.47	F8 IV	5660	-	-0.07	UMa stream star
π Ori	30652	3.19	+0.45	F6 V	6380	4.5	+0.10	
γ Ser	142860	3.85	+0.48	F7 V abw	6300	4.0	-0.24	old disk star
γ Pav	203608	4.22	+0.49	F6 V	6000	4.35	-0.62	old disk star
o Aql	187691	5.11	+0.55	F8 V	6150	4.40	+0.12	
-	140283	7.20	+0.49	F9 V wl	5730	4.0	-2.15	Halo dwarf
Moon (Sun)	-	-	-	-	5770	4.43	0.00	chief standards for detailed
Ceres (Sun)	-	6.85	-	-	5770	4.43	0.00	analyses for F, G and K stars
	44594	6.60	+0.66	G3 V	-	-	-	Hardrop's solar twin
	81809	5.38	+0.64	G2 V	-	-	-	Mihala's solar twin
VB 73	28344	7.85	+0.60	G1 V	5900	4.50	+0.12	Hyades cluster star
VB 64	28099	8.12	+0.66	G2 V ?	5768	4.50	+0.15	" " "
VB 92	28805	8.66	+0.74	G8 V ?	5540	4.50	+0.16	" " "
	1835	6.39	+0.66	G2 V	5774	4.50	+0.19	Hyades group star
	10307	4.95	+0.62	G1.5 V	5860	4.38	-0.03	
α Cent A	128620	-0.01	+0.71	G2 V	5760	4.38	+0.26	π = 0."750
	76151	6.00	+0.67	G3 V	5600	4.40	-0.02	The position in the HR diagram implies He-rich
85 Peg	224930	5.75	+0.67	G3 V	5200	4.35	-0.80	Old disk star
-	20630	4.83	+0.68	G5 V	5660	4.45	+0.08	
μ Cas	6582	5.17	+0.69	G5 Vp	5230	4.40	-0.70	Old disk star
31 Aql	182572	5.16	+0.77	G8 IV	5660	4.20	+0.40	Metal rich G-subgiant
Grmb 1830	103095	6.45	+0.75	G9 Vp	5040	4.60	-1.40	Halo dwarf star
ε Vir	113226	2.83	+0.94	G8 IIIab	4940	2.70	0.00	
β LMi	90537	4.21	+0.90	G9 IIIab	5090	3.00	+0.21	
36 Oph A	155886	4.31	+0.86	KO V	5090	4.60	-0.01	
β Gem	62509	1.14	+1.00	KO IIIb	4850	2.5	0.00	
γ Tau	27371	3.65	+0.99	KO IIIab CN1	4990	2.6	+0.10	
α Cent B	128621	1.33	+0.88	K1 V	5250	4.73	+0.20	
-	190404	7.28	+0.82	K1 V	4990	4.50	-0.20	
36 Oph B	155885	4.33	+0.86	K1 V	5090	4.60	+0.09	
α Boo	124897	-0.04	+1.23	K1 IIb CN-1	4270	2.0	-0.50	Old disk giant
γ Lep B	38392	6.15	+0.94	K2 V	-	-	-	UMa stream star
μ Leo	85503	3.88	+1.22	K2 IIIb CN1 Ca1 Ba-1	4460	2.40	+0.07	Chief SMR candidate
-	122563	6.20	+0.90	K2 p	4500	1.00	-2.70	Halo giant
-	219134	5.56	+1.01	K3 V	4670	4.50	0.00	
36 Oph C	156026	6.34	+1.16	K5 V	4380	4.70	-0.06	
61 Cyg A	201091	5.21	+1.18	K5 V	4380	4.50	-0.07	
α Tau	29139	0.85	+1.54	K5 III	3920	1.20	-0.10	
61 Cyg B	201092	6.03	+1.37	K7 V	3880	4.60	-0.10	
β And	6860	2.06	+1.58	MO IIIa	3250	-	+0.10	
	36395	7.97	+1.47	M1.5 V	3630	4.80	+0.60	
	204961	8.67	+1.46	M1 V	-	-	-	
	35601	7.35	+2.20	M1 Ib	4000	0.7	-0.24	

DISCUSSION

GRIFFIN, R. E. M.: In the pictures of reticon spectra which you showed, we could see some apparent emission spikes. Could you please explain their origin?

CAYREL: The spikes are caused by cosmic rays. They are even stronger in the Pleiades spectrum, but before the cosmic ray "events" become really disturbing you have gained at least three magnitudes in your your abundance work and I consider this a crucial advantage.

GRIFFIN, R. E. M.: If they are artifacts of the reticon and you cannot predict their positions, how can you be sure that they do not coincide with absorption lines? You say that you can measure confidently lines down to 2 mÅ, but how do you know there is not an artificial -6 mÅ line sitting on top of it? In my experience at the McDonald Observatory these spikes are observed in spectra of bright stars and they do not disappear by flat-fielding.

CAYREL: Yes, but though you may have spikes, you can observe much fainter stars with the Reticon.

GRIFFIN R. E. M.: It also appears that they are in the same place for several stars.

ARDEBERG: From your spectrograms I am quite sure that you have used the same background correction for all the spectrograms displayed. I think that you simply have to invest more time in your background measurements; then your spikes will diminish.

CAYREL: Yes. We did not use the same background corrections at CFHT. At CFHT four flat field exposures are taken after each stellar spectrum, having the same exposure level as the stellar spectrum. What I have shown is the ratio of the stellar exposure to the average of the four flat fields.

HEINTZE: For one observed H-alpha profile there are several combinations of effective temperature and log g that produce such a profile. How did you disentangle these two quantities?

CAYREL: We know that the surface gravity of the solar Hyades dwarfs is 4.5. They might be slightly more massive than the Sun. The microturbulent velocity is 1.0 km/sec. We got the effective temperature by comparison of observed H-alpha profiles with theoretical profiles. The corresponding uncertainty is less than 10°K.

GUSTAFSSON: You have obtained an impressive accuracy in the effective temperature determinations; however, it should be noted that these temperatures refer to the "effective temperature labels" of the models. For your abundance analysis the high accuracy in determining this label temperature is the relevant and important one, but one should warn people not to believe that your T_{eff}'s, as defined in the normal way related to the stellar surface flux, could be without systematic errors due to failures in the models of even more than 50°K.

CAYREL: We measured the depression in the line at 2.5 and 5.0 Å from the line center for the Hyades stars and the Sun. We calculated the ratio and compared with the grid of models. The Sun was compared with vB 64 or vB 73 and the temperatures agree extremely well between different spectra taken for the same star (typically ± 15°K).

GARRISON: One has to be as careful with Reticon spectra as we have learned to be with photographic spectra. Too many observers use the Reticon as a black box and believe everything that comes out, whether properly treated or not. Properly used, it is a beautiful tool, as you have said.

It will be nice to be able to use new detectors in the blue. With all the reports of spectroscopic binaries in the Hyades we cannot be sure that your H-alpha results translate to compare with blue results exactly for any particular star.

The signal-to-noise ratio quoted for photographic spectra does not tell the whole story in comparing with S/N for Reticons. At the MK workshop in Toronto (1983) this whole problem was brought up several times and it was agreed that Reticon spectra of S/N = 100 are not equivalent to photographic spectra. Millward's spectra of S/N = 1000 were much more comparable and even Morgan agreed that they could be classified using the MK process.

I wish people would not use the Moon as a source of solar spectrum. Unless the spectrograph is perfect, the use of an extended source will cause scattered light and will also fill the collimator differently. The result is that the H-lines look weaker; the metal lines also get filled in, making the spectrum look metal-weak. One thus concludes that the Sun is cooler and more metal-weak than it really is.

I wish people would not quote the compilations of spectral types, such as Jaschek, but rather use them as guides to get back to the original sources. Your types are Morgan's, not Jaschek's.

Finally, the case of vB 64 is an interesting one, with an interesting history. Hardorp has used it as evidence that the Sun is cooler than other stars classified G2 V. I took a good spectrum and found it to be close to that of the Sun, about G2 - G3, which is different from the type of G6 given by Morgan in his study of Hyades stars. After investigation, I found that he had classified his spectrum correctly, but that Hiltner had taken the wrong star! Thus his type of G6 V was not for vB 64 but for some other star. It is unfortunate that Hardorp concluded that the solar type was in error instead of trying to find out why the results were different.

CAYREL: Well, it is exciting to use a new technique. As to using the Jaschek catalogue, I trust it and do not have time to go back to the original sources. But the reader can always go to the Jaschek Catalogue and find the original sources. As for our use of the Moon, observations go much more quickly than with asteroids. We do have spectra of Ceres, and they agree very well with the Moon. As for vB 64, I will use G2 V for its spectral type.

ADELMAN: It is possible to obtain similar abundances from the Fe II lines in the optical and in the ultraviolet in B and A stars (Lekrone and Adelman, in preparation - Pi Cet and Nu Cap). One can increase the signal-to-noise ratio obtainable from photographic spectrograms by co-adding spectrograms. This is a useful technique when one needs a large spectral region provided one is working on bright stars. The process of co-addition should be checked by obtaining high signal to noise spectra of selected spectral regions with Reticons or CCD's.

The older elemental analysis of most B- and A-type stars suffers from systematic errors in the gf values of the atomic species used to determine the microturbulent velocity. This usually leads to errors in the microturbulence and in the elemental abundances. Whose Fe I gf values did you use?

CAYREL: May's. Blackwell has not yet measured 4 ev excitation Fe I lines (typically weak lines in our spectral range).

ADELMAN: The best Fe I gf values are those of Blackwell and his collaborators. Dr. J. R. Fuhr, National Bureau of Standards, has prepared a revision of the NBS critical compilation of gf values for Fe I lines. It will appear as part of a forthcoming volume on gf values of iron peak elements. Another useful source of Fe I gf values are the recalibration of Corliss and Bozman values by Cowley and Corliss (1983 MNRAS 203, 651). The recent gf values published in Astron. Astrophys. Suppl based on solar lines are also helpful (See, e.g., Gurtovenko, E. A. and Kostik, R. I. 1982, Astron. Astrophys. Suppl. 49, 193.). Some additional comments on gf values and atomic data are contained in my paper with C. R. Cowley (1983 QJRAS 24, 393). Let me note in closing that Dr. Fuhr and Dr. William C. Martin of the National Bureau of Standards have been most helpful to me and many other astronomers in giving advice on gf values, damping parameters, atomic energy levels, atomic line lists and other topics in atomic physics.

CAYREL: Thank you very much for your helpful comments. I have only one complaint about this symposium, happily, and that is that there are no papers on oscillator strengths. You have spoken for Dr. Blackwell, who is not here.

WALKER: Some of the Reticon spectra we have seen today have not been fully reduced. We have no difficulty achieving the high S/N per diode discussed and there is no problem with spikes in sea-level observations. CFHT spectra do have some cosmic ray events. For a

critical discussion of the performance and calibration of our Reticon
systems, see my paper in the Eighth Symposium on Photoelectronic Imaging
Devices which is now in press.

GRIFFIN: I do not suppose that Madame Cayrel would expect me to
subscribe to her remarks concerning the comparison between Reticon and
photographic spectra! Without wishing to denigrate Reticons, I would
like to suggest that the comparison is not so one-sided as it has been
made to appear. Since much of my work is photoelectric, I feel that it
is in order for me to say that a good deal of unwarranted prejudice has
developed against photography in recent years. People seem to have been
so carried away with Reticons, and we have got so used to seeing the
words "high-resolution, low-noise" adhering to every reference to
Reticon spectroscopy - often with the judicious use of initial capital
letters! - that an impression almost of sacredness now surrounds
everything to do with Reticons. The facts do not altogether merit that
impression. For one thing, reticons are small. This smallness means
that either the wavelength coverage is very short, or the binning of the
spectrum is coarse, or both. Another comment I would make concerns
signal/noise ratios. Few users of Reticons refrain from quoting
signal/noise ratios (usually very high ones), presumably because they
are very easy to calculate. They do not usually bother to tell us
whether the quoted ratios are per bin, per resolution element or what.
In any case, although the ratios are made to appear large, their
significance seems relatively small. They are probably derived from
inferred photon counts in the continuum and do not represent the true
errors. There is an example of a Reticon spectrum in one of the poster
papers: it shows absorption lines with equivalent widths of just a few
milliangstroms and looks superficially to be a very nice tracing, but it
does also show apparent emission lines with equivalent widths ranging up
to about 6 mÅ. Such features are readily recognized as artifacts where
they are seen on the continuum, but in a complicated spectrum some of
them must inevitably compromise the profiles of absorption lines.

CAYREL: The occurrence of "spikes" is a very mild nuisance, because they
are rare and do not reappear at the same location on several spectra.
The basic advantage of a Reticon over the photographic plate is the much
higher quantum efficiency of the solid state detector, which is
conservatively at least 50 times the one of a good photographic plate.
So, if the main interest is to reach fainter stars with a decent S/N
ratio (such as globular cluster stars) one is compelled to use Reticon
or photocounting devices, even if their wavelength coverage is smaller.

DETERMINATION OF STELLAR ROTATIONAL VELOCITIES

Arne Slettebak

Perkins Observatory
Ohio State and Ohio Wesleyan Universities

ABSTRACT. The three basic methods for measuring axial rotation of stars had been suggested before the beginning of this century. These are (1) Modulation of starlight due to dark or bright areas on a rotating star; (2) Distortions in the radial velocity curves of eclipsing binary systems; and (3) Line profile analysis. Research in each of these areas is reviewed.

1. INTRODUCTION

"If I may give my own opinion to a friend and patron, I shall say that the solar spots are produced and dissolve upon the surface of the Sun and are contiguous to it, while the Sun, rotating upon its axis in about one lunar month, carries them along . . . ." So wrote Galileo (Drake 1957) in his letter of May 4, 1612 to the wealthy Augsburg merchant and enthusiastic amateur of science, Mark Welser. Here we have the first determination of the rotation period of a star. But over 300 years were to pass before the next determination of a stellar rotational velocity. It is interesting, however, that the three basic methods for measuring stellar rotation had all been suggested, although not yet carried out, before the end of the last century.

As early as 1667, Bouillaud suggested that stellar variability was a direct consequence of axial rotation, the rotating star showing alternately its bright (unspotted) and dark (spotted) hemispheres, an idea which was further explored by Cassini, Fontenelle, and Miraldi (cf. Brunet 1931). More recently, Pickering (1881), in a study of variable stars, concluded that "The most natural explanation of the variation of a star of short period is that it is due to its rotation around its axis". Furthermore, "The difference in brightness of the two sides of a star may be due to spots like those of our Sun, to large dark patches, or to a difference in temperature". Pickering erred in trying to apply these ideas to close binary systems and Cepheid variables, but in recent years, periodic variations in light due to starspots and stellar plages have been used to determine rotation periods for certain classes of stars. Line profile analysis was first

D. S. Hayes et al. (eds.), Calibration of Fundamental Stellar Quantities, 163–184.
© 1985 by the IAU.

introduced in 1877, when Abney suggested that the effect of a star's rotation on its spectrum would be to broaden all of the lines and that ". . . other conditions being known, the mean velocity of rotation might be calculated." The third basic method for determining stellar rotational velocities was apparently first suggested by Holt (1893): ". . . in the case of variable stars, like Algol, where the diminution of light is supposed to be due to the interposition of a dark companion, it seems to me there ought to be a spectroscopic difference between the light at the commencement of the minimum phase, and that of the end, inasmuch as different portions of the edge would be obscured. In fact, during the progress of the partial eclipse, there should be a shift in the position of the lines; and although this shift is probably very small, it ought to be detected by a powerful instrument." This predicted distortion of the radial velocity curve of an eclipsing binary, sometimes called the "Rossiter Effect", was indeed detected less than 20 years later, and resulted some years after in the first actual measurements of stellar rotation since the observations of Galileo.

The aforementioned three methods of determining stellar rotational velocities will be discussed in this review. I should like to emphasize that the review will be limited to observational work only -- there exists an enormous literature concerning the theoretical aspects of stellar rotation which will not be included here. Furthermore, it would be impossible even to include all of the references to measurements of stellar rotation in a paper of this length. Therefore, whenever possible, I have referred to review papers with extensive bibliographies of earlier work rather than repeat long lists of references. Some of these, which have been very helpful in preparing this paper, are by Struve (1945), Huang and Struve (1960), Kraft (1969, 1970), and Moss and Smith (1981). In addition, two books on stellar rotation have appeared: The proceedings of IAU Colloquium No. 4 (ed. Slettebak, 1970a) and "Theory of Rotating Stars" (Tassoul, 1978). The latter is a superb contribution, with many references to both observational and theoretical papers.

2. METHODS FOR THE DETERMINATION OF ROTATIONAL VELOCITIES

2.1. Rotational Modulation

Although stellar disks generally cannot be seen beyond the sun, a rotating non-uniform disk should produce periodic changes in light which, if detectable, would permit determination of the period of rotation. Perhaps the first convincing detections of this kind were by Kron (1947, 1952), who found evidence in the light curves of the eclipsing binaries AR Lac and YY Gem for a patchy, non-uniform surface of the stellar components. Subsequent work generally divides into two categories: (1) photometry of a variable continuous spectrum as starspots appear and disappear on the rotating disk, and (2) measurements of the variation in strength of emission lines (e.g., Ca II H and K) which arise in plages in a rotating chromosphere. A large literature

has developed in the last decade or so, and I will discuss only
selected objects and papers.

2.1.1. <u>BY Dra Stars</u>. These are defined as UV Cet flare stars of spectral types dKe and dMe which, outside of periods of flaring, show
periodic variations in light of several tenths of a magnitude, attrib-
utable to large, cool starspots on their surfaces.

Following Chugainov's (1966) discovery of periodic light variability in Popper's flare star HDE 234677, Krzeminski and Kraft (1967)
using UBV photometry found similar variations in brightness in three
dKe-dMe stars and ". . . agree with Chugainov that the most promising
model appears to be the rotational modulation of a star with a
non-uniform distribution of surface brightness". Krzeminski (1969)
reached the same conclusion, finding equatorial rotational velocities
of 10 and 15 km/sec for two of his stars. Upper limits to flare star
rotational velocities were estimated by Gershberg (1970) to be 25-30
km/sec, based on flare statistics and Hα emission widths in four UV Cet
stars. Bopp and Evans (1973) developed a model for the spots on BY
Dra, based on 1965 and 1966 photometry, showing that "The spots must be
large, covering up to 20 percent of the stellar hemisphere". Observations by Vogt (1975) in 1973 for this star were found to be inconsistent with a thermal dark spot model but explainable in terms of Ca
II emission regions or plages. More recently, Pettersen (1980)
observed sinusoidal variations in the magnitude of EV Lac outside of
flares, interpreted these as due to intensity modulations from a
photospheric spot group, and estimated an equatorial rotational
velocity of 4.2 km/sec for this dM4.5e star. Many more references can
be found in the review paper by Vogt (1983).

2.1.2. <u>RS CVn Binaries and FK Com Stars.</u> RS CVn binaries have been
defined by Hall (1976) as "binaries with orbital periods between 1 day
and 2 weeks, with the hotter component F-G V-IV, and with strong H and
K emission seen in the spectrum outside eclipse". In 1972, Hall
suggested that a model in which ". . . a region of tremendous sunspot
activity darkens one side of the cool star", assuming synchronous
rotation, can reproduce the many photometric complications observed in
RS CVn. Recent studies of starspots and rotations of the components of
RS CVn systems include those by Ramsey and Nations (1980), Vogt (1981),
Guinan et al. (1982), and Fekel (1983), plus the references therein.

Another class of stars with photometric and spectral properties
similar to those of the RS CVn binaries and BY Dra stars are the FK Com
stars. These are apparently single G2-K0 giants with unusually large
rotational velocities for their types (v sin i ≃ 100 km/sec), whose
photometric variability can best be interpreted in terms of starspots
(cf. Bopp and Stencel 1981; Holtzman and Nations 1984; Dorren et al.
1984).

2.1.3. <u>Photometric Variations of Pre-Main-Sequence and Pleiades K
Stars.</u> Four pre-main-sequence K stars in the direction of the Taurus

dark cloud complex were monitored with UBVRI photometry by Rydgren and
Vrba (1983) and found to show quasi-sinusoidal light variations "apparently due to the presence of large starspots". The periods, which
range from 1.9 to 4.1 days, correspond to rotational velocities of 75
to 20 km/sec. Van Leeuwen and Alphenaar (1983) observed 19 late G and
early K-type members of the Pleiades cluster, finding all to be
variable and 12 with semi-regular light curves like those of the BY Dra
stars. If the variations are assumed to be due to rotational modulation, the corresponding rotational velocities are consistent with
v sin i values of 75-150 km/sec found spectroscopically for two of the
stars.

2.1.4. <u>Chromospheric Variations in Lower Main-Sequence Stars</u>. Wilson
(1978) was the first to study the variation with time of the
chromospheric activity in main-sequence stars. He measured fluxes at
the centers of the Ca II H and K lines in 91 F5-M2 stars for time
intervals of 9-11 years, but stated that "the data points are too few
to establish short-term periodicities such as rotational modulation".
Stimets and Giles (1980) analyzed his data for periodicities using an
autocorrelation technique, however, and determined rotational periods
for 10 stars, ranging between 2 and 37 days. Meanwhile, new H and K
flux observations of 47 lower main sequence and eight evolved stars
were made over a nearly continuous 14-week observing run and discussed
by Vaughan et al. (1981) and Baliunas et al. (1983). These investigators ". . . find rotation rates easily for the main-sequence stars
with strong emission or those later than about spectral type K0. With
this technique, rotation rates can be measured precisely for the first
time for equatorial velocities as slow as 1 km/sec, and independently
of the aspect of the rotation axis".
        Radick et al. (1983) studied the photometric variability of solar-
type stars. Although their time resolution was not sufficient to
obtain rotation periods, they point out ". . . the possible ability of
photometric observations to measure stellar rotation should not be
overlooked. Continuum photometry probably is sensitive mainly to dark
spots, whereas the chromosphere diagnostics characterize the surrounding plage regions. However, for the Sun, spots are confined, both
individually and as groups, to a much narrower range of latitudes than
are the plage regions. Accordingly, spot-derived rotation periods may
be more precise".
        Lambert and O'Brien (1983) detected a marked variation of the
chromospheric He I D_3 line at 5876 Å in the spectrum of the G5V star
κ Cet, which they attribute to rotational modulation. The rotational
velocity of κ Cet is estimated to be 3.9 km/sec on the basis of an 8.5
day period and an assumed radius $R_* \simeq 0.9\ R_\odot$.
        Ultraviolet studies of F, G, and K main-sequence stars with the
IUE satellite by Blanco et al. (1979), Hallam and Wolff (1981), and
Boesgaard and Simon (1982) have also revealed periodic variations in a
number of chromospheric lines, which these investigators attribute to
rotational modulation.

2.1.5. <u>Ap Stars</u>. It was Babcock (1949) who first considered ". . .the alternative hypothesis that the spectrum variables of type A are stars in which the magnetic axis is more or less highly inclined to the axis of rotation and that the period of magnetic and spectral variation is merely the period of rotation of the star". Deutsch's (1952, 1956) discovery of an inverse relation between the periods of Ap stars and the widths of their absorption lines supported this suggestion and led him to conclude that ". . . in the spectrum variables we observe the rotation of A stars that exhibit intensely magnetic areas, within which the peculiar line strengths are produced". Many studies of the spectrum and magnetic variables of type A exist, including several symposia devoted to these objects (e.g., Cameron 1967; Weiss et al. 1976; Liège Astrophys. Colloq. 23, 1981). A catalog of observed periods for Ap stars, with an extensive bibliography, was recently published by Catalano and Renson (1984).

2.1.6. <u>Be Stars</u>. Hutchings (1970) found a period of 0.7 days in the peak separation and V/R ratio of the double emission profiles of Hγ and Hβ in the spectrum of the B0 IVe star γ Cas, which he attributed to rotation. His observations show considerable scatter, however, and have not been confirmed as yet. Slettebak and Snow (1978) searched for rotational modulation of the Si IV and Mg II resonance doublets during 64 nearly-continuous hours of <u>Copernicus</u> observations of γ Cas but found no evidence for a rotation period. Most recently, Walker et al. (1979) reported velocity variation in the He I 6678 Å line in the spectrum of the rapidly-rotating O9.5 V star ζ Oph which they suggested to be consistent with non-unformities in stellar surface brightness being carried across the line of sight by rotation. In a later paper (Walker et al. 1981), however, they state that the observed features move too rapidly and linearly to be due to irregularities on the stellar surface itself, and suggest that an interpretation in terms of non-radial pulsations coupled with rotation cannot be ruled out. The latter interpretation was supported by a recent study of ζ Oph by Vogt and Penrod (1983). At this time, therefore, no unassailable evidence for rotational modulation has been found for Be stars.

2.2. Distortions in the Radial Velocity Curves of Eclipsing Binary Systems.

Following Holt's (1893) prediction, the first observational evidence for a rotational effect in the radial velocity curve of an eclipsing binary system was presented by Schlesinger in 1909 for the Algol variable δ Lib. In his words, "The rotation of the bright star has another consequence in certain parts of the orbit. In general we obtain light from the whole disk and the observed velocity is equal to that of the center of the star. Just before and just after light minimum, however, this is not the case; before minimum the bright star is moving away from us and part of its disk is hidden by the dark star. The part that remains visible has on the whole an additional motion away from us on account of the rotation; the observed velocity will therefore be greater than the orbital. On the other hand just after minimum the circumstances are reversed so that the observed velocity is less than the orbital". Forbes (1911) suggested that this effect could be used to measure equatorial rotational velocities, but Schlesinger (1911) replied that because of unknown limb-darkening effects, ". . .it would appear to be well-nigh hopeless, with present-day appliances at least, to attempt to determine . . . the rate of rotation from observational material alone". Schlesinger (1913) found the effect again in the radial velocity curve of λ Tau, as did Hellerich (1922) for a number of Algol-type variables. The first actual measurement of the rotational effect was by Rossiter (1924), who found a total range of 26 km/sec in the brighter component of β Lyr. He pointed out in this work that this rotational effect ". . . does not represent the value of the velocity of rotation at the limb of the star but represents only a fractional part of it. The magnitude of the effect depends on how much the light from the unbalanced visible limb is able to displace the apparent center of the line from the point where it would normally be measured". Soon thereafter, McLaughlin (1924, 1926) discussed the effect in Algol and in λ Tau, while Plaskett (1926) showed it to be present in 21 Cas.

In an important paper, Struve and Elvey (1931), assuming a rotational velocity derived from line broadening of 60 km/sec for Algol, showed that asymmetrical line profiles computed as the star goes in and out of eclipse agreed well with observed line profiles at those phases. They were able to derive the rotational effect in the radial velocities of Algol and show that it agreed well with McLaughlin's (1924) observations, thereby establishing the rotational velocity at about 60 km/sec.

McLaughlin (1933, 1934) computed "rotation factors" (the distances of the center of light of the luminous area from the center of the disk, expressed in units of the radius of the disk) from the photometric elements of four eclipsing binary systems and applied them to the observed rotational distortions in the radial velocity curves to obtain v sin i's for the principal components. He obtained 42 km/sec for Algol, 42 km/sec for λ Tau, 60 km/sec for δ Lib, and 200 km/sec for α CrB. These values may be compared with v sin i's determined from line-profile analysis of the same stars (Uesugi and Fukuda 1982): Algol, 55 km/sec; λ Tau, 85 km/sec; δ Lib, 75 km/sec; and α CrB, 135 km/sec. Kopal (1942), who included limb-darkening in his analysis, obtained v sin i = 48 km/sec from McLaughlin's (1934) radial velocity data for Algol -- somewhat closer to the value from line-profile analysis. Struve (1944) found a very large rotation effect in the partial phases of the eclipse of the B8 component of U Cep, from which he concluded that its equatorial velocity of rotation is 200 km/sec. After correcting the radial velocity measurements for asymmetries displayed by the Balmer lines, Hardie (1950) found a rotation effect of 250 km/sec for the same star, whereas Olson (1968) obtained v sin i = 310 km/sec from line profile widths. Many other examples of radial velocity curve distortions due to rotation exist: Struve (1950) estimated that the phenomenon had been observed in about 100 systems by 1949. A recent study is by Wilson and Twigg (1980).

2.3. Line Profile Analysis

2.3.1. Profile Fitting and Line-Width Measurements.

After Abney's (1877) suggestion, Schlesinger (1909) seems to have been the first to call attention to rotational line broadening. Noting that "In close spectroscopic binaries like δ Lib the periods of rotation of the two stars are doubtless the same as the time of revolution", he suggested that "This rotation will introduce a Doppler effect that will broaden the lines in the spectrum to a considerable extent. One of the limbs of the bright star (in the δ Lib system) is approaching us at the rate of 35 km/sec while the other is receding at the same rate .... These (including orbital velocity changes) are sufficient to account for the general character of the spectrum as we see and photograph it".

Shapley and Nicholson (1919) investigated spectral line profiles in a rotating or pulsating star in connection with Cepheid variables, and showed that an undarkened rotating stellar disk would broaden an infinitely sharp line into a semi-ellipse. Adams and Joy (1919), investigating the short-period variable W UMa, suggested that the unusually broad spectral lines in that system are due ". . . mainly to the rotational effect in each star, which may cause a difference of velocity in the line of sight of as much as 240 km/sec between the two limbs of the star". Estimates of rotational velocities of μ^1 Sco and V Pup from line widths were made by Maury (1920) and (for V Pup) found to agree satisfactorily with computed values by Hellerich (1922).

In an important paper, Shajn and Struve (1929) showed that fast rotational velocities predominate in short-period spectroscopic binaries and developed a graphical method of computing rotationally-broadened line profiles. The resulting "dish-shaped" line contours were observed by Elvey (1929) in a number of broad-lined single stars, with such a ". . . marked similarity . . . with the theoretical contours of rapidly rotating stars obtained by Shajn and Struve . . . that we feel safe in stating that the stars . . .are rotating rapidly". Elvey (1930) followed up this work with a study of rotational broadening of the Mg II 4481 line in 59 O, B, A, and F stars, using the Shajn-Struve graphical method. Instead of using line profiles in the spectrum of the Moon as non-rotating lines, as had been done by Shajn and Struve, Elvey chose profiles from sharp-lined stars of early type. "The stellar disk is divided into sections parallel with the axis of rotation, each section receiving a weight equal to its area expressed as a fraction of that of the total disk. The original contour (i.e., from a sharp-lined star) multiplied by its weight is assigned to each section and displaced by the amount corresponding to the velocity of rotation in the line of sight for the section, and the sum of the contours is taken. The result is the contour of the line for the rotating star. Comparison with observed line profiles then gives v sin i, the component of the rotational velocity in the line of sight". This technique was later used by many investigators.

Struve (1930) summarized the evidence for "Broad and shallow absorption lines in stellar spectra (being) due to axial rotation" and was the first to show that ". . . rotational speed is a function of spectral type, the fastest rotations occurring in the earliest types".

This conclusion was strengthened by a number of statistical studies of line widths in O, B, A, and F stars by Westgate (1933a, 1933b, 1934).

Some 15 years passed before the next flurry of interest in stellar rotation. Using the Shajn-Struve graphical method but including the effects of limb darkening, Slettebak (1949, 1954, 1955, 1956) and Slettebak and Howard (1955) investigated stellar rotation in some 700 stars across the HR diagram using mostly moderate dispersions. Huang (1953) measured line widths in the spectra of 1550 O-G stars, while Herbig and Spalding (1955) made visual estimates of line widths in 656 F0-K5 stars and compared them with line widths in standard stars to obtain v sin i's. A number of high-dispersion studies concentrated on evolved stars. Thus, Oke and Greenstein (1954) used the Shajn-Struve graphical method to study rotation in A, F, and G giant stars; Abt (1957, 1958) compared observed and computed line profiles in high-luminosity A-F stars; Rosendhal (1970) studied line profiles in B and A supergiants; and Danziger and Faber (1972) analyzed stellar rotation among A-F evolved stars. The latter study is interesting in that entire spectral regions, rather than individual lines, were mathematically broadened and compared with observed spectra to obtain rotational velocities. The aforementioned investigations of evolved stars appear to show that solid-body rotation applies to stars which have not expanded too greatly whereas some form of differential rotation applies for the larger, more luminous stars. Other studies of stellar rotation during the 1960's include those of Walker and Hodge (1966), Palmer et al. (1968), and Buscombe (1969). The above plus other references are listed in a review article by Slettebak (1970b).

Lists of v sin i's determined from line profile measurements or visual estimates of line broadening of early-type stars were published by a number of investigators during the past decade or so. Most of the following are high-resolution studies:

    Abt and Moyd (1973): late A-type dwarfs
    Balona (1975): southern O and B stars
    Buscombe and Stoeckley (1975): O, B, and A main-sequence stars
    Conti and Ebbets (1977): O-type stars
    Day and Warner (1975): sharp-lined B stars
    Dworetsky (1974): A0 stars
    Wolff and Preston (1978): late B-type stars
    Wolff et al. (1982): early B-type stars.

A system of 217 bright northern and southern standard stars of types O9-F9 for rotational velocity determinations was established by Slettebak et al. (1975), based on comparisons of theoretical rotationally-broadened profiles with observed profiles from photoelectric scans and coudé spectrograms.

Struve (1930) and Westgate (1934) had shown that ". . .appreciable rotation disappears in the middle F's". High-resolution spectra are therefore required to show measurable line broadening in the later-type stars. Kraft (1967a) was able to study rotation in solar-type stars by employing coudé spectrograms of 4.5-5 Å/mm, giving him an estimated limiting resolution in v sin i of about 6 km/sec. More recently, Soderblom (1982) compared calculated line profiles with observed, high-resolution, echelle spectrograms for solar-type stars. He found that v sin i's as low as 1.5 km/sec and differences in v sin i of < 0.3 km/sec can be discerned for stars rotating as slowly as the Sun. Additional studies of late-type stars using the same techniques were made by Vogt et al. (1983) of BY Dra stars and by Soderblom et al. (1983) of Pleiades K dwarfs.

The determination of rotational velocities of very rapidly rotating stars, as for example the Be stars, involves a number of additional factors. In extreme cases, the stars can no longer be regarded as spherical; gravity darkening is likely to play an important role; and the assumption (inherent in the Shajn-Struve graphical method) that the flux profile of a sharp-lined star can be used to approximate the non-rotating intensity profile at various places on the disk of a rotating star is no longer valid. A crude attempt to incorporate these effects into a line-profile analysis of rapidly-rotating B and Be stars was made by Slettebak (1949). Later papers, using more sophisticated methods, include those by Collins and Harrington (1966), Hardorp and Strittmatter (1968), Stoeckley (1968), Vilhu and Tuominen (1971), Hardorp and Scholz (1971), Hardorp and Strittmatter (1972), Norris and Scholz (1972), and Collins (1974). The latter investigation, in which rotationally-broadened line profiles for O9-F8 main-sequence stars were calculated using the ATLAS model-atmosphere program and taking shape distortion and gravity darkening into account, served as the basis for the aforementioned system of standard stars for rotational velocity determinations (Slettebak et al. 1975).

2.3.2. Fourier Analysis of Line Profiles. Carroll (1933) was the first to suggest the application of Fourier analysis to spectral line profiles for stellar rotational velocity determinations. Since the rotationally-broadened profile is obtained by a convolution, it is advantageous to take the Fourier transform of the observed profile, which is the product of the transforms of the non-rotating profile and of the rotational-broadening function. Comparison of the zeroes of the latter transform with the zeroes of the observed profile allows the determination of v sin i. Carroll and Ingram (1933) applied this method to line profiles published by Elvey (1929, 1930) in several O and B stars and found good agreement with his estimated v sin i's. Later, Colacevich (1937) and Wilson (1969) also applied Carroll's method, the latter to obtain an upper limit of 3.5 km/sec for the rotation velocity of Arcturus.

The introduction of photoelectric scanners to obtain accurate line profiles has made it possible to extend Carroll's method from a location of zeroes to a fitting of the entire transform, thereby adding weight to the measurement. This was first done by Gray (1973) and is the basis of a number of papers by Gray and by M. Smith, primarily on the rotation of G and K stars, plus individual studies of Vega, Arcturus, and Procyon. The method is described and discussed in review articles by Gray (1976), Smith and Gray (1976), Gray (1978), Smith (1979), and Gray (1980a). Other recent investigations of stellar rotation using Fourier transform techniques include those by de Jager and Neven (1982) on Procyon; Ebbets (1979) on O-type stars; and Vogel and Kuhi (1981) on pre-main-sequence stars. In addition to obtaining rotational velocities, many of the aforementioned authors have also measured turbulent velocities as a part of their Fourier analyses, as will be discussed in Section 2.3.4.

Variations of Fourier transform methods have been proposed by several authors. Thus, Deeming (1977) uses a Bessel transform technique in which the Fourier transform is multiplied by a Bessel weighting function and integrated over Fourier frequency, while Milliard et al. (1977) employ the "L_2 norms" of the rotation profile to derive v sin i's for Sirius and Vega.

2.3.3. <u>Other Methods of Line Profile Analysis</u>. Using a PEPSIOS interferometer, Kurucz et al. (1977) observed the profile of an intrinsically narrow Ba II line at 6496.9 Å in the spectrum of Sirius interferometrically and obtained a projected rotational velocity v sin i of 16 km/sec.

Following a method due to Griffin (1967) for measuring radial velocities, Benz and Mayor (1981) use the cross-correlation between the spectrum of a cool star and an appropriate mask located in the focal plane of the spectrograph to obtain stellar rotational velocities. Whereas the position of the correlation dip depends on the radial velocity of the star, the width of the dip depends on the width of the absorption lines chosen for the correlation, and can be used for v sin i measurements. The authors find good agreement between their rotational velocities obtained with the CORAVEL spectrometer and values obtained using Fourier transform techniques.

2.3.4. <u>Rotation Versus Other Line-Broadening Agents</u>. All spectrum lines suffer thermal Doppler and collisional broadening to various degrees, as well as microturbulent broadening and Zeeman broadening in certain types of stars. Fortunately most of these are either small in normal stars or can be avoided (e.g., Stark-broadened Balmer lines are not suitable for rotational velocity determinations).

One broadening agent which competes significantly with rotation in broadening spectrum lines, particularly in early-type stars of high luminosity, however, is macroturbulence (defined as turbulence with elements as large as the thickness of the effective photospheric layers, thereby causing no change in line strengths). Huang and Struve (1960) summarized the earlier work on the recognition and separation of macroturbulence from rotation. Statistical studies (Huang and Struve 1954; Slettebak 1956) had suggested the existence of macroturbulence in early-type stars but it proved to be impossible to distinguish unambiguously from line profile analysis of a single star between rotation and various types of macroturbulence (Huang and Struve 1953; Slettebak 1956; Abt 1958; Rosendhal 1970).

Interest in this problem was revived in the mid-1970's by the application of Fourier analysis to line profiles by Gray, Smith and others. These investigators found (cf. the review paper by Smith and Gray [1976]) that high-resolution, low-noise data make it possible to distinguish, at high frequencies, between the Fourier transform profiles arising from rotation versus those of macroturbulence. They

also find evidence for a model of macroturbulence which involves only
radial and tangential streams, in contrast to the earlier isotropic
Gaussian models. A recent paper by Gray (1984a) points out that Zeeman
broadening may also be a significant component of the line broadening
in late-type dwarf stars, and comments on the separation of rotation,
macroturbulence, and Zeeman effect from line profiles. Other recent
attempts to separate macroturbulence from rotation using Fourier
analysis include those by de Jager and Neven (1982), who studied
Procyon; Ebbets' (1979) study of O-type stars, and Soderblom's (1982)
work on solar-type stars.

2.3.5. The Separation of v from i. The determination of stellar
rotational velocities from line profile analysis generally yields only
the projected (along the line of sight) velocity, v sin i. Information
about the true equatorial rotational velocity can be deduced for some
special cases, however. Thus, if a large sample of stars of a given
spectral type is studied, for example (cf. Slettebak 1966), those stars
with the largest observed rotational velocities are presumably being
viewed essentially equatorially (sin i = 1) and the observed rotational
velocity v sin i must be very close to the equatorial velocity v. The
Be stars represent another interesting example. Struve's (1931)
rotational model for Be stars suggests that all Be stars of a given
type rotate at a very rapid (near critical) rate and that the observed
line broadening for a given star depends only upon its inclination
angle i. There is statistical evidence for this view (cf. Struve 1951;
Slettebak 1976, 1982) -- if the model is adopted, the equatorial
velocity for the sample plus inclination angles for individual stars
can, in principle, be obtained.

   Many investigators have given evidence for a random orientation in
space of the axes of rotation in rotating stars. Under these circumstances it is possible, using statistical methods, to obtain a
relationship between the mean observed v sin i and the mean true v for
a sample of stars (cf. Chandrasekhar and Münch 1950).

   The problem of separating v from i for an individual star remains,
however. As long as the star is spherically symmetrical, only v sin i
can be determined. Very rapid rotation may result in deviations from
spherical symmetry, however, as we have seen, and also in gravity darkening and variations in the line profile across the disk of the star.
The integrated line profile will then depend upon the inclination and,
in principle, v and i can be determined separately. This was first
attempted by Stoeckley (1968), who computed rotationally-broadened
profiles of several lines and compared them with observed profiles in
five rapidly- rotating B and A stars. Later, Hutchings (1976),
Hutchings and Stoeckley (1977), and Hutchings et al. (1979) made use of
the observed difference in ultraviolet versus visual line widths for
rapidly-rotating stars to attempt the separation of i and v for a
number of O and B stars. Hutchings and his collaborators assumed
spherical stars in their calculations, however, and did not include the
variation of the line profile with temperature and gravity. In a more
detailed treatment, Sonneborn and Collins (1977) computed
rotationally-distorted, gravity-darkened models which include the

latitude dependence of T_e and g in their line profiles. Their results were qualitatively similar to those of Hutchings but they predict a smaller ultraviolet-to-visual line width variation. More recently, Ruusalepp (1982) attempted to separate v and i from a study of He I 4471 and Mg II 4481 line widths in a number of B-type stars. Carpenter et al. (1984) have also investigated rotational velocities (for later B-type and A-type stars) as determined from ultraviolet versus visual line profiles, using IUE spectra. They enumerate the various errors and uncertainties which are inherent in such an analysis and suggest that ". . . attempts to separate v from i from line profiles must be carried out with great caution and, indeed, it is not clear to us that such a separation is even feasible."

An interesting special case occurs where v is obtained by means of rotational modulation and v sin i from spectroscopic measurements. This would permit a determination of i and, for a sufficiently large sample, a check on the distribution of i.

3. DIFFERENTIAL ROTATION

No direct information regarding differential rotation (that is, surface rotation as a function of latitude) exists for stars other than the Sun. Slettebak (1949) computed the effects of a solar-type differential rotation on the He I 4026 line profile of a rapidly-rotating model star, using the Shajn-Struve graphical method. The computed profile was deeper and narrower than that for the rigid-body model, but the differences so slight that no meaningful comparison with observed line profiles was possible. Huang (1961) derived a general formula to describe the geometrical broadening of a line profile by differential rotation, but an analytical solution exists only under certain physical conditions. Later, Stoeckley (1968) did attempt to detect differential rotation in five rapidly-rotating B and A main-sequence stars using line profile analysis, and suggested that his objects appeared to be either solid-body rotators or differential rotators in the sense opposite to the solar case. Again, there is a question whether observed line profiles are sufficiently accurate to support such conclusions.

The possibility also exists of detecting differential rotation by means of light variations of rotating stars. Thus, Hall (1972) suggested that a region of tremendous starspot activity darkens one side of the cool component of the RS CVn system and that differential rotation, like that observed in the Sun, then produces the migration of the wave-like distortion in the light curve. Vogt (1975) found period changes in the BY Dra light curves, indicating the presence of differential rotation. Gray (1977, 1982) searched for differential rotation in line profiles of both A-type and F-type stars using Fourier transform techniques, with negative results. Bruning (1981) suggested a numerical model based on Fourier analysis of line profiles to detect differential rotation in late-type stars. He points out that while modulation studies may also be used to determine the amount of stellar differential rotation, the Fourier method retains a certain advantage over K-line index studies since not only the amount but also the sense of differential rotation may be determined. Hallam and Wolff (1981) observed periodic variations in the ultraviolet chromospheric fluxes of H I, Si II, and Mg II in six main-sequence F, G, and K stars, but with period and time dependence of the modulated spectral flux not identical from one ionic species to another in the same star. They attribute this to differential stellar rotation. LaBonte (1982, 1984), however, tried to detect solar differential rotation using data sets analogous to stellar observations (magnetic flux and 2.8 GHz flux) and was not able to do so. The problem of detecting stellar differential rotation is obviously a difficult one.

4. ACCURACY OF ROTATIONAL VELOCITY DETERMINATIONS

4.1. Rotational Modulation

This method of determining stellar rotational velocities has the advantage of being independent of the inclination angle i. It is also very sensitive, allowing rotation rates as slow as 1 km/sec to be measured (Baliunas et al. 1983). On the other hand, the radius of the star must be known in addition to the period of modulation in order to calculate the equatorial rotational velocity. For most stars, the radius must be calculated from the black-body assumption with the effective temperature estimated from the spectral type and the luminosity corrected for interstellar reddening. Including uncertainties in these quantities plus errors in the periods of their light curves, Rydgren and Vrba (1983), for example, estimate errors of $\pm$ 16 percent of their derived rotational velocities.

4.2. Eclipsing Binary Radial Velocity Curve Distortions

As stated previously, the determination of equatorial rotational velocities using this method depends upon the "rotation factors" for the time of observation; i.e., how much of the eclipsed star is showing and what is the distribution of light across the disk. These, in turn, will depend upon the elements of the light curve and the assumed limb darkening, among other things. McLaughlin (1933) found v = 200 km/sec for α CrB in this way, but admitted that this value is too large relative to the width of the K-line for that star (Uesugi and Fukuda [1982] list a rotational velocity of 135 km/sec for α CrB, derived from line profile analysis). On the other hand, Kopal (1942), who included limb-darkening in his analysis, claimed a mean error of $\pm$ 2 km/sec for his equatorial rotational velocity of Algol of 48 km/sec (Uesugi and Fukuda [1982] give 55 km/sec from line profiles). It would seem that definitive analyses of rotational distortions in eclipsing binary radial velocity curves to obtain rotational velocities should be carried out, in the manner of Struve and Elvey (1931), using line profiles rather than visual radial velocity measures. The computed line profiles should be based on models which include the effects of limb and gravity darkening, possible shape distortion, and profile changes across the stellar disk. Comparisons with observed line profiles would then yield rotational velocities. Only systems with good photometric solutions should be used and those with known gaseous streams avoided, since the latter may affect the line profiles (Hardie 1950).

4.3. Line Profile Analysis

As was stated very clearly by Abt (1962), the observational precision required and the importance of competing line-broadening agents in addition to stellar rotation varies across the HR diagram. Thus, in A-type main-sequence stars, where the mean v sin i is 125-150 km/sec, the rotational line broadening is likely to dominate other mechanisms such as thermal, microturbulent, or Zeeman broadening. In supergiants, on the other hand, rotational broadening may be comparable to microturbulent and/or macroturbulent broadening.

Careful determinations of v sin i from line widths or line profiles, using moderate-to-high dispersions, seem to agree from one investigator to the next to within about 10 percent. An exception is the determination of the largest rotational velocities (v sin i > 200 km/sec), where the uncertainties may be as large as 15-20 percent of the estimated value (Slettebak et al. 1975). This is due to a number of ambiguities which are introduced when the star rotates close to its critical velocity for which the centrifugal force at the star's equator balances the gravitational force.

A very rough rule for the limit of detection has been stated by a number of investigators (e.g., Treanor 1960; Kraft 1965): the minimum rotational velocity that can be resolved in km/sec is numerically comparable to the dispersion employed in Å/mm.

With regard to visual estimates of line broadening versus measurements from tracings, the opinion stated by Kraft (1967b) seems a valid one: ". . . no real increase in accuracy is obtained from tracings, provided that the visual estimates are based on a sufficiently large number of standards well distributed in spectral type. . . The reason is clear: on a tracing one considers one or two lines only which may be subject to photometric irregularities, whereas visual inspection allows the examination of a number of spectral features at once." This general point-of-view is the basis for the establishment of a system of standard rotational velocity stars by Slettebak et al. (1975).

The internal errors quoted for rotational velocities determined using Fourier transform methods are quite low. Thus, Gray (1980b, 1981a, 1981b) finds mean errors of 0.3-0.4 km/sec for his v sin i's for Vega, Arcturus, and Procyon. Comparison with other determinations show significant differences, however. Gray (1980b), for example, obtains v sin i = 23.4 km/sec for Vega while Milliard et al. (1977) find 18 km/sec. For Procyon, Gray (1981b) lists 2.8 km/sec and De Jager and Neven (1982) 10.0 km/sec. Another example is Sirius: using Fourier transform methods, Smith (1976) finds v sin i = 17 km/sec and Milliard et al. (1977) obtain 11 km/sec, while Kurucz et al. (1977) using an interferometer, find 16 km/sec. The latter write: ". . . we feel that Fourier transform methods should be used with caution unless damping constants, blending, and the continuum level are well determined." In an application of Fourier analysis to A-type stars rotating between 100 and 300 km/sec, Gray (1980c) estimates the internal errors for eleven of his stars to be ". . . no worse than $\pm$ 5 km/sec." A comparison with Slettebak et al. (1975) shows that nearly all of his v sin i's fall within $\pm$ 10 percent deviation lines.

According to Milliard et al. (1977): ". . . (Fourier transform methods) make use of the whole set of frequency points available in the observed profile, from core to wings; in this sense they are more adequate to obtain v sin i from high resolution and high signal-over-noise data, than methods based on the measure of simply one parameter in the line (usually the half-width)." As Moss and Smith (1981) point out, however, "The main difficulty with this procedure (Fourier analysis) in its original form is that it assumes that there is a unique 'non-rotating profile' that represents the local profile equally at all points on the stellar disk. In practice, the intrinsic profile is likely to vary over the disk, and the de-convolution procedure will yield an unbroadened profile which does not correspond to any real line profile produced by the star. . . . What is required and is now being done is to calculate the transforms by integrations over the disk." Gray (1984b) points out, however, that ". . . in practice, the thermal-microturbulence profile is sufficiently small compared to the rotation-macroturbulence portions that a very approximate calculation is adequate." He suggests furthermore that while disk integrations to combine the effects of rotation and macroturbulence are feasible, the thermal-microturbulence part cannot be done until the small center-to-limb variations seen for the Sun are understood.

5. CATALOGUES AND SELECTED LISTS OF ROTATIONAL VELOCITIES

The first general catalogue of rotational velocities was published by Boyarchuk and Kopylov (1964) and included 2558 stars. A few years later, Uesugi and Fukuda (1970) presented v sin i's for 3951 stars, as did Bernacca and Perinotto (1970, 1971) and Bernacca (1973) for a total of 3074 stars. The most recent list is the revised catalogue of Uesugi and Fukuda (1982) which reviews 11,460 v sin i determinations from 102 sources to give rotational velocities for 6472 stars. All of the aforementioned catalogues have extensive lists of references to the individual papers.

In closing, it may be useful to reference a few lists of rotational velocities of objects which have not been explicitly mentioned earlier in this review. In a number of cases, no up-to-date review papers exist and only the most recent review is cited.

> Open clusters: Abt (1970). Individual papers by Abt, Levato, and others since 1970 may be found in the literature.
> Binary stars: Slettebak (1963); Van den Heuvel (1970); Weis (1974); Levato (1975).
> Ap stars: Preston (1970); Abt et al. (1972).
> Be stars: Slettebak (1982).
> White dwarfs: Greenstein et al. (1977).
> Horizontal-branch stars: Peterson et al. (1983); Peterson (1983).

ACKNOWLEDGEMENTS

I am indebted to a number of persons for comments and suggestions in writing this review: Helmut Abt's discussion of the Rossiter effect in an unpublished (1962) review was very helpful; J. L. Tassoul called my attention to the early ideas on rotational modulation; and George Collins, Dave Gray, and Robert Smith all contributed valuable comments. Finally, I would like to acknowledge my debt to Otto Struve, whose encouragement many years ago was of enormous importance to someone just starting in the field.

REFERENCES

Abney, W. de W.   1877, Monthly Notices Roy. Astron. Soc. 37, p. 278.

Abt, H. A.   1957, Astrophys. J. 126, p. 503.

Abt, H. A.   1958, Astrophys. J. 127, p. 658.

Abt, H. A.   1962, Stellar Rotation (review), unpublished.

Abt, H. A.   1970, in IAU Colloq. No. 4, Stellar Rotation ed. A. Slettebak, (Dordrecht: Reidel), p. 193.

Abt, H. A. and Moyd, K. I.   1973, Astrophys. J. 182, p. 809.

Abt, H. A., Chaffee, F. H., and Suffolk, G.   1972, Astrophys. J. 175, p. 779.

Adams, W. S. and Joy, A. H.   1919, Astrophys. J. 49, p. 186.

Babcock, H. W.   1949, Observatory 69, p. 191.

Baliunas, S. L., Vaughan, A. H., Hartmann, L., Middelkoop, F., Mihalas, D., Noyes, R. W., Preston, G. W., Frazer, J., and Lanning, H.   1983, Astrophys. J. 275, p. 752.

Balona, L. A.   1975, Mem. Roy. Astron. Soc. 78, p. 51.

Benz, W. and Mayor, M.   1981, Astron. Astrophys. 93, p. 235.

Bernacca, P. L.   1973, Contrib. Oss. Astrofis. Asiago, Univ. Padova, No. 294.

Bernacca, P. L. and Perinotto, M.   1970, Contrib. Oss. Astrofis. Asiago, Univ. Padova, No. 239.

Bernacca, P. L. and Perinotto, M.   1971, Contrib. Oss. Astrofis. Asiago, Univ. Padova, No. 249.

Blanco, C., Catalano, S., and Marilli, E.   1979, Nature 280, p. 661.

Boesgaard, A. M. and Simon, T.   1982, in Second Cambridge Workshop on Cool Stars, Stellar Systems, and the Sun, Smithsonian Astrophys. Obs., Special Report 392, Vol. II, p. 161.

Bopp, B. W. and Evans, D. S.   1973, Monthly Notices Roy. Astron. Soc. 164, p. 343.

Bopp, B. W. and Stencel, R. E.   1981, Astrophys. J. 247, p. L 131.

Boyarchuk, A. A. and Kopylov, I. M.   1964, Publ. Crimean Astrophys. Obs. 31, p. 44.

Brunet, P.   1931, L'introduction des theories de Newton en France au XVIII siecle (Geneva: Slatkine Reprints, 1970).

Bruning, D. H.   1981, Astrophys. J. 248, p. 274.

Buscombe, W.   1969, Monthly Notices Roy. Astron. Soc. 144, p. 1.

Buscombe, W. and Stoeckley, T. R.   1975, Astrophys. Space Sci. 37, p. 197.

Cameron, R. C. (ed.): 1967   The Magnetic and Related Stars
    (Baltimore: Mono Book Corp).
Carpenter, K. G., Slettebak, A., and Sonneborn, G.   1984, Astrophys.
    J. 286, p. 741.
Carroll, J. A.   1933, Monthly Notices Roy. Astron. Soc. 93, p. 478.
Carroll, J. A. and Ingram, L. J.   1933, Monthly Notices Roy. Astron.
    Soc. 93, p. 508.
Catalano, F. A. and Renson, P.   1984, Astron. Astrophys. Suppl. 55,
    p. 371.
Chandrasekhar, S. and Münch, G.   1950, Astrophys. J. 111, p. 142.
Chugainov, P. F.   1966, IAU Comm. 27 Information Bull. Var. Stars
    No. 122.
Colacevich, A.   1937, Mem. Astron. Soc. Italy 10, p. 163.
Collins, G. W., II,   1974, Astrophys. J. 191, p. 157.
Collins, G. W., II and Harrington, J. P.   1966, Astrophys. J. 146,
    p. 152.
Conti, P. S. and Ebbets, D.   1977, Astrophys. J. 213, p. 438.
Danziger, I. J. and Faber, S. M.   1972, Astron. Astrophys. 18, p. 428.
Day, R. W. and Warner, B.,   1975, Monthly Notices Roy. Astron. Soc.
    173, p. 419.
Deeming, T. J.   1977, Astrophys. Space Sci. 46, p. 13.
De Jager, C. and Neven, L.   1982, Astrophys. Space Sci. 84, p. 297.
Deutsch, A. J.   1952, IAU Trans. 8, p. 801.
Deutsch, A. J.   1956, Publ. Astron. Soc. Pacific 68, p. 92.
Dorren, J. D., Guinan, E. F., and McCook, G. P.   1984, Publ. Astron.
    Soc. Pacific 96, p. 250.
Drake, S.   1957, Discoveries and Opinions of Galileo (Garden
    City: Doubleday Anchor), p. 102.
Dworetsky, M. M.   1974, Astrophys. J. Suppl. 28, p. 101.
Ebbets, D.   1979, Astrophys. J. 227, p. 510.
Elvey, C. T.   1929, Astrophys. J. 70, p. 141.
Elvey, C. T.   1930, Astrophys. J. 71, p. 221.
Fekel, F.   1983, Astrophys. J. 268, p. 274.
Forbes, G.   1911, Monthly Notices Roy. Astron. Soc. 71, p. 578.
Gershberg, R. E.   1970, in IAU Colloq. No. 4,   Stellar Rotation , ed.
    A. Slettebak (Dordrecht: D. Reidel), p. 249.
Gray, D. F.   1973, Astrophys. J. 184, p. 461.
Gray, D. F.   1976, The Observation and Analysis of Stellar Photo-
    spheres (New York: Wiley), Chaps. 2 and 17.
Gray, D. F.   1977, Astrophys. J. 211, p. 198.
Gray, D. F.   1978, in High Resolution Spectroscopy , Proc. 4th
    Trieste Colloq. on Astrophysics, ed. M. Hack (Trieste: Trieste
    Astrophysical Obs.), p. 268.
Gray, D. F.   1980a, in IAU Colloq. 51, Stellar Turbulence , ed. D. F.
    Gray and J. L. Linsky (Heidelberg: Springer-Verlag), p. 75.
Gray, D. F.   1980b, Publ. Astron. Soc. Pacific 92, p. 154.
Gray, D. F.   1980c, Publ. Astron. Soc. Pacific 92, p. 771.
Gray, D. F.   1981a, Astrophys. J. 245, p. 992.
Gray, D. F.   1981b, Astrophys. J. 251, p. 152.
Gray, D. F.   1982, Astrophys. J. 258, p. 201.
Gray, D. F.   1984a, Astrophys. J. 281, p. 719.

Gray, D. F.   1984b, private communication.
Greenstein, J. L., Boksenberg, A., Carswell, R., and Shortridge,
    K.   1977, Astrophys. J. 212, p. 186.
Griffin, R. F.   1967, Astrophys. J. 148, p. 465.
Guinan, E. F., McCook, G. P., Fragola, J. L., O'Donnell, W. C.,
    Tomczyk, S., and Weisenberger, A. G.   1982, Astron. J. 87, p. 893.
Hall, D. S.   1972, Publ. Astron. Soc. Pacific 84, p. 323.
Hall, D. S.   1976, in IAU Colloq. 29, Multiple Periodic Variable
    Stars , ed. W. S. Fitch (Dordrecht: D. Reidel), p. 287.
Hallam, K. L. and Wolff, C. L.   1981, Astrophys. J. 248, L 73.
Hardie, R. H.   1950, Astrophys. J. 112, p. 542.
Hardorp, J. and Scholz, M.   1971, Astron. Astrophys. 13, p. 353.
Hardorp, J. and Strittmatter, P. A.   1968, Astrophys. J. 153, p. 465.
Hardorp, J. and Strittmatter, P. A.   1972, Astron. Astrophys. 17,
    p. 161.
Hellerich, J.   1922, Astron. Nach. 216, p. 277.
Herbig, G. H. and Spalding, J. F., Jr.   1955, Astrophys. J. 121,
    p. 118.
Holt, J. R.   1893, Astron. and Astro-Physics 12, p. 646.
Holtzman, J. A. and Nations, H. L.   1984, Astron. J. 89, p. 391.
Huang, S. -S.   1953, Astrophys. J. 118, p. 285.
Huang, S. -S.   1961, Astrophys. J. 133, p. 130.
Huang, S. -S. and Struve, O.   1953, Astrophys. J. 118, p. 463.
Huang, S. -S. and Struve, O.   1954, Ann. d'Ap. 17, p. 85.
Huang, S. -S. and Struve, O.   1960, in Stellar Atmospheres , ed.
    J. L. Greenstein (Chicago: University of Chicago Press), p. 321.
Hutchings, J. B.   1970, in IAU Colloq. No. 4, Stellar Rotation ,
    ed. A. Slettebak (Dordrecht: D. Reidel), p. 283.
Hutchings, J. B.   1976, Publ. Astron. Soc. Pacific 88, p. 5.
Hutchings, J. B. and Stoeckley, T. R.   1977, Publ. Astron. Soc.
    Pacific 89, p. 19.
Hutchings, J. B., Nemec, J. M., and Cassidy, J.   1979, Publ. Astron.
    Soc. Pacific 91, p. 313.
Kopal, Z.: 1942, Astrophys. J. 96, p. 399.
Kraft, R. P.   1965, Astrophys. J. 142, p. 681.
Kraft, R. P.   1967a, Astrophys. J. 150, p. 551.
Kraft, R. P.   1967b, Astrophys. J. 148, p. 129.
Kraft, R. P.   1969, in Stellar Astronomy , ed. H. -Y. Chiu, R. L.
    Warasila, and J. L. Remo (New York: Gordon and Breach), Vol. 1,
    p. 317.
Kraft, R. P.   1970, in Spectroscopic Astrophysics , ed. G. H. Herbig
    (Berkeley: University of California Press), p. 385.
Kron, G. E.   1947, Publ. Astron. Soc. Pacific 59, p. 261.
Kron, G. E.   1952, Astrophys. J. 115, p. 301.
Krzeminski, W.   1969, in Symposium on Low Luminosity Stars , ed.
    S. S. Kumar (New York: Gordon and Breach), p. 57.
Krzeminski, W. and Kraft, R. P.   1967, Astron. J. 72, p. 307.
Kurucz, R. L., Traub, W. A., Carleton, N. P., and Lester, J. B.   1977,
    Astrophys. J. 217, p. 771.
LaBonte, B. J.   1982, Astrophys. J. 260, p. 647.
LaBonte, B. J.   1984, Astrophys. J. 276, p. 335.

Lambert, D. L. and O'Brien, G. T.   1983, Astron. Astrophys. 128,
     p. 110.
Levato, H.   1975, Astron. Astrophys. Suppl. 19, p. 91.
Liège Astrophys. Colloq. No. 23: 1981, Upper Main Sequence CP Stars
     (Liège: Univ. of Liège Astrophys. Inst.).
Maury, A. C.   1920, Harvard Ann. 84, p. 157.
McLaughlin, D. B.   1924, Astrophys. J. 60, p. 22.
McLaughlin, D. B.   1926, Pop. Astron. 34, p. 624.
McLaughlin, D. B.   1933, Publ. Obs. Univ. Michigan 5, p. 91.
McLaughlin, D. B.   1934, Publ. Obs. Univ. Michigan 6, p. 3.
Milliard, B., Pitois, M. L., and Praderie, F.   1977, Astron.
     Astrophys. 54, p. 869.
Moss, D. and Smith, R. C.   1981, Rep. Prog. Phys. 44, p. 831.
Norris, J. and Scholz, M.   1972, Astron. Astrophys. 17, p. 182.
Oke, J. B. and Greenstein, J. L.   1954, Astrophys. J. 120, p. 384.
Olson, E. C.   1968, Publ. Astron. Soc. Pacific 80, p. 185.
Palmer, D. R., Walker, E. N., Jones, D. H. P., and Wallis,
     R. E. 1968, Roy. Obs. Bull. No. 135, p. E 385.
Peterson, R. C.   1983, Astrophys. J. 275, p. 737.
Peterson, R. C., Tarbell, T. D., and Carney, B. W.   1983,
     Astrophys. J. 265, p. 972.
Pettersen, B. R.   1980, Astron. J. 85, p. 871.
Pickering, E. C.   1881, Proc. American Acad. Arts Sci. 16, p. 257.
Plaskett, J. S.   1926, Publ. Dominion Astrophys. Obs. 3, p. 247.
Preston, G. W.   1970, in IAU Colloq. No. 4, Stellar Rotation
     ed. A. Slettebak, (Dordrecht: Reidel), p. 254.
Radick, R. R., Lockwood, G. W., Thompson, D. T., Warnock, A. III,
     Hartmann, L. W., Mihalas, D., Worden, S. P., Henry, G. W., and
     Sherlin, J. M.   1983, Publ. Astron. Soc. Pacific 95, p. 621.
Ramsey, L. W. and Nations, H. L.   1980, Astrophys. J. 239, L 121.
Rosendhal, J. D.   1970, Astrophys. J. 159, p. 107.
Rossiter, R. A.   1924, Astrophys. J. 60, p. 15.
Ruusalepp, M.   1982, in IAU Symp. 98, Be Stars , ed. M. Jaschek and
     H. -G. Groth (Dordrecht: D. Reidel), p. 303.
Rydgren, A. E. and Vrba, F. J.   1983, Astrophys. J. 267, p. 191.
Schlesinger, F.   1909, Publ. Allegheny Obs. 1, p. 123.
Schlesinger, F.   1911, Monthly Notices Roy. Astron. Soc. 71, p. 719.
Schlesinger, F.   1913, Publ. Allegheny Obs. 3, p. 23.
Shajn, G. and Struve, O.   1929, Monthly Notices Roy. Astron.
     Soc. 89, p. 222.
Shapley, H. and Nicholson, S. B.   1919, Comm. Nat. Acad. Sci.
     Mt. Wilson Obs. 2, p. 65.
Slettebak, A.   1949, Astrophys. J. 110, p. 498.
Slettebak, A.   1954, Astrophys. J. 119, p. 146.
Slettebak, A.   1955, Astrophys. J. 121, p. 653.
Slettebak, A.   1956, Astrophys. J. 124, p. 173.
Slettebak, A.   1963, Astrophys. J. 138, p. 118.
Slettebak, A.   1966, Astrophys. J. 145, p. 126.
Slettebak, A. (ed.)   1970a, IAU Colloq. No. 4, Stellar Rotation
     (Dordrecht: D. Reidel).

Slettebak, A.   1970b, in IAU Colloq. No. 4, Stellar Rotation ,
     ed. A. Slettebak (Dordrecht: D. Reidel), p. 3.
Slettebak, A.   1976, in IAU Symp. No. 70, Be and Shell Stars ,
     ed. A. Slettebak (Dordrecht: D. Reidel), p. 123.
Slettebak, A.   1982, Astrophys. J. Suppl. 50, p. 55.
Slettebak, A. and Howard, R. F.   1955, Astrophys. J. 121, p. 102.
Slettebak, A. and Snow, T. P., Jr.   1978, Astrophys. J. 224, L 127.
Slettebak, A., Collins, G. W., II, Boyce, P. B., White, N. M., and
     Parkinson, T. D.   1975, Astrophys. J. Suppl. 29, p. 137.
Smith, M. A.   1976, Astrophys. J. 203, p. 603.
Smith, M. A.   1979, Publ. Astron. Soc. Pacific 91, p. 737.
Smith, M. A. and Gray, D. F.   1976, Publ. Astron. Soc. Pacific 88,
     p. 809.
Soderblom, D. R.   1982, Astrophys. J. 263, p. 239.
Soderblom, D. R., Jones, B. F., and Walker, M. F.   1983,
     Astrophys. J. 274, L 37.
Sonneborn, G. H. and Collins, G. W., II.   1977, Astrophys. J. 213,
     p. 787.
Stimets, R. W. and Giles, R. H.   1980, Astrophys. J. 242, L 37.
Stoeckley, T. R.   1968, Monthly Notices Roy. Astron. Soc. 140, p. 121.
Struve, O.   1930, Astrophys. J. 72, p. 1.
Struve, O.   1931, Astrophys. J. 73, p. 94.
Struve, O.   1944, Astrophys. J. 99, p. 222.
Struve, O.   1945, Pop. Astron. 53, p. 201 and 259.
Struve, O.   1950, Stellar Evolution (Princeton: Princeton Univ.
     Press), p. 125.
Struve, O.   1951, in Astrophysics , ed. J. A. Hynek (New York:
     McGraw-Hill), p. 89.
Struve, O. and Elvey, C. T.   1931, Monthly Notices Roy. Astron.
     Soc. 91, p. 663.
Tassoul, J. -L.   1978, Theory of Rotating Stars (Princeton:
     Princeton University Press).
Treanor, P. J.   1960, Monthly Notices Roy. Astron. Soc. 121, p. 503.
Uesugi, A. and Fukuda, I.   1970, Mem. Fac. Science Kyoto Univ. Ser.
     Phys. Astrophys. Geophys. Chem. 33, p. 205.
Uesugi, A. and Fukuda, I.   1982, Revised Catalogue of Stellar
     Rotational Velocities , Dept. of Astron., Kyoto Univ. Also,
     Centre de Données Stellaires, Strasbourg, France.
Van den Heuvel, E. P. J.   1970, in IAU Colloq. No. 4, Stellar
     Rotation, ed. A. Slettebak, (Dordrecht: Reidel), p. 178.
Van Leeuwen, F. and Alphenaar, P.   1983, in Activity in Red-Dwarf
     Stars , ed. P. B. Byrne and M. Rodono (Dordrecht: D. Reidel),
     p. 189.
Vaughan, A. H., Baliunas, S. L., Middelkoop, F., Hartmann, L. W.,
     Mihalas, D., Noyes, R. W., and Preston, G. W.   1981, Astrophy. J.
     250, p. 276.
Vilhu, O. and Tuominen, I. V.   1971, Astron. Astrophys. 13, p. 136.
Vogel, S. N. and Kuhi, L. V.   1981, Astrophys. J. 245, p. 960.
Vogt, S. S.   1975, Astrophys. J. 199, p. 418.
Vogt, S. S.   1981, Astrophys. J. 247, p. 975.

Vogt, S. S.   1983, in Activity in Red-Dwarf Stars , ed. P. B. Byrne
    and M. Rodono (Dordrecht: D. Reidel), p. 137.
Vogt, S. S. and Penrod, G. D.   1983, Astrophys. J. 275, p. 661.
Vogt, S. S., Soderblom, D. R., and Penrod, G. D.   1983, Astrophys. J.
    269, p. 250.
Walker, G. A. H. and Hodge, S. M.   1966, Publ. Dominion Astrophys.
    Obs. 12, p. 401.
Walker, G. A. H., Yang, S., and Fahlman, G. G.   1979, Astrophys. J.
    233, p. 199.
Walker, G. A. H., Yang, S., and Fahlman, G. G.   1981, in Proc.
    Workshop on Pulsating B Stars, Nice Obs., ed. M. Auvergne et al.
    p. 261.
Weis, E. W.   1974, Astrophys. J. 190, p. 331.
Weiss, W. W., Jenkner, H., and Wood, H. J. (eds.)   1976, IAU Colloq.
    No. 32, Physics of Ap Stars (Vienna: Univ of Vienna Obs.).
Westgate, C.   1933a, Astrophys. J. 77, p. 141.
Westgate, C.   1933b, Astrophys. J. 78, p. 46.
Westgate, C.   1934, Astrophys. J. 79, p. 357.
Wilson, A.   1969, Monthly Notices Roy. Astron. Soc. 144, p. 325.
Wilson, O. C.   1978, Astrophys. J. 226, p. 379.
Wilson, R. E. and Twigg, L. W.   1980, in IAU Symp. No. 88, Close
    Binary Stars , ed. M. J. Plavec, D. M. Popper, and R. K. Ulrich
    (Dordrecht: D. Reidel), p. 263.
Wolff, S. C. and Preston, G. W.   1978, Astrophys. J. Suppl. 37,
    p. 371.
Wolff, S. C., Edwards, S., and Preston, G. W.   1982, Astrophys. J.
    252, p. 322.

DISCUSSION

POLIDAN: I would like to comment on a program we are pursuing with the
Voyager UV spectrometers.   G. J. Peters, R. Stalio and I are comparing
the far-UV (912-1700 Å) flux distributions of a sample of Be stars with
sharp-lined stars of similar (ground based) spectral types.   Our
preliminary conclusions are that there is no significant difference
between the FUV flux distributions of the rapidly rotating stars and the
sharp-lined stars.   This would imply that these Be stars are not
rotating at velocities high enough to exhibit gravity darkening
effects.

PETERS: Non-radial pulsation may prove to be much more of a problem in
the determination of rotational velocities from line profiles than we
have recognized thus far.   An increasing number of early B stars are
being shown to be such non-radial pulsators and the effect on the line
profiles is certainly a function of the pulsational phase (i.e., the
deduced vsini may be dependent on the time of observation).
Furthermore, there is some evidence that an individual star can abruptly
change its pulsational modes. As an example of variable absorption
lines in a non-radial pulsator, consider recent IUE observations of the
active B2 IVe star Mu Cen (Peters 1984 PASP, submitted).   The

"photospheric" absorption lines were, on the average, considerably broader when H alpha emission was present than during the star's quiescence. Therefore, the deduced vsini from the "photospheric" lines in this "standard" star contained in the reference list of Slettebak et al. (1975) will indeed vary. I suggest that in view of the implied importance of non-radial pulsations that rotational velocity standards be checked for constancy regularly over an extended period of time.

WALKER: I would like to support Dr. Peters' remarks about the effect of non-radial pulsations and the variations in line width. In our study of Spica, published in 1982, we showed that the formal measure of rotational velocity based on line profiles was almost double the old value measured by Struve and others when the star was known to be a Beta Ceph-type variable. Now it has apparently gone into a higher mode and no photometric variability has been detected for over a decade.

Although the interpretation of the "blips" moving through the line profiles is model dependent, there is a consistent acceleration associated with them which must be associated with the rotational velocity and I feel confident that ultimately this will be a useful technique for the determination of rotational velocity.

RAKOS: We have used Fourier methods introduced by Gray on high resolution ultraviolet spectra from IUE. The agreement between our results and the published values was very good.

PARTHASARATHY: How significant is the effect of gravity darkening?

SLETTEBAK: Gravity darkening effects on colors and spectrum lines in rotating stars become important only for values of the fractional angular velocity larger than about 0.9, according to the work of Collins.

LACY: I would like to point out that when the "dip" profile of a radial velocity spectrometer such as CORAVEL is used to estimate the rotational velocity of a double-lined binary, careful attention must be paid to the additional broadening produced by the blending of some of the oppositely displaced features of each component's spectrum. This effect is not present in single star observations but it is significant in most double-lined binaries in all phases and must be taken into account. I believe there has already been at least one analysis of CORAVEL results on YY Gem where this was not done.

MEASURING STARS WITH HIGH ANGULAR RESOLUTION: RESULTS FROM NARRABRI
OBSERVATORY

R. Hanbury Brown

Chatterton Astronomy Department
University of Sydney

ABSTRACT. The stellar intensity interferometer at Narrabri Observatory
was used to measure the angular diameters of 32 stars in the spectral
range 05f to F8 and brighter than about V = +2.6. The results were
combined with measurements of absolute flux and, in a few cases
parallax, to find the emergent flux, effective temperature, radius and
luminosity of the stars. The methods used to calibrate these measurements and their uncertainties are discussed. It is shown that in
measuring emergent flux and effective temperature the major uncertainties were most frequently in the measurements of angular size.

1.   INTRODUCTION

     The stellar intensity interferometer at Narrabri Observatory
(Hanbury Brown et al. 1967a,b, Hanbury Brown 1974) consisted of two
large mosaic reflectors (6.7m in diameter) mounted on trucks which moved
around a circular railway track 188 m in diameter. The reflectors were
controlled by a computer assisted by automatic photoelectric star-
guiding; they followed a star in azimuth by moving around the track,
and in elevation by tilting about a horizontal axis. The separation
between the reflectors, the baseline, could be pre-set anywhere in the
range 10-188 m and their movement was controlled so that this baseline
was always normal to the direction of the star. At the focus of each
reflector the light from the star passed through a narrow-band (~ 10 nm)
interference filter and was then focussed on to the cathode of a
photomultiplier. High-frequency (10-100 MHz) fluctuations in the anode
currents of these photomultipliers were carried by cables to a central
control building where the time-average of their cross-product, or
correlation, was measured in an electronic correlator. The angular
diameter of a star was found by measuring this correlation c(d) as a
function of the baseline length, d.
     The principal programme of the Observatory was to measure the
angular diameters of 32 stars in the spectral range 05f to F8 and with
magnitudes brighter than V = +2.6. This program took 6 years to
carry out (May 1965-February 1972).

D. S. Hayes et al. (eds.), Calibration of Fundamental Stellar Quantities, 185–192.

2.   DATA REDUCTION

2.1 Correcting for Zero-drift of the Electronic Correlator

One of the most difficult technical problems in developing the stellar intensity interferometer in the early 1960's was to reduce the zero-drift in the output of the electronic correlator. This zero-drift was monitored as follows. In the intervals between observations the phototubes were illuminated by small lamps which were adjusted to give exactly the same light flux as the star. The light from these two lamps was completely uncorrelated and so the mean output of the correlator was a measure of the zero-drift. Experience showed that this drift did not change significantly over a period of 3 days and so the observed correlation $c(d)$ from a star was corrected by an amount D equal to the mean rate of zero-drift for a period of 1½ days before and after the observation, to find the corrected value of correlation $c_o(d)$, where

$$\overline{c_o(d)} = \overline{c(d)} - \overline{D} \tag{1}$$

2.2 Normalizing the Observed Correlation for Variations of Gain and Light Flux

The measured values of correlation were normalised for variations in the gain of the correlator and phototubes and in the light flux from the star as follows. The gain of the correlator (G) was measured before and after every night's observation with an uncertainty of 1 per cent. The measured values of correlation together with the two phototube anode currents were printed out by the correlator every 100 s.

The final mean value of correlation $c_N(d)$, for a 100 s observation, weighted by the square of the signal/noise ratio in each observation, and normalized for variations in the gain of the phototubes and correlator was found from,

$$\overline{c_N(d)} = \overline{c_o(d)}/(G) \cdot \overline{i_1, i_2} \tag{2}$$

where i_1, i_2 are the mean phototube anode currents due to the star, found by subtracting from the total observed anode currents the components due to the night sky and moon light. Near new moon these corrections were about 1 per cent for stars of magnitude +1 and were never allowed to exceed 10 per cent.

2.3 Uncertainty in the Normalised Correlation

The principal source of uncertainty in the normalized correlation was the statistical fluctuation (noise) in the output of the correlator. This uncertainty had two components, the uncertainty in the mean correlator output $c(d)$ during the actual run on the star and the uncertainty in the measured value of D the mean drift. Both of the uncertainties were found by statistical analysis of the 100 s observations and were combined to find the final uncertainty in the normalized correlation $\overline{c_N(d)}$.

There were a number of other minor uncertainties which were taken into account. These included possible small losses of correlation due to misalignment of the baseline and uncertainties in the measurement of the phototube anode currents and correlator gain. Finally we allowed for the possibility that there were unknown sources of unwanted correlation from the night sky, for example, due to Cerenkov radiation from cosmic rays. To set limits to any such false correlation we observed the bright star β Crucis for 55 hours with a baseline (154 m), so long that any correlation due to the star itself would have been negligible. No significant correlation was observed. We also exposed the instrument to areas of sky without any bright stars and, again, no significant correlation was observed.

The overall uncertainty in the final value of normalized correlation was found by combining all these uncertainties as though they were random and statistically independent.

2.4 The Determination of Angular Diameter

It can be shown that the correlation $\overline{c_N(d)}$ varies with the length of the baseline (d) as,

$$c_N(d) \quad \propto \quad \Delta_\lambda \; \Gamma_\lambda^2(d) \tag{3}$$

where Δ_λ is the *partial coherence factor* and takes account of the fact that stars of large angular diameter are partially resolved by the very large reflectors themselves. The *correlation factor* $\Gamma_\lambda^2(d)$ is a function of the angular size θ of the star, and the effective wavelength of the light λ and the baseline length d; for a circular disc of uniform surface brightness and angular diameter θ_{UD} it is given by,

$$\Gamma_\lambda^2(d) \quad = \quad 2[J_1(x)/x]^2 \tag{4}$$

where $x = \pi d \theta_{UD}/\lambda_o$ and it is assumed that θ_{UD} is so small that $\Delta_\lambda \approx 1$. For a limb-darkened star the shape of this theoretical curve is not significantly altered, although the scale is changed.

For each baseline the final value of mean correlation was found by combining the observations on different nights weighted according to the square of their signal/noise ratios. A curve was then fitted to these points using a computer. For most stars, where $\Delta_\lambda \approx 1$, this curve was given by equation (4), but for some stars with larger angular sizes the observed points were fitted to a theoretical curve using equation (3). The results of this analysis gave the correlation at zero-baseline $\overline{c_N(0)}$ and the angular diameter of the equivalent uniform disc θ_{UD} together with their r.m.s. uncertainties. The final uncertainties in both $\overline{c_N(0)}$ and θ_{UD} took into account the uncertainties in fitting the observations to a theoretical curve and included the uncertainties in the zero-level of the correlator.

These values of θ_{UD} (the angular diameter of an equivalent uniform
disc) were then corrected for limb-darkening to find θ_{LD} the true
angular diameter for each star. The ratio θ_{LD}/θ_{UD} was computed for each
star by comparing the intensity distribution across the disc of the star,
as predicted by an appropriate model atmosphere, with the intensity
distribution for an uniform disc. The correction factor for limb-
darkening was small and had a mean value of about 4 per cent; the
uncertainties in this factor introduced an additional uncertainty into
the final true angular diameter of about 0.5 per cent.

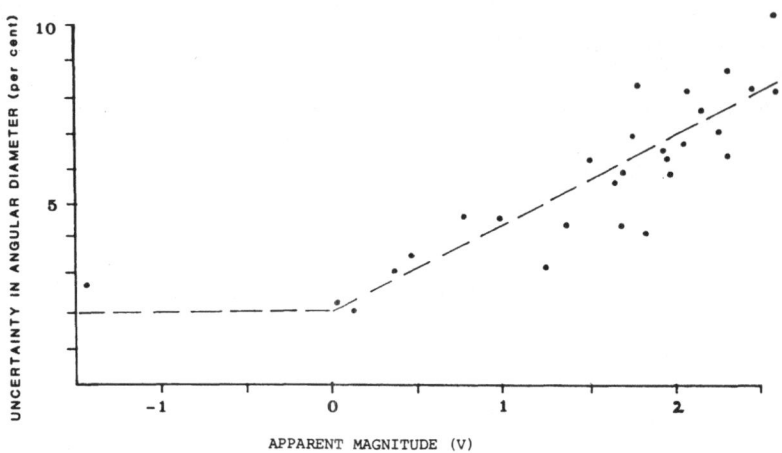

Fig. 1. The uncertainty ($\sigma\theta_{LD}$) in the measured values of angular
diameter as a function of the apparent (V) magnitude of the 32 stars.

2.5 Uncertainties in the Final Angular Diameters

Fig. 1 shows the final uncertainties ($\sigma\theta_{LD}$) in the measured angular
diameters of the 32 stars as a function of their apparent V magnitudes.
Three of the stars have been omitted from this diagram (α Car, δ CMa,
γ Vel) because they presented special problems. The broken line
suggests the "average" relationship between uncertainty and apparent
magnitude. The results are scattered about this line because, for any
one star, the uncertainty is not a simple function of brightness but
also depends on spectral type and on the length of time for which the
star was observed. Most of the exposures lay in the range 40 h for
the brighter stars, to a maximum of 120 h for the fainter stars. The
figure shows that for stars brighter than about V = 0 the uncertainty
did not decrease but remained constant at about 2 per cent. It was
determined by the combined uncertainties in, the gain of the correlator
($\sim$ 1½ per cent), the photo-tube anode currents ($\sim$ 1 per cent), misalignment of the base-line ($<$ 1 per cent) and in the corrections for
limb-darkening ($\sim$ 0.5 per cent). These errors were taken to be random
and statistically independent and did not, as far as we could find,
introduce any significant systematic errors into the measurement of

angular size.

     For stars fainter than V = 0 the uncertainty increased and was
largely due to the combined effects of the statistical fluctuations
(noise) and zero-drift in the output of the electronic correlator. It
was a function of brightness and increased from 2 per cent at V = 0 to
about 8 per cent at V = +2.5.

3.   INTERPRETATION OF RESULTS

3.1 The Detection of Multiple Stars

     In finding the properties of these 32 stars it was important
to know whether or not any particular star was single or multiple. It
was not always possible to find this out from the existing optical
evidence, because the interferometer was capable of resolving binary
or multiple stars that had not previously been detected. If a star is
binary, and if the angular separation of the two components is resolved
at the shortest basline, the apparent value of the zero-baseline
correlation $c_N(0)$ will be reduced relative to that expected from a
single star of the same total brightness by the factor,

$$(I_1^2 + I_2^2)/(I_1 + I_2)^2 \qquad\qquad\qquad (5)$$

where I_1 and I_2 are the intensities of the two components, and it is
assumed that the individual stars are themselves unresolved.
     For 27 of the 32 stars measured in this program, the measured
value of $c_N(0)$ was not significantly less than unity, which implied
that any companion stars which they might have were at least 2.5 magnitudes fainter. It follows that any correlation due to such companions
was less than 1 per cent of that from the bright stars themselves, and
could not have introduced significant errors into the measurements of
angular size. Five of the stars were clearly multiple, and their
values of $c_N(0)$ were used to estimate the brightness of their companions.
While these companions did not introduce appreciable errors into the
measurements of angular size, their presence had to be taken into account
in finding emergent fluxes and effective temperatures.
     In the case of one binary star (α Vir) we made an elaborate series
of observations (Herbison-Evans et al. 1971) which, when combined with
spectroscopic data, yielded all the parameters of the orbit including
the inclination, eccentricity and angular size of the semi-major axis.
They also yielded the brightness ratio of the two components, the
emergent flux, the effective temperature and the luminosity of the
primary component. An interesting feature of this work was that the
distance of the star (84±4 pc) was found by combining the measured values
of the physical size and angular diameter of the semi-major axis.

3.2 Finding the Emergent Fluxes and Effective Temperatures

     The emergent flux (F) at the surface of a star is related to the
integrated flux (f) outside the Earth's atmosphere by,

$$F = 4f/\theta_{LD}^2 \tag{6}$$

The measurement of these integrated fluxes is discussed in detail by
Code et al. (1976). Briefly, they were found by integrating the flux
found in 5 wavelength bands extending from the ultra-violet to the infrared. Measurements of the ultra-violet flux in the range 110-350 nm
were carried out by the satellite observatory OAO-2, and were calibrated
by Aerobee rocket flights and synchrotron radiation from the University
of Wisconsin electron storage ring. Measurements in the range 330-808
nm were carried out by Davis and Webb (1974) using ground-based
telescopes, and were calibrated in absolute units by reference to the
absolute spectrophotometry of α Lyr by Oke and Schild (1970). Measurements at wavelengths greater than 808 nm were taken from the infra-red
photometry of Mitchell and Johnson (1969) and Johnson et al. (1966) and
were also calibrated by the absolute spectrophotometry of Oke and
Schild (1970).

For stars of spectral type earlier than B3 the integrated flux was
corrected for the absence of observations at wavelengths shorter than
110 nm by using the LTE statistically blanketed models of Kurucz et al.
(1972). A generous estimate of the uncertainty in this correction was
made: 20 per cent for T_e < 25 000 K and 50 per cent for
T_e > 25 000 K.

Two other corrections were made to the integrated fluxes. Firstly,
the values for 5 stars in our list were corrected for the presence of
companions; the largest correction was 21 per cent (γ Vel). The
integrated fluxes were also corrected for extinction which was significant for only 10 of the stars; the color excesses for these stars,
together with the extinction curve, is given by Code et al. (1976).

The effective temperature of a star (T_e) is defined by,

$$T_e = (F/\sigma)^{\frac{1}{4}} \tag{7}$$

where σ is the Stefan-Boltzmann constant. The effective temperatures
of the 32 stars were calculated from this equation using the final
values of the integrated flux (F).

3.3 Uncertainties in the Emergent Fluxes and Effective Temperatures

Fig. 2(a) is a histogram of the uncertainties in the final values of
effective temperature (T_e) for the 32 stars. From equation (7) it can
be seen that they are compounded of the uncertainties in the measured
values of angular diameter ($\sigma\theta_{LD}/2$) and in the measured values of
integrated flux ($\sigma f/4$). Fig. 2(b) illustrates the relative contributions by these two quantities to the final uncertainty in F. It shows
that for the majority of stars, 20 out of 32, the major contribution to
the uncertainty in T_e, and hence of F, was due to the measurements of
angular size. However, for 11 of the stars the major uncertainty was
due to the measurements of integrated flux. All of these 11 stars were
of spectral type earlier than AO, and this results reflects the
difficulty of measuring the integrated flux of very hot stars.

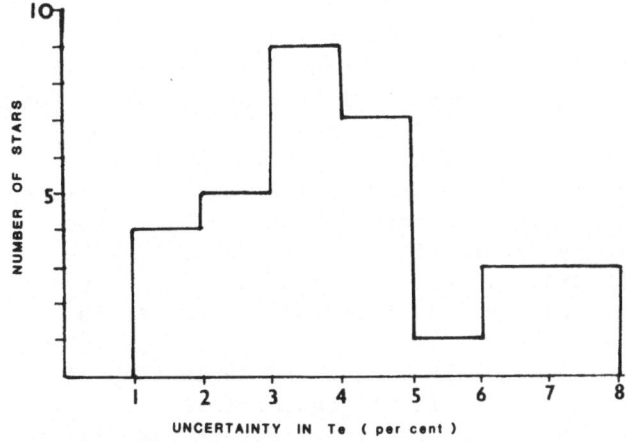

Fig. 2(a) The distribution of the uncertainties in the effective temperatures (T_e) of the 32 stars.

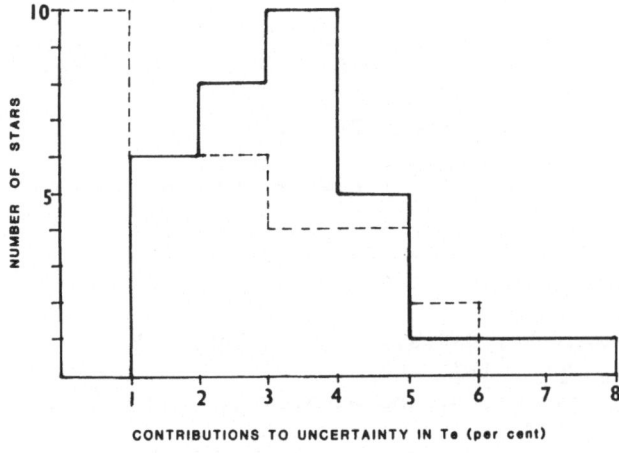

Fig. 2(b) A comparison of the contributions to the uncertainties in the effective temperatures (T_e) due to uncertainties: (solid line) in the angular diameters ($\sigma\theta_{LD}/2$): (broken line) in the integrated fluxes ($\sigma f/4$).

It should be noted that the measurements of F and T_e are wholly empirical with two qualifications. Firstly, model atmospheres were used to make small corrections for limb-darkening. Secondly, for the hottest stars the flux at wavelengths shorter than 110 nm was estimated using model atmospheres. For most stars this use of models accounted for only small corrections.

3.4   Stellar Luminosities and Radii

Only 12 of 32 stars have reliable trigonometrical parallaxes while the distance of α Vir was found, as we have already noted, in a separate program. For these 13 stars it was possible to compute luminosities and radii. For 10 of the stars the uncertainty in the trigonometric parallax was significantly greater than that in the measured angular size. Thus the uncertainties in the radii and luminosities of these stars were due, predominantly, to the large uncertainties in their parallaxes.

4.   CONCLUSION

The Stellar Intensity Interferometer at Narrabri Observatory was used to measure the apparent angular diameters of 32 stars brighter than V = +2.6 in the spectral range 05f to F8. These measurements were combined with measurements of integrated flux to find the emergent fluxes and effective temperatures of these stars. The resulting values are almost wholly empirical, only minor corrections having been made by using model atmospheres. It is interesting to note that, for the majority of these stars, the accuracy of the emergent flux and effective temperature, found in this way, was limited by the uncertainties in the measurement of angular size.

The results were also used to find the radius and luminosity of the few stars for which there are reasonable trigonometrical parallaxes. For almost all these stars the accuracy of the final result was limited by the uncertainty in the trigonometrical parallax.

REFERENCES

Code, A.D., Davis, J., Bless, R.C. and Hanbury Brown, R. 1976,
     Astrophys.J., **203**, 417.
Hanbury Brown, R. 1974, The Intensity Interferometer, (Taylor and
     Francis, London).
Hanbury Brown, R., Davis, J. and Allen, L.R. 1967a, Mon. Not.
     R. Astron. Soc., **137**, 375.
_____1967b, ibid., **137**, 393.
Herbison-Evans, D., Hanbury Brown, R., Davis, J. and Allen, L.R.
     1971, Mon. Not. R. Astron. Soc., **151**, 161.
Johnson, H.L., Mitchell, R.I., Iriarte, B. and Wisniewski, W.Z. 1966,
     Comm. Lunar and Planet. Lab., **4**, 99.
Kurucz, R.L., Peytremann, E. and Avrett, E.H. 1972, Blanketed Model
     Atmospheres for Early Type Stars. (Smithsonian Press,
     Washington, D.C.).
Mitchell, R.I. and Johnson, H.L. 1969, Comm. Lunar and Planet. Lab.,
     8, 1.
Oke, J.B. and Schild, R.E. 1970, Astrophys. J., **161**, 1015.

Discussion of this paper occurred after the paper by McAlister.

MEASURING STARS WITH HIGH ANGULAR RESOLUTION: CURRENT STATUS AND FUTURE PROSPECTS

John Davis

Chatterton Astronomy Department
University of Sydney

ABSTRACT. The current state of knowledge of angular diameters of stars
is reviewed and, based on this review and the requirements for the
determination of surface fluxes, effective temperatures, radii and
masses, targets of sensitivity, angular resolution and accuracy for
future programs of stellar angular diameter measurements are established. Long baseline interferometry is the only technique with the
potential to meet all the targets. The necessary improvements in
sensitivity, angular resolution and accuracy are promised by the
approach adopted in the modern Michelson stellar interferometer under
development at the University of Sydney and the prototype instrument,
which is currently nearing completion, is briefly described to
illustrate how the atmospheric and mechanical problems which have
inhibited the development of amplitude interferometry may be overcome
using modern technology. This program together with the developments
taking place at CERGA lead to the conclusion that the prospects for
contributions by high angular resolution measurements to the determination of fundamental stellar quantities during the next decade are
excellent.

1. INTRODUCTION

Since the completion of the program of stellar angular diameter
measurements with the Narrabri intensity interferometer, which has been
described in the preceding paper (Hanbury Brown 1985), the Chatterton
Astronomy Department at the University of Sydney has been engaged in a
program to develop a stellar interferometer with increased sensitivity,
resolution and accuracy. The primary objective is to develop an instrument capable of contributing to the determination of fundamental stellar
quantities, namely emergeant fluxes and effective temperatures, radii
and hence luminosities, and masses. Additional studies would include
limb-darkening, stellar rotation, extended atmospheres, emission-line
stars and Cepheids.

This paper reviews the current status of angular diameter measurements, discusses the targets for future angular diameter determination

D. S. Hayes et al. (eds.), Calibration of Fundamental Stellar Quantities, 193–208.
© *1985 by the IAU.*

programs and assesses the potential of available techniques. The
modern Michelson stellar interferometer under development at the
University of Sydney is described briefly to illustrate how the targets
might be met. Since high angular resolution studies of binary stars
are discussed by McAlister (1985) in this symposium the discussion is
restricted to single stars. However, it is emphasized that the
University of Sydney program is just as concerned with the observation
of binary stars (Davis 1983) since for at least the primary components
of double-lined spectroscopic binaries it is possible to determine all
the gross properties (i.e. mass, radius and luminosity) as demonstrated
with the intensity interferometer for α Vir (Herbison-Evans et al. 1971).

2.   CURRENT KNOWLEDGE OF THE ANGULAR DIAMETERS OF STARS

    Discussion in this section is restricted to direct measurements of
angular diameters by interferometry or occultations. A survey of the
literature reveals that angular diameters have been measured for some
150 stars with more than half (~ 90) being determined from lunar
occultations. However, many of the published angular diameters are of
little interest because of the poor accuracy of their determination.
Figure 1 shows the distribution by spectral class of stars whose
angular diameters have been determined with published uncertainties
better than   (a) ±20% (111 stars),   (b) ±10% (82 stars) and   (c) ±5%
(32 stars). In preparing these histograms a weighted mean angular
diameter was derived for each star for each observational technique
(i.e. for lunar occultations, for speckle interferometry etc.). Where
a star had been observed by more than one technique, measurements from
the different techniques were not combined and only the most accurate
weighted mean angular diameter was taken for the purposes of Figure 1.
Combination of measurements from different techniques would only affect
the positions of 2 late type stars in Figure 1.
    Figure 1 has two distinct ranges. For spectral types earlier than
G0 the data is dominated by results from the intensity interferometer
whereas for spectral types later than G0 the majority of the data is
from lunar occultations with contributions from the two telescope
interferometer at CERGA (I2T), from the amplitude interferometer and
from speckle interferometry. All the data for classes G, K and M are
for evolved stars.
    The data in Figure 1 have been sorted according to the published
uncertainties in the angular diameters. Independent checks on the
reliability of the data are clearly desirable and ideally the results
for individual stars determined by different techniques should be
compared. Unfortunately, because of the limitations of the different
techniques, there are only a few cases where such comparisons can be
made and the relevant data are listed in Table I. Mira type stars have
been excluded. α Ori has also been omitted because of the apparent
variation of angular diameter with wavelength and time (White (1980).
    Except for ρ Per and α Her where the speckle and amplitude inter-
ferometry results are inconsistent the agreement is generally as good
as would be expected. Detailed intercomparison of the results from the

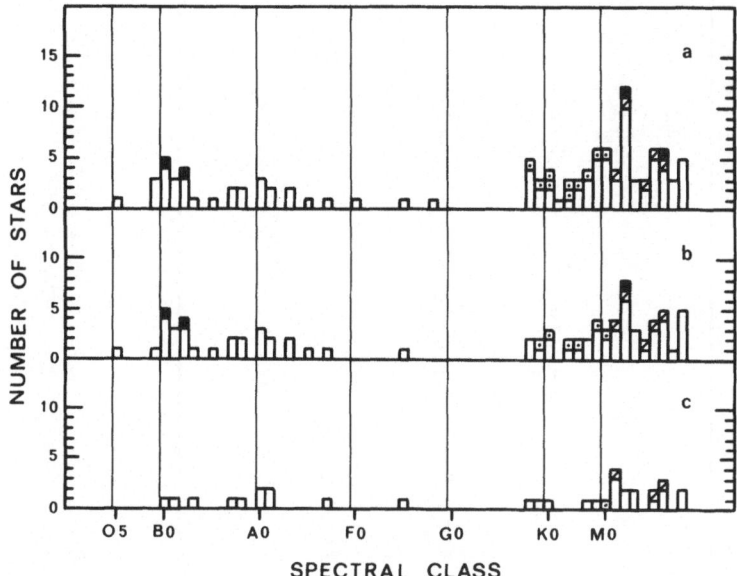

Fig. 1. The number of stars with measured angular diameter as a function of spectral class. The distributions are for angular diameters determined to better than ±20% (a), ±10% (b) and ±5% (c).

For spectral classes O, B, A and F: □ represents measurements by intensity interferometry and ■ by Jovian occultations. For spectral classes G, K and M: □ represents measurements by lunar occultations, ⊡ by CERGA I2T, ▨ by speckle interferometry and ■ by amplitude interferometry.

different techniques is not warranted because of the small amount of data but Table I serves to emphasize the desirability of obtaining more data, especially for spectral classes earlier than K.

An alternative check on the reliability of data can be obtained by intercomparison of results by the same technique but by different observers with different equipment. This approach is possible for lunar occultations and speckle interferometry although it is limited to late spectral types as shown by the spectral class coverage of the techniques shown in Figure 1. In the case of lunar occultations Ridgway et al. (1980) have compared the error estimates for 66 measurements of 24 stars with the dispersion of measurements for each star. Expressing the deviations from the mean for each star in units of the expected standard deviation, and excluding Mira-type stars, they found excellent agreement between the observed distribution and an appropriately normalized normal distribution for 53 out of 57 observations. The 4 discrepant measurements were for 2 stars and it was concluded that while these results may be the result of real variations it is possible that occasionally (∼ 2 out of 57) the occultation technique may be subject to errors which do not reveal themselves in the analysis.

The speckle interferometer results for 3 stars measured by different observers are listed in Table II. α Ori has been omitted for the

Table I. Angular diameters for stars measured by more than one technique.

BS	Star	MK	Angular Diameter of Equivalent Uniform Disk (milliseconds of arc)				
			Intensity Interferometer	Lunar Occultation	Speckle Interferometer	CERGA I2T	Amplitude Interferometer
921	ρ Per	M4 II-III			29±2[1]		17±2[2]
1457	α Tau	K5 III		20.0 ±0.1[3]*	24±3[1]		19±2[2]
2286	μ Gem	M3 III		12.1 ±0.1[4-8]*		14.2±0.7[9]	15±2[2]
3982	α Leo	B7 V	1.32±0.06[10]	1.38±0.06[11]*			
5340	α Boo	K2 IIIp			20±2[12-14]*	21.0±1.2[9]	24±2[2]
6134	α Sco	M1-2 Iab		41 ±1[15]	42±2[12]		42±5[2]
6406	α Her A	M5 II			29±1[1,12,16]*		42±3[2]
6705	γ Dra	K5 III				8.3±0.6[9,17]*	13±5[2]
8775	β Peg	M2 II-III			16±2[12]		19±1[2]

* Weighted mean of published values.

References: 1. Welter & Worden 1980
2. Currie et al. 1976
3. Ridgway et al. 1982
4. White 1974
5. Ridgway et al. 1974
6. Dunham et al. 1975
7. Nelson 1975
8. Beavers et al. 1981
9. Di Benedetto & Conti 1983
10. Hanbury Brown et al. 1974
11. Radick 1981
12. Gezari et al. 1972
13. Worden 1976
14. Blazit et al. 1977
15. Evans 1957
16. Worden 1975
17. Faucherre et al. 1983

TABLE II. Intercomparison of angular diameters determined by speckle
           interferometry.

BS	Star	MK	Angular Diameters (milliseconds of arc)		
911	α Cet	M2 III	14±5(400) [1]* 25±2(550) [2]		
5340	α Boo	K2 IIIp	18±4(450) [1] 22±3(500) [3] 18±5(500) [1]	19±6(420) [4]	
6406	α Her A	M5 II	29±1(510) [2] 31±3(500) [3] 26±3(520) [2] 25±3(550) [2]	30±1(510) [5] 28±3(520) [5]	

*The mean wavelength of observation in nm is given in parenthesis after
each result.
References: 1. Blazit et al. 1977
             2. Welter & Worden 1980
             3. Gezari et al. 1972
             4. Worden 1976
             5. Worden 1975

----

same reason it was omitted from Table I.

     Welter and Worden (1980) have commented on the descrepancy between
the values for α Cet and noted that the star is close to the limit of
their technique. The results for α Boo and α Her A are in reasonable
agreement but again there is insufficient data to warrant detailed
discussion.
     The intercomparison of results between techniques and within
techniques, although restricted by the limited amount of data, can be
summarized as follows. Although a small number of discordant results
exist, the agreement between independent observations of a star is
generally in accord with the published error estimates.
     Where intercomparison of results is not possible, reliance has to
be placed on the internal consistency of data from repeated observations,
preferably obtained with modified equipment configurations, and on the
efforts taken to eliminate systematic errors from the data. This is
particularly true for the results from the intensity interferometer and
this is discussed in detail by Hanbury Brown (1985) in the preceding
paper.

3. FUTURE ANGULAR DIAMETER MEASUREMENTS

     Fig. 1 shows that existing measurements of angular diameters with
an accuracy better than ±5% are spread very thinly across the spectral
classes. There is a real paucity of data for spectral classes from late
A to late G even if the accuracy is relaxed to ±20%. The situation is

even worse if only the main sequence is considered since there are no
main-sequence stars later than F5 in Figure 1. However, it should be
noted that the Sun (G2 V) and the eclipsing binaries YY Gem (MO.5V) and
CM Dra (M4V) have not been included. Angular diameters derived for
these eclipsing binaries from trigonometric parallaxes have uncertainties in the ±5-10% range.

Let us examine the sensitivity, angular resolution and accuracy
required to add substantially to our knowledge of stellar surface fluxes
and effective temperatures for all spectral types via angular diameter
determinations.

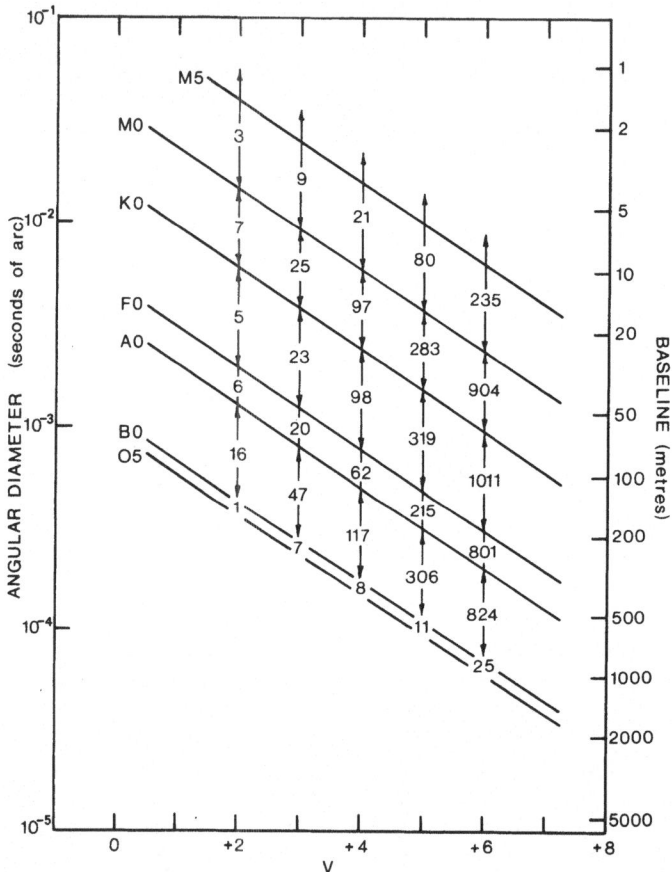

Fig. 2. Distribution of stars with respect to apparent brightness
and angular size. The diagonal lines represent mean relationships
for the spectral types shown and the figures give the total number of
stars in the Catalogue of Bright Stars (Hoffleit 1964) brighter than
a given V magnitude with declination < +30° for each spectral group
(e.g. there are 215 A-type stars with V < +5). The right-hand ordinate
gives the interferometer baseline required for resolution ($|\gamma|^2 = 0.5$)
at a wavelength of 550 nm.

Figure 2 shows the distribution with respect to brightness and angular diameter of stars brighter than V = +6 and with declination < +30°. The figure suggests that angular diameters for a large sample of stars could be determined with an instrument capable of measuring stars to V = +4. This is true for the middle range of spectral classes but a closer examination of the statistics reveals that for early O and for M main-sequence stars greater sensitivity is required. To measure hot O stars it is necessary to reach at least V = +6 and, in order to obtain a reasonable sample, V(limit) should exceed +7. For main-sequence stars of spectral class MO and later it is also necessary to have V(limit) > +7. In fact, the V(limit) required to measure main-sequence stars increases towards later spectral classes and exceeds +9.5 at M5.

It can be seen from Figure 2 that an angular resolution of ~ 2 x 10^{-4} arcseconds would be sufficient for measuring all except the hottest stars but for these an angular resolution of ~ 4 x 10^{-5} arcseconds is required.

If a measured angular diameter is to be used to determine the emergent flux at the stellar surface (F) and effective temperature (T_e) then the aim should be to measure the angular diameter (θ) with an accuracy such that its contribution to the uncertainty in F is no greater than that from the integrated absolute flux received from the star (f). F, f, θ and T_e are related by

$$F = 4f/\theta^2 = \sigma T_e^4 \qquad (1)$$

where σ is the Stefan-Boltzmann constant. From equation (1) it can be seen that the target is to make $2\sigma(\theta) \leq \sigma(f)$ where $\sigma(\theta)$ and $\sigma(f)$ are the fractional uncertainties in θ and f.

Figure 3 is a plot of $2\sigma(\theta)$ and $\sigma(f)$ against $(B-V)_o$ for the 32 stars measured with the intensity interferometer. The data is taken from Code et al. (1976). For $(B-V)_o \leq -0.15$ the uncertainties are generally well matched largely because of the increasing uncertainty in the integrated fluxes for the hotter stars. However, for $(B-V)_o \gtrsim -0.15$, the uncertainty in the integrated flux is of the order of ±4% whereas $2\sigma(\theta)$ ranges from ~ ±4% up to ~ ±28%. On the basis of the evidence shown in Figure 3 future angular diameter measurements should aim at reducing $\sigma(\theta)$ to $\lesssim$ ±2%.

Blackwell et al. (1979) have also set a target accuracy of about 2% in θ and 1% in T_e for their infrared flux method which would be commensurate with that of the oscillator strengths being measured at Oxford.

The targets to be aimed at in future angular diameter determinations thus include V(limit) > +7 for spectral classes earlier than MO increasing to V(limit) > +9.5 for M5 main-sequence stars, an angular resolution of 2 x 10^{-4} arcseconds decreasing to 4 x 10^{-5} arcseconds for a sample of hot O stars, and generally an accuracy of $\leq$ ±2%.

The difficulties involved in reaching these targets are formidable and have led Barnes and Evans (Barnes & Evans 1976; Barnes et al. 1976; Barnes et al. 1978) to develop an empirical relationship between visual surface brightness and color index based on existing angular diameter

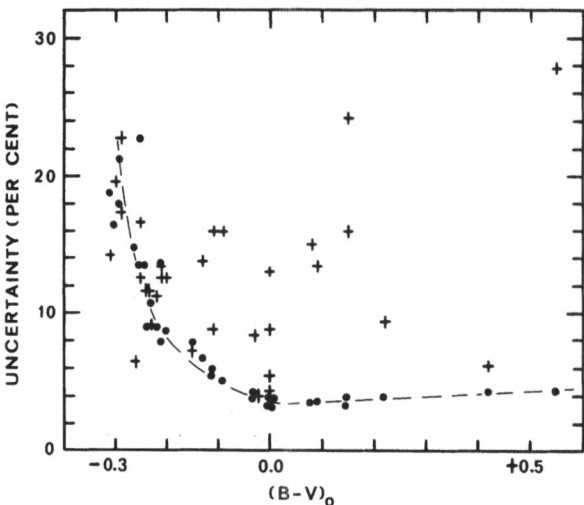

Fig. 3. The uncertainty in integrated flux received from a star (●) and twice the uncertainty in angular diameter (+) as a function of $(B-V)_0$ for the 32 stars measured with the intensity interferometer. The broken line indicates the trend of the uncertainties in the fluxes.

measurements. While it is not fundamental in the sense that it is calibrated with empirical data it has nevertheless found numerous applications (Barnes et al. 1978) in the absence of direct angular diameter measurements. An important feature of the relationship is the fact that it can be applied to stars which are below the current sensitivity of techniques for direct angular diameter determinations. Its usefulness would be extended by improvements in calibration which would be brought about by angular diameter determinations of improved accuracy. This is particularly true in the range of spectral classes A to G where the visual surface brightness relationship is least well defined.

The infrared flux method has been devised by Blackwell & Shallis (Blackwell & Shallis 1977; Blackwell et al. 1978; Blackwell et al. 1979) with the stated intention of providing a fundamental technique that will give angular diameters and effective temperatures for selected stars of virtually any spectral type to a magnitude limit of at least +7 or +8 with an accuracy at least equal to that attainable by other methods and which is completely independent of results from other methods. They have shown (Blackwell et al. 1979) that their results are insensitive to the model atmosphere used, at least for $T_e \lesssim 8000$ K, and that the interstellar extinction is not a significant problem. So far the method has given angular diameters that are larger by 3.8% on average than those measured with the intensity interferometer. This discrepancy may be the result of errors arising from the absolute infrared flux calibration which is crucial to the success of the infrared flux method. The cause of the discrepancy needs to be resolved and Blackwell and his colleagues are endeavoring to improve the infrared flux calibration. It is also important in this context to improve the spectral coverage and accuracy of direct angular diameter measurements to provide further crosschecks on the results.

The targets for future angular diameter measurements have been discussed in the context of surface fluxes and effective temperatures. Stellar radii, and hence luminosities, can also be determined from angular diameter measurements for stars of known parallax. The usefulness of this application is currently limited by the accuracy of existing trigonometric parallaxes. The radii of stars of known trigonometric parallax can only be obtained with an accuracy better than ±20% for spectral classes AO and later (Davis 1979a). From one site there are some 100 stars brighter than V = +10 with trigonometric parallaxes known to better than ±20% but only ~ 12 better than ±5%. Thus the targets set above are more than adequate to match existing parallaxes and, in fact, will be adequate in almost all cases for radii determined from parallaxes determined with the astrometric satellite HIPPARCOS (mean error 0.002 arcseconds (Høg 1978)). This conclusion is also true for the determination of masses of single stars by the method proposed by Blackwell et al. (1978) since it too involves parallaxes.

High angular resolution observations of suitable double-lined spectroscopic binaries will also enable stellar masses and radii to be determined. The case of binary stars is discussed by McAlister (1985), but the targets discussed here are appropriate for this application (Davis 1983).

4. THE DIRECT MEASUREMENT OF ANGULAR DIAMETERS

Of the techniques available for the direct measurement of stellar angular diameters only long baseline interferometry has the potential to meet all the targets set in the previous section although baselines in excess of 1 km will be required to achieve an angular resolution of 4×10^{-5} arcseconds at optical wavelengths as can be seen from Figure 2.

The limitation in angular resolution of the lunar occultation technique restricts it to the measurement of luminous cool stars and it is here that it has made a major contribution as shown by Figure 1. Although the technique is capable of measuring angular diameters down to the 1-2 milliseconds of arc range (Ridgway 1979) the accuracy deteriorates below ~ 3 milliseconds of arc and, with only a few notable exceptions, accurate angular diameter measurements will remain restricted to luminous stars of late G, K and M spectral classes.

The angular resolution of amplitude and speckle interferometry using a single telescope is restricted to ~ 15 milliseconds of arc by the diameters of existing telescope apertures. Although the angular diameters of 9 stars have been determined by each technique with an accuracy better than ±20%, in the cases of 4 of the speckle results and 7 of the amplitude results the angular diameters have been determined with greater accuracy by other methods. This is shown by the surviving results appearing in Figure 1. Even with the next generation telescopes

under consideration these techniques will not make a significant
contribution to stellar angular diameter determinations.

There are several possible approaches to long baseline interferometry but since intensity interferometry has been by far the most
successful it would seem to be the obvious approach to pursue. A
possible design for a very large intensity interferometer with increased
sensitivity, resolution and accuracy has been discussed by Hanbury Brown
(1974) and Davis (1975). It appears that the resolution and accuracy
targets given in Section 3 could be reached with an intensity interferometer but that an increase in sensitivity of ~ 80 times over the
Narrabri instrument, which had a limiting blue magnitude B(limit)
~ +2.5 (Hanbury Brown et al. 1974), is all that can be achieved without
great difficulty and prohibitive cost. There is no doubt that a very
large intensity interferometer could be built and that it would work
but a modern amplitude interferometer offers the possibility of greater
sensitivity at lower cost. If reasonable parameters are assumed it can
be shown that it would be necessary to use apertures of the order of
30 m in an intensity interferometer in order to achieve the same
sensitivity as an amplitude interferometer with 10 cm diameter apertures.

The development of amplitude interferometers has been inhibited by
two main areas of difficulty in the past. One is associated with the
severe requirements of mechanical stability imposed by the wavelength
of light and the other is the problem of making accurate measurements
in the presence of the disruptive effects of atmospheric turbulence on
the incoming wavefronts. The problems imposed by the atmosphere can
be avoided by an interferometer in space (see for example Labeyrie
1978; Stachnik & Labeyrie 1984) but since the space proposals are still
at an early stage of development and are not likely to contribute
measurements of angular diameters within at least the next decade they
will not be discussed here. The use of lasers, and modern control,
detection and data handling techniques offers the possibility of over-
coming the mechanical and atmospheric problems in a ground-based
amplitude interferometer and this, together with the sensitivity
advantage of an amplitude interferometer, are the reasons why long-
baseline optical interferometers currently under development are all
amplitude instruments.

A large aperture amplitude interferometer is being developed at
CERGA (Labeyrie 1978) and small aperture amplitude interferometers are
being developed at the Universities of Sydney (Davis 1979b) and
Maryland (Liewer 1979). In the CERGA interferometer the Coudé images
from two telescopes, whose apertures are large compared with Fried's
atmospheric coherence diameter r_0 (Fried & Mevers 1974), are relayed
to a central station where fringes appear in the superposed images
providing the optical paths are matched within the tolerances set by
the coherence length of the light. The prototype instrument (I2T)
using 25 cm diameter aperture telescopes has been successfully used out
to baselines of ~ 40 m to determine $|\gamma|$ for bright stars with
V < 3.1 (Faucherre et al. 1983) ($|\gamma|$ is the modulus of the complex
degree of coherence (Born & Wolf 1964) equal to Michelson's fringe
visibility). Work is in progress to develop telescopes with aperture
diameters of 1.52 m (Labeyrie 1978) for use in this type of instrument.

In the small aperture approach the apertures are restricted to
$\lesssim r_o$ (which has a value of ~ 10 cm at a good site) and the effects
of atmospheric turbulence are then removed in real time from the
measurement of $|\gamma|$. How this is done in the instrument being developed
at the University of Sydney will be described in the following section.

The relative merits of the large aperture and small aperture
approaches have been outlined by Davis (1976). It is not yet clear
which will prove most suited to stellar angular diameter determinations
although it is to be hoped that both will be successful so that stars
may be observed in common to provide crosschecks. The large aperture
approach is expected to have greater sensitivity but a small amplitude
interferometer, because it incorporates removal of atmospheric
effects in real time from the measurement of $|\gamma|$, should be capable
of achieving greater accuracy.

5. A SMALL APERTURE MODERN MICHELSON STELLAR INTERFEROMETER

A small aperture modern Michelson stellar interferometer is under
development at the University of Sydney and a prototype instrument is
currently nearing completion in the grounds of the Australian National
Measurement Laboratory at West Lindfield, near Sydney.

The prototype instrument represents a logical step following the
work of Twiss and Tango with the Monteporzio 2 m amplitude interferometer (Tango 1979a). The principles and theory of a small aperture
amplitude interferometer have been discussed by Tango and Twiss (Tango
1979b; Tango & Twiss 1980) and they apply directly to the prototype
instrument. Since the design considerations for the prototype (Davis
1979b) and, recently, a more detailed description of the design and
layout (Davis 1984) have been published only a very brief summary of
the key features will be given here.

The prototype interferometer has a fixed 11.4 m north-south
baseline and coelostats mounted on plinths at each end of the baseline
to steer starlight into a central laboratory. This laboratory houses
the optical system of the interferometer which includes a path length
compensator to equalise the light paths from the star to the beamsplitter
where the beams from the two sides of the instrument are combined.

5.1 The Elimination of Atmospheric Problems

The effect of atmospheric turbulence is to produce time dependent
deformations of the wavefronts reaching the coelostats which will reduce
the observed visibility $|\gamma|$ below the true value. The atmospheric
effects are circumvented by restricting the effective apertures of the
coelostats to $\lesssim r_o$ (the maximum primary beam diameter is 10 cm) so
that 'flat' sections of wavefront are selected and also by reducing
the mean tilts of the wavefronts using optical servo systems. These
steps produce a significant reduction in the loss in $|\gamma|$ (Tango &
Twiss 1980) and the residual loss can be estimated from auxilliary
measurements. For this reason an auxilliary interferometer is being
developed to provide real time measurements of r_o which will be used

to select the effective aperture size and to provide a correction
factor for the residual losses due to wavefront curvature. The loss
arising from uncompensated wavefront tilt fluctuations and shot noise
in the tilt correcting servo signals, which will cause the tilt
correcting mirrors to dither randomly, can be estimated from the noise
power of the servo error signal (Tango & Twiss 1980). This procedure
has been used with very good results in the Monteporzio interferometer
(Tango 1979a).

The tilt corrected wavefronts from the two arms of the interferometer are arranged to interfere at nominally zero angle in a beamsplitter so that, in the absence of aberrations, the complementary
output beams are uniformly illuminated. The irradiances of the beams
are proportional to $(1 + |\gamma| \cos\phi)$ and $(1 - |\gamma| \cos\phi)$ where ϕ is a
randomly varying phase angle resulting from atmospheric optical path
length fluctuations and path equalisation errors. The sampling time
τ of the photon counting detectors measuring the irradiance in the
two beams is made short so that changes in ϕ are not significant during
τ which is expected to be in the range 1-10 ms. The photocounts
registered by the two detectors in τ are processed by a correlator
which basically measures the square of the difference between the two
signals. The data are integrated for a total observing time T and
analysed to yield $2|\gamma|^2 <\cos^2\phi>$, where the brackets indicate an average
over T. If the phase is a uniform random variable $2|\gamma|^2 <\cos\phi>$
becomes $|\gamma|^2$, the square of the fringe visibility. Provision has been
made to introduce a path difference of $\lambda/4$ between the two interfering
beams halfway through the sampling period τ so that the correlator
gives $2|\gamma|^2 <\cos^2\phi + \sin^2\phi>$ instead of $2|\gamma|^2 <\cos^2\phi>$ so that it will not
be necessary to assume that the phase is a uniform random variable
to obtain $|\gamma|^2$.

Tango and Twiss (1980) have shown that irradiance fluctuatons due
to scintillation will have a quite negligible effect on the measure-
ment of $|\gamma|^2$ in a small aperture amplitude interferomter.

5.2 The Elimination of Mechanical Problems

In order to provide the required mechanical stability the component
parts of the prototype interferometer are all mounted on massive
reinforced concrete plinths anchored in a monolithic layer of sandstone
approximately 1 m below ground level. Seismometer tests carried out
prior to site selection established that ambient vibration levels were
acceptable.

Particular care has been taken in the design and construction of
the coelostat mounts to minimize vibration and path length changes
with orientation. The only other component mechanically driven during
an observation is a truck carrying a retroreflector in the optical
path length compensator (OPLC) and this is the most critical component
of the interferometer. This is not simply because it has to match the
optical paths from the star to within a tolerance set by the optical
bandwidth but because it also has to meet a particularly severe requirement regarding the smoothness of its motion. Not only has the reflector
to move at the correct velocity but any error or irregularity in the

velocity in the sampling time τ will result in a loss of coherence and an error in the measured $|\gamma|^2$. The tolerances for the OPLC to keep losses in $|\gamma|^2 \leq 1\%$ have been established by Tango and Twiss (1980) and the OPLC for the prototype interferometer, which has been designed and built to meet these tolerances, has been described by Davis (1984).

5.3 The Prototype Program

The prototype interferometer is nearing completion and it will be used to establish the accuracy and reliability that can be achieved in measurements of $|\gamma|^2$ through the atmosphere. In developing the design of the interferometer considerable effort has been made to keep individual sources of error in the measurement of $|\gamma|^2$, both atmospheric and mechanical, to $< \pm 1\%$ so that, at least for bright stars, an accuracy of $\leq \pm 2\%$ in $|\gamma|^2$ can be achieved. An accuracy of ± 0.02 in $|\gamma|^2$ measured at the optimum baseline $d_o \simeq 1.3 \times 10^{-4} \lambda/\theta$, where λ is the wavelength of observation in nm and θ is the angular diameter in arcseconds, corresponds to an accuracy of $\sim \pm 2.5\%$ in the angular diameter. The limiting accuracy is ultimately expected to be set by the tilt-correcting servo and this will deteriorate for fainter stars. The faintest magnitude at which $\pm 2\%$ accuracy can be achieved depends on how well the corrections for residual losses due to atmospheric effects can be established but at this stage it appears that V(limit) will be in the range +7 to +8 (Tango 1979b). The accuracy will deteriorate rapidly below V = +8 to +9.

The prototype program is the foundation for the major high angular resolution interferometer that we plan to build and the component parts of the prototype instrument have been designed to become the heart of the long baseline instrument. The preliminary specification includes baselines extending eventually to a kilometer or more with the same limiting magnitude and accuracy as for the prototype interferometer. This would enable angular sizes down to $\sim 5 \times 10^{-5}$ arcseconds to be measured with an accuracy of the order of $\pm 2\%$ to a V(limit) of +7 to +8.

Apart from the increase in baseline lengths and the corresponding increase in the effective length of the OPLC there are two additional features which will be included in the major instrument. Firstly, for baselines in excess of ~ 50 m it will be necessary to use an automatic fringe tracking system (Davis 1984) and a system such as that developed by Shao and Staelin (1980) will be incorporated. Secondly, the single spectral channel detectors used in the prototype instrument for the determination of $|\gamma|^2$ will be replaced by photon counting linear array detectors covering a range of the spectrum. This will enable simultaneous observations to be made in spectral lines as well as in multiple channels in the continuum.

A search for a suitable site with stable rock near the surface and large enough to accommodate either east-west and/or north-south baselines of at least 1 km in length is under way.

6.   CONCLUSIONS

The targets for future angular diameter determinations suggested in Section 3, namely V(limit) > +7, angular resolution of ~ 4 x 10^{-5} arc-seconds and an accuracy of $\leq$ ±2%, appear to be within the reach of a ground based small aperture amplitude interferometer of the type being developed at the University of Sydney. They may also be within the reach of the large aperture approach being developed by Labeyrie and his colleagues at CERGA.

The importance of observing stars by more than one technique is stressed. In due course it will be desirable to establish a list of reference stars visible from both the Northern and Southern Hemispheres, including stars occulted by the Moon, for comparison and calibration purposes.

The prospects for contributions by high angular resolution measurements to the determination and calibration of fundamental stellar quantities during the next decade are excellent.

ACKNOWLEDGEMENTS

The University of Sydney stellar interferometer program is supported by the Australian Research Grants Scheme, the University of Sydney Research Grants Committee and the Science Foundation for Physics within the University of Sydney.

REFERENCES

Barnes, T.G. & Evans, D.S. 1976, Mon. Not. R. Astron. Soc., 174, 489.

Barnes, T.G., Evans, D.S. & Moffett, T.J. 1978, Mon. Not. R. Astron. Soc., 183, 285.

Barnes, T.G., Evans, D.S. & Parsons, S.B. 1976, Mon. Not. R. Astron. Soc., 174, 503.

Beavers, W.I., Cadmus, R.R. & Eitter, J.J. 1981, Astron. J., 86, 1404.

Blackwell, D.E., Petford, A.D. & Shallis, M.J. 1978, in Colloquium on European Satellite Astronomy, Padua, 1978 June 5-7, p.223.

Blackwell, D.E. & Shallis, M.J. 1977, Mon. Not. R. Astron. Soc., 180, 177.

Blackwell, D.E., Shallis, M.J. & Selby, M.J. 1979, Mon. Not. R. Astron. Soc., 188, 847.

Blazit, A., Bonneau, D., Koechlin, L. & Labeyrie, A. 1977, Astrophys. J., 214, L79.

Born, M. & Wolf, E. 1964, Principles of Optics, Oxford: Pergamon Press.

Code, A.D., Davis, J., Bless, R.C. & Hanbury Brown, R. 1976, Astrophys. J., 203, 417.

Currie, D.G., Knapp, S.L., Liewer, K.M. & Braunstein, R.H. 1976, University of Maryland Technical Report, #76-125.

Davis, J. 1975, in Multicolor Photometry and the Theoretical HR Diagram, ed. A.G.D. Philip & D.S. Hayes, Dudley Observatory Report No.9,p.199.

Davis, J.   1976, Proc. Astron. Soc. Australia, 3, 26.
Davis, J.   1979a, in High Angular Resolution Stellar Interferometry
      (IAU Colloquium No. 50), ed. J. Davis & W.J. Tango, p.1.1.
      Chatterton Astronomy Department, University of Sydney.
Davis, J.   1979b, in High Angular Resolution Stellar Interferometry
      (IAU Colloquium No. 50), ed. J. Davis & W.J. Tango, p.14.1.
      Chatterton Astronomy Department, University of Sydney.
Davis, J.   1983, in Current Techniques in Double and Multiple Star
      Research (IAU Colloquium No. 62), ed. R.S. Harrington &
      O.G. Franz, p.191. Lowell Observatory Bulletin No. 167.
Davis, J.   1984, in Indirect Imaging, ed. J.A. Roberts, p.125.
      Cambridge University Press.
Di Benedetto, G.P. & Conti, G. 1983, Astrophys. J., 268, 309.
Dunham, D.W., Evans, D.S. & Vogt, S.S. 1975, Astron. J., 80, 45.
Evans, D.S. 1957, Astron. J., 62, 83.
Faucherre, M., Bonneau, D., Koechlin, L. & Vakili, F. 1983, Astron.
      Astrophys., 120, 263.
Fried, D.L. & Mevers, G.E. 1974, Applied Optics, 13, 2620 (Correction
      in Applied Optics, 14, 2567).
Gezari, D.Y., Labeyrie, A. & Stachnik, R.V. 1972, Astrophys. J.,
      173, L1.
Hanbury Brown, R. 1974, The Intensity Interferometer. London:
      Taylor & Francis.
Hanbury Brown, R. 1985, in Calibration of Fundamental Stellar
      Quantities (IAU Symposium No. 111), ed. D. S. Hayes, L. E.
      Pasinetti and A. G. D. Philip, (Reidel, Dordrecht), p. 185.
Hanbury Brown, R., Davis, J. & Allen, L.R. 1974, Mon. Not. R. Astron.
      Soc., 167, 121.
Herbison-Evans, D., Hanbury Brown, R., Davis, J. & Allen, L.R. 1971,
      Mon. Not. R. astron. Soc., 151, 161.
Hoffleit, D. 1964, Catalogue of Bright Stars, Yale University
      Observatory.
Høg, E. 1978, in Colloquium on European Satellite Astronomy, Padua,
      1978 June 5-7, p.7.
Labeyrie, A. 1978, Ann. Rev. Astron. Astrophys., 16, 77.
Liewer, K.M. 1978, in High Angular Resolution Stellar Interferometry
      (IAU Colloquium No. 50), ed. J. Davis & W.J. Tango, p.8.1.
      Chatterton Astronomy Department, University of Sydney.
McAlister, H. A. 1985, in Calibration of Fundamental Stellar
      Quantities (IAU Symposium No. 111), ed. D. S. Hayes, L. E.
      Pasinetti and A. G. D. Philip, (Reidel, Dordrecht), p. 97.
Nelson, M.R. 1975, Astrophys. J., 198, 127.
Radick, R.R. 1981, Astron. J., 86, 1685.
Ridgway, S.T. 1979, in High Angular Resolution Stellar Interferometry
      (IAU Colloquium No. 50), ed. J. Davis & W.J. Tango, p.6-1.
      Chatterton Astronomy Department, University of Sydney.
Ridgway, S.T., Jacoby, G.H., Joyce, R.R., Siegel, M.J. & Wells, D.C.
      1982, Astron. J., 87, 1044.
Ridgway, S.T., Jacoby, G.H., Joyce, R.R. & Wells, D.C. 1980, Astron.
      J., 85, 1496.
Ridgway, S.T., Wells, D.C. & Carbon, D.F. 1974, Astron. J., 79, 1079.

Shao, M. & Staelin, D.H. 1980, Applied Optics, 19, 1519.
Stachnik, R.V. & Labeyrie, A. 1984, Sky & Telescope, 67, 205.
Tango, W.J. 1979a, in High Angular Resolution Stellar Interferometry
      (IAU Colloquium No. 50), ed. J. Davis & W.J. Tango, p.13.1.
      Chatterton Astronomy Department, University of Sydney.
Tango, W.J. 1979b, in High Angular Resolution Stellar Interferometry
      (IAU Colloquium No. 50), ed. J. Davis & W.J. Tango, p.12.1.
      Chatterton Astronomy Department, University of Sydney.
Tango, W.J. & Twiss, R.Q. 1980, in Progress in Optics XVII, ed.
      E. Wolf, p.239. Amsterdam: North Holland.
Welter, G.L. & Worden, S.P. 1980, Astrophys. J., 242, 673.
White, N.M. 1974, Astron. J., 79, 1076.
White, N.M. 1980, Astrophys. J., 242, 646.
Worden, S.P. 1975, Astrophys. J., 201, L69.
Worden, S.P. 1976, Pub. Astron. Soc. Pacific, 88, 69.

Discussion of this paper occurred after the paper by McAlister.

THE ROLE OF SPACE OBSERVATIONS IN THE CALIBRATION OF FUNDAMENTAL
STELLAR QUANTITIES

Arthur D. Code

Washburn Observatory
University of Wisconsin

ABSTRACT. This paper summarizes the current status of ultraviolet
spectrophotometry with emphasis on the instrumental characteristics
unique to space observations and on the application of existing data to
the calibration of stellar properties. The currently available data
bases will be briefly reviewed. When combined with ground based data,
ultraviolet observations provide information on effective temperatures
and bolometric corrections for early type stars and on the nature of the
intervening interstellar medium. The ultraviolet measurements are
sensitive to chemical composition differences and provide a powerful
tool in discussion of stellar evolution in composite systems. This
review concludes with a brief discussion of future directions in
instrumentation and analysis.

1.   INTRODUCTION

     The traditional advantages of carrying out astronomical observations
from space are associated with the absence of significant terrestrial
atmospheric effects. Thus the sky is darker, the entire electromagnetic
spectrum is observable and the "seeing" is always excellent. In
particular the ability to extend observations to the X-ray, ultraviolet
and the far infra-red have led to many new and exciting advances. The
high energy region of the spectrum provides data on the final stages of
stellar evolution and on the interaction of close binary systems, while
infra-red observations enable astronomers to penetrate the dense clouds
surrounding stellar systems in the process of forming. In this review,
however, attention will be primarily directed to a discussion of stellar
measurements of the ultraviolet flux. Ultraviolet spectrophotometry is
simply an extension of classical stellar astronomy to shorter wavelengths. As such, the techniques and applications are in general
similar to those of ground based observations. There are, however,
certain characteristics of the instrumentation and of the data which
are unique to this spectral region. In the following section we shall
describe the current problems relating to photometric calibration.

D. S. Hayes et al. (eds.), Calibration of Fundamental Stellar Quantities, 209–224.
© *1985 by the IAU.*

2. PHOTOMETRIC CALIBRATION

The bulk of the radiation from early-type stars is emitted at wavelengths shorter than 3300 Å in the vacuum ultraviolet, observable only from above the terrestrial atmosphere. The fraction of the flux radiated in the visual and ultraviolet regions of the spectrum of normal stars as a function of effective temperature is illustrated in Figure 1. For temperatures above 13000 °K most of the flux is radiated at wavelengths shortward of 3300 Å, while above about 42000 °K the major fraction of the flux is in the Lyman continuum shortward of 912 Å. Clearly any comprehensive discussion of the properties of stars of spectral types O8 through B8 must include data on the flux distribution in the vacuum ultraviolet. Main sequence and giant stars in this spectral interval represent the youngest population in a stellar system. The main sequence age is indicated by the dashed curve and the right hand scale on Figure 1. Any main sequence star with more than 50% of the flux in the ultraviolet has a life-time less than one galactic rotation. On the other hand many highly evolved stars such as sdO stars, WDA precursors, and planetary nuclei are characterized by temperatures in excess of 40000 °K. For these objects we cannot expect to measure the total flux distribution since the interstellar hydrogen is opaque near 912 Å for even the closest stars. We can, however, derive important information on their structure by extending the observation to the Lyman limit. Moreover, many of these objects are close enough and hot enough that observation in the extreme ultraviolet, (EUV), can be made. Observations of these high surface gravity hot stars have recently played a role in the calibration of ultraviolet fluxes.

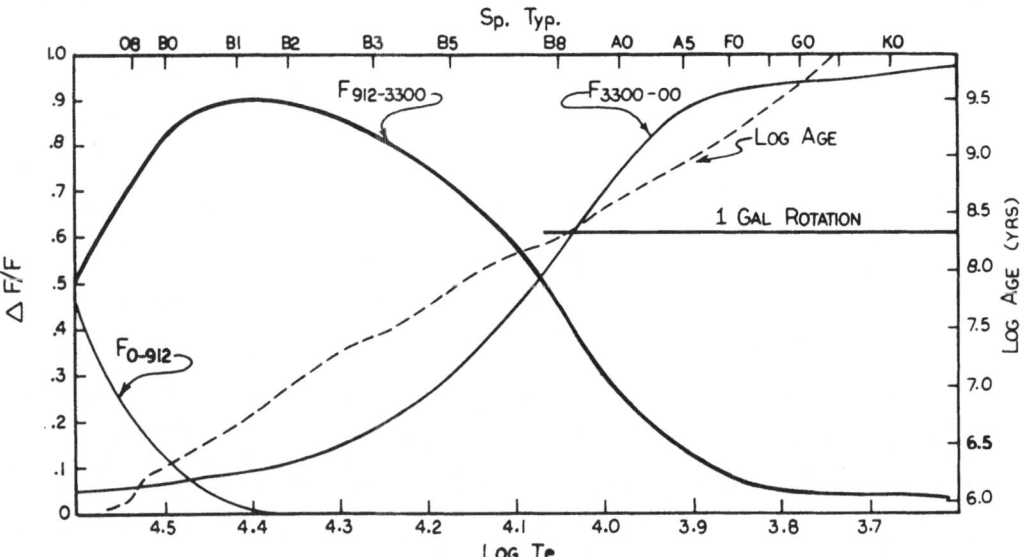

Fig. 1. Fractional flux vs effective temperature in the wavelength intervals 0-912 Å, 912-3300 Å and 3300-∞. Dashed curve and right hand scale is the logarithm of main sequence life-time.

The absolute energy calibration in the ultraviolet presents several unique problems. Most standard sources (and all thermal sources) have exceedingly little energy in the ultraviolet and a great deal at longer wavelengths, making scattered light a very difficult problem. Moreover all measurements must be carried out in a vacuum environment. This presents no problem in principle but is time consuming and complicates the procedures. Suitable transmission optics are limited and reflectivities are generally poorer in the ultraviolet. The fundamental absolute calibration efforts in the optical and infra-red have been based upon blackbody sources. For practical temperatures, thermal radiation is not suitable, however, in the UV. For example, a tungsten ribbon filament lamp operated at 2850 °K drops by 20 magnitudes between 5500 Å and 1500 Å.

In general calibration of flight instruments have been carried out using black receivers or calibrated detectors. Synchrotron radiation provides an ideal fundamental source for calibration of these secondary standards. Several synchrotron storage rings are now available as radiation sources. For calibration purposes, only a very small beam current is required to provide radiation of the desired intensity and spectral distribution. In calibrations carried out at the University of Wisconsin, it was found practical to operate at currents corresponding to approximately 100 electrons with energies of 240 Mev. The total number of electrons could then be counted. It is possible to measure the radiation from a single electron. Thus by allowing the beam to decay, it is possible to count the electrons one by one as they are "kicked" out of the beam. From these data and the theory of synchrotron

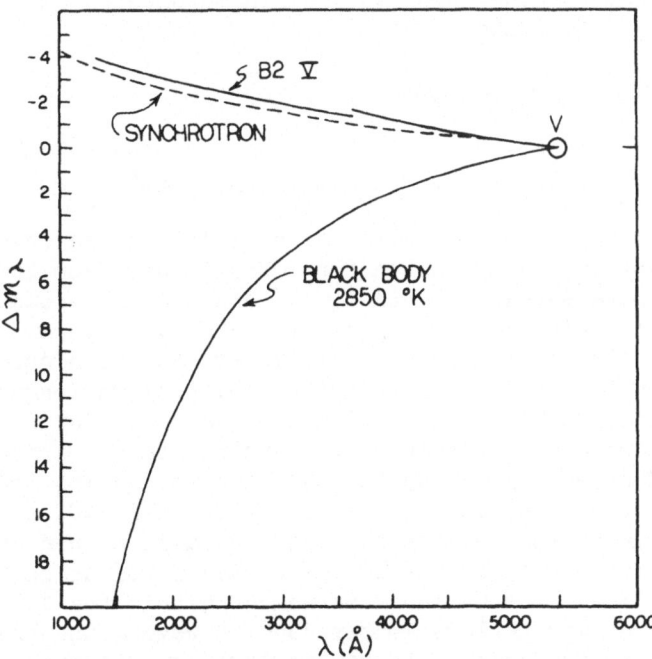

Fig. 2. Spectral distribution of a B2 V star, 240 Mev. synchrotron radiation and a 2850 °K black-body.

radiation the number of ergs per second per unit wavelength interval
intercepted by the detector is easily determined. The synchrotron
radiation from such storage rings has spectral distribution similar to a
B5 star. Figure 2 shows the spectral distribution between 1000 Å and
5500 Å for a black body at 2850 °K, a 240 Mev synchrotron spectrum and
the spectrum of a B2 V star. The flux is given in magnitudes per unit
wavelength interval normalized to the V magnitude. The advantage of
synchrotron radiation as a fundamental standard is obvious. Currently
most calibration efforts are carried out by using photodiodes that are
referenced to synchrotron calibrations carried out in the US by the
National Bureau of Standards (Canfield, Johnston and Madden 1973).
Despite the availability of these standard diodes the absolute energy
calibration, particularly shortward of about 2000 Å, is a difficult
process.

Most precision ultraviolet spectrophotometry has been obtained with
the OAO-2, TD1, ANS, Copernicus and IUE satellite observatories. Of
these only IUE is still providing data and the IUE calibration is
generally adopted as the standard. Before discussing the current status
of the IUE calibration, however, it is helpful to comment on the quality
of data obtained by these earlier satellites. Koornneef et al. (1980)
have compared OAO-2, ANS and TD1 photometric data for some 531 stars.
In general the absolute calibrations longward of 1800 Å agreed to within
10%. At 1550 Å the OAO-2 calibration was about 20% brighter than the
other determinations. When these three independent sets of data were
reduced to a common absolute energy calibration the agreement was
generally better than 0.1 magnitudes. The photometric system adopted
for these data expresses the absolute flux in magnitudes in accordance
with the following definition.

$$m_\lambda = -2.5 \log F_\lambda - 21.10$$

The constant -21.10 was chosen so that a constant flux per unit
wavelength interval would have a magnitude approximately equal to the
visual magnitude of the star.

The basis for the absolute calibration of IUE is the observed flux
for η UMa. The absolute flux chosen for η UMa is based on the results
obtained by the above satellites, Apollo 17 and sounding rocket results.
The procedure is described by Bohlin et al. (1980). Figure 3 shows the
ratio of absolute fluxes for the several systems to the adopted IUE
scale. The system is essentially the OAO-2 calibration longward of
2000 Å and the Johns Hopkins calibration shortward of 2000 Å. In order
to reduce the random errors a number of standard stars observed by OAO
and TD1 were used after correcting the data for the ratios indicated in
the figure. With the exception of the high flux near 1500 Å shown by
the OAO data the agreement between all these data is within about 10%.
In addition comparison with a line blanketed LTE model atmosphere
(T_{eff} = 17000 °K, log g = 4.0) shows agreement within the 10% level.
This IUE calibration was revised in May 1980, however, the basis of the
calibration remained the same. The new calibration utilized more data
and incorporated some refined reduction algorithms. The current IUE
calibration is described by Bohlin and Holm (1980). Since that time the

sensitivity of the detectors has decreased and a redundant detector
employed. In the final reduction of IUE data instrument signatures
resulting from a variety of causes, temporal changes, temperature
corrections, background non-linearities etc., will be more clearly
defined and we can expect significant enhancement of the data.

The absolute calibration of IUE is based on standard stars and not
any measurement of the instrument response function. Currently the flux
of these standard stars is known to about 10%. For some applications
this accuracy is insufficient and efforts to improve the flux
calibration continue.

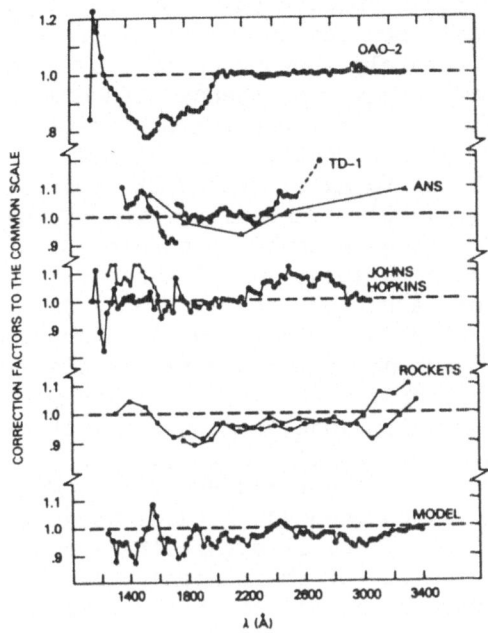

Fig. 3. Ratio of absolute fluxes on the IUE scale to those from
         different sources as given by Bohlin et al. (1980).

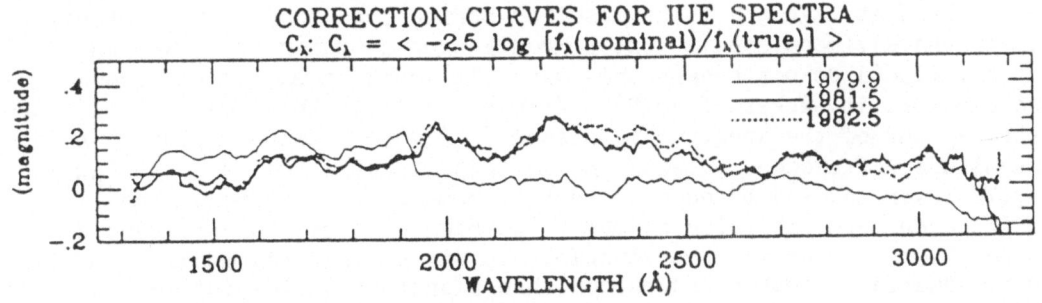

Fig. 4. Correction curves for IUE spectra based on fitting observations
         of DA white dwarfs to models (Finley et al. 1984).

One direction that calibration efforts has taken is to compare the measurements of hot subluminous stars to model atmospheres. Finley, Basri and Bowyer (1984) presented a self-consistent recalibration of IUE based on observations of hot DA white dwarfs. By averaging color differences between observed spectra of 13 objects and appropriate models they found the correction curves shown in Figure 4. The differences are large--amounting to as much as 0.2 magnitudes at 2200 Å. It was also found that the calibration corrections changed with time, although these time variations are somewhat different than those reported in the IUE Newsletters. Finley et al. believe that the model atmosphere fluxes are reliable to about 2%, that the physics employed is well understood and since these objects have very little interstellar reddening the DA white dwarfs provide the best available means of calibrating UV fluxes. Ultimately, however, the flux calibration must be based upon measurements and not consistency with models.

In the extension of the flux calibration from 1200 Å to 912 Å the hot subluminous stars have also recently been invoked. Ultraviolet observations of stars with the Voyager 1 and 2 UV spectrometers were reported by Holberg et al. (1982). The calibration of the flight instruments were based on reference to NBS photodiodes. They found that their fluxes from 1200 A down to about 1050 Å were in good agreement with the earlier calibrations of Carruthers et al. (1981) and the earlier Johns Hopkins calibration of Brune et al. (1979). Shortward of 1050 Å the Voyager fluxes were 60% higher and the Brune fluxes 60% lower than Carruthers. They then recalibrated the Voyager fluxes based on the high temperature, high surface gravity models of Wesemael et al. (1980) using the measured flux of HZ43. This recalibration showed good agreement with other star-model comparisons. Polidan and Holberg (1985) report in this volume on stellar fluxes between 912 Å and 1200 Å, based on the revised Voyager 1 calibration. They find that the spectra of the sdO star BD +28 4211 and central star of the planetary nebula NGC 246 fit a simple power law, $\lambda^{-3.78}$ from the visual to the Lyman limit. These results are interesting but a bit perplexing. The asymptotic limit as $T \to \infty$ is a slope of -4. To maintain a slope of -3.78 over this entire spectral range it is necessary that the opacity in the atmosphere vary in such a manner as to look to progressively higher temperatures towards the ultraviolet. Curiously a simple gray atmosphere temperature distribution with T_{eff} = 50000 °K and hydrogen opacity only does fit the relation, and indeed at high gravities electron scattering becomes unimportant. Unfortunately more sophisticated models such as the grid of Wesemael et al. (1980) do not behave this way. It would appear to require a temperature in excess of 100000 °K to approximate the power law although the red end of the spectrum is steeper. Figure 5 shows a plot of the magnitude differences between the infinite temperature -4 power law and a -3.78 power law and three representative model atmospheres. The differences are small for a 100000 °K model but the appropriateness of a power law spectrum is questionable. It may be that the modeling of DA white dwarfs is better understood than planetary nuclei but we return to the point that the ultimate calibration must involve absolute energy measurements. Both the NRL group and Johns Hopkins group have reported on new rocket measurements at the recent Seventh VUV Calibration Workshop.

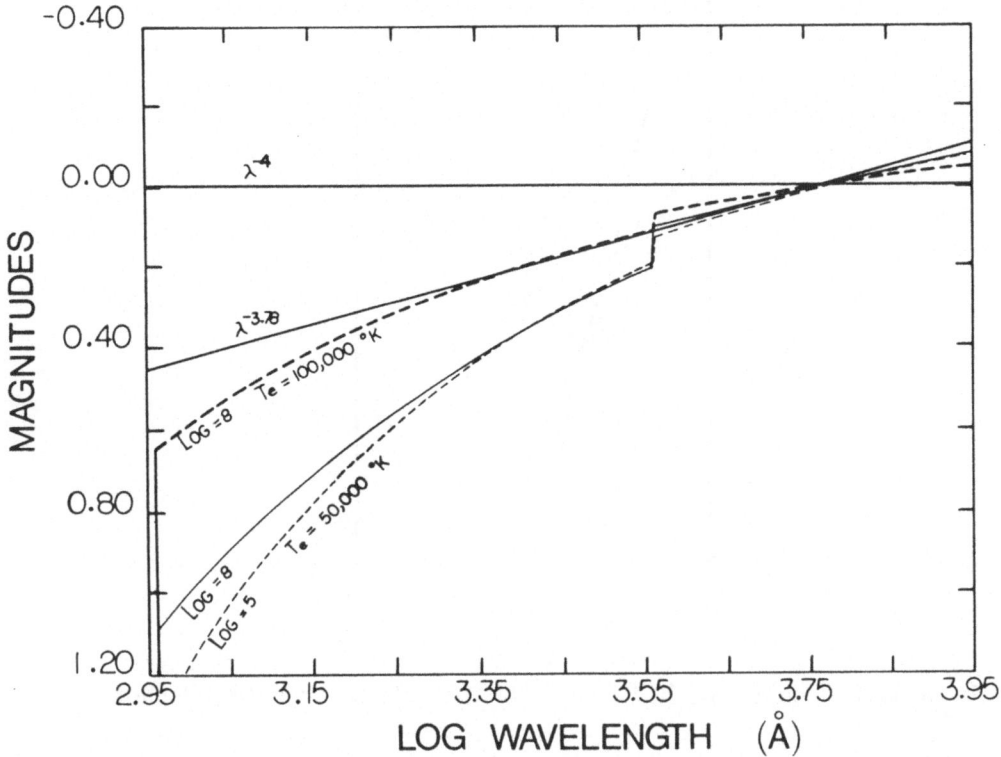

Fig. 5. Difference in magnitude between the flux from an infinite
         temperature black-body and (a) a power law of slope -3.78,
         and hydrogen model atmospheres for (b) upper dashed curve
         T_e = 100000 °K, log g = 8.0, (c) solid curve, T_e = 50000 °K
         log g = 8.0, (d) lower dashed curve T_e = 50000 °K, log g = 5.0.

Woods, Feldman and Bruner (1984) presented preliminary data agreeing
with the revised Voyager calibration within ±20% Å in the 1600 Å to
1400 Å and the 1200 Å to 1000 Å region but falling below the Voyager
results by nearly a factor of two shortward of 1000 Å. The possibility
of stellar variability at these wavelengths must be kept open when
comparing measurements at different epochs. This requires a larger data
set than currently available. In any event recent calibration efforts
appear to be converging. The discussion of hot subdwarfs is an
interesting astrophysical problem in its own right and it is to the
application of ultraviolet spectrophotometry that we turn our attention.

3. STELLAR OBSERVATIONS IN THE ULTRAVIOLET

     The extension of stellar observations into the vacuum ultraviolet
provides large photometric baselines which substantially increase the
sensitivity to temperature, gravity and chemical composition variations.

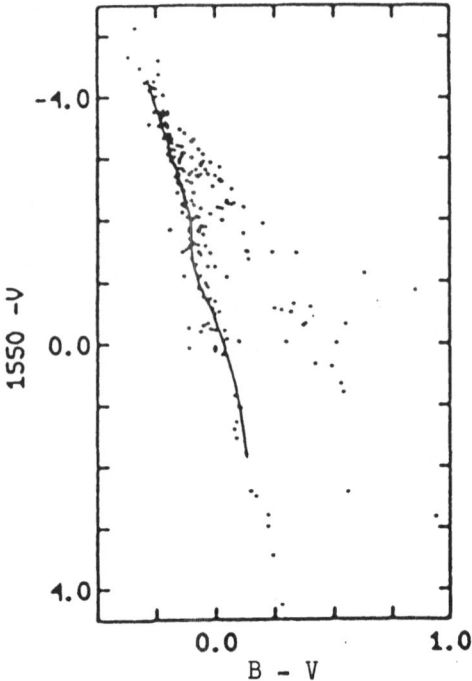

Fig. 6. A 1550-V color vs B-V color for 238 stars from OAO-2 photometry.

Figure 6 shows a typical color-color diagram in which the B-V color
index is plotted against a color index between 1150 Å and the V band
pass, m(1550)-V. For little redened early type main sequence stars the
slope of this relation is $\nabla(1550-V)/\nabla(B-V) = 12.7$. As such the colors
of comparable accuracy offer an order of magnitude greater
discrimination between stellar types. For the data sets described above
the photometric accuracy between systems is the order of 0.1 magnitudes
while the internal accuracies are significantly higher, of the order of
a few hundredths of a magnitude and thus comparable to ground-based
photometry. In principal higher photometric accuracy should be
attainable from space observations. The environment is benign, the
instruments stable over long periods of time and there are no
atmospheric extinction corrections or scintillation to contend with.
This is particularly advantageous in the study of rapid time variations.
        The ultraviolet is also more sensitive to interstellar extinction.
The average interstellar extinction at 2400 Å is 5 times as great as the
visual extinction and about 6 times the visual at 1200 Å. Fortunately
the interstellar extinction curve is non-linear and displays a
pronounced bump at 2200 Å corresponding to about 7 times the visual
extinction. The measurement of the 2200 Å bump is frequently used to
provide a determination of interstellar extinction. Unfortunately there
is a significant variation from the average extinction for some objects
(c.f. Meyer and Savage 1981). It is worth noting that in their
discussion of the IUE calibration Finley et al. point out that their
correction curve has a pronounced peak at 2200 Å that would lead one to
assume that an unreddened star had a B-V color excess of 0.08 magnitudes.

There now exists a substantial set of data on the spectra of stars
in the ultraviolet of both high and low spectral resolution. Thus from
TD1, Copernicus and the IUE high resolution mode many studies have been
carried out on interstellar and stellar lines of light elements
previously unavailable from observations in the visual. From these
investigations we have learned a great deal about the structure of the
interstellar medium and have come to recognize the non-thermal nature of
the upper atmosphere of B stars and the importance of mass loss. From
these data and insight we have new tools to apply to the task of
calibrating fundamental stellar quantities. Let me simply cite two
examples in which the addition of ultraviolet data has played a role.

The theory of stellar structure provides us with an interpretation
of the evolution and present structure of stars which traditionally is
depicted in the form of an HR diagram. The connection between these
theoretical HR diagrams and the stellar parameters of the observer has
not always been direct. Appeal is often made to the concept of
effective temperature and the bolometric correction. In the last decade
we have been able to make empirical determinations of these parameters
relatively independent of theory. The ability to observe most of the
electromagnetic spectrum provides not only details on the spectral
distribution but a reliable determination of the integrated flux. From
the integrated flux we may determine bolometric corrections or
alternatively, given a reliable parallax, the total stellar luminosity.
Based on the definition of effective temperature the total integrated
flux at the earth is

$$f = (\theta^2/4)\ T^4$$

where θ is the angular diameter subtended at the earth by the star.
Significant progress has also been made in the measurement of angular
diameters and with this information empirical effective temperatures can
be found. This provides the framework within which other less direct
methods may be compared and extended to a larger number of stars.

Due to the large number of atomic transitions occurring in the
ultraviolet, this spectral region is sensitive to chemical composition.
At high resolution, both weak lines and the resonance lines of abundant
elements can be studied. At low resolution, line blends and line
blanketing provide information on the metallicity. An example is shown
in Figure 7 which compares the composite IUE spectra of a sample of
Population I stars and field horizontal-branch stars with an average
effective temperature of 8500 °K. The log of the flux is normalized to
zero at 5500 Å. The lower spectrum is that of the Population I stars
showing both stronger lines and a depressed ultraviolet continuum due to
line blanketing. A color-color plot employing magnitude near 1500 Å
provides a sensitive index of metallicity in the temperature range from
about 7000 °K to 10000 °K. A comparison of HB stars with Population I
standards has been described by Huenemoerder et al. (1984).

By combining measurements in the Balmer continuum (ultraviolet) and
the Paschen continuum (visible) we should expect to be able to define
photometric parameters whose principal components are sensitive to
specific physical parameters such as temperature, reddening, gravity and

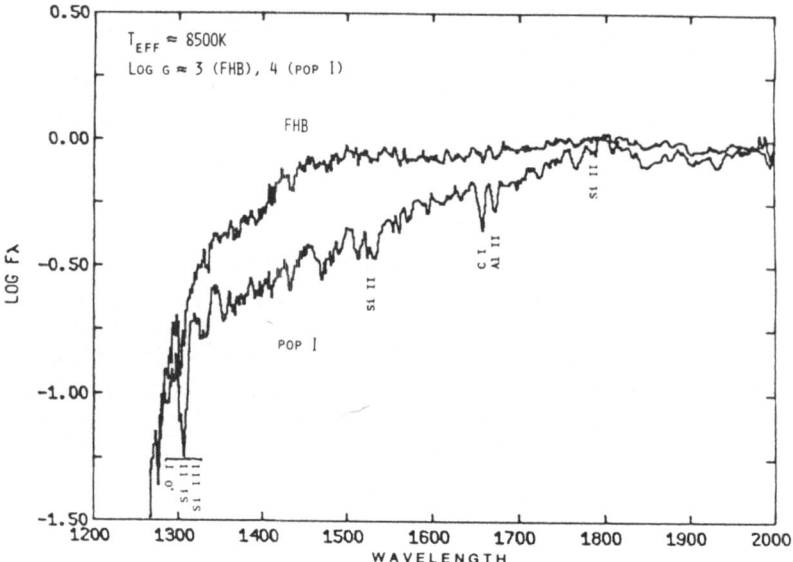

Fig. 7. Ultraviolet spectral distribution of average FHB star, upper
curve and Population I star at T_e = 8500 °K. The curves are
normalized at 5500 Å.

metallicity. The calibration of such indicators provides a useful tool
in the study of composite systems such as globular clusters and galaxies.
        One valuable resource for reviewing the scope of current activity
in ultraviolet astronomy are the conference reports of the several NASA
and ESA IUE Conferences.

4. ULTRAVIOLET OBSERVATIONS OF COMPOSITE STELLAR SYSTEMS

        The extension of the spectral range into the ultraviolet provides
additional leverage in separating the components of multiple stellar
systems. By way of a simple example Figure 8 shows the spectrum of
Antares. It is of course primarily the spectrum of the companion (α Sco
B), a B2.5 V star rather than the M1 Ib standard. There are a number of
systems in which the difference is far less extreme for which UV data
provides the decisive criteria. Among the composite stars contained in
the ANS Ultraviolet Photometry catalog of point sources (Wesselius
et al. 1982) the star 58 Per presents an interesting case. It has been
variously classified as K4 III + A3 V, G5 Ib-II + A3 or simply G8 II.
The ANS photometry contradicts these assignments. The 2200 Å channel
suggest a color excess of about 0.2 magnitudes. Figure 9 shows the
observed magnitudes in the five ANS channels along with the magnitudes
at U, B and V. The upper solid curve is the result of applying the
average galactic extinction law for a B-V color excess of 0.22
magnitudes. This envelope can be fit by two components with flux
distributions similar to a G8 II star and a B 2.5 V star. On the basis

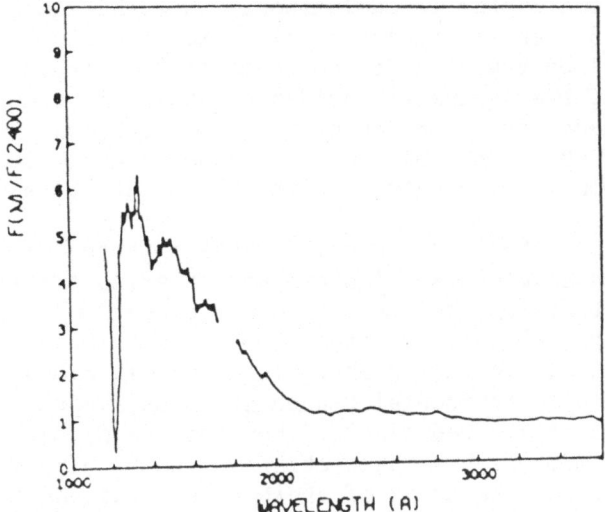

Fig. 8. Ultraviolet spectrum of Antares. The early type, B2.5 V,
companion dominates.

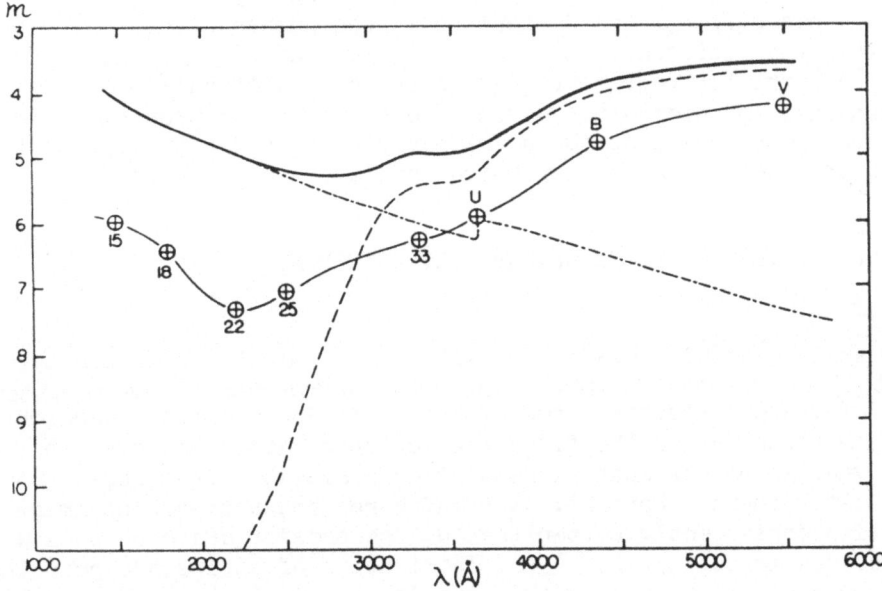

Fig. 9. Spectral distribution of possible components of the star 58 Per.
Circled crosses are UBV and ANS observed magnitudes. Upper solid
curve corresponds to observations corrected for E(B-V)=0.22 mag.
reddening. Dashed curve is the flux for a G8 II star and the
dash-dot curve that of a B2.5 V star. The two curves are
normalized to yield the corrected total magnitudes.

of the ANS photometry one would conclude that either the G star is more
luminous than a II or the B star is subluminous. Several IUE spectra
have been obtained of 58 Per and Harmer et al. (1983) conclude that the
components are G8 I and B5 V with a color excess of 0.3 magnitudes.

Another example of the leverage provided by extending observations
into the ultraviolet comes from IUE observations of classical Cepheid
variables. Apparently about one third of the Cepheids have blue
companions such as S Mon whose spectrum in the UV is that of a B4 V
star.

The extension of the study of composite systems to the integrated
light from objects such as globular clusters and galaxies provides
useful data for characterizing the population and evolution of these
objects. In general, the ultraviolet spectra of these sources is
dominated by the relatively few hot stars. In a typical globular
cluster a single faint blue horizontal-branch star contributes as much
light at 1500 Å as 1000 bright red giants. For the nearby systems
observations of single members have been obtained in the ultraviolet as
well as in the optical. On the basis of these investigations the
integrated properties can be understood in terms of age and chemical
composition. The integrated properties of globular clusters can then be
extended to more distant clusters and to members of other galaxies. A
review of some of the data on the ultraviolet spectra of globular
clusters has been given by Code (1983).

Ultraviolet observations of galaxies have shown that many systems
with similar spectra in the optical have distinctly different spectra in
the UV. This reflects the difference in the number of hot stars and
presumably the difference in the current rate of star formation. Galaxy
evolution currently represents one of the more difficult problems facing
the classical approaches to solving the cosmological problem. The
extension of observations to the ultraviolet provides an important tool
in studying the stellar and dynamical evolution of galaxies.

5. FUTURE DEVELOPMENTS IN ULTRAVIOLET ASTRONOMY

I shall conclude this review with some comments on future directions
in satellite instrumentation. Many astronomers regard the forthcoming
launch of Space Telescope a new milestone in astronomical research. This
2.4 meter telescope is the first stellar observatory designed to take
full advantage of the spatial resolution achievable from above the
earth's atmosphere. Optical, near infra-red and ultraviolet imagery with
resolution consistently better than 0.1 seconds of arc open up a wide
range of new investigations. The fabrication of optics and pointing
systems capable of providing this precision have been a major challenge
and this challenge appears to have been met. Space Telescope is
different from past missions in a number of ways. It is to be launched
by the shuttle and accessible to the shuttle for refurbishment and
maintenance. Communication to and from the spacecraft will be through
the new Tracking and Data Relay Satellite System (TDRSS). The quantity
and diversity of data obtained require an extensive suite of dedicated

hardware and software. The high resolution, in particular, provided by the Wide Field/Planetary camera and the Faint Object Camera present challenges in image processing that will have significant impact in all areas of astronomical data reduction. Perhaps most important to the astronomer is the fact that this unique space observatory is to be operated in a unique manner. The scientific responsibility for the operation of Space Telescope has been placed in the hands of the user community through the establishment of the Space Telescope Science Institute.

Space Telescope and other planned missions will, among other things, enrich our base of fundamental stellar quantities. We can anticipate an order of magnitude precision in parallaxes and proper motions and orbit determinations. Angular diameters from occultation measurements and from the space equivalent of speckle interferometry will add to our meager list. The gain to be expected from interferometric measurements in space is spectacular. The signal-to-noise ratio for such measurements varies linearly with the "coherency time", limited to less than 0.1 seconds due to seeing for ground-based observations; while in space times of the order of 1000 seconds are practical. The ability to observe stellar objects fainter than 26th magnitude will extend the luminosity function and explore the lower end of the main sequence in clusters in our own and other galaxies. Infra-red satellites will also make their contributions to fundamental astronomy and to a better determination of the birth rate function. Other instruments carried in the shuttle bay will provide data on the polarization of stars in the ultraviolet, and on the EUV spectrum of nearby stars. The shuttle may provide a means of carrying out absolute energy calibrations of high fidelity.

To exploit these opportunities, however, places a significant demand on current ground based telescopes. First the facilities in space for the near future will be limited and if an investigation can be done from the ground it will not have a very high priority in space. Secondly the preparation for carrying out space observations and the interpretation and exploitation of the results often requires ground based observing. Another related problem that I will conclude this review with is that of calibrating these advanced space flight instruments. The basis for photometric calibration of the scientific instruments on Space Telescope will be standard stars. For the most part current standards are too bright for some of the focal plane instruments. The instruments span a decade of the electromagnetic spectrum and yield imagery at flux levels unattainable from the ground. I believe the topic of calibration techniques for Space Telescope would be a suitable topic for discussion at this symposium and hope that there are some here that are prepared to contribute to that discussion.

In this review I have tried to indicate the scope of current stellar research in ultraviolet spectrophotometry. I have not strived for completeness, others here might well have chosen different topics to emphasize. The theme that runs throughout this text, however, is the unity of observations across the electromagnetic spectrum. Space observations eliminates the artifical boundary established by the atmospheric cut-off. Given facilities such as Space Telescope, ultraviolet astronomy, at least, disappears as a separate discipline.

REFERENCES

Bohlin, R.C., Holm, A.V., Savage, B.D., Snijders, M.A.J., and
    Sparks, W.M. 1980, Astron. Astrophys., 85, 1.
Bohlin, R.C., and Holm, A.V. 1980, NASA IUE Newsletter No. 10, 37
Brune, W.H., Mount, G.H., and Feldman, P.D. 1979, Astrophys. J.,
    227, 884.
Canfield, L.R., Johnston, R.G., and Madden, R.P. 1973, Appl. Optics,
    12, 1611.
Carruthers, G.R., Heckathorn, H.M., and Opal, C.B. 1981, Astrophys. J.,
    243, 855.
Code, A.D. 1983, Adv. Space Res., 2, 119.
Finley, D.S., Basri, G., and Bowyer, S. 1984, to be published in
    Proceedings of the Third Goddard IUE Symposium.
Harmer, D.L., Stickland, D.J., Lloyd, C., Harmer, C.F.W., Pike, C.D.,
    and Corft, D. 1983, Mon. Not. R. Astron. Soc., 204, 927.
Holberg, J.B., Forrester, W.T., and Shemansky, D.E. 1982, Astrophys. J.,
    257, 656.
Huenemoerder, D.P., de Boer, K.S., and Code, A.D., 1984, Astron. J.,
    89, 851.
Koornneef, J., Meade, M.R., Wesselius, P.R., Code, A.D., and
    van Duinen, R.J. 1982, Astron. Astrophys. Suppl., 47, 314.
Meyer, D.M., and Savage, B.D. 1981, Astrophys. J., 248, 545.
Polidan, R.S. and Holberg, J.B. 1985, in IAU Symposium No. 111:
    Calibration of Fundamental Stellar Quantities, eds. D.S. Hayes,
    L.E. Pasinetti and A.G. Davis Philip (Reidel: Dordrecht), p. 479.
Wesemael, F., Auer, L.H., Van Horn, H.M., and Savedoff, M.P. 1980,
    Astrophy. J. Suppl., 43, 159.
Wesselius, P.R., van Duinen, R.J., De Jonge, A.R.W., Aalders, J.W.G.,
    Luinge, W., and Wildeman, K.J. 1982, Astron. Astrophys. Suppl.,
    49, 427.
Woods, T.N., Feldman, P.D. and Bruner, G.H. 1984, Bull. Amer. Astron.
    Soc., 16, 492.

DISCUSSION

POLIDAN: I have two comments. First I agree with you completely that a definitive far-UV calibration does not yet exist. As stated in our paper (elsewhere in this volume) we feel that some of the differences between the existing calibrations are due to intrinsic variability in the reference stars observed. As discussed by many of the speakers at this symposium the proper choice of reference stars is critical to any fundamental system. Our observations of widespread flux variability in B stars at wavelengths less than 1150 Å would appear to make them unreliable as flux standards. It is primarily for this reason that we suggest that sub-luminous stars are more suitable as UV flux calibration standards. (See Polidan and Holberg in this volume for additional reasons.) Secondly, I will remark on a clarification of the techniques used to arrive at the revised Voyager calibration. The model atmosphere was used only to set the zero point of the flux calibration. The relative sensitivity function used was still the pre-launch value. While I agree with you that this is not a fundamental calibration, the situation in the far-UV was, as you have stated, confused. A working calibration had to be assumed. The fact that recent rocket flights have supported this assumed "calibration" have been quite encouraging.

BOHLIN: The Space Telescope project is the most expensive project ever undertaken by mankind for pure scientific research (except possibly the pyramids of Egypt). The ST calibration for the various photometric, geometric and polarimetric modes will be done using standard astronomical sources after launch. Therefore, the preparation of the data for the best standard targets must be completed during the remaining time before launch. There are considerable efforts currently being expended to define proper standards. We at the Space Telescope Science Institute invite the active participation of all calibration experts in defining the most appropriate ST standard sources.

GUSTAFSSON: I think the models we produce are quite up to date, but therefore necessarily very primitive descriptions of the quite complex systems that stars seem to be. Therefore, if somebody would rely on our models for calibrating the fundamental properties of his spectrophotometric system I am afraid I would stop calculating them, just because I would have difficulties sleeping at night.

ROUNTREE: I think it is time to lay to rest the ghost of Antares as a case of an MK standard that has a "different spectral type" in the ultraviolet. In fact, the M supergiant standard is Antares A, while the B star seen in the UV is Antares B. This is not a classification problem – we are seeing two different stars, not one star with two spectral types. There may be subtle problems that arise in using MK standards in the ultraviolet, but this is not one of them.

CODE: Yes, of course. I did not mean to imply a "different spectral type" but simply to illustrate what an actual composite spectrum of these type components would look like in the ultraviolet.

CAYREL: Have your hot subdwarfs all the same chemical composition? If not, how can you use them as standards?

CODE: I don't know the chemical composition of the hot subdwarfs. The spectra provide little information on chemical composition and there are various theoretical evolutionary scenarios for these stars. On the other hand, the flux is little affected by chemical composition and of course a standard star is simply a standard candle independent of chemical composition.

STELLAR ABSOLUTE FLUXES AND ENERGY DISTRIBUTIONS FROM 0.32 to 4.0 μm

D. S. Hayes

Kitt Peak National Observatory
National Optical Astronomy Observatories[1]

ABSTRACT. The absolute fluxes and energy distributions of stars
are the foundation of the calibration of fundamental effective
temperatures and bolometric corrections. In this paper I will review
recent progress in the calibration of absolute fluxes and energy
distributions in the visual and IR parts of the spectrum. In the
visual, the calibration of the absolute flux and energy distribution
of Vega has settled down well, and the remaining difficulties include
the lack of a worldwide common list of brighter secondary standard
stars, the lack of enough satisfactory fainter secondary standard
stars and the possibility of variability in Vega. In the IR, the
process of arriving at a dependable and accurate calibration, and
of linking it to commonly used photometric systems, is in its infancy.
A final, and rather special problem, is the question of the calibration
of the Sun. The Sun is a special case both because it is so well
studied astrophysically and because its extreme brightness makes
it very difficult to calibrate photometrically. Some progress has
recently been made on the calibration of the absolute flux and energy
distribution of the Sun, and I will discuss this work.

1. INTRODUCTION

     I am concerned here with the measurement of the absolute flux
and energy distribution of the stars within that part of the spectrum
which includes thermal radiation from the apparent surface of the
star. In terms of the calibration of fundamental stellar quantities,
the apparent total flux, f, radiated by a star, is related to the
effective temperature, T_{eff}, and angular diameter, $\theta = (2R/d)$, of
the star, through the equation:

$$f = (\theta^2/4)\sigma T^4_{eff}. \tag{1}$$

---

[1] Operated by the Association of Universities for Research in
Astronomy, Inc., under contract with the National Science Foundation.

D. S. Hayes et al. (eds.), Calibration of Fundamental Stellar Quantities, 225–252.
© *1985 by the IAU.*

The apparent bolometric magnitude, m_{bol}, is related to the apparent
total flux through the equation:

$$m_{bol} = -2.5 \log_{10} f + C = V + B.C., \tag{2}$$

where V is the apparent visual magnitude in the Johnson UBV system,
B.C. is the bolometric correction, and the zero point constant, C,
is determined by reference to the Sun:

$$m_{bol,*} - m_{bol,o} = -2.5 \log_{10}(f_*/f_o). \tag{3}$$

The measured quantities in these equations are the apparent total
fluxes, the angular diameters, and the V magnitudes of the Sun and
stars. The measurement of the V magnitudes of the stars is not a
major contributor to the errors here, so I will not discuss it further.
The V magnitude of the Sun is discussed below, and the measurement
of angular diameters is discussed by Hanbury Brown (1985) and Davis
(1985) in this symposium.

The quantity which I have been calling the "apparent total flux"
is the integral over wavelength (or frequency) of the apparent
monochromatic flux, f_λ (or f_ν). In fact, we do not measure the
apparent total flux because of the nature of our detectors and the
transmission of the Earth's atmosphere, and what is actually done
is to measure the apparent monochromatic flux at a number of
wavelengths and to perform the integral numerically. The measurement
of the apparent monochromatic flux of a star divides naturally into
three wavelength ranges: a) the UV, with wavelengths shortward of
the atmospheric cutoff at about 0.32 μm; these measurements must
be made from above the atmosphere, b) the "visual," with wavelengths
between the atmospheric cutoff and about 1.0 μm, and c) the IR, with
wavelengths longer than 1.0 μm. The UV is discussed by A. D. Code
(1985) in this symposium.

I will further separate the measurement of the apparent
monochromatic flux into two parts: the measurement of the absolute
monochromatic flux, which is measured at some standard wavelength,
such as 5000 or 5556A, and the measurement of the absolute energy
distribution, which is the apparent monochromatic flux normalized
to the standard wavelength. I emphasize the term "absolute" to
distinguish it from conventional relative photometry, in which the
measurement of the program stars is referred to one or more standard
stars. We make measurements of the absolute monochromatic flux and
energy distribution for only a limited number of stars, which become
the standard stars. For traditional and practical reasons the star
Vega (Alpha Lyrae = HR 7001 = HD 172167) is the primary standard
star. A number of other bright early-type stars have been defined
as secondary standard stars; the fluxes and energy distributions
of these stars have, for the most part, been determined through careful
measurements relative to Vega. In some cases the secondary standards
have been measured absolutely. For the sake of brevity, I will

hereafter refer to the absolute monochromatic flux and absolute energy distribution as the "flux" and "energy distribution" of the star. There will be no confusion since there is no longer any reason to consider the total flux, and since the term "absolute" is to be understood during the entire discussion. That is, I will specifically refer to the relative flux (or energy distribution), if that is what I mean.

In addition to their being the basis of the determination of fundamental effective temperatures and bolometric corrections, absolute fluxes and energy distributions are very important because of what can be learned by fitting the calculated energy distributions and fluxes from model atmospheres to the observations. Firstly, the degree of fit can be used to diagnose problems with the models and to improve the physics and the method of calculation. Secondly, the fitting of the model energy distributions to observations can be used to obtain values for fundamental stellar parameters such as effective temperatures and surface gravities. Although the values so obtained are not fundamental determinations, they can be very valuable as a supplement to the fundamental results. The fundamental measurements of effective temperatures are limited to only a few stars because of the paucity of well-measured angular diameters, particularly for certain regions of the HR diagram. There are similar limits to the numbers of fundamental determinations of surface gravities. The non-fundamental results can thus be a valuable supplement if they are properly calibrated by reference to such fundamental determinations as do exist. Reviews of this subject may be found in a number of places in the literature. For a discussion of the fundamental determination of effective temperatures and bolometric corrections, see Hayes (1978). For a discussion of the interplay between the calibration of energy distributions and the understanding of the physics in model atmospheres, see Mihalas (1975). For a discussion of the determination of fundamental parameters of stars through the use of model atmospheres, see Gustafsson and Graae-Jørgensen (1985), in this symposium, and references therein. The model atmospheres appear to fit the observations best for A-type and late B-type stars. For a discussion of these and related problems for the A-stars, see Wolff (1983), and references therein.

I will discuss several measurements of the absolute flux and energy distribution of Vega in the "visual" range; these "modern" measurements now agree very well. I will also discuss what has been done in the IR between 1 and 4 μm; here the agreement is not so good. I will discuss the secondary standard stars briefly later on. I will also discuss another bright and important star: the Sun. The Sun is a special case; it is measured absolutely and without any reference to any other star, except in rare (and generally unsuccessful) cases. The measurement of the energy distribution and flux of the Sun is very difficult because of its extreme brightness, and yet it is very important because this extreme brightness (along with its large angular size) has made possible

very detailed astrophysical investigations which we wish to relate
to other stars. The interagreement of some of the various energy
distributions and fluxes for the Sun which have been published is
good, while others disagree strongly. Nevertheless, there does seem
to be good reason to prefer the monochromatic flux distribution by
Neckel and Labs (1984, 1985). I will discuss this situation, below.

2. THE ENERGY DISTRIBUTION AND FLUX OF VEGA

The _absolute_ measurement of the flux and energy distribution
of Vega is carried out by comparing Vega with a _terrestrial_ source
of radiant energy whose monochromatic flux is known. The absolute
measurement of the flux and energy distribution of Vega or some
secondary standard is often called a _calibration_, or even "absolute
calibration" of the flux and energy distribution, and I will often
use this term here. In relative photometry, a star is compared with
a standard star in such a way as to minimize the difference between
the method or circumstances of measuring the two stars. In the case
of comparing Vega and a terrestrial source, the success of the
comparison depends upon making the measurement of the star and the
standard source as nearly the same as possible, or in accounting
for the differences. Two geometries have been used: 1) the most
common geometry used in the "visual" is to place the standard source
a few hundred meters from the telescope such that the telescope may
be pointed at the source and the measurement made in the same way
as for a star; 2) the most common geometry used in the IR involves
placing the standard source in the dome and introducing the light
from it into the optical system _after_ the telescope. In the first
case, the optical system is the same for the star and source, except
for a generally small differential vignetting due to the fact that
the source will not be at optical infinity for the telescope. The
difficulty is that there will be significant atmospheric extinction
between the source and the telescope when the distance is large enough
to place the source near enough to optical infinity to satisfy the
condition that the source and star be measured in the same way. In
the second case, obviously, the optical system will not be the same
for the source and the star, and the effects of the different optical
components and geometry must be carefully evaluated.

The terrestrial standard sources have been of two types: 1)
a blackbody, and 2) a tungsten striplamp, operated at a specified
current. The blackbodies which are used for this purpose have a
small chamber surrounded by a pure metal whose melting point is used
to define the temperature of operation; a small hole in the chamber
is the source of blackbody radiation. The blackbodies used for
astronomical calibrations usually operate at the copper-point or
platinum-point (primary gold-point blackbodies are too expensive
and generally too large to use in typical observatory locations).
The platinum-point is preferable because its temperature is higher
(2042°K), which gives more light in the UV. Copper-point (1358°K)

and gold-point (1338°K) blackbodies are difficult to use below 4000Å. On the other hand, platinum-point blackbodies are more difficult to construct and their melting point has not been well known until recently. The striplamp must be calibrated at a standards laboratory; most often the calibration is done by comparing the striplamp with a gold-point blackbody. In the "visual," both striplamps and blackbodies have been used successfully, whereas in the infrared blackbodies have been used exclusively.

The process of carrying out a measurement of the energy distribution and flux of a star can be broken into three parts: 1) the standardization, including the provision of a terrestrial standard source whose monochromatic flux is known as a function of wavelength with adequate accuracy, 2) the comparison, involving the decision on which geometry to use, the determination of horizontal extinction (if necessary), the determination of the effects of any optical components which are not the same for the measurement of the star and the terrestrial standard source, etc., and 3) the photometry of the star and source, including photometric or spectrophotometric system and the determination of the (vertical) atmospheric extinction. All three of these parts must be done well if the final result is to come out well. In my discussion of the measurements which are found in the literature, I will discuss these three parts, as appropriate.

I will consider data resulting from six calibrations (in parentheses I give an abbreviation): 1) Hayes and Latham (H&L) (1975), 2) Tüg, White and Lockwood (TWL) (1977); 3) Terez and Terez (T&T) (1979), 4) Kharitonov, et al. (KHAR) (1980), 5) Terez (1982) and 6) Arkharov and Terez (A&T) (1982). In the case of H&L, the data represent the result of a discussion and combination of data from three calibrations: the measurement of the energy distribution between 3200 and 10870Å at Lick Observatory by Hayes (1970), the measurement of the monochromatic flux at 31 wavelengths between 3300 and 10800Å at Palomar Mt. by Oke and Schild (1970), and the measurement of the fluxes at 6800, 8090 and 10400Å and the energy distribution between 7100 and 10800Å at the Mt. Hopkins Observatory by Hayes, Latham and Hayes (1975). It should be emphasized that the discussion by H&L is vital to the use of these calibrations, because H&L correct the original data for errors in the treatment of horizontal extinction, and also correct it to the International Practical Temperature Scale (IPTS) of 1968, to which all the other calibrations are referred. The calibration TWL gives the monochromatic flux distribution at 90 wavelengths from 3200 to 9040Å; it was done at Lowell Observatory. The calibration by KHAR gives the monochromatic flux distribution at 23 wavelengths from 3200 to 7500Å; it was done at Alma-Ata. The calibrations reported by T&T, Terez and A&T were done during the Ararat Expedition of the Main Astronomical Observatory. They are unusual in that the standard source was located in the dome; in the other cases the standard sources were placed from about 200m to about 1100m from the telescope. The calibration by T&T also includes data

taken at the Crimean Astrophysical Observatory and reports fluxes
at seven wavelengths, of which I have used only the flux at 5556Å.
The wavelength coverage of Terez and A&T will be discussed, below.

My objective here is twofold: one aspect is to compare the
data in order to show the present status of our knowledge of the
flux and energy distribution of Vega, and the other aspect is to
combine the data to create a "mean" flux and energy distribution
which may be used with greater confidence than any one of the original
calibrations. I will combine the data in such a way as to derive
a continuous energy distribution for Vega. This has not been done
in the past, generally, because at wavelengths near strong lines
and in the Balmer and Paschen confluences the details of wavelength
setting accuracy and relative bandpass size have made the use and
comparison of the data difficult, and the accuracy low. These problems
will remain present with this new energy distribution, so it must
be used with caution, particularly in the regions of the Balmer and
Paschen confluences. There is now a demand for continuous energy
distributions because of the increasing use of array detectors, and
this continuous energy distribution will be useful for calibrating
them. There is now an increasing need for continuous energy
distributions of stars for synthetic photometry (Hayes 1975, Buser
1978a, 1978b, Buser and Kurucz 1978, and Buser and Kurucz 1985),
and continuous energy distributions are needed for a large number
of stars for this purpose. They must be calibrated against Vega,
so the present continuous energy distribution will provide the basis
for improved results in this field. As an example, I use synthetic
photometry below in the discussion of the energy distribution of
the Sun. It turns out that doing the first aspect, comparing the
original calibrations, requires doing the second beforehand, so I
will next explain the combination of the data to form a "mean" flux
and energy distribution.

Each calibration has used a different bandpass and a different
set of wavelengths for their measurements, and this fact makes
comparing them difficult. The bandpasses range from 10 to 100Å,
and the set of wavelengths does not cover the spectrum continuously,
with the exception of Terez and A&T, and TWL in certain pieces. Often
the comparison is performed by interpolating with a smooth or linear
curve to a common set of wavelengths, ignoring the differential
line-blocking effects. The proper way to perform the interpolation
is to use a continuous spectrum with a resolution several times better
than the smallest bandpass to be considered. I do not have such
a spectrum of Vega available, but, fortunately, Terez and A&T report
the data continuously at 25Å steps over the entire wavelength range.
The bandpass was also 25Å, which is wider than ideal, but it will
do. In order to combine the different calibrations, I have
interpolated (with approximate allowance for the relative bandpass)
their reported wavelengths in the data given by Terez and A&T. From
this I determine a correction to the data by Terez and A&T; I then
interpolate in the correction to make it continuous with wavelength.

This correction is applied to the data by Terez and A&T to produce a continuous energy distribution which represents the energy distribution of the calibration being considered. I have then formed a weighted mean energy distribution of Vega, using the continuous energy distribution for each calibration, and the weights given in Table I in the last three columns. The final weighted mean continuous energy distribution, in terms of the relative magnitude of the monochromatic flux per unit wavelength interval, is given in Table 2. Note that I have taken the standard wavelength for normalization to be 5000Å.

With the final weighted mean continuous energy distribution of Vega in hand, we can now compare the different calibrations. I have again interpolated at the wavelengths of each calibration, and formed the differences (calibration minus weighted mean) for each calibration at its natural wavelength set. These differences are shown in Fig. 1 (3300-7500Å) and Fig. 2 (7000-10500Å). If we remember that good relative photoelectric spectrophotometry is characterized by observational errors on the order of 0.01 mag. (std. dev.), and also remember that the results shown in Figs. 1 and 2 include possible systematic errors characteristic of absolute calibrations, then we can conclude that the agreement shown here is superb. In particular, the agreement between 4000 and 8500Å shows that the standardization of these five calibrations is excellent. We have included here, a) tungsten striplamps calibrated in Heidelberg (Lick), Washington D.C. (Palomar) and Leningrad (Alma Alta, Crimea and Ararat Expedition), b) copper-point blackbodies following an NBS design (Palomar, Mt. Hopkins, Lowell), and c) a platinum-point blackbody of original design (Lowell), which also has been compared with a gold-point blackbody. One of the copper-point blackbodies (Lowell) was also compared with a gold-point blackbody with excellent results. Note, that Hayes, Oke and Schild (1970) directly compared the striplamps used in the Lick calibration with the striplamp used in the Palomar calibration, and found excellent agreement. There are some signs of problems, here: Terez departs significantly in the UV below 3400Å, KHAR departs significantly at 4000 and at 7000Å, and TWL departs significantly around 5900A and 8700-8800Å. In the case of the departures by Terez in the UV, they were recognized by the author and he had no explanation; neither do I. The departures by TWL around 5900Å appear to involve the end of the range of an order-separation filter; perhaps low signal levels or the leakage of extraneous light are the problem. The departures by TWL at 8700-8800Å are probably due to mismatching of wavelengths and bandpasses near high-order Paschen lines, and may very well be artifacts of my comparison process. The data shown in the two figures was constructed in the two wavelength ranges 3300-7500Å and 7000-10500Å because Terez and A&T report their data split in this way. In order to see if there is any systematic shift between the two pieces, I have compared the continuous weighted mean against H&L in Fig. 3 for the full wavelength range. Clearly, the agreement is excellent, and there is no evidence of a systematic shift of the "red" and "blue" pieces greater than reasonable observational error.

    I next consider the flux of Vega. There are five calibrations
to consider, including the same authors as represented above. H&L
include the flux measurements by Oke and Schild (1970) made at Palomar;
Oke and Schild measured the monochromatic flux at all 31 wavelengths
but report a result at 5556Å; this value is used here. H&L also
include fluxes by Hayes, Latham and Hayes (1975) made at Mt. Hopkins
at wavelengths of 6800, 8090 and 10400Å. The energy distributions
by Hayes (1970) and by Oke and Schild (1970) were used to derive
a weighted mean flux at 5556Å. TWL also measured the monochromatic
flux at all of their wavelengths, but also report a final flux
measurement for 5556Å. T&T report flux measurements made at seven
wavelengths; I use here a value for 5556Å which is the mean of their
values from observations at the Crimean Astronomical Observatory
and at the Ararat Expedition of the Main Astronomical Observatory.
The calibrations KHAR and Terez report fluxes for 5556Å, although
this was not one of the wavelengths at which fluxes were measured.
These results, in ergs/cm^2/sec/Å, are given in Table I, and the weights
used in calculating the mean are given in the following column headed
by the letter "f." The formal error (std. dev.) is only about 1.5%,
which is excellent agreement for six absolute flux measurements. In
Table II, the energy distribution is normalized at 5000Å, whereas
in Table I I derive the flux at 5556Å. Combining these two sets
of data allows calculating the flux per unit wavelength interval
at 5000A to be 4.65 x 10^{-9} ergs/cm^2/sec/Å.

    I would like to summarize the results of the discussion of the
flux and energy distribution of Vega in the "visual" in the following
way. Let us consider the usual observational errors found in good
spectrophotometry; these are, as stated above, about 0.01 mag. The
measurements of the absolute flux and absolute energy distribution
of Vega involve much of the same observational errors as normal
spectrophotometry. They involve, in addition, possible systematic
errors, which can be of any size. I will characterize the efforts
of a series of calibration measurements as mature when there is a
statistically useful number of calibrations and the systematic
agreement is on the order of the internal error, as is true for Vega
in the "visual." In the case of the IR for Vega there appear to
be systematic errors several times the size of the photometric errors,
and in the case of the "visual" for the Sun there are not enough
calibrations; in neither case can the accumulated calibrations be
said to be "mature."

    In the case of the IR between 1.0 and 4.0 μm, there are fewer
calibrations to consider: 1) Walker (1969), 2) Selby, et al. (1983)
and 3) Blackwell, et al. (1983). Walker's calibration was carried
out at the Agassiz Station of the Harvard College Observatory, in
Massachusetts. The blackbody was mounted in the adaptor between
the photometer and the telescope, and was operated at a temperature
of 402K. Measurements were made at wavelengths of 1.06, 1.13, 1.63

and 2.21 μm, with "equivalent widths" of .077, .114, .173 and .271 μm, respectively. The calibrations by Selby, et al. and Blackwell, et al. were carried out at Tenerife in 1980 and 1981, respectively with essentially the same equipment. The standard source was a furnace mounted between the telescope and the photometer, and the 1980 and 1981 observations differed with respect to the methods used to control the intensity of the furnace relative to the star. Observations were made at 2.20 and 3.80 μm in 1980, and 1.24, 2.20, 3.76 and 4.6 μm in 1981. The halfwidths were .034, .054, .145 and .323μm at the 1981 wavelengths, respectively. The furnace was calibrated against a standard blackbody. The calibration is ultimately traced back to the National Physical Laboratory, Teddington.

In each case, only a few wavelengths have been calibrated, and they are not wavelengths used in any commonly used system except where they are close to wavelengths in the standard JHKL system; in the latter case the bandpass is narrower, even where the wavelength is close to one of the effective wavelengths of JHKL. Because the calibrations are few and their wavelengths widely spaced, the approach used for the "visual" range is not appropriate. In order to have a reference spectrum for Vega I have used an ATLAS model (Kurucz 1979) which fits the "visual" energy distribution well. The model I have used is the (9400, 3.95, 0.00) model proposed by Kurucz to be a good fit to the Vega energy distribution. The fit to my new weighted mean is good, as can be seen in Fig. 4. The discrepancy between 4000 and 5000Å is disturbing, and would be interesting to investigate further. It is not my purpose, here, to discuss model atmosphere energy distributions, so I will pass it by. Except for that region, however, the fit of the ATLAS model is good, systematically. I use the IR energy distribution of this model for reference in Fig. 5, in which the IR flux calibrations are shown. I should emphasize that the IR measurements are made and reported as individual flux measurements, rather than as a flux plus an energy distribution. I show in Fig. 5 the weighted mean flux value given in Table III(the point is labelled "Hayes (1985)"). I also show in Fig. 5 a point for the flux calibration at 1.04μm from the Mt. Hopkins calibration by Hayes, Latham and Hayes (1975). This point is part of the data combined and reported by H&L, but is separated out and presented individually here. Note that in Fig. 5, the scale of the ordinate is coarser by a factor of two than used in the previous four figures. Clearly, the agreement is not nearly as good as in the "visual" range, and the amount of data far less. In this case, the systematic errors are significantly larger than the internal errors, and the number of calibrations are few, so I would characterize the situation as "immature."

Also shown in Fig. 5 are points representing three non-absolute calibrations. The non-absolute calibrations constructed in recent years use one of two basic assumptions: a) that the Sun has infrared colors similar to one or more solar analog stars; this assumption plus the solar absolute calibration in the infrared allows calibrating

the stars, and b) that the infrared calibration can be obtained from
a model atmosphere fitted to the visual energy distribution of Vega
or other stars. Hayes (1979b) constructs a calibration using both
of these bases and compares against the other absolute and non-absolute
calibrations available up to that time. Wamsteker (1981) uses the
solar-analog approach, and Koorneef (1983), as part of a critical
homogenization of JHKLM photometry, has constructed a calibration
which is very close to Wamsteker's, but which is based upon a constant
color temperature for a star with zero color-indices. These three
non-absolute calibrations are significant here because they are
attempts to calibrate the JHKLM photometry, which is the closest
we have to a standard system for spectrophotometry in the IR. Each
one presents the flux for zero magnitude in this system; I have assumed
V = +0.03 mag. and zero color indices for Vega in calculating the
values shown in Fig. 5. The agreement between these calibrations
is about as good (or as poor) as between the absolute calibrations
discussed above. If a mean of the absolute calibrations were to
be taken, it would not be well represented by any one of the three
non-absolute calibrations. Overall, Koorneef's appears to be the
closest, and is within roughly 0.05 mag. of such a mean. Clearly,
more work needs to be done on the IR calibration of Vega (or other
appropriate stars).

3. THE FLUX AND ENERGY DISTRIBUTION OF THE SUN

     In principle, the measurement of the flux and energy distribution
of the Sun is very similar to such measurements for Vega or any other
star, but, in practice, its extreme apparent brightness (compared
to the brightest of other stars) plus the fact that it is an extended
source make the measurements especially difficult. The extreme
apparent brightness and large angular size of the Sun also provide
for some great opportunities for detailed astrophysical investigations.
We would, of course, like to be able to compare the Sun with other
stars in terms of measurements which are made commonly on other stars,
such as the effective temperature and bolometric correction, which
depend, as described above, upon measurements of the flux and energy
distribution. One should note that, although the measurements of
the flux and energy distribution are difficult, the angular diameter
can be measured with an accuracy far better than for any other star.
Since the accuracy with which the angular diameter is measured is
the primary determinant of the accuracy with which the effective
temperature is measured (Hanbury Brown 1985, Davis 1985), the result
is that the effective temperature is better known for the Sun than
for any other star.

     As is true for other stars, the flux and energy distribution
are also important for comparison with model atmospheres of the Sun;
this case is very important because the models may be compared with
other observations with a detail which cannot be achieved for other
stars. Because the Sun can be observed so well in other ways, it

is particularly important that the models be a good fit, and that
means that it is particularly important that the energy distribution
be well measured. The present status of model atmospheres for
solar-type dwarfs is discussed by Gustafsson and Graae-Jørgensen
(1985) in this symposium.

The measurement of the flux and energy distribution of the Sun
have been the object of much effort in recent decades, but the result
has been a number of highly discordant results. There has been an
active controversy about whether making observations from a
high-altitude aircraft improves the measurements. The assertion
by the proponents is that atmospheric extinction is the major
contributor to systematic errors in the ground-based observations;
the alternate assertion is that the difficulties of doing the
standardization and the comparison will dominate because of the
environment in the aircraft and the restricted time available in
which to do the observations. In fact, the restricted time available
in which to do the observations makes the measurement of what
atmospheric extinction there is (and it is not negligible) more
difficult. The results seem to bear out the proponents of the
ground-based measurements. I do not wish to review all the recent
measurements nor to go through this controversy in detail, because
this effort has been undertaken by myself and many others already,
and the results have been published (Makarova and Kharitonov 1972,
1976; Neckel and Labs 1973, Labs 1975, Pierce and Allen 1977, Hayes
1979a, Hardorp 1980 and Taylor 1984a). I am most interested in
discussing the recent publication by Neckel and Labs (1984; see also
Neckel 1984 and Neckel and Labs 1985), which gives the monochromatic
flux continuously with wavelength from 3300 to 12500Å, with bandpasses
(and wavelength steps) of 10Å (3300-6300A), 20Å (6300-8700Å) and
50A (8700-12500Å). This work is based upon ground-based results;
the primary basis being measurements of the intensity of the center
of the solar disc made from the Jungfraujoch Scientific Station.
The standard source was a blackbody. This investigation demonstrates
the special demands made upon attempts to calibrate the solar spectrum;
since the original measurements the authors have spent considerable
effort on obtaining the data needed to determine the flux from the
entire solar disc, based upon the intensity of the center. Their
most recent efforts involve new limb-darkening and high-resolution
FTS spectrum measurements made at the Kitt Peak National Observatory.

A comparison and averaging of solar data in a manner like that
used above for Vega in the "visual" region is not appropriate in
the solar case, because of the large scatter in the solar data. I
note that the aircraft data by Arveson, et al. (1969), corrected
for a revised lamp calibration reported by Duncan (1969), is compared
with an earlier version of the Labs and Neckel (1968, 1970) data
by Labs (1975) and by Hardorp (1980); the comparison shows good
agreement from 4000 to 8000Å and from 1 to 2μm. Recent measurements
of the monochromatic flux of the Sun at ten wavelengths between 4100
to 10100Å, made at Mauna Kea (Shaw and Fröhlich 1979) and from a

stratospheric balloon (Fröhlich and Wehrli 1981) agree with a
preliminary version (Neckel and Labs 1981) of the new data by Neckel
and Labs excellently - with a standard deviation of 1.2% (Fröhlich
1983; Neckel 1984). Interference filters of typically 70Å bandpass
were used with silicon diode detectors. The radiometers at eight
wavelengths were calibrated against a tungsten striplamp which had,
in turn, been calibrated by the NBS. At two wavelengths, the
radiometers were calibrated at the Physikalisch-Meteorologisches
Observatorium, World Radiation Center, Davos, Switzerland, by using
dye lasers as intermediate standards, referenced to an electrical
cavity radiometer (Fröhlich 1983, Shaw 1982).

The agreement described above with the data by Arvesen, et al.
(1969) and by Fröhlich and his collaborators, plus the concensus
of the discussions of older data cited earlier, leads me to conclude
that the new data by Neckel and Labs is probably accurate to something
like ±1-2% (std. dev.) over the entire wavelength range covered (and
perhaps better). The fact that there are no other calibrations of
the solar monochromatic flux as a function of wavelength which cover
the entire wavelength range with a resolution and continuity comparable
to theirs means that one cannot be as confident as in the case of
Vega. Thus, I have tried to make other comparisons which might test,
if only roughly, the systematic accuracy of the new Neckel and Labs
data.

The first test I have performed is to compare the N&L data with
the energy distribution of two "solar analogs" which are calibrated
with respect to my new energy distribution of Vega. I have chosen
the double star system 16 Cyg A & B (HR 7503 and 7504), which has
been analysed by Perrin and Spite (1981), who concluded from high
dispersion spectra covering 4300-6000Å that 16 Cyg B was
"indistinguishable" from the Sun in terms of effective temperature,
surface gravity and chemical composition, and that 16 Cyg A was
"somewhat hotter," with a "smaller gravity." They used spectra of
the Moon for their solar reference. I have used the scans by Taylor
(1984a), covering 3288-7000Å continuously with passbands of 49Å
(3288-5304Å), 32Å (5248-6182Å) and 100Å (6050-6950Å), and corrected
from the calibration of Vega by Hayes and Latham (1975) to that of
this paper. I have made a very rough allowance for the difference
in passband sizes of the data for Vega, 16 Cyg A & B and the Sun,
but I am clear that bandpass mismatches represent a major difficulty
in the comparison I have made here. I smoothed the data by Neckel
and Labs roughly to Taylor's bandpasses and interpolated to Taylor's
passband centers. The differences (16 Cyg A minus the Sun) and (16
Cyg B minus the Sun) are shown in Figure 6. The agreement is here
very good systematically, but there are problems which lead me to
recommend that this comparison be carried through more carefully
and for more stars. The excursion of about 0.06 mag. at about 3500Å
is somewhat disturbing, as are the "waves" in the data through the
rest of the spectral range, but these effects may well be due to
the problems of matching bandpasses and wavelengths. In any case,

considering the number of steps involved, the agreement does show that there is a meaningful degree of coherence between the calibrations of the Sun and Vega.

The next comparison is, in some respects, weaker yet, but also shows to a useful degree the coherence between the calibrations of the Sun and Vega. I have calculated synthetic values of V and (B-V) for the Sun using a method which I have described earlier (Hayes 1975, 1979a). This method involves convolving the response functions of the B and V filters as recommended by Ažusienis and Straižys (1966) with the Neckel and Labs monochromatic flux distribution of the Sun. The transformation coefficients in (B-V) were determined by fitting synthetic values of (B-V) with observed ones for energy distributions of sample spectral types from B0 to M0 given by Straižys and Sviderskienė (1972); the latter were converted to the calibration of Hayes and Latham (1975) which is for this purpose indistinguishable from the calibration derived in this paper. The observed mean colors for each spectral type are from Johnson (1966). I have shown (Hayes 1975) that one must use such a wide range of spectral type in order to obtain a trustworthy value for the transformation coefficients unless one is concerned with a very narrow range of spectral type. The zero-point in V was determined from the energy distribution and flux of Vega, itself, derived earlier in this paper. The synthetic values of V and (B-V) for the Sun and Vega are given in Table III. The value of (B-V) for Vega of -0.016 mag. is a good indication of the maximum systematic error which one can expect from this method, when good energy distributions are used. Thus, I would associate an error of about 0.02 mag. with the final synthetic values of V and (B-V) of the Sun, -26.75 and +0.661 mag. respectively.

I must compare the synthetic photometry with observations of the Sun, and this clearly is the weak point of the comparison, because direct photometric measurements of the Sun are very difficult because of its extreme brightness, compared to the stars for which UBV photometers were designed to measure. There have been a number of determinations of the apparent visual brightness of the Sun, but I find only three which have been made photoelectrically: that by Nikonova (1949), transformed to the V-magnitude scale by Martynov (1960), that by Stebbins and Kron (1957), corrected by myself for an error in the treatment of horizontal extinction (see Hayes and Latham 1975) and that by Gallouët (1964). Their values are summarized in Table IV. Similarly, there are only a few photoelectric determinations of (B-V) for the Sun. Stebbins and Kron (1957), measured color in the six-color system of Stebbins and Whitford; I have transformed their results into (B-V) (Hayes 1979) and do not find the use of the six-color system and the need to transform it into the UBV system a significant problem in this context. Additionally, there are the measurements by Gallouët (1964), Preski (1970) and Tüg and Schmidt-Kaler (1982). Their results are also summarized in Table IV. The mean values for these observations are -26.75±0.06 and +0.661±0.03 mag.

The exact agreement of the values of V and (B-V) at the ends of the last two paragraphs is accidental, of course, but the fact that the agreement is good is an indication of a significant degree of coherence in the calibrations of the Sun and Vega. I wish also to point out that the interagreement of the photometric observations, which is the basis of the error figures I attach to the means, is not nearly as good as the internal errors quoted by the authors. For example, all four measurements of (B-V) quote internal errors of 0.01 mag., and yet the range is 0.06 mag! Clearly, there are significant systematic errors in these measurements. One can reduce the systematic error in the mean of a series of such measurements if there are enough of them and if they are all really measuring the same thing; in this case the averaging of only four measurements does not guarantee that the mean is free of significant systematic error. On the other hand, the fad of determining the value of (B-V) of the Sun from spectroscopic measurements misses the point: our objective is to determine the photometric behavior of the Sun!

I would like to conclude this section by recalling my earlier characterization of the calibration of Vega in the "visual" range as "mature," whereas I concluded that the calibration of the calibration of Vega in the IR is yet "immature." In the case of the calibration of the Sun in the "visual" range, the calibration is yet immature, even though the new calibration of the Sun by Neckel and Labs is probably as accurate as the calibration of Vega! The reason for my characterization of the calibration of the Sun as "immature" is that there are not a large enough number of calibrations which agree at a level close to their internal errors. Thus, we do need more excellent calibrations of the Sun, in addition to what we have.

4. THE POSSIBILITY OF VARIABILITY IN VEGA

Since Vega is used as the primary standard star for the measurements of stellar energy distributions and fluxes, it is important to consider the possibility that it is a variable star. As Batten (1985a) puts it, a standard star should be: "constant in the characteristic for which it has been chosen as a standard, within the smallest attainable errors of measurement." There have been reports in the literature for over 50 years of observations of variable brightness, spectrum and radial velocity for Vega. A useful summary of the history of these reports has been published by Wisniewski and Johnson (1979); their concern about this topic was spurred by their apparent discovery of emission lines in the near infrared spectrum (Johnson and Wisniewski 1978). These emission lines have not been confirmed by other observers (Barker et al. 1978; Griffin and Griffin 1978), and their relevance to the use of Vega as a spectrophotometric standard is purely circumstantial. The earliest observations of brightness variations include those by Guthnick, who built the first successful photoelectric photometer,

and who used the new photometric technique to observe Vega from 1915
to the 1930's. He reported variations with an amplitude of a few
hundredths of a magnitude over characteristic times of variation
of hours to months (Guthnick 1918, 1930a, 1930b, 1930c, 1931). It
must be remarked, on the one hand, that these observations should
not be rejected solely because of their age. On the other hand,
they should be treated with considerable caution, because the
photometric equipment and technique used were primitive and the
observing site marginal for photometry. The amplitudes he reports
cannot be much larger than his internal errors. More recently, the
long series of UBV observations by H. L. Johnson and his collaborators
shows residuals larger than some other bright stars, and if interpreted
as evidence of variability, then the amplitude would be several
hundredths on a magnitude (Johnson 1980; Wisniewski and Johnson 1979).
Kharitonov, et al. (1980) have performed absolute energy distribution
and flux measurements on eight secondary standard stars similar to
that discussed above for Vega. They have cross-compared the
observations, made during 1977 and 1978, of all the stars, including
Vega, and found evidence of variability of Vega on the order of 0.02
to 0.05 mag. Kozyreva, Moshkalev and Khaliullin (1981) have reviewed
some of the literature, and have reported their own observations.
These are WBVR photoelectric photometry made at an altitude of 3
km, covering three months during 1980. They say that the observations
"showed no brightness variations significantly exceeding the
measurement error (σ = 0.006 mag.)," but the data "indicates the
possibility of (quasi) periodic microvariability of Vega with an
amplitude of ∿0.02 mag. and a period (characteristic time) of about
an hour." Their mean value for the V magnitude (V = 0.034 mag.)
agrees well with the results in the literature. Fernie (1981) made
photoelectric observations on 14 nights over four months in 1980.
He used a mask on the telescope only for observations of Vega. On
one night Vega appeared to be brighter by 0.041 mag; two other nights
had brightenings of about 0.015 mag. On the remaining nights the
star was constant to about 0.006 mag. Glushneva (1983b) reports
that Sperauskas in 1983 described photoelectric observations covering
three seasons during which variations did not exceed 0.01 mag.
Finally, I can report unpublished IR observations made at Kitt Peak
which also do not show evidence of variability. R. R. Joyce has
made 10 JHKL measurements of the difference in brightness of Vega
and γ Lyr over the 3 1/2 year period from October, 1980 to March,
1984, and finds an overall standard deviation of 0.007 mag.
Measurements have been made in a nearly monochromatic photometric
system which has 13 wavelengths between 1.04 and 4.0 μm by the author
and R. F. Wing, S. T. Ridgway, R. R. Joyce and C. P. Rinsland (Hayes,
et al. 1980, Hayes, et al. 1983). Twenty-nine scans of Vega were
made on 27 dates between December, 1979 and November, 1982, and were
reduced in a network with 46 other stars. Because of the way the
observations and reductions were made I cannot give a precise value
for the limit on the variations of Vega, but they must be less than
0.01 mag. In summary, there is some evidence for low-amplitude
variations in the brightness of Vega, but it results from the

less-controlled or older observations; the more recent observations with the most appropriate observational techniques do not show variability on a scale which would be important, here. I should also note that the six absolute flux measurements which I discussed above cover a period of time of over ten years and have a standard deviation of only 1.5%.

As noted above, there have also been reports of radial-velocity variations; some of these and the history are discussed by Wisniewski and Johnson (1979). Clearly, evidence of pulsation would be relevant here, but the evidence is far from definitive. In fact, reports of unpublished observations, given at this symposium by Batten (1985b) and Walker (1985) indicate that no variability of the radial velocity is present on a scale which would be significant, here.

My overall conclusion is that the evidence for variability in Vega is not strong enough to indicate a need for a program to find and begin observing a substitute primary spectrophotometric standard star. I think the evidence is strong enough that we should be aware of the possibility of variability, so that we make observations and encourage our colleagues to make observations which will help decide the issue. Photometric observations of this type are needed for many of the brightest stars, and I would like to encourage the photometrists in the audience and the readership to undertake them, if they are so inclined.

5. THE SECONDARY STANDARD STARS

My concern in this Section is the availability of secondary standard stars which can be used when Vega is not visible or is too bright. The secondary standard stars which are in use today are, for the most part, standards for energy distribution measurements but not for fluxes; as noted above fluxes for stars other than Vega and the Sun are usually obtained by use of the V magnitude relative to that of Vega. I will henceforth only consider standards for measurements of energy distributions. A secondary standard star should have an energy distribution measured with a photometric accuracy which is close to that of Vega; one finds stars in the literature which are used as standards which are simply taken from one or more of the many catalogues of stars with measured energy distributions. Certainly, without a critical evaluation of energy distributions from several sources this procedure is very dangerous. I can recommend here only stars which are well-measured several times and critically evaluated as secondary standards.

After reviewing the lists of secondary standards to be found in the literature, one can conclude that while there are some truly useful lists of such stars available, there is not enough unity to make such lists universally valuable. For example, observers in the Soviet Union mostly use secondary standards from lists containing seven or eight bright stars (Kharitonov and Glushneva 1978, Kharitonov, et al. 1980, Voloshina, Glushneva and Shenavrin 1980 and Glushneva and Ovchinnikov 1982). Observers in the Western countries have mostly used the 11 secondary standards proposed by Breger (1976). Taylor (1984b) has recently published a list of 16 secondary standards which include, and supersede, Breger's. I have made a preliminary attempt to compare the Soviet and Western lists, but there are only two stars in common (α Leo and η UMa) in the "blue" spectral region. The agreement between the energy distributions by Taylor and by Glushneva and Ovchinnikov (1982) appears to be very good. In the "red" γ Ori is common, as well, and the agreement between the energy distributions by Taylor and by Voloshina, Glushneva and Shenavrin (1980) is not so good, especially at the longer wavelengths.

I have not carried out the comparison above in any greater detail because of the lack of an adequate number of overlapping stars, and because the wavelength sets are so different that a detailed comparison using continuous spectra would be needed. I recommend strongly that the Soviet and Western observers include each other's secondary standard stars in their observing programs, so that this comparison can be carried out properly. Having made the comments above, I can call attention to the very useful combined list of secondary standards published by Glushneva (1983a) and the supplementary list by Burnashev (1984).

In addition to the bright secondary standard stars discussed in the previous paragraphs, there is an intensifying need for faint standards. An early list of such standards is by Stone (1974, 1977); the stars are between 10th and 12th mag. The recent publication of four secondary standards by Oke and Gunn (1983) is very important; the stars are F subdwarfs between 8th and 10th mag., and they have been very carefully calibrated against Vega. The energy distributions are continuous, as well. Hayes and Philip (1984) have compared their observations of five stars observed at Palomar using BD +17° 4708, Oke and Gunn's "primary" secondary standard, with their observations at Kitt Peak and Cerro Tololo using Breger's secondary standard stars; the comparison shows excellent agreement. This means that between 3400 and 6800A the standards by Oke and Gunn are on the same system as those by Breger. Hayes and Philip (1985) have made a comparison which indicates that Taylor's energy distributions will give, if anything, improved agreement. Other lists of fainter secondary standards include those by Stone and Baldwin (1983) and Baldwin and Stone (1984) for the southern sky. There is no overlap with any other lists of faint secondary standards, so comparison is impossible. Hayes and Philip (1985) also give a list of faint (7th to 12th mag.) and fainter (15th to 16th) secondary standards. Another list is that by Ipatov (1983), which includes 10 stars between 7th and 9th mag. There is only one star in common with any other list.

The situation in the IR is similar, in some respects, to that for the bright stars in the "visual." First, I note that the absolute

calibrations discussed above do not reproduce any standard combination
of bandpasses and wavelengths. Second, however, it should be noted
that there is not one JHKL system, but several, since a number of
observatories are using instrumentally defined systems and their
own sets of standard stars (Glass 1973, 1974a, 1974b, Wamsteker 1981,
Elias, et al. 1982, Allen and Cragg 1983 and Joyce, Probst, and Guetter
1984). Clearly, a true spectrophotometric system in the IR is needed;
we have one in process at Kitt Peak (Hayes, et al. 1980, Hayes, et
al. 1983) in which 47 stars have been observed at 13 nearly
monochromatic wavelengths between 1.04 and 4.0 μm. The publication
of this system is waiting upon the completion of an absolute
calibration.

CONCLUSIONS

The calibration of the energy distribution of Vega has matured
in recent years, and the mean energy distribution and flux given
in this paper can be recommended as having an accuracy on the order
of 1.0 to 1.5% over the wavelength range 3300 to 10500Å. On the
other hand, the secondary standards need more work, in that more
overlap between the various lists (both bright and faint) in use
is badly needed. The calibration of the IR, and the availability
of secondary standard stars in the IR, is yet immature, and I recommend
more effort in this wavelength range. The calibration of the energy
distribution of the Sun, again, is probably now quite accurate, and
is apparently quite coherent with the new energy distribution of
Vega, but the lack of a number of co-equal calibrations leads to
a lack of confidence which would be best remedied by having more
such calibrations.

REFERENCES

Allen, D.A. and Cragg, T.A. 1983, Mon. Not. R. Astron. Soc., 203,
    777.
Arkharov, A.A. and Terez, È.I. 1982, preprint.
Arvesen, J.C., Griffin, R.N., Jr., and Pearson, D.B., Jr. 1969, Appl.
    Optics, 8, 2215.
Ažusienis, A. and Straižys, V. 1966, Bull. Vilnius Obs., No. 16,
    3.
Baldwin, J.A. and Stone, R.P.S. 1984, Mon. Not. R. Astron. Soc.,
    206, 241.
Barker, E.S., Lambert, D.L., Tomkin, J. and Africano, J. 1978, Publ.
    Astr. Soc. Pacific, 90, 514.
Batten, A.H. 1985a, in: IAU Symposium No. 111: Calibration of
    Fundamental Stellar Quantities, ed. D.S. Hayes, L.E. Pasinetti
    and A.G. Davis Philip, (Reidel: Dordrecht), p. 3.
Batten, A.H. 1985b, Discussion following paper by McAlister (1985).
Breger, M. 1976, Astrophys. J. Suppl., 32, 1.
Blackwell, D.E., Leggett, S.K., Petford, A.D., Mountain, C.M. and

Selby, M.J. 1983, Mon. Not. R. Astron. Soc., 205, 897.

Burnashev, V. 1984, Std. Star Newslett., No. 4, 5.

Buser, R. 1978a, Astron. Astrophys., 62, 411.

Buser, R. 1978b, Astron. Astrophys., 62, 425.

Buser, R. and Kurucz, R.L. 1978, Astron. Astrophys., 70, 555.

Buser, R. and Kurucz, R.L. 1985, in: IAU Symposium No. 111: Calibration of Fundamental Stellar Quantities, ed. D.S. Hayes, L.E. Pasinetti and A.G. Davis Philip, (Reidel: Dordrecht), p. 513.

Code, A.D. 1985, in: IAU Symposium No. 111: Calibration of Fundamental Stellar Quantities, ed. D.S. Hayes, L.E. Pasinetti and A.G. Davis Philip, (Reidel: Dordrecht), p. 209.

Davis, J. 1985, in: IAU Symposium No. 111: Calibration of Fundamental Stellar Quantities, ed. D.S. Hayes, L.E. Pasinetti and A.G. Davis Philip, (Reidel: Dordrecht), p. 193.

Duncan, C.H. 1969, GSFC Report No. X-713-69-382 (Greenbelt).

Elias, J.H., Frogel, J.A., Matthews, K. and Neugebauer, G. 1982, Astron. J., 87, 1029.

Fröhlich, C. 1983, Appl. Optics, 22, 3928.

Fröhlich, C. and Wehrli, C. 1982, in Proceedings, Third Scientific Assembly of IAMAP, Hamburg, 1981; The Symposium on the Solar Constant and the Spectral Distribution of Solar Irradience, (Boulder).

Gallouët, L. 1964, Ann. d'Astrophys., 27, 423.

Glass, I.S. 1973, Mon. Not. R. Astron. Soc., 164, 155.

Glass, I.S. 1974a, Mon. Not. Astron. Soc. S. Africa, 33, 53.

Glass, I.S. 1974b, Mon. Not. Astron. Soc. S. Africa, 33, 71.

Glushneva, I.N. 1983a, Inform. Bull. CDS, No. 24, 7.

Glushneva, I.N. 1983b, Std. Star Newslett., No. 3, 10.

Glushneva, I.N. and Ovchinnikov, S.L. 1983, Soviet Astron., 26, 548.

Griffin, R. and Griffin, R. 1978, Publ. Astr. Soc. Pacific, 90, 518.

Gustafsson, B. and Graae-Jørgensen, U. 1985, in: IAU Symposium No. 111: Calibration of Fundamental Stellar Quantities, ed. D.S. Hayes, L.E. Pasinetti and A.G. Davis Philip, (Reidel: Dordrecht), p. 303.

Guthnick, P. 1918, Veroff. Berlin-Babelsberg, II, 91.

Guthnick, P. 1930a, Vierteljahrschrift den Astro. Gesellshaft, 51, 79.

Guthnick, P. 1930b, Sitzungsberichte den Akad. Wiss. Berlin (Phys. Math. Klasse), I, 3.

Guthnick, P. 1930c, Sitzungsberichte den Akad. Wiss. Berlin (Phys. Math. Klasse), I, 495.

Guthnick, P. 1931, Sitzungsberichte den Akad. Wiss. Berlin (Phys. Math. Klasse), II, 22.

Hanbury Brown, R. 1985, in: IAU Symposium No. 111: Calibration of Fundamental Stellar Quantities, ed. D.S. Hayes, L.E. Pasinetti and A.G. Davis Philip, (Reidel: Dordrecht), p. 185.

Hardorp, J. 1980, Astron. Astrophys., 91, 221.

Hayes, D.S. 1970, Astrophys. J., 159, 165.

Hayes, D.S. 1975, in: Multicolor Photometry and the Theoretical HR Diagram, ed. A.G. Davis Philip and D.S. Hayes, (Dudley Obs. Rept. No. 9), p. 309.

Hayes, D.S. 1978, in: <u>IAU</u> <u>Symposium</u> <u>No</u>. <u>80</u>, <u>The</u> <u>HR</u> <u>Diagram</u>, ed. A.G. Davis Philip and D.S. Hayes, (Reidel: Dordrecht), p. 65.

Hayes, D.S. 1979a, in: <u>Problems</u> <u>of</u> <u>Calibration</u> <u>of</u> <u>Multicolor</u> <u>Photometric</u> <u>Systems</u>, ed. A.G. Davis Philip, (Dudley Obs. Rept. No. 14), p. 223.

Hayes, D.S. 1979b, in: <u>Problems</u> <u>of</u> <u>Calibration</u> <u>of</u> <u>Multicolor</u> <u>Photometric</u> <u>Systems</u>, ed. A.G. Davis Philip, (Dudley Obs. Rept. No. 14), p. 297.

Hayes, D.S., Joyce, R.R., Ridgway, S.T., Rinsland, C.P. and Wing, R.F. 1980, <u>Bull</u>. <u>Amer</u>. <u>Astr</u>. <u>Soc</u>., 12, 837.

Hayes, D.S. and Latham, D.W. 1975, <u>Astrophys</u>. <u>J</u>., 197, 593.

Hayes, D.S., Latham, D.W. and Hayes, S.H. 1975, <u>Astrophys</u>. <u>J</u>., 197, 587.

Hayes, D.S., Oke, J.B. and Schild, R.E. 1970, <u>Astrophys</u>. <u>J</u>., 162, 361.

Hayes, D.S. and Philip, A.G.D. 1985, in: <u>IAU</u> <u>Symposium</u> <u>No</u>. <u>111</u>: <u>Calibration</u> <u>of</u> <u>Fundamental</u> <u>Stellar</u> <u>Quantities</u>, ed. D.S. Hayes, L.E. Pasinetti and A.G. Davis Philip, (Reidel: Dordrecht), p. 469.

Hayes, D.S. and Philip, A.G.D. 1985, <u>Astrophys</u>. <u>J</u>. <u>Suppl</u>., 53, 759.

Hayes, D.S., Wing, R.F., Ridgway, S.T., Joyce, R.R. and Rinsland, C.P. 1983, <u>Std</u>. <u>Star</u> <u>Newslett</u>., No. 2, 9.

Ipatov, A.P. 1982, <u>Soviet</u> <u>Astron</u>., 26, 368.

Johnson, H.L. 1966, in: <u>Ann</u>. <u>Rev</u>. <u>Astron</u>. <u>Astrophys</u>., 4, 193.

Johnson, H.L. 1980, <u>Rev</u>. <u>Mex</u>. <u>Astron</u>. <u>Astrofis</u>., 5, 25.

Johnson, H.L. and Wisniewski, W.Z. 1978, <u>Publ</u>. <u>Astr</u>. <u>Soc</u>. <u>Pacific</u>, 90, 139.

Joyce, R.R., Probst, R.G. and Guetter, H.H. 1984, <u>Bull</u>. <u>Amer</u>. <u>Astron</u>. <u>Soc</u>., 16, 497.

Kharitonov, A.V., Tereshchenko, V.M., Knyazeva, L.N. and Boiko, P.N. 1980, <u>Soviet</u> <u>Astron</u>., 24, 168.

Kharitonov, A.V., Tereshchenko, V.M., Knyazeva, L.N. and Boiko, P.N. 1980, <u>Soviet</u> <u>Astron</u>., 24, 417.

Kharitonov, A.V. and Glushneva, I.N. 1978, <u>Soviet</u> <u>Astron</u>., 22, 284.

Koorneef, J. 1983, <u>Astron</u>. <u>Astrophys</u>., 128, 84.

Kozyreva, V.S., Moshkalev, V.G. and Khaliullin, Kh.F. 1981, <u>Soviet</u> <u>Astron</u>., 25, 705.

Kurucz, R.L. 1979, <u>Astrophys</u>. <u>J</u>. <u>Suppl</u>., 40, 1.

Labs, D. 1975, in: <u>Problems</u> <u>in</u> <u>Stellar</u> <u>Atmospheres</u> <u>and</u> <u>Envelopes</u>, ed. B. Baschek, W.H. Kegel and G. Traving (Springer-Verlag, Berlin), p. 1.

Labs, D. and Neckel, H. 1968, <u>Z</u>. <u>für</u> <u>Astrophys</u>., 69, 1.

Labs, D. and Neckel, H. 1970, <u>Solar</u> <u>Phys</u>., 15, 79.

Makarova, Ye.A. and Kharitonov, A.V. 1972, <u>Distribution</u> <u>of</u> <u>Energy</u> <u>in</u> <u>the</u> <u>Solar</u> <u>Spectrum</u> <u>and</u> <u>the</u> <u>Solar</u> <u>Constant</u>, Nauka Publishing House, Moscow, (English Translation: NASA Technical Translation TT F-803, National Aeronautics and Space Administration, Washington, D.C., June 1974).

Makarova, Ye.A. and Kharitonov, A.V. 1976, <u>Soviet</u> <u>Astron</u>., 19, 585.

Martynov, D.Ya. 1960, <u>Soviet</u> <u>Astron</u>., 3, 633.

McAlister, H.A. 1985, in: <u>IAU</u> <u>Symposium</u> <u>No</u>. <u>111</u>: <u>Calibration</u> <u>of</u>

Fundamental Stellar Quantities, ed. D.S. Hayes, L.E. Pasinetti and A.G. Davis Philip, (Reidel: Dordrecht), p. 97.

Mihalas, D. 1975, in: *Multicolor Photometry and the Theoretical HR Diagram*, ed. A.G. Davis Philip and D.S. Hayes, (Dudley Obs. Rept. No. 9), p. 241.

Neckel, H. 1984, *Space Sci. Rev.*, 38, 87.

Neckel, H. and Labs, D. 1973, in: *IAU Symposium No. 54, Problems of Calibration of Absolute Magnitudes and Temperatures of Stars*, ed. B. Hauck and B.E. Westerlund, (Reidel: Dordrecht), p. 149.

Neckel, H. and Labs, D. 1981, *Solar Phys.*, 74, 231.

Neckel, H. and Labs, D. 1984, *Solar Phys.*, 90, 205.

Neckel, H. and Labs, D. 1985, in: *IAU Symposium No. 111: Calibration of Fundamental Stellar Quantities*, ed. D.S. Hayes, L.E. Pasinetti and A.G. Davis Philip, (Reidel: Dordrecht), p. 473.

Nikonova, E.K. 1949, *Izv. Krymsk. Astrofiz. Observ.*, 4, 114.

Oke, J.B. and Gunn, J.E. 1983, *Astrophys. J.*, 266, 713.

Oke, J.B. and Schild, R.E. 1970, *Astrophys. J.*, 161, 1015.

Perrin, M.N. and Spite, M. 1981, *Astron. Astrophys.*, 94, 207.

Pierce, A.K. and Allen, R.G. 1977, in: *The Solar Output and its Variation*, ed. O.R. White (Colorado Assoc. U. Press, Boulder), p. 169.

Preski, R.J. 1970, *Goodyear Aerospace Corp. Memorandum*, SP-7276.

Selby, M.J., Mountain, C.M., Blackwell, D.J., Petford, A.D. and Leggett, S.K. 1983, *Mon. Not. R. Astron. Soc.*, 203, 795.

Shaw, G.E. 1982, *Appl. Optics*, 21, 2006.

Shaw, G.E. and Fröhlich, C. 1979, in *Solar-Terrestrial Influences on Weather and Climate*, ed. B. McCormac and B. Seliga (Reidel: Dordrecht), p. 69.

Stebbins, J. and Kron, G.E. 1957, *Astrophys. J.*, 126, 226.

Stone, R.P.S. 1974, *Astrophys. J.*, 193, 135.

Stone, R.P.S. 1977, *Astrophys. J.*, 218, 767.

Stone, R.P.S. and Baldwin, J.A. 1983, *Mon. Not. R. Astron. Soc.*, 204, 347.

Straižys, V. and Sviderskienė, Z. 1972, *Bull. Vilnius Obs.* No. 35, 3.

Taylor, B.J. 1984a, *Astrophys. J. Suppl.*, 54, 167.

Taylor, B.J. 1984b, *Astrophys. J. Suppl.*, 54, 259.

Terez, È.I. 1982, preprint.

Terez, G.A. and Terez, È.I. 1979, *Soviet Astron.*, 23, 449.

Tüg, H., White, N.M. and Lockwood, G.W. 1977, *Astron. Astrophys.*, 61, 679.

Tüg, H. and Schmidt-Kaler, T. 1982, *Astron. Astrophys.*, 105, 400.

Voloshina, I.B., Glushneva, I.N. and Shevarin, V.I. 1980, *Soviet Astron.*, 24, 576.

Walker, G.A.H. 1985, private communication.

Walker, R.G. 1969, *Phil. Trans. R. Soc. London, A*, 264, 209.

Wamsteker, W. 1981, *Astron. Astrophys.*, 97, 329.

Wisniewski, W.Z. and Johnson, H.L. 1979, *Sky and Tel.*, 57, 4.

Wolff, S.C. 1983, *The A-Stars: Problems and Perspectives* (NASA: Washington, D.C.).

TABLE I

WEIGHTS AND FLUXES FOR CALIBRATION OF VEGA

Calibration	flux, E-9	f	Weights 3300-7500	7000-9040	9040-10500
HAYES AND LATHAM (1975)	3.39	2	2	2	2
TÜG, ET AL. (1977)	3.47	1	1	1	—
TEREZ AND TEREZ (1979)	3.42	1	—	—	—
KHARITONOV, ET AL. (1980)	3.54	1	1	—	—
TEREZ (1982)	3.44	1	1	—	—
ARKHAROV AND TEREZ (1982)	—	—	—	1	1
Mean	3.44 ± 0.05				

TABLE III

BV SYNTHETIC PHOTOMETRY OF THE SUN AND VEGA

	$(B-V)_{SYN}$[*]	V_{SYN} + CONST	V_{OBS}	V_{SYN}
VEGA	− 0.016 MAG	− 7.738 MAG	+ 0.03 MAG	—— MAG
SUN	+ 0.661	−34.515	——	−26.75 MAG

[*]$(B-V)_{SYN}$ = 1.00(b−v) + 1.09 MAG. (HAYES 1975), FOR THE
RESPONSE FUNCTIONS BY AŽUSIENIS AND STRAIŽYS (1966).

TABLE IV

DIRECT PE DETERMINATIONS OF V, (B−V) OF THE SUN

SOURCE	V (MAG)
NIKONOVA (1949)[*]	−26.81 ± 0.05
STEBBINS AND KRON (1957) HOR. EXT. CORR. BY HAYES	−26.75 ± 0.03
GALLOUËT (1964)	−26.70 ± 0.01
MEAN	−26.75 ± 0.06

	(B−V) (MAG)
STEBBINS AND KRON (1957)[+]	+ 0.627 ± 0.01
GALLOUËT (1964)	+ 0.68 ± 0.01
TÜG AND SCHMIDT-KALER (1982)	+ 0.686 ± 0.01
PRESKI (1970)	+ 0.65 ± 0.01
MEAN	+ 0.661 ± 0.03

[*]TRANSFORMED TO V-MAG BY MARTYNOV (1960).
[+]TRANSFORMED TO (B−V) BY HAYES (1979).

TABLE II

ENERGY DISTRIBUTION OF VEGA

λ	$Maq(f_\lambda)$	λ	$Maq(f_\lambda)$	λ	$Maq(f_\lambda)$	λ	$Maq(f_\lambda)$
3300	.358	5100	.064	6900	1.049	8700	1.706
3325	.378	5125	.079	6925	1.063	8725	1.860
3350	.391	5150	.095	6950	1.069	8750	1.899
3375	.393	5175	.110	6975	1.081	8775	1.762
3400	.395	5200	.126	7000	1.088	8800	1.717
3425	.406	5225	.139	7025	1.101	8825	1.835
3450	.419	5250	.154	7050	1.115	8850	1.940
3475	.430	5275	.167	7075	1.129	8875	1.810
3500	.439	5300	.183	7100	1.141	8900	1.762
3525	.442	5325	.192	7125	1.156	8925	1.762
3550	.445	5350	.206	7150	1.171	8950	1.774
3575	.453	5375	.221	7175	1.181	8975	1.885
3600	.458	5400	.234	7200	1.191	9000	2.102
3625	.463	5425	.249	7225	1.201	9025	2.102
3650	.459	5450	.264	7250	1.212	9050	1.925
3675	.455	5475	.278	7275	1.226	9075	1.859
3700	.452	5500	.296	7300	1.241	9100	1.872
3725	.414	5525	.311	7325	1.252	9125	1.898
3750	.265	5550	.324	7350	1.264	9150	1.995
3775	.105	5575	.333	7375	1.273	9175	2.102
3800	-.060	5600	.349	7400	1.281	9200	2.220
3825	-.191	5625	.368	7425	1.293	9225	2.238
3850	-.342	5650	.384	7450	1.305	9250	2.118
3875	-.454	5675	.399	7475	1.316	9275	1.953
3900	-.438	5700	.415	7500	1.327	9300	1.912
3925	-.555	5725	.426	7525	1.340	9325	1.925
3950	-.469	5750	.441	7550	1.349	9350	1.939
3975	-.415	5775	.456	7575	1.364	9375	1.953
4000	-.595	5800	.471	7600	1.372	9400	1.953
4025	-.680	5825	.479	7625	1.387	9425	1.981
4050	-.669	5850	.498	7650	1.395	9450	2.024
4075	-.502	5875	.513	7675	1.404	9475	2.117
4100	-.364	5900	.529	7700	1.421	9500	2.256
4125	-.460	5925	.545	7725	1.429	9525	2.371
4150	-.581	5950	.561	7750	1.437	9550	2.372
4175	-.568	5975	.575	7775	1.453	9575	2.184
4200	-.553	6000	.592	7800	1.462	9600	2.070
4225	-.541	6025	.606	7825	1.470	9625	2.024
4250	-.520	6050	.619	7850	1.478	9650	2.010
4275	-.487	6075	.630	7875	1.487	9675	2.024
4300	-.393	6100	.642	7900	1.496	9700	2.024
4325	-.205	6125	.657	7925	1.514	9725	2.055
4350	-.208	6150	.672	7950	1.523	9750	2.085
4375	-.332	6175	.687	7975	1.533	9775	2.101
4400	-.404	6200	.702	8000	1.542	9800	2.116
4425	-.391	6225	.717	8025	1.551	9825	2.133
4450	-.375	6250	.733	8050	1.560	9850	2.132
4475	-.357	6275	.742	8075	1.570	9875	2.133
4500	-.340	6300	.754	8100	1.580	9900	2.148
4525	-.316	6325	.769	8125	1.589	9925	2.165
4550	-.272	6350	.779	8150	1.599	9950	2.217
4575	-.280	6375	.785	8175	1.620	9975	2.272
4600	-.270	6400	.793	8200	1.630	10000	2.349
4625	-.256	6425	.810	8225	1.640	10025	2.477
4650	-.240	6450	.824	8250	1.650	10050	2.596
4675	-.219	6475	.839	8275	1.662	10075	2.389
4700	-.199	6500	.859	8300	1.673	10100	2.329
4725	-.179	6525	.957	8325	1.683	10125	2.271
4750	-.157	6550	1.064	8350	1.695	10150	2.251
4775	-.140	6575	1.109	8375	1.705	10175	2.233
4800	-.119	6600	.992	8400	1.717	10200	2.215
4825	-.034	6625	.908	8425	1.739	10225	2.233
4850	.134	6650	.920	8450	1.750	10250	2.233
4875	.129	6675	.932	8475	1.762	10275	2.249
4900	.011	6700	.944	8500	1.774	10300	2.248
4925	-.049	6725	.953	8525	1.798	10325	2.267
4950	-.035	6750	.965	8550	1.810	10350	2.266
4975	-.015	6775	.978	8575	1.835	10375	2.283
5000	.000	6800	.991	8600	1.861	10400	2.282
5025	.017	6825	1.006	8625	1.810	10425	2.301
5050	.034	6850	1.022	8650	1.751	10450	2.300
5075	.048	6875	1.034	8675	1.705	10475	2.318
						10500	2.318

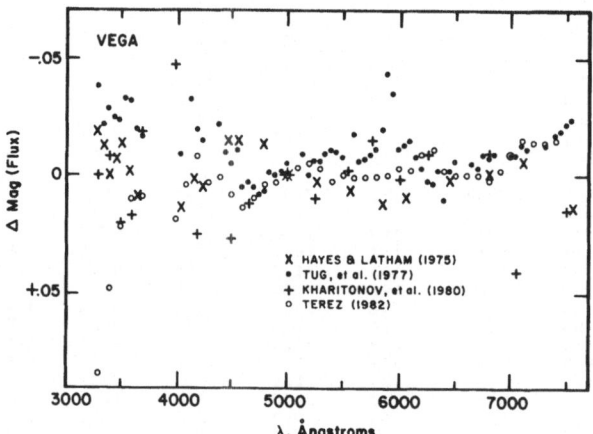

Fig. 1 The energy distribution of Vega over the wavelength range
3300 – 7500Å, referred to the weighted mean, for four sources.

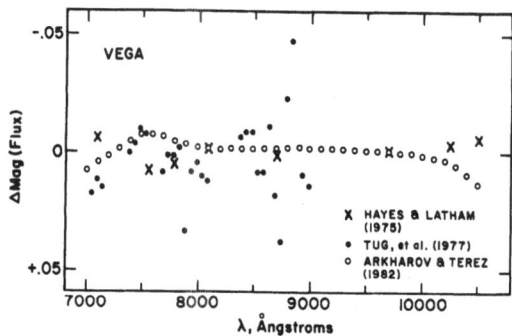

Fig. 2 The energy distribution of Vega over the wavelength range
7000 – 10500Å, referred to the weighted mean, for three sources.

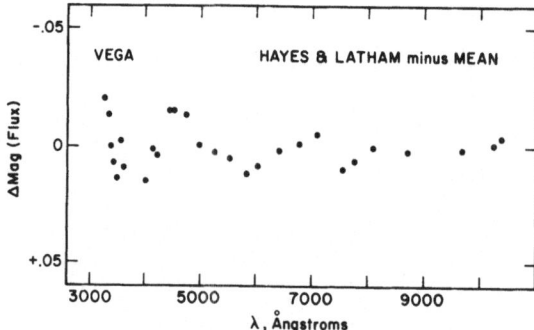

Fig. 3 The energy distribution of Vega over the wavelength range
3300 – 10500Å by Hayes and Latham (1975), compared to the weighted
mean.

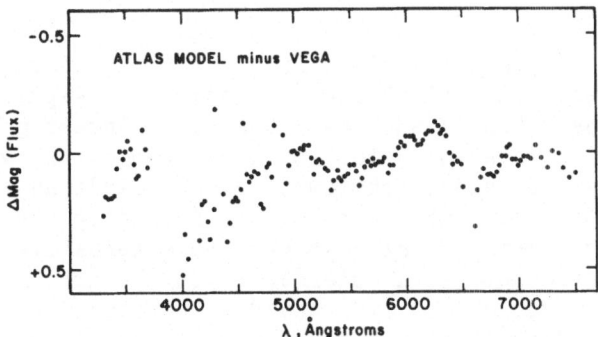

Fig. 4 An ATLAS model atmosphere (Kurucz 1979) fitted to the weighted mean over the wavelength range 3300 - 10500Å.

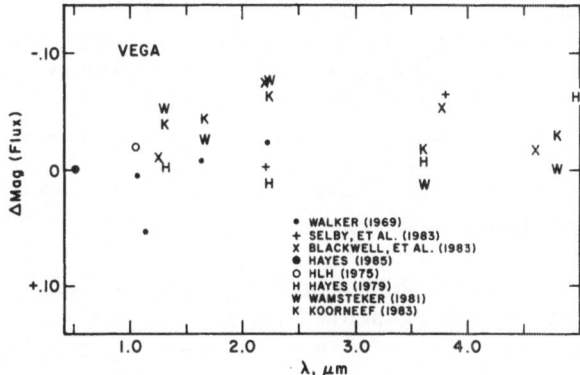

Fig. 5 The energy distribution of Vega over the wavelength range 0.55 - 4.6 μm, referred to the ATLAS model atmosphere shown in Fig. 4, for five sources.

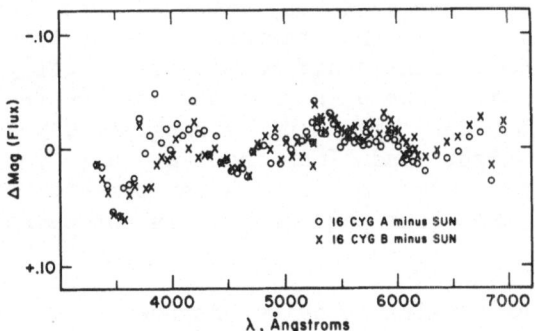

Fig. 6 The energy distribution of BS 7503 (16 Cyg A) and BS 7504 (16 Cyg B) (Taylor 1984, standardized to the weighted mean energy distribution of Vega) over the wavelength range 3300 - 7000Å, compared with the energy distribution of the Sun by Neckel and Labs (1984).

DISCUSSION

GLUSHNEVA: I want to begin the discussion with two comments. The first one concerns the possible brightness variation of Alpha Lyrae. We see that much effort has been made by a number of investigators to obtain reliable calibration data in a wide spectral range. If we delete Alpha Lyrae from the list of spectrophotometric standards as a variable star all of our spectrophotometric catalogues, including several thousand stars, lose their basis. Really we have no alternative to Alpha Lyrae as a reliable spectrophotometric standard. On the other hand, if Alpha Lyrae is really variable it would increase the errors of spectrophotometric data if a star is compared with Alpha Lyrae directly or by means of a secondary which is compared directly with Alpha Lyrae. So the importance of photometric observations of possible brightness variations of this star in the future is obvious.

My second comment concerns the reliability of the calibration in the infrared. It can be demonstrated that when we use monochromatic fluxes at the I, J, and K bands of the Johnson system for the determination of effective temperatures we find a systematic difference in the J and K determinations which must be taken into account if we use the calibration by Johnson. But if we use another calibration, for example the recent calibration data by Koorneef, the dependence of temperature on color becomes stronger.

ADELMAN: Was the Atlas model whose predictions you showed optimized for fit to the Vega calibration presented in this paper? If so, what T_{eff} and log g were used for this model and how was log g determined?

HAYES: No. I used the (9400, 3.95, 0.00) model which Kurucz proposed several years ago as a good fit to Vega.

GARRISON: I have two comments on the determination of the (B-V) color of the Sun by Tüg and Schmidt-Kaler, compared with the others. The internal errors they quote cannot represent the systematic errors. The correction for an ideal pinhole diffraction is probably not absolutely known to 1% and their pinhole is probably not perfect. Also, their observations of the Sun are made during the day and the stars for transformation at night and I am not at all convinced that the transformation can be made to only a few percent under these extreme conditions. I find that most photometry does not successfully transform even at night!

Secondly, their value is quite extreme and I can quite clearly state that stars with (B-V) of 0.69 do NOT have the same line spectrum as the Sun at 1 - 2 Ångstroms resolution. I am not saying that they are wrong, just that the difference is significant. I agree that the Tüg and Schmidt-Kaler determination is elegant and very interesting. I only question the relationship of systematic and internal errors.

HAYES: For all four of the direct measurements of $(B-V)_0$ the authors quote an internal error of 0.01 mag., and yet the spread is from 0.63 to 0.69 mag. Thus, there must be significant systematic errors in these measurements, and we do not really know which measurements are the most seriously affected. With regard to the disagreement between the result of Tüg and Schmidt-Kaler and the typical spectroscopic behavior, I agree. But I must say that the degree to which the Sun is atypical photometrically for stars of its spectral type is the interesting question.

GARRISON: With regard to your comment about using more precise photometric systems, there are people working on it. Erik Olsen in Denmark has observed many thousands of early G stars on the Strömgren system and is currently working on the mid-G stars. Chmielewski has used the Geneva system to infer the color of the Sun.

HAYES: Yes. So the next step is to make direct observations of the Sun in these systems.

BESSELL: With regard to secondary standards there are three glaring unfilled needs. We need V = 16 mag. DC stars from 0.34 to 1.1 microns for photon counting spectrophotometry. We need V = 7 - 9 mag. G and K extreme metal-deficient ([Fe/H] = -2.0) stars for CCD spectrophotometry and we need K = 3 - 5 mag solar-like dwarfs for J, H, K, L (1.2 - 4.0 micron) spectrophotometry. A and B stars with large continuum discontinuities and strong hydrogen lines are very unsuitable as spectrophotometric standards for work from 0.3 - 5 microns.

HAYES: I agree. In cases 1 and 3 work is in progress on such standards.

MILLWARD: I would just like to point out that Vega was one of our trigonometric parallax standards for the H-gamma-luminosity calibration, but had to be excluded from the group as it was found to be one magnitude too luminous for its H-gamma-equivalent width. So, in this sense it is anomalous.

HAYES: Yes. I believe this effect has been known for some time.

GALGANI: I would like to make a comment. Calibrations are important not only for applications, as discussed here, but also for general physics. Indeed there is a problem of internal consistency. Take for example the case of isochromatics; a blackbody is observed at a fixed frequency but at various known temperatures. The fluxes so obtained should fit Planck's law. I studied this problem of the internal consistency for blackbodies in the last two years and found that the situation is quite striking. Only last year a very good result was found by Quinn and Martin (one part over a thousand) but only for the global emission (determination of the Stefan-Boltzmann constant). If one, instead, looks at the spectrum for a relevant range of the variable $x = h\nu/KT$; one finds that essentially no data were published after 1921 (Rubens and

Michel) and that the data fit Planck's law within 3% (three standard deviations). My point is then that if in the calibrations one finds a consistency better than 3%, then one should publish this as an interesting result in general physics.

APPROACHES TO PHOTOMETRIC CALIBRATIONS

F. Rufener

Geneva Observatory

ABSTRACT. In this review an examination will be made of the experimental conditions which must be satisfied by multicolor photometric systems if the observational parameters are to be correlated unequivocally with physical quantities of the star. An inventory of error sources disturbing such calibrations contains: instability of the natural system, procedures of observation and reduction to outside the atmosphere and correlation with a standard system. The requirements desirable for the preparation of lists of standard stars, for the definition of the pass bands, for the "orthogonality" of the calibrated parameter, for the definition of the domain of validity of the calibration are listed. Some remarks are made on the concepts of "open" or "closed" calibrations.

1. INTRODUCTION

The evaluation of physical quantities by means of photometric observations often gives the impression of being a simple process, requiring little equipment and telescope time. When we need to know the color excess of a star of known spectral type, one UBV measurement is sufficient. The question is more critical, however, if the spectral type is unknown, and even more so if we want to determine other properties of the star also. We are faced by the hazards which may affect each step of the following process:

$$* \rightarrow \begin{array}{c} \text{photometric} \\ \text{observation} \end{array} \rightarrow \text{standardization} \rightarrow \text{calibration} \rightarrow \begin{array}{c} \text{physical} \\ \text{quantities} \end{array}$$

The legitimate concern about the use of the latest available calibration sometimes causes one to neglect the precautions which are essential for the previous stages. My intention is to insist here on the importance of these precautions, which are all the more necessary if the number of filters is large and if the object is to evaluate several physical quantities characteristic of the star.

D. S. Hayes et al. (eds.), Calibration of Fundamental Stellar Quantities, 253–269.
© *1985 by the IAU.*

We recall here that each photometric measurement is a direct measurement of the integral over the energy flux $E(\lambda)$ of the star filtered by a more or less wide passband $\phi_i(\lambda)$, and that the signal is recorded at ground-level through the absorbing atmospheric screen $A(\lambda, Fz, t)$. Therefore, the measurement made at time t through an air mass Fz corresponds to a complicated function

$$m_{i,z,t} = -2.5 \log \int E_t(\lambda)\phi_i(\lambda)A(\lambda,Fz,t)d\lambda$$

If an accurate measurement of $m_{i,z,t}$ is to be subjected to a calibration, its value outside the atmosphere must be determined:

$$m_{i,o,t} = -2.5 \log \int E_t(\lambda)\phi_i(\lambda)d\lambda$$

which can only be significant if, on one hand, the profile of the passband is well defined and conserved and, on the other hand, the reduction to outside the atmosphere is properly carried out.

The opinions which I express in the following pages are the result of twenty-five years of confronting the difficulties of ground-based photoelectric multicolor photometry. This is why many of my references will relate to the Geneva photometry which I have been practising continually since its creation in 1960; besides, it is the only photometry I know really well.

2. THE NATURAL SYSTEM

The natural system, or instrumental system, corresponds to the photoelectric responses of the equipment used by the photometrist during the observations. The natural system is therefore the product of the chromatic responses of all the reflecting, absorbing or even diffusing elements encountered along the optical path, beginning at the entrance of the telescope, multiplied by the response of the detector which is generally a photomultiplier. The whole equipment presents an analogical response which varies with time. Indeed, the reflectivity of the teles-cope mirrors, or the transmission of the elements in the photometer as well as of the filters can evolve with time and vary under the influence of external factors which are mainly governed by temperature. The same is also true for the detector. Quite often the observer has very little information concerning the natural system he is using. Is it stable? Is it sufficiently close to the definition of the standard system to be sensitive in the same manner to features in the stellar spectra? The answer to these questions must not remain vague. Let us begin with the question of stability and detail a few precautions.

The permanence of the system and the continuity of its use are tokens of stability. One must however remember that the aging of the reflective coating of the mirrors is rapid, and that it is also chromatic.

Freshly aluminized mirrors undergo a more rapid decrease of their
reflectivity in the near ultraviolet than in the visible (Hass, 1955;
Rufener, 1968). As it is desirable to use mirrors which are as clean
as possible one could prefer a careful washing to a new aluminium
coating. Our experience with a mirror protected by a wide-band interference coating of magnesium fluoride has been very satisfactory.
This coating has been washed frequently without suffering any significant alteration. The chromatic properties of the coating have remained
stable. If the cleanliness of the mirrors and other elements is bene-
ficial to signal intensity, it is equally important to the reduction of
diffused light. This diffused light increases the sky background and
can in certain cases be chromatic. We have noted several times that it
decreases rapidly with increasing distance to the optical axis. It is
then necessary to introduce small corrections when we change the
diameters of the diaphragms which define the measuring field. A further
advantage of continued use of a natural system lies in the possibility
of monitoring its evolution and of identifying the cause of each change
noted.

We must insist here on the essential role of temperature regulation
in the conservation of the passbands. Most absorbing glasses which are
used to define the passbands show a variation of their cut-off wavelength with temperature. It is not uncommon to observe an effect of the
order of $\Delta\lambda = 2\Delta T$ (λ in Å, T in °K). On the other hand, the chromatic
response of the photomultiplier (PM) also varies with temperature. One
generally observes a redward displacement of the sensitivity threshold
of a cathode with increasing temperature. The maximum of the sensitivity
function often decreases with increasing temperature. We therefore have
to stabilize the temperatures of all the elements which influence the
response curves of the natural system (Young, 1974a). In most photometers, the temperature of the photocathode is lowered radically so as
to reduce the thermionic emission which uselessly increases the dark
current. This is hazardous when the refrigeration is not regulated but
only set to be as low as possible. This situation places the photocathode in a temperature field with a high gradient, all the more so
since the entrance window is often heated to prevent condensation. The
true temperature of the cathode is then very badly defined. It is by
far preferable to have a true temperature regulation of the PM around
a sufficiently low value so that thermionic emission would not be
troublesome, but at which the thermal gradients are moderate and well
localized by the construction of the insulating mounting. A second
temperature regulation will be necessary for stabilizing the filters
and possibly other technical elements related with the measurement of
the PM current.

Several cases are reported, in the literature, of photometers having
shown a sensitivity to the influence of disturbing fields such as
magnetic, electric or gravitational fields (Young, 1974b). This type of
subtle disturbance is often difficult to test systematically.

Magnetic fields act mainly on the PM; they can modify the distribution
and the orientation of the initial velocities of the photoelectrons
leaving the photocathode. Perturbations of gain and chromatic response
have been reported (Rufener, 1966). A good protection is insured by
properly installing a shielding of high magnetic permeability.
Perturbations due to electric fields are difficult to eliminate
completely. They enter at several levels. The stabilization of the
voltages is generally sufficient and allows one to obtain stable PM
gains as well as a permanence of the characteristics of the system which
measures the photoelectric current. On the other hand, the effects of
electromagnetic disturbances which propagate in the environment of the
photometer, or in the power network, are more often than one may suspect
the causes of instrumental instabilities. In particular, when a photon
counting system is used to analyze the PM current, the frequency and
origin of the disturbing pulses are not always recognized. The effects
of gravity have also caused a few surprises which can, however, be
prevented by a rigid construction of the photometer and by tests made
with a stable calibration source which can be moved with the photometer.
The use of an internal reference source has sometimes been proposed, but
the stability of such a source can also present problems (Peytremann,
1964).

3.   REPRODUCTION OF THE PASSBANDS

     This is truly a subject on which most users of photometers do not
have any direct influence. It is however of the utmost importance that
the natural system imitates as well as possible the standard system.
To make this possible it is therefore at first necessary that the standard system be well defined by a description of the elements used and
by precise laboratory measurements. This question will be taken up later. When we wish to build a new photometer, we should choose components
which have the same specifications as those which created the original
natural system. These components are often no longer produced, having
been replaced by new ones with better performance according to catalogs.
This is the beginning of a series of compromises. In the past few years,
the availability of new photocathodes which are highly sensitive in the
red has become a source of many temptations. The use of photovoltaic,
two-dimensional detectors such as the CCD is being generalized. How can
one conserve the definition of the passbands? A scrutiny of colored
glass catalogs suggests some solutions. One must, however, be very mis-
trusting; a combination defined on the basis of transmission and response curves given in the catalogs of the manufacturers is only a first
approximation which must be controlled by laboratory measurements.
It is certain that precise measurements made through badly reproduced
passbands will later on be the cause of much uncertainty and dispersions
which will affect in a very troublesome manner the significance of these
measurements. A detailed example of this kind of difficulty has been
presented by Olsen (1983) in the introduction of his catalog.

We can already foresee unpleasant surprises due to the redward
extension of the spectral responses of the new detectors which will be
either badly, or not at all, taken into account. Two instructive
examples of transformations and comparisons figure in the articles by
Graham (1982) and especially by Landolt (1983). They reveal the
difficulty of obtaining UBV measurements by means of an RCA 31034 PM
with a Gallium-Arsenide cathode. The lack of definition of a passband
cannot be corrected by a simple transformation made with the help of
a few standard stars. One must actually maintain the content of spectral
information which is summed up by the passband of the natural system.
Figure 1 shows an example of a mismatch between two manufacturings of
interference filters which were used for intermediate band photometry.
The localization of a few important lines shows the order of magnitude
of the potential problems.

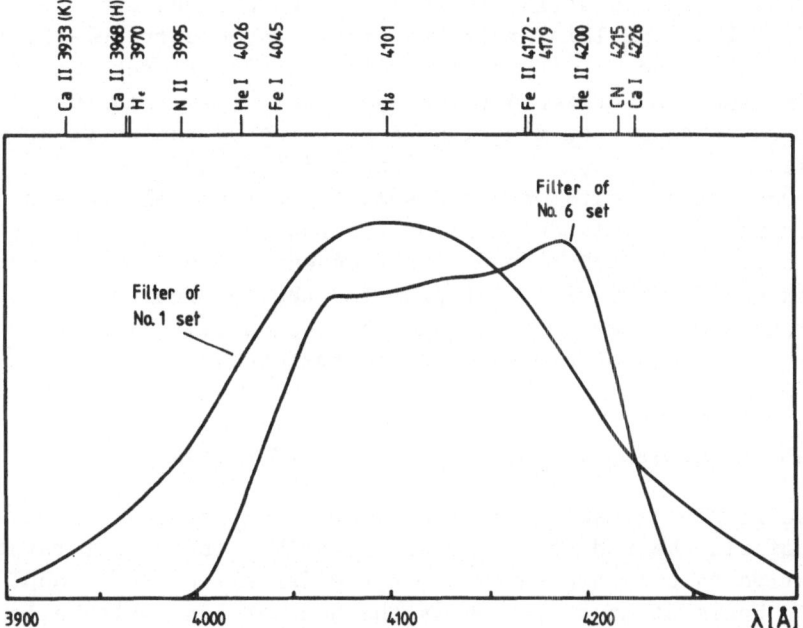

Fig. 1

Optical responses of two interference filters which
have been used for intermediate band photometry.
Differential effects uncorrectable by the standard
transformation are to be expected if one considers
the locations of some important lines.

4. PROCEDURES OF SIGNAL MEASUREMENT

Whether we measure the photoelectric current with a DC amplifier
or with an amplifier adapted to photon counting, there is no real
advantage in using large time-constants or long integration times. This
filtering of high-to-medium frequencies (f > 0.1 Hz) often only serves
to mask anomalies of the signal and to average them with the interesting
part of the signal. A large variety of disturbances have time
characteristics which correspond to frequencies > 10 Hz. We can mention
disturbances of the electric field, scintillations due to cosmic
radiation or to natural radioactivity as well as several atmospheric
phenomena such as lightning, and some of the effects related to
turbulence and scintillation. Bartholdi et al. (1984) have shown the
advantage of rapidly sampling the signal while applying statistical
tests calculated in real-time. This technique, which is made feasible
by available microcomputers, allows one to recognize the anomalies
of the observed time distribution of incident photons compared with the
Poisson distribution. This procedure proved to be particularly interesting for judging the quality of recordings of low fluxes. It allows one,
in certain cases, to filter out the anomalies. This leads us to recall
again the advantages resulting from a planning of the observations
which makes plentiful use of the differential method. On one hand,
simultaneous measurements or rapid sampling through all passbands
reduce the perturbations due to slow fluctuations of atmospheric
transparency. On the other hand, the frequent use of comparison stars
from an extended and varied list promotes the intercomparison of the
observations and allows a monitoring, or even a filtering-out, of the
very slow variations of atmospheric transparency.

5. PROCEDURES OF REDUCTION TO OUTSIDE THE ATMOSPHERE

It would not be useful to undertake in this context a detailed
analysis of all the reduction methods described in the literature. They
are as varied as they are numerous in the details of their applications.
The common basis of most of them is the Bouguer line which allows the
evaluation of the magnitude outside the atmosphere m_o, and of the
atmospheric extinction coefficient k, if several observations m_{zi} of
the same star are made at different air-masses F_{zi}. A linear regression
determined by least squares is applied to the relation

$$m_{zi} = m_o + k F_{zi}$$

One must insist on the fact that this model is rather unrealistic since
it assumes that three hypotheses are satisfied: constancy of the star,
stability of the response of the photometer, constancy and isotropy of
the atmospheric extinction during the whole period of acquisition of
the observations (5 to 7 hours). One can, in practice, hope to select
a stable star.

The photometer must be able to satisfy the desired requirements of
stability; the third hypothesis, however, only has a chance of being
satisfied accidentally. Nikonov (1952) and Young & Irvine (1967) have
proposed methods which account for variability of the extinction.
Rufener (1964) adopts a less restrictive form of the third hypothesis:
the extinction can be slowly variable but it remains isotropic during
the whole period of acquisition of the measurements. By grouping into
quasi-simultaneous pairs the measurements of an ascending extinction
star (M) with those of a descending one (D), it is possible to
calculate for these stars their magnitudes outside the atmosphere and
to obtain the instantaneous extinction at the time of each observation.
The remaining measurements of the night are then reduced by interpolation. This method, called M + D, has been applied in the Geneva
photometry for   twenty years each time extinction was measured. This
more realistic model allows a better understanding of the observations
and explains the sometimes misleading results obtained through the use
of the oversimplified Bouguer method. Figure 2 shows these effects in
the frequently encountered case of a slow decrease of extinction during
the night. When we do not wish to devote the necessary time to measure
extinction by the M + D method, we use mean extinction coefficients.
A planning of observations, which imposes a small dispersion of air
masses around a predetermined value for the night (the so-called
constant air mass night), allows an evaluation of the necessary
corrections due to drifts of extinction to be made by readjusting the
zero point with the help of a sufficient number of comparison stars.
The reduction to outside the atmosphere is complicated by the effects
due to the width of the passbands. The color of the star and the air
mass along the line of sight modify the effective wavelength and
consequently also the actual extinction coefficient. Among the number
of solutions proposed to take this effect into account, the best
approach is that of King (1952). Put into practice by Rufener (1964)
it leads to a development into a series centered on the mean wavelength
(λ_o) of the passband. A magnitude outside the atmosphere is then
expressed by

$$m_o = m_z - F_z[k(\lambda_o) + \alpha + \beta C + \gamma F_z]$$

α, β and γ are coefficients which can be calculated as soon as one has
a good knowledge of the profiles of the passbands and of the mean
extinction. C is a color index which describes the energy distribution
of the star. This way of proceeding thus distinguishes itself from the
method which consists in introducing an extinction $k_1 + k_2$ C, whose
coefficients k_1 and k_2 are determined empirically. We see that for
evaluating atmospheric extinction as well as for compensating the
effects due to the width of the passbands, we may choose between more or
less perfect methods which are nothing but more or less accurate models
of reality. It seems important to me to conserve for a given photometric
system the method adopted by its initiators. Thus, if that procedure
causes systematic errors, these would at least not become randomized by
the choices of subsequent observers.

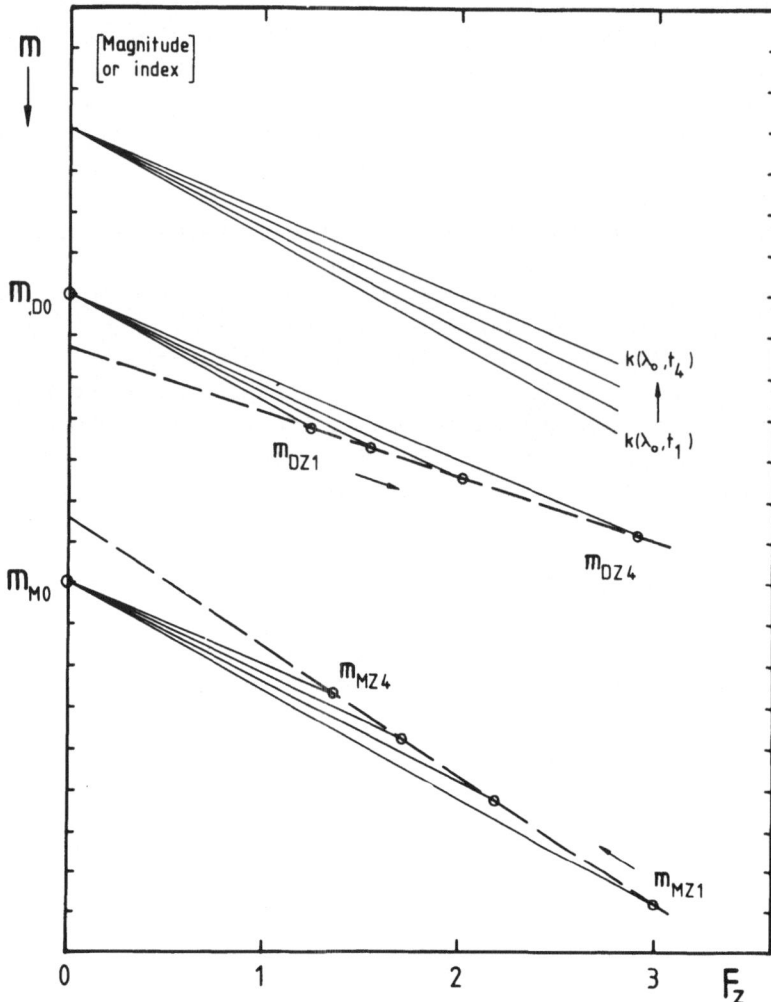

Fig. 2

Simulation of a night with variable extinction. The
adopted decreasing values (0.260, 0.240, 0.220, 0.200)
of the instantaneous extinction coefficient are shown at
the top of the figure. The m_{Mo} and m_{Do} correspond to the
true values outside the atmosphere of the stars M and D
(these can be unknown). The real observations made at the
times t_i would be close to the synthetic values m_{MZi} and
m_{DZi} represented here by open circles. The two Bouguer
lines which one would be tempted to adopt are represented
by the broken lines. The M + D method described in the
text allows one to determine the values m_{Mo} and m_{Do} and the
i values $k(\lambda_o, t_i)$. Tick marks of the ordinate scale
correspond to 0.1 mag.

6. CORRELATION OF THE OBSERVATIONS WITH A STANDARD

In the best case, where the natural system is close to the standard one, we confirm that a linear transformation is sufficient to transform the observations of any given night into standardized measurements. Let C^s_{1-2} and C^n_{1-2} be a color index in the standard and natural system, respectively.

$$C^s_{1-2} = aC^n_{1-2} + b$$

Rufener (1968) has formulated the approximations to the coefficients

$$a = \varepsilon(1+ \frac{\Delta\lambda}{\lambda_2 - \lambda_1}) = \frac{\lambda_2\lambda_1}{\lambda^s_2\lambda^s_1} \frac{(\lambda^s_2 - \lambda^s_1)}{(\lambda_2 - \lambda_1)}$$

with λ_2, λ_1 the mean wavelengths of the natural system

λ^s_2, λ^s_1 the mean wavelengths of the standard system

$$\Delta\lambda = (\lambda^s_2 - \lambda^s_1) - (\lambda_2 - \lambda_1)$$

$$\varepsilon = \frac{\lambda_1\lambda_2}{\lambda^s_1\lambda^s_2}$$

$$b = \Delta\Phi_{1-2}+(\Phi_1-\Phi_2) [1-\varepsilon- \frac{\Delta\lambda}{\lambda_2-\lambda_1}] - \Delta k_{1-2}\bar{F}_z \cdot \varepsilon(1+ \frac{\Delta\lambda}{\lambda_2-\lambda_1})$$

$$\cong \Delta\Phi_{1-2} - \Delta k_{1-2} \cdot \bar{F}_z$$

with Φ_1, Φ_2 the magnitudes of the passbands of the natural system $\Phi = -2.5 \log \int \phi(\lambda)d\lambda$

and $\Delta\Phi_{1-2} = (\Phi^s_1-\Phi^s_2) - (\Phi_1-\Phi_2)$ the difference between the

passbands of the standard and the natural system.

We notice that the coefficient "a" only depends on changes of mean wavelengths between the standard and natural systems. The term "b" is more complicated; it represents the zero point of the transformation. The main term reflects the change in magnitude of the passbands while the second term involves the error in the extinction coefficient used (Δk_{1-2}). The factor $\bar{F}_z$ expresses the mean air mass used during the observations of the given night. We see here the advantage of a limited dispersion of the individual values of F_z if it is our aim to estimate the term Δk_{1-2} and to monitor its evolution during the night.

The influence of changes in the wavelength is of the second order for this term b. In the unfavorable case where the natural and standard systems differ seriously, the transformation is no longer linear; moreover, it is no longer unique. Stars reddened or not by interstellar matter, evolved to different degrees, of various chemical compositions, etc.... would each have to be treated by different transformations; by overstatement, the relation between the natural system and the standard system becomes in a certain sense a photometric diagram. There is no good solution in this case; Young (1974c) who examined a few approaches was not able to draw a conclusion.

7. NETWORKS OF COMPARISON AND STANDARD STARS

We must first explain the distinction implied by this title. In view of applying the principle of differential measurement as conveniently as possible, the photometrist must have at his disposal a collection of comparison stars. These must be in sufficient number, well distributed over the sky, present a good spread in V magnitudes, easy to identify, without troublesome neighboring stars. They have to be thoroughly intercompared so that the probability of finding a variable among them, even of small amplitude, is practically zero. On the other hand, this same photometrist needs a network of standard stars. These must cover the whole HR diagram, the whole range of reddening by interstellar matter and all populations. They should be measured from the origin of the system onwards, and thus be able to guarantee the conservation of the system. Here also, convenience of use requires a distribution over the whole sky and a strong intercomparison. In practice, the same stars could serve as comparison and standard stars. We must note however that in the case of the comparison stars we would be tempted to eliminate all microvariables (supergiants, CP stars, extremely red stars etc....). This choice would not be desirable from the point of view of the standard stars which have to represent all types. When we establish these collections, we can distinguish between the standardization of the colors (indices) and of the magnitude (V) by using different weightings. One common but risky practice has to be avoided, namely that of choosing for the one or other purpose stars from a compilation catalog. It is truly necessary to use stars chosen in lists of primary or secondary standards and, as a last resort, stars which have been at least strongly intercompared and measured frequently.

8. GLOBAL TREATMENT

Several authors have proposed a global treatment of the reduction to outside the atmosphere and of the correlation with a standard. This attitude is the result of the desire to make better use of all the available elements of information with the help offered by computers.

Harris et al. (1981), Popper (1982) and Manfroid and Heck (1983, 1984)
have contributed with insight to present this new orientation, with all
its advantages and drawbacks. As Popper (1982) points out, the benefit
lies less in the mixing of both problems but rather in the better use
made of the constancy of certain parameters during several nights of a
same series, thereby often allowing to improve their estimate. Let us
note, however, that this type of treatment will not be better than is
allowed by the model chosen for the interpretation. There is the danger
that a global solution will lead the observer to increasingly neglect
the strictness of the planning of his observations for each night. The
above authors do stress, however, the importance of the latter. The
use of a more complicated model, which takes into account the variation
of extinction during the night, could also be incorporated into the
global treatment. The optimization by least squares of the adopted
model may render more difficult the separation of extinction anomalies
from variations of the natural system due to accidental circumstances
which it is important to recognize. In Geneva we treat both problems
consecutively; to maintain homogeneity, we do not intend to unify the
treatment. My lack of experience does not allow me to judge the extent
of a possible gain that might be achieved.

9.   DESCRIPTION OF THE PASSBANDS

This question presents at least two aspects. First, the physical
definition of the photometric system which usually figures in the first
publications describing it. This technical aspect is important as soon
as the question of reproducing the system arises. Of greater importance
for the user is the exact description of the passbands which characterize the standard system. One may choose for example to present, as a
function of wavelength with steps of maybe 25 to 50 $\overset{\circ}{A}$, the response of
each passband to an equienergetic flux expressed in units proportional
to photons per second. It is difficult to calibrate this description by
direct measurements with a primary spectrophotometric standard. We
should at least dispose of an indirect calibration so that by filtering
the spectrophotometric distributions of stars considered as secondary
standards we reproduce, by numerical integration, the corresponding
indices in the system considered. For some systems it will be necessary
to apply, before the comparison is made, a normalization fixed at the
origin of these systems. The calibration of the passbands will then
depend on the absolute reference calibration adopted for the spectrophotometric distributions of the secondary calibration stars. D.S. Hayes
will certainly discuss this question during this symposium (Hayes
1985). It would be desirable for these stellar standards to have a
continuous spectrophotometric description from 3000 to 10,000 $\overset{\circ}{A}$,
including the effects of the lines. It would also be useful to dispose,
among these secondary standards, of stars covering the whole HR diagram
and, if possible, also strongly reddened ones as well as representatives
with extreme chemical compositions.

Regarding data presently available we may mention Straižys &
Sviderskienè (1972) and more recently Glushneva (1982) and Gunn &
Stryker (1983). The type of approach briefly described above has been
made in particular by Hayes (1975) and by Buser (1978). A variant which
allows to optimize this calibration procedure by distinguishing the
optical response of each passband from the electrical response of the
detector has been proposed and applied by Rufener & Maeder (1971) to
the passbands of the Geneva photometry.

As soon as an adequate description of the passbands has been
realized by means of a variety of secondary standards for which a
consensus exists regarding their absolute calibration, numerous
comparisons between stellar model atmospheres and observations can then
be considered by confronting the synthetic photometry with the actual
measurements. The reliability of the inferences made depends then
obviously on the exactness of the description of the standard passbands.

10.   CHOICE OF THE PARAMETER TO BE CALIBRATED

It is often a heuristic approach which leads to the selection of
a photometric parameter P presenting a variation destined to be
correlated to a given physical parameter x_c. Before calibrating this
correlation it is preferable to check the "orthogonality" of the
future parameter P, or at least to seek the best formulation, in view
of obtaining the highest independence of P relatively to variations of
the other physical quantities (x_i) of the star and of the interstellar
medium. In other words, it is desirable to optimize the definition of
the parameter P in such a manner that one obtains a maximum for
$\partial P/\partial x_c$ and a minimum for $\partial P/\partial x_i$ for all $i \neq c$. This task is not trivial;
it can be undertaken via a geometrical fine analysis of the n-dimensio-
nal hyperspace corresponding to the n independent color indices of a
photometry. An interesting approach consists in selecting first as
coordinates combinations of indices which are reddening-free but
sensitive to intrinsic differences of the spectral energy distribution.
In the Geneva photometry, for example, such a reddening-free space was
used by Cramer and Maeder (1979) who defined three orthogonal parameters
X, Y and Z which are correlated with the effective temperature, the
surface gravity and the spectral peculiarity of B-type stars
respectively. Such an analysis requires that the photometric system
used has already been applied to the greatest possible variety of
stars with distinct properties, so that this optimization can be validly
carried out and verified.

We must recall here that the number of discernable independent
parameters cannot exceed the number of passbands of the photometry;
it is always smaller by one unit.

On the other hand, the number of physical quantities and evolutionary
or random circumstances which can influence the spectral energy
distribution is large; therefore a certain danger always exists in
hastily applying a correlation with x_c which proves to be more or less
parallel with one x_i, or with a combination of the latter.

We can be tempted to orient the choice of a photometric parameter
in a prospective manner by using synthetic photometry applied to a set
of spectrophotometric continua, or even to a grid of model atmospheres,
filtered by the passbands. This procedure might seem alluring; it has
however only rarely contributed to the perfecting of an optimized parameter. There is, among the criteria for choosing a parameter, what we
may call its resolution or its sensitivity. This is an evaluation of
the following type:

$$\Delta x_c = \frac{\partial x_c}{\partial P} \cdot 2\sigma_P$$

where σ_P is the standard deviation characteristic of the experimentally
obtained parameter P.

11. DISTINCTION BETWEEN "OPEN" AND "CLOSED" CALIBRATIONS

The establishment of a calibration can be undertaken within
extremely different contexts and lead to opposed practical choices. This
will clearly influence the resulting performance. By schematizing
extreme circumstances, I propose to confront two cases, the one defined
as "open", the other as "closed". To the first belong the UBVRI and
uvby,β photometries and to the latter correspond, for example, the
Geneva photometry and probably also that of Walraven. We tabulate below
the options which distinguish both orientations.

Nature and origin of the data	Circumstances for a calibration open	closed
multiband photometric system	Popular system, frequently copied. Great variety of natural systems	Stable experimental system, under single supervision
Acquisition of the observations	Heterogeneous procedures of measurement and reduction to outside the atmosphere	Complete homogeneity of the method of treatment
Available data	Compilation and means of results dispersed in the literature	One single source of compilation

Origin of the physical reference parameters	Collection of estimates having met with a wide consensus	Possibly, physical parameters taken from a small number of primary standards
Form of the calibration	Mean standard correlation, tabulated and immediately accessible	More confidential calibration, requiring a complex critical analysis
Number of persons contributing to this condition	Often more than 100	Maybe less than 10
Performance	Rather rough possibilities of discrimination	Limit of discrimination finer by a factor 2 to 3. Could lead to the estimate of physical parameters without publication of the intermediary stages

We note, in the case of the Geneva photometry, the application by M. Grenon of the completely "closed" case with the elaboration of a series of algorithms to determine the T_{eff}, the absolute magnitude, the chemical composition (Fe/H) and consequently the photometric parallax of F, G, K, M-type stars. This calibration is applied to the individual measurements before they are averaged. The n estimates of the physical parameters corresponding to the n measurements, and their subsequent mean with its standard deviation, are shown to be more favorable for interpretation than the estimate based on the mean of the photometric measurements above. This way of analyzing the individual measurements is illustrated by the examples of Table I; it reveals the cases for which ambiguities are to be feared and also provides a concrete appreciation of the sensitivity attained.

12. CONCLUSIONS

To conclude, I would like to insist once again on the preliminary requirements which seem to me to be most important for the elaboration or use of photometric calibrations with a sufficiently high guarantee of security and resolution.

Insure the greatest experimental rigor to define the natural system, systematically maintain the procedures of reduction to outside the atmosphere and of correlation with a reliable and homogeneous standard.

Insert the new measurements in an extended collection of reference stars for which a consensus exists, or will exist, regarding the values of the characteristic physical quantities.

Precise description of the standard passbands and of the possible normalizations (zero point) necessary for the observed color indices to correspond to the synthetic photometry of the best available spectrophotometric continua, which are themselves compared with the best absolute primary standards.

Checking of the degree of orthogonality of the chosen photometric parameters, together with a precise definition of the domain of application and possible resolution.

ACKNOWLEDGEMENTS

I would like to express my thanks to Noël Cramer for the English version of this text and for his assistance in the preparation of the figures. My thanks also to E. Teichmann for her careful reading and typing.

REFERENCES

Bartholdi, P., Burnet, M. and Rufener, F. 1984, Astron. Astrophys., 134, 290.
Buser, R., 1978, Astron. Astrophys., 62, 411.
Cramer, N. and Maeder, A., 1979, Astron. Astrophys., 78, 305.
Glushneva, I. N. 1982, Spectrophotometry of Bright Stars, (Nauka, Moscow)
Graham, J.A., 1982, Publ. Astron. Soc. Pac., 94, 244.
Gunn, J.E. and Stryker, L.L., 1983, Astrophys. J. Suppl., 52, 121.
Harris, W.E., Fitzgerald, M.P. and Reed, B.C., 1981, Publ. Astron. Soc. Pac., 93, 507.
Hass, G., 1955, Journ. Opt. Soc. Amer., 45, 945.
Hayes, D.S., 1975, in Multicolor Photometry and the Theoretical HR Diagram, eds. A.G.D. Philip & D.S. Hayes, (Dudley Obs. Rep.) 9, 309.
Hayes, D.S. 1985, in IAU Symposium No. 111: Calibrations of Fundamental Stellar Quantities, eds. D.S. Hayes, L.E. Pasinetti and A.G. Davis Philip (Reidel: Dordrecht), p. 225.
King, I., 1952, Astron. J., 57, 253.
Landolt, A.U., 1983, Astron. J., 88, 439.
Manfroid, J. and Heck, A., 1983, Astron. Astrophys., 120, 302.
Manfroid, J. and Heck, A., 1984, Astron. Astrophys., 132, 110.
Nikonov, V.B., 1952, Izw. Krim Obs., 9, 41.
Olsen, E.H., 1983, Astron. Astrophys. Suppl., 54, 55.
Peytremann, E., 1964, Publ. Obs. Genève, Ser. A, 69.
Popper, D.M., 1982, Publ. Astron. Soc. Pac., 94, 204.
Rufener, F., 1964, Publ. Obs. Genève, Ser. A, 66.
Rufener, F., 1966, Publ. Obs. Genève, Ser. A, 72.
Rufener, F., 1968, Publ. Obs. Genève, Ser. A, 74.

Rufener, F. and Maeder, A., 1971, Astron. Astrophys. Suppl., **4**, 43.
Straižys, V. and Sviderskienė, Z., 1972, Bull. Vilnius Astron. Obs., **35**.
Young, A.T., 1974a, in Astrophysics. Methods of Experimental Physics,
    Vol. 12, part A, ed. N. Carleton (Academic Press: New York)
    pp. 46 and 105.
_____1974b, ibid., pp. 60, 119 and 137.
_____1974c, ibid., p. 190.
Young, A.T. and Irvine, W.M., 1967, Astron. J., **72**, 945.

T A B L E   I

No. HD	TS	m_V	$B_2{-}V_1$	T_{eff}	M_V	πphot.	Fe/H
20 807	G2V	--	.365	5766	5.16	.095	- .21
		5.223	.357	5715	5.15	.097	- .24
		5.220	.365	5773	5.06	.093	- .12
discarded measure		5.217	.345	[5830	4.75	.081	.02]
		5.225	.359	5747	5.10	.094	- .19
		5.225	.358	5726	5.17	.098	- .26
		5.229	.354	5762	5.06	.092	- .19
		5.226	.359	5739	5.15	.097	- .24
		5.216	.356	5789	4.94	.088	- .08
		5.219	.353	5781	4.97	.089	- .12
mean parameters		5.223	.358	5755	5.08	.094	- .18
standard deviation		.004	.004	25	.08	.004	.06
99 492	K2V	7.564	.617	4847	6.28	.055	.41
		--	.608	4845	6.21	.052	.26
discarded measure		7.556	.630	[4651	6.59	.064	---]
		7.566	.610	4899	6.07	.050	.61
		7.573	.627	4850	6.32	.056	.48
		7.582	.613	4913	6.08	.050	.64
mean parameters		7.568	.618	4871	6.19	.053	.48
standard deviation		.010	.009	33	.11	.003	.16

DISCUSSION

ARDEBERG: I find your definition of closed and open photometric systems a quite provocative one. May I propose an equally provocative definition? An open photometric system is a system open to use by the astronomical community, whereas a closed system is one closed off from such use. I think that we should take care not to create systems bent into themselves but rather systems that can solve pending astrophysical problems. I suggest as the best approach to apply, using your terminology, open systems with reduction methods as closed as possible.

RUFENER: I agree with your last sentence. The distinction proposed between open and closed calibration processes underlies two alternative approaches which I hope can be complimentary in their usefulness.

PAPOULAR: Our experience in near- and mid-infrared photometry closely confirms the statement by Dr. Rufener to the effect that good photometry requires differential measurements and intercomparisons between program star, standard star and sky to be made at a high rate and with small integration times. Now, photometers in big observatories are usually under the responsibility of the staff. Would it not be worthwhile, therefore, for IAU officials to help us induce these members of the staff into modifying their acquisition procedures in the directions suggested above?

RUFENER: Often it is difficult to transform existing equipment. We can hope that new photometers will permit such practices.

CALIBRATION IN TEMPERATURE OF PHOTOMETRIC PARAMETERS

B. Hauck

Institute of Astronomy, University of Lausanne

1. INTRODUCTION

It is very easy to use a Planck distribution to show that a color index is a temperature parameter, but it is more difficult to calibrate such a color index in terms of temperature because only a few T_{eff} values are determined. A pioneering work is that of Kuiper (1938), who derived a temperature scale according to spectral type (from A0 to M2 for dwarfs and from G0 to M8 for giants) and a Becker index. The first study giving a relation between T_{eff} and a photoelectric color index is that of Popper (1959) in which the author derives a relation between T_{eff} and B-V for the A and F stars and for G8 to K5 (dwarf and giant) stars. On this occasion, Popper shows the relation between temperature parameters of two photometric systems, R-I from the six-color system of Stebbins and Whitford, and B-V. This work was followed by a quantity of others from numerous authors in various systems, one of the most important being that of Johnson (1966). The purpose of the present study is not to review all the relations to be found in the literature but to define a set of stars which can be used to determine a calibration of a photometric parameter in terms of T_{eff}.

2. DETERMINATION OF T_{eff}

Böhm-Vitense (1981) has published a very important and complete study of the T_{eff} determination to which the reader may be referred. Böhm-Vitense classes the various methods to determine T_{eff} into direct methods (mainly Code et al. 1976), semi-direct and indirect methods. Semi-direct methods are those of Pottasch et al. (1979) for O stars and those of Underhill et al. (1979) for O and B stars. In both cases angular radii θ are obtained with the help of model atmospheres calculations. Indirect methods are of various kinds: a) comparison of energy distribution (in the visible and/or in the UV) with model energy distribution methods applicable to O, B, A and F stars, b) ionization equilibrium (O stars),

D. S. Hayes et al. (eds.), Calibration of Fundamental Stellar Quantities, 271–284.
© *1985 by the IAU.*

c) Balmer lines (G stars), d) synthetic colors (B, A, F, G stars),
e) spectrum analvsis (F, G, K stars).

Since the publication of the review paper of Böhm-Vitense, some new
studies have been published and mention may be made of that of Malagnini
et al. (1982) who use for A and F stars a method described by Morossi
and Crivellari (1980) for O stars. Malagnini et al. compare UV distributions published in the UV Bright Stars Spectrophotometric Catalog (S2/68
experiment, Jamar et al. 1976) with Kurucz models. The same procedure is
also applied by Malagnini et al. (1983) to B5-A stars. Malagnini et al.
(1984) give a relation based on 81 B5-F7 non-supergiant stars between
$R = \log \dfrac{F_{1965}}{F_{5445}}$ and T_{eff} derived from the two previous studies:
$\theta_{eff} = -0.266R + 0.557.$

Böhm-Vitense (1982) also proposes an indirect method to determine
effective temperatures of late A and early F stars from the UV flux
obtained in the S2/68 experiment.

Adelman and Pyper (1983) have published T_{eff} values for eleven stars
using the same approach as in their earlier studies by comparing energy
distributions with Kurucz models.

3. STANDARD STARS

The present purpose is to define a set of standard stars to calibrate
various photometric parameters in terms of T_{eff}. There are many papers
giving T_{eff} but it is not possible to use all the stars. Reddened stars
must be avoided since only a small number of photometric parameters,
such as Q in the UBV system and X in the Geneva system, are independent
of interstellar reddening. Unreddened or only slightly reddened stars are
proposed. A method proposed by Cramer (1982) based on the X and Y parameters to select unreddened stars has been used for O and B stars. Companions are often a source of "pollution" and unresolved visual binaries
must be excluded. In addition we must consider only stars belonging to
the same luminosity class and having the same chemical composition.

The first study in which such stars may be found is that of Code, et
al. (1976). It is the only one giving T_{eff} values determined with direct
knowledge of the star's diameter. Semi-direct and indirect methods must
be used because Code et al. give T_{eff} values only for B and A stars.
To complete the sample, data were taken from Adelman (1978), Adelman
et al. (1980, 1983), Hayes (1978), Underhill et al. (1979), Perrin et
al. (1977) and Oinas (1974). Considering first the main-sequence stars,
Table I contains those proposed as standard. We have thus selected 104
stars from B9 to K2. The column contents are as follows: (1) name,
(2) BS number, (3) HD number, (4) θ_{eff}, (5) T_{eff}, (6) B2-V1, (7) B-V,

(8) source of T_{eff}, then the various systems in which the star is measured. This information is extracted from the General Catalogue of Photometric Data which is in course of preparation. Table II gives the same information (except B-V) for the giant stars from Oinas' list that are measured in the Geneva system. The number of stars in each system is given in Table III. The first column of this table gives the identification numbers of systems, these also appearing in the last part of each line of Tables I and II.

4. CALIBRATION OF B2-V1 AND B-V

In this section the stars in Table I are used to calibrate B2-V1 (from the Geneva system) and B-V (from the Johnson and Morgan UBV system) for main-sequence stars with a solar chemical composition. The data are plotted in Figures 1 and 2 respectively. A relatively large dispersion is to be noted near both B2-V1=-.150 and B-V=.00 which could be due to a small gravity effect. If we plot the synthetic colors calculated by North and Hauck (1979) for the Kurucz models (1979) we remark this effect clearly, which is maximum near A0. Both linear and polynomial fittings were used for each diagram, giving the following relations:

$$\theta_{eff}=0.917 + 2.306(B2-V1) \qquad -.300 \leq B2-V1 \leq -.160 \qquad (1)$$
$$\pm .014 \pm 0.057$$

$$\theta_{eff}=0.632 + 0.640(B2-V1) \qquad -.160 < B2-V1 \leq .730 \qquad (2)$$
$$\pm .002 \pm 0.006$$

$$\theta_{eff}=0.573 + 1.512(B-V) \qquad -.22 \leq B-V \leq -.05 \qquad (3)$$
$$\pm .007 \pm .049$$

$$\theta_{eff}=0.536 + 0.514(B-V) \qquad -.05 < B-V \leq 1.20 \qquad (4)$$
$$\pm .003 \pm .005$$

$$\theta_{eff}=0.643 + 0.567(B2-V1) -0.597(B2-V1)^2 + \qquad (5)$$
$$+ 4.406(B2-V1)^3 - 9.033(B2-V1)^4 + 5.970(B2-V1)^5$$

$$\theta_{eff}=0.525 + 0.807(B-V) -1.442(B-V)^2 + 3.194(B-V)^3 - \qquad (6)$$
$$- 3.208(B-V)^4 + 1.153(B-V)^5$$

Relation (2) was checked for the main-sequence K stars for which Oinas (1974) and Lambert (1977) have determined the temperature and good agreement was found. We may therefore assume that our calibration is valid from B9 to K2.

The relations obtained are valid for stars on the main sequence with normal of solar chemical composition. Many metal-deficient stars are to be found in the paper of Perrin et al. (1977) and if these stars are

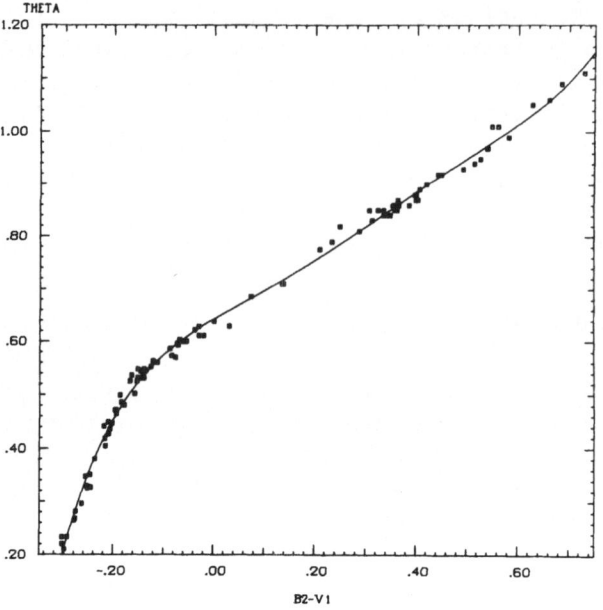

Fig. 1. θ_{eff} vs B2-V1 for the main-sequence standard stars. The full line is the fitted polynomial function.

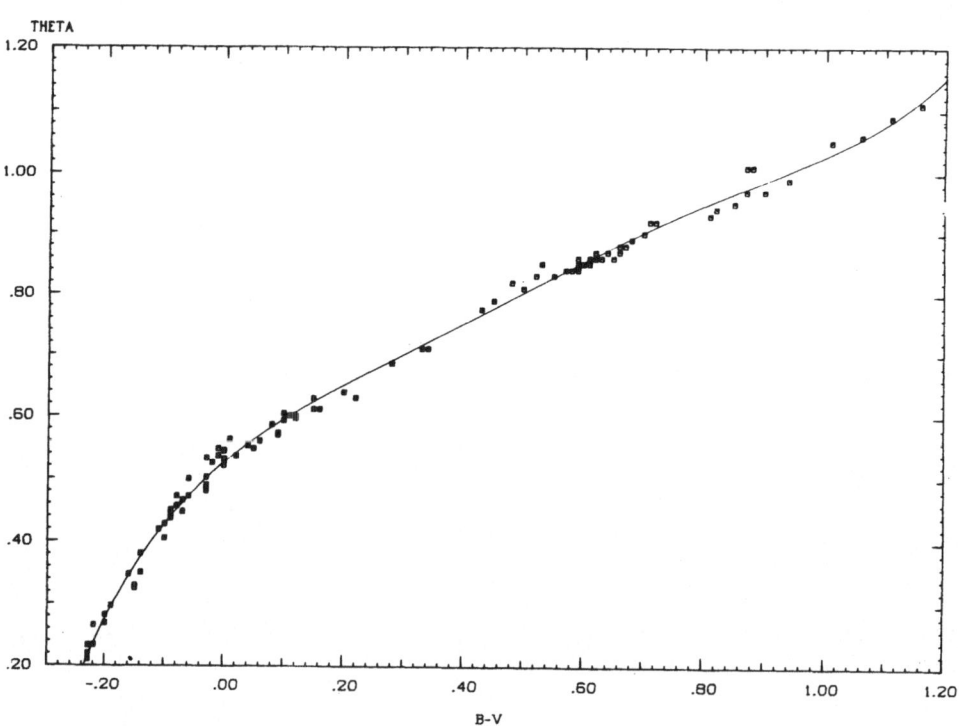

Fig. 2. Same as Fig. 1 for θ_{eff} vs B-V

plotted in a θ_{eff} vs B2-V1 diagram there is a clear blanketing effect, all these stars being shifted to the left. But B2-V1 can be corrected (Hauck 1973) and the corrected values bring about a good fitting of these stars with the relation for the normal stars.

For the giant stars hotter than the Sun, we have noticed that only the few for which the temperature is known do not show any deviation from the relation for main-sequence stars. Using the data Oinas (1974) has published for some K0-K3 giants we obtain the following relation:

$$\theta_{eff} = 0.634 + 0.548 \ (B2-V1) \tag{7}$$

5. OTHER CALIBRATIONS

In the previous section we have seen that the stars proposed as standard can be used to calibrate B2-V1 and B-V in terms of effective temperature. These stars are also measured in some other systems and many calibrations could be determined. We can also use relations between B2-V1 and some other photometric parameters established by Meylan et al. (1981). ьut it should be remembered that these relations can be sensitive to gravity and blanketing effects.

Various calibrations can be found in **Golay**'s book (1974) and in the proceedings of the workshop on Problems of Calibration of Multicolor Photometric Systems (A.G. D. Philip, ed. 1979). Among those that have been published more recently I would mention Cramer's (1984) study of the B stars, in which we found the following relation:

$$\log T_{eff} = 4.586 - 1.038 \ X + 1.094 \ X^2 - 0.646 \ X^3 + 0.139 \ X^4$$

where $X = 0.3788 + 1.376 \ U - 1.2162 \ B1 - 0.8498 \ B2 - 0.1554 \ V1 +$
             $+ \ 0.8450 \ G$

is a temperature parameter (in the Geneva system) for the hot stars independent of reddening.

Stone (1983) has published the following relation for extremely metal-deficient red giants:

$$T_{eff} = 7101 - 3768 \ (V-R)_o + 848 \ (V-R)_o^2$$

while Straižys et al. (1982) give tabulated relations between $\log T_{eff}$ and the indexes U-P, P-V, X-Y, Y-Z, Z-V, Z-S of the Vilnius system.

Another approach to the determination of T_{eff} may now be mentioned. Some authors have calculated synthetic colors from model atmospheres and then constructed a photometric diagram (color or parameter vs color or

parameter) with loci of constant T_{eff} and log g. Releya and Kurucz
(1978) calculated theoretical uvby and UBV colors from the Kurucz grid
of model atmospheres for O, B, A, F and G stars, while Philip and
Releya (1979) used these synthetic colors to propose three large-scale
$(c)_0$ vs $(b-y)_0$ grids calculated for [Fe/H] values of 0.0, -1.0 and -2.0.
These authors claim rms errors of the calculated values of ±0.2 in log g
and ± 250 K in T_{eff}. Philip and Egret (1980) applied these grids to the
whole uvbyβ catalogue (Hauck and Mermilliod, 1980). Synthetic colors
for the RGU system have been published by Buser (1978), while Buser and
Kurucz (1978) give synthetic UBV colors and derive an effective temperature scale as a function of $(B-V)_0$, $(U-B)_0$ and $(U-V)_0$. North and
Hauck (1979) have also published synthetic colors in the Geneva system
from Kurucz models and also given an effective temperature scale as a
function of B2-V1, while Lub and Pel (1977) have published a set of
theoretical two-color diagrams in the VBLUW system calculated from the
first set of Kurucz models (1975).

Theoretical colors in various photometric systems (UBVR, Geneva,
uvby, DDO, gnkmf, Uppsala, 13-color (Arizona)) were calculated for the
G and K giant stars by Bell and Gustafsson (1978, 1979) using
their own set of model atmospheres (Gustafsson et al. 1975, Bell
et. al. 1976)

6. TEMPERATURES FOR THE Ap AND Am STARS

Several attempts have been made to determine effective temperatures,
one being that of Babu and Shylaja (1981), who determined θ_{eff} for 125 Ap
and Am stars. Their method is semi-direct, since they use flux distributions in the λλ4000-7800 wavelength region and Mihalas models. Shallis
and Blackwell (1979) determined effective temperatures for five Ap stars
using a semi-direct method (Blackwell et al. 1977, 1979) in the IR
range. Floquet (1981) derived effective temperatures for 69 mostly cool
Ap stars taking the intensity of the CaII line as a temperature indicator. More recently Lanz (1984) has determined effective temperature
for six Ap and six He-weak stars, employing the Shallis and Blackwell
method. We thus have a lot of effective temperatures for Ap and Am
stars but the basic problem is that color indexes are generally affected by the peculiar characteristics of the star. B2-V1 is not affected
in the case of Am stars (Hauck and Van't Veer, 1970) but B-V is affected and also b-y, but to a lesser extent. In the case of Ap stars,
many photometric parameters are affected because one filter is in the
spectral interval of one of the continuum depressions, mainly that at
λ 5200 Å. B2-V1, B-V and b-y are strongly affected, but one of the less
affected photometric parameters is B2-G (see Gerbaldi et al., 1974) and
recently Lanz (1984) obtained a calibration of $(B2-G)_0$ in terms of
effective temperature for the Ap stars.

ACKNOWLEDGEMENTS

   I am very grateful to Dr. P. North for his assistance with the computer programming and many stimulating discussions. I am also grateful to Mrs. M. Mermilliod, Mrs. E. Bertinotti and Messrs. D. Lechaire, R. Hachadourian and E. Pfister for their help at various stages of the preparation of the manuscript and to Mrs. B. Wilhelm for its final realization. This work was supported by the Swiss National Foundation for Scientific Research.

REFERENCES

Adelman S.J.   1978, *Astrophys. J.* 222, 547
Adelman S.J., Pyper D.M., White R.E.   1980, *Astrophys. J. Suppl.* 43,491
Adelman S.J., Pyper D.M.   1983, *Astrophys. J.* 266, 732
Babu G.S.D., Shylaja B.S.   1981, *Astrophys. Sp. Sci.* 79, 243
Bell R.A., Gustafsson B.   1978, *Astron. Astrophys. Suppl.* 34, 229
Bell R.A., Gustafsson B.   1979, *Astron. Astrophys.*   74, 313
Bell R.A., Eriksson K., Gustafsson B., Nordlund A.   1976, *Astron. Astrophys. Suppl.* 23, 37
Blackwell D.E., Shallis M.J.   1977, *Monthly Notices Roy. Astron. Soc.* 180, 177
Blackwell D.E., Shallis M.J., Selby M.J.   1979, *Monthly Notices Roy. Astron. Soc.* 188, 847
Böhm-Vitense E.   1981, *Ann. Rev. Astron. Astrophys.* 19, 295
Böhm-Vitense E.   1982, *Astrophys. J.* 255, 191
Buser R.   1978, *Astron. Astrophys.* 62, 411
Buser R., Kurucz R.L.   1978, *Astron. Astrophys.* 70, 555
Code A.D., Davis J., Bless R.E., Hanbury-Brown R.   1976, *Astrophys. J.* 203, 417
Cousins A.W.J.   1980, *South Afr. Astr. Obs. Circ.* 1, 166
Cramer N.   1982, *Astron. Astrophys.* 112, 330
Cramer N.   1984, *Astron. Astrophys.* 132, 283
Floquet M.   1981, *Astron. Astrophys.* 101, 176
Gerbaldi M., Hauck B., Morguleff N.   1974, *Astron. Astrophys.* 30, 105
Golay M.   1974, *Introduction to Astronomical Photometry*, Reidel, Dordrecht
Gustafsson B., Bell R.A., Eriksson K., Nordlund A.   1975, *Astron. Astrophys.* 42, 407
Hauck B.   1973, in *Problems of Calibration of Absolute Magnitudes and Temperature of Stars*, B. Hauck and B.E. Westerlund (eds.), Reidel, Dordrecht, p. 117
Hauck B., Mermilliod M.   1980, *Astron. Astrophys. Suppl.* 40, 1
Hauck B., Van't Veer, C.   1970, *Astron. Astrophys.* 7, 219
Hayes D.S.   1978, in *The HR Diagram*, A.G.D. Philip and D.S. Hayes (eds.), Reidel, Dordrecht, p. 65
Jamar C., Macau-Hercot D., Monfils A., Thompson G.I., Houziaux L., Wilson R.   1976, *UV Bright-Star Spectrophotometric Catalogue*, ESA SR-27

Jasniewicz G.   1982, *Astron. Astrophys. Suppl.* 49, 99
Johnson H.L.   1966, *Ann. Rev. Astron. Astrophys.* 4, 193
Kuiper G.P.   1938, *Astrophys. J.* 88, 429
Kurucz R.L.   1975, in *Multicolor Photometry and the Theoretical HR
  Diagram*, A.G.D. Philip and D.S. Hayes (eds.), *Dudley Obs. Rep.* 9, 271
Kurucz R.L.   1979, *Astrophys. J. Suppl.* 40, 1
Lambert D.L.   1977, *Astrophys. J.* 217, 508
Lanz T.:   1984, submitted to *Astron. Astrophys.*
Lub J., Pel J.W.   1977, *Astron. Astrophys.* 54, 137
Magnenat P.:   1977, *Inf. Bull. CDS* 13, 95
Malagnini M.L., Faraggiana R., Morossi C., Crivellari L.   1982,
  Astron. Astrophys. 114, 170
Malagnini M.L., Faraggiana R., Morossi C.   1983, *Astron. Astrophys.*
  128, 375
Malagnini M.L., Morossi C., Faraggiana R.   1984, in *The MK Process
  and Stellar Classification*, R. F. Garrison (ed.), David Dunlap
  Obs., Toronto, p. 321.
Mermilliod J.-C.   1983, *Inf. Bull. CDS* 25, 79
Mermilliod J.-C., Mermilliod M.   1984, in preparation
Meylan G.   1982, *Astron. Astrophys. Suppl.* 47, 483
Meylan G., Hauck B.   1981, *Astron. Astrophys. Suppl.* 46, 281
Morguleff N., Gerbaldi M.   1975, *Astron. Astrophys. Suppl.* 19, 189
Morossi C., Crivellari L.   1980, *Astron. Astrophys. Suppl.* 41, 299
Nicolet B.   1975, *Astron. Astrophys. Suppl.* 22, 239
Nicolet B.   1978, *Astron. Astrophys. Suppl.* 34, 1
Nicollier C., Hauck B.   1978, *Astron. Astrophys. Suppl.* 31, 437
North P.   1980, *Astron. Astrophys. Suppl.* 41, 395
North P., Hauck B.   1979, in *Problems of Calibration of Multicolor
  Photometric Systems*, A.G. Davis Philip (ed.), *Dudley Obs. Rep.* 14, 183
Oinas V.   1974, *Astrophys. J. Suppl.* 27, 391
Perrin M.N., Hejlesen P.M., Cayrel de Strobel G., Cayrel R.   1977,
  Astron. Astrophys. 54, 779
Philip A.G.D.   1979, *Proceedings of Problems of Calibration of Multi-
  color Photometric Systems, Dudley Obs. Rep. No. 14*
Philip A.G.D., Egret D.   1980, *Astron. Astrophys. Suppl.* 40, 199
Philip A.G.D., Releya L.J.   1979, *Astron. J.* 84, 1743
Popper D.M.   1959, *Astrophys. J.* 129, 647
Pottasch S.R., Wesselius P.R., van Duinen R.J.   1979, *Astron. Astrophys.*
  77, 189
Python M.   1979, *Astron. Astrophys. Suppl.* 38, 463
Releya L.J., Kurucz R.L.   1978, *Astrophys. J. Suppl.* 37, 45
Rufener F.   1981, *Astron. Astrophys. Suppl.* 45, 207
Shallis M.J., Blackwell D.E.   1979, *Astron. Astrophys.* 79, 48
Stone R.P.S.   1983, *Publ. Astron. Soc. Pacific* 95, 27
Straižys V., Kurilienè G., Jodinskienè E.   1982, *Vilnius Obs. Bull.*
  60, 44
Underhill A.B., Divan L., Prévot-Burnichon M.L., Doazan V.   1979,
  Monthly Notices Roy. Astron. Soc. 189, 601

Table I – Main sequence standard stars

Name	BS	HD	θ_e	T_e	B2-V1	B-V	ref	Photometric systems
GAM PEG	39	886	.233	21600	-.304	-.23	2	1 4 6 8 9 12 13 14 18 19 21 23
THE AND	63	1280	.560	9000	-.113	.06	3	1 4 6 8 12 13 14 18 21 23 26
54 PSC	166	3651	.949	5309	.526	.85	6	1 4 6 8 12 13 14 19 21 23
		3765	.990	5093	.582	.94	6	1 4 6 9 13 14 19
PHI2 CET	235	4813	.810	6223	.286	.50	6	1 2 8 12 13 14 54
	244	5015	.850	5929	.306	.53	6	1 8 9 13 14 18 19 23
ALF ERI	472	10144	.347	14510	-.257	-.16	1	1 2 8 9 13 14 23 54
	483	10307	.870	5794	.362	.62	6	1 4 6 8 9 12 13 14 18 19 54
	511	10780	.930	5420	.492	.81	6	1 4 8 13 14 18 21 23
58 AND	620	13041	.600	8400	-.055	.12	2	1 4 8 13 14 18 23
PI CET	811	17081	.380	13250	-.237	-.14	2	1 2 4 8 11 13 14 18 19 23 54
41 ARI	838	17573	.405	12450	-.216	-.10	3	1 4 8 12 13 14 18 19 23
IOT PER	957	19373	.850	5929	.334	.61	6	1 2 6 8 9 12 13 14 18 19 23 54
KAP CET	996	20630	.890	5662	.405	.68	6	1 2 6 8 12 13 14 18 19 23
17 ERI	1070	21790	.436	11550	-.207	-.09	3	1 2 4 8 13 14 18 23 54
51 TAU	1197	24167	.636	7900	.001	.20	2	1 13 23
64 TAU	1331	27176	.686	7350	.073	.28	2	1 4 6 8 11 12 13 14 18 21 23
PI3 ORI	1330	27819	.611	8250	-.030	.15	3	1 4 6 8 11 12 13 14 18 23
	1543	30652	.790	6383	.233	.45	6	1 4 6 8 9 12 13 14 18 19 23 54
	1614	32147	1.060	4753	.661	1.06	6	1 2 6 8 12 13 14 19 54
LAM AUR	1639	32608	.622	8102	-.038	-.07	2	1 2 6 8 9 12 13 14 18 23
RHO AUR	1729	34411	.860	5861	.361	.63	6	1 6 8 11 12 13 14 18 19 23
GAM ORI	1749	34759	.329	15300	-.254	-.15	2	1 6 13 23
DZE LEP	1790	35468	.234	21580	-.294	-.22	1	1 4 6 8 9 12 13 14 18 19 21 23 26 54
	1998	38678	.593	8500	-.072	.10	2	1 2 6 8 13 14 19 54
134 TAU	2010	38899	.465	10850	-.195	-.07	2	1 4 6 8 13 14 18 19 23 26
CHI1 ORI	2047	39587	.840	5998	.347	.59	6	1 4 8 9 11 12 13 14 18 19 23
ETA LEP	2085	40136	.710	7100	-.134	.33	2	1 4 8 9 11 12 13 14 18
	2209	42818	.533	9450	-.145	.03	3	1 8 13 16 23
	2244	43445	.472	10689	-.197	-.08	5	1 4 8 13 14 18 23 54
GAM GEM	2421	47105	.544	9260	-.146	.00	1	1 4 6 8 9 12 13 14 18 19 23
21 LYN	2818	58142	.525	9600	-.168	-.02	4	1 8 13 14 18 23
64 GEM	2857	59037	.600	8400	-.061	.11	2	1 13 14 18 23 26
27 LYN	3173	67006	.548	9200	-.140	.05	2	1 4 8 13 14 18 23
16 PUP	3192	67797	.326	15460	-.246	-.15	5	1 2 8 13 18 23 54

Table I – (continued)

Name	BS	HD	θ_e	T_e	B2-V1	B-V	ref	Photometric systems
LAM CNC	3268	70011	.489	10300	-.184	-.03	3	1 4 23
CNC	3348	71906	.486	10375		-.03	3	1 4 13 23 26
GAM CNC	3449	74198	.536	9400	-.278	-.02	3	1 4 8 14 18 23
ETA HYA	3454	74280	.268	18792		-.20	5	1 2 6 8 11 12
RHO1 CNC	3522	75732	.969	5200	.539	.87	6	1 6 13 14 19
BET CAR	3685	80007	.545	9240	-.136	.00	1	1 2 4 8 9 13 14 19 23 54
26 UMA	3699	82621	.563	8950	-.122	-.01	3	1 4 8 13 14 18 23 26
KAP HYA	3849	83754	.327	15398	-.250	-.15	5	1 2 4 8 13 18 23 54
	3881	84737	.860	5861	.362	.62	6	1 2 6 8 13 14 18 23
20 LMI	3951	86728	.880	5728	.399	.66	6	1 4 6 12 13 14 21 23
ALF LEO	3982	87901	.419	12040	-.217	-.11	7	1 4 6 8 9 13 14 18 19 21 23
BET SEX	4119	90994	.350	14400	-.247	-.14	4	1 4 8 13 14 18 23 54
37 UMA	4141	91480	.710	7100	-.137	.34	4	1 4 8 13 14 21 23 26
47 UMA	4277	95128	.860	5861	.364	.61	6	1 4 6 13 23 26
BET UMA	4295	95418	.525	9600	-.154	-.02	4	1 4 6 8 9 13 18 21 23
THE LEO	4359	97633	.548	9200	-.152	-.01	3	1 4 8 9 12 13 14 18 23 26
61 UMA	4496	101501	.919	5483	.449	.72	6	1 4 6 8 12 13 18 19
BET LEO	4534	102647	.570	8850	-.077	.09	1	1 2 4 6 9 11 12 13 14 18 19 21
BET VIR	4540	102870	.831	6067	.311	.55	6	1 2 4 8 11 12 13 14 18 19 21
GAM UMA	4554	103287	.531	9500	-.151	.00	3	1 4 6 9 13 18 21 23
DEL CRU	4656	106490	.220	22906	-.304	-.23	5	1 2 4 8 11 13 19 23 54
DEL UMA	4660	106591	.586	8600	-.088	.08	4	1 4 6 13 14 18 21 23
GAM MUS	4773	109026	.325	15506	-.253	-.15	5	1 4 8 13 23 54
BET CVN	4785	109358	.860	5861	.352	.59	6	1 4 6 8 9 12 14 18 19 21 23
14 CVN	4943	113797	.456	11050		-.08	3	1 14
BET COM	4983	114710	.840	5998	.334	.57	6	1 4 6 8 9 11 12 13 14 18 19 21 23 45
		115043	.850	5929	.356	.60	6	1 4 8 12 13 21 23
59 VIR	5011	115383	.850	5929	.323	.59	6	1 6 11 12 13 14 19 23 54
ALF VIR	5056	116658	.211	23930	-.300	-.23	1	1 2 6 8 12 13 18 19 21
80 UMA	5062	116842	.611	8250	-.020	.16	3	1 4 6 13 18 21 23
70 VIR	5072	117176	.919	5483	.442	.71	6	1 4 6 8 9 11 12 13 14 18 19 23
	5169	119765	.521	9675		.00	3	1 4 14 21 23
ETA UMA	5191	120315	.296	17000	-.264	-.19	4	1 4 6 8 11 12 13 18 21 23 26 54
TAU VIR	5264	122408	.604	8350	-.069	.10	3	1 2 4 11 12 13 14 18 21 23 26 54
	5332	124683	.502	10050	-.158	-.03	3	1 2 4 11 13 23

Table I - (continued)

Name	BS	HD	θ_e	T_e	B2-V1	B-V	ref	Photometric systems
45 BOO	5422	127304	.480	10500	-.179	-.03	4	1 4 13 14 21 23 26
	5568	131977	1.090	4624	.685	1.11	6	1 2 6 8 9 12 13 14 19 54
	5634	134083	.775	6500	.209	.43	4	1 4 6 8 9 13 14 18 19 21 23 26
LAM SER	5868	141004	.850	5929	.360	.60	6	1 2 4 6 8 11 12 13 14 18 19 21
GAM SER	5933	142860	.819	6152	.248	.48	6	1 2 4 6 8 11 12 13 14 18 19 21
		145675	.969	5200	.540	.90	6	1 2 12 13 14 19 23 26
18 SCO	6060	146233	.860	5861	.385	.65	6	1 2 4 12 13 14 18 19
12 OPH	6171	149661	.941	5358	.514	.82	6	1 4 6 8 9 12 13 14 54
		156026	1.110	4539	.730	1.16	6	1 6 8 12 13 14 54
ALF OPH	6556	159561	.628	8020	-.030	.15	1	1 4 6 8 13 14 18 19 21 23
MU ARA	6585	160691	.900	5598	.419	.70	6	1 2 8 12 13 14 18 19 21 23
GAM OPH	6629	161868	.552	9125	-.126	.04	3	1 2 4 6 8 11 12 13 19 21 45 54
	6806	166620	1.010	4999	.548	.87	6	1 4 6 8 9 12 13 14 19 21 23
EPS SGR	6879	169022	.533	9460	-.142	-.03	1	1 2 6 8 9 13 14 19 23 54
ALF LIR	7001	172167	.531	9490	-.152	.00	7	1 4 6 8 9 12 13 14 18 19 21 23 26
SIG SGR	7121	175191	.265	18987	-.280	-.22	5	1 2 4 8 13 23 54
LAM AQL	7236	177756	.442	11414	-.219	-.09	5	1 8 9 13 14 18 21 23 45 54
16 SYG	7503	186408	.870	5794	.398	.64	6	1 4 6 8 9 12 13 18 21 23
	7504	186427	.870	5794	.402	.66	6	1 4 6 8 9 12 13 18 21 23
ALF AQL	7557	187642	.629	8010	.031	.22	1	1 4 6 8 9 13 14 19 21 54
	7722	192310	1.010	4989	.561	.88	6	1 2 6 8 12 13 14 19 26
	7783	193664	.840	5998	.344	.58	6	1 6 13 21 45
ALF PAV	7790	193924	.282	17880	-.277	-.20	1	1 2 4 6 8 13 19 23 54
ALF DEL	7906	196867	.472	10681	-.196	-.06	5	1 4 6 11 14 18 19 23
IOT AQR	8418	209819	.447	11284	-.203	-.07	5	1 2 4 6 8 13 14 18 23 26 45
SIG AQR	8573	213320	.499	10100	-.187	-.06	2	1 2 4 8 9 13 14 18 23
ETA AQR	8597	213998	.449	11218	-.211	-.09	5	1 4 8 13 14 18 23 54
	8607	214279	.596	8450	-.074	.12	2	1 4 13 21 23
OMI PEG	8641	214994	.536	9400	-.164	-.01	2	1 2 6 8 13 14 18 23 26
ALF PSA	8728	216956	.573	8800	-.085	.09	1	1 2 8 9 13 14 19 23 54
51 PEG	8729	217014	.880	5728	.396	.67	6	1 4 6 8 9 11 12 13 14 18 19 21 23
	8872	219134	1.051	4797	.628	1.01	6	1 4 6 9 12 13 14 18 19 21 45
	8853	219623	.831	6067		.52	6	1 4 14 21 23
IOT AND	8965	222173	.427	11800	-.210	-.10	2	1 4 8 13 14 18 21 23

Table II – K0-K3 giant standard stars

Name	BS	HD	θ_e	T_e	B2-V1	ref	Photometric systems
ALF ARI	617	12929	1.084	4650	.797	8	1 4 6 8 9 12 13 14 18 21
GAM TAU	1346	27371	0.998	5050	.645	8	1 4 8 11 12 13 19 23
EPS TAU	1409	28305	0.998	5050	.678	8	1 4 6 8 9 11 12 13 16 19 21 45
THE1 TAU	1411	28307	0.979	5150	.628	8	1 4 6 8 9 11 12 13 18 19 21 45
NU AUR	2012	39003	1.061	4750	.785	8	1 8 9 13 14 18
MU LEO	3905	85503	1.084	4650	.840	8	1 8 9 12 13 14 18 45
NU UMA	4377	98262	1.200	4200	1.023	8	1 4 8 9 12 13 14 18 21
ALF SER	5854	143333	1.061	4750	.797	8	1 4 13 14 19 23
11 CEP	8317	206952	1.050	4800	.754	8	1 8 12 13 14 18
	8924	221148	1.029	4900	.727	8	1 12 13 14 19 21 54

Sources for tables I and II.

1. Code et al. (1976)
2. Adelman (1978)
3. Adelman et al. (1980)
4. Adelman and Pyper (1983)

5. Underhill et al. (1979)
6. Perrin et al. (1977)
7. Hayes (1978)
8. Oinas (1974)

Table III - Number of standard (n_1=m.s., n_2=giant) stars measured in some photometric systems

System	Source of data	n_1	n_2
1 UBV	Nicolet (1978), Mermilliod J.C. (1983)	103	10
2 UBV CAPE	Nicolet (1975)	31	-
4 UVBY	Hauck and Mermilliod M. (1980)	94	6
6 UVBGRI Stebbins, Whitford	Nicollier and Hauck (1978)	60	3
8 UBVRI Johnson	Mermilliod J.C. and Mermilliod M. (1984)	82	8
9 JKLMN Johnson	Mermilliod J.C. and Mermilliod M. (1984)	45	6
11 WUBVL Walraven	Python (1979)	19	3
12 DDO	Meylan (1982)	45	8
13 UBVB1B2V1G Geneva	Rufener (1981)	99	10
14 UBV Eggen	Mermilliod J.C. and Mermilliod M. (1984)	72	7
18 13-COLOR	Magnenat (1977)	63	7
19 RI Kron and Eggen	Jasniewicz (1982)	47	5
21 UPXYVTZ Vilnius	North (1980)	42	5
23 β	Hauck and Mermilliod M. (1980)	89	2
26 Cl ...C8 Barbier, Morguleff	Morguleff and Gerbaldi (1975)	19	-
45 CMT1T2V Canterna	Mermilliod J.C. and Mermilliod M. (1984)	10	3
54 RI Cousins	Cousins (1980)	39	1

DISCUSSION

PAPOULAR: Is there any hope of extending this effort to M giants and supergiants?

HAUCK: The problem is not with the colors but with effective temperatures and at the present time we need more effective temperature determinations for these stars.

CODE: What is the basis of the effective temperatures for metal deficient stars?

HAUCK: Perrin et al. (1977) have obtained the effective temperatures from detailed analysis based on model atmosphere calculations.

POPPER: Effective temperature is defined in terms of bolometric flux. This quantity is known only for stars with measured angular diameters determined, which are, unfortunately, very few in number. Temperatures derived by comparing synthetic spectra with observations should, perhaps, be termed "hypothetical" effective temperatures. This distinction, usually overlooked, has been pointed out many times in the past. It is typical for those who interpret IUE observations, for example, to quote temperatures from models without reference to the calibration of these models against ultraviolet observations of stars with known effective temperatures from angular diameter measurements.

TOBIN: To reinforce Dr. Popper's comment, can I say that I would classify the Underhill et al. method as indirect, rather than semi-direct, as what it really does when strictly applied is to compare the dereddened value of (monochromatic, visual or infrared flux/observed integrated flux) with model values, and it is just as dependent on the Kurucz models as anyone else's "effective temperatures".

CODE: To follow up the remarks by Popper, I wish to remind you that another assumption the concept of effective temperature is that the star is spherical and the atmosphere thin. For hot main sequence stars and supergiants this may not be true and some other concept must be substituted for effective temperature.

THE CALIBRATION OF PHOTOMETRIC DETERMINATIONS OF ABUNDANCE

V. Straižys

Vilnius Astronomical Observatory

ABSTRACT. Photometric abundance determinations using blanketing in the ultraviolet and violet spectral regions as well as the intensity of strong lines and bands is the most important tool in the study of chemical composition of faint and distant stars. However, this method requires the calibration of different photometric quantities in terms of abundance at given temperatures and luminosities. Two principal methods of calibration are in use. One of them uses stars with abundances determined from high dispersion spectra by curve of growth or model atmosphere analysis of the spectral lines. The other one uses synthetic spectra based on model atmospheres which are integrated to imitate the narrow band photometric indices. This paper will summarize the results of the application of both of these methods to calibrate a number of the important photometric systems. Most attention will be given to the late-type stars, which demonstrate the strongest photometric abundance effects.

1. INTRODUCTION

The photometric method of metallicity determination is very attractive since it has the folowing advantages in comparison with spectroscopic methods: (1) much fainter limiting magnitudes, (2) a much faster rate of collecting observational data, (3) the ability to work in clusters and in other crowded fields, (4) the parallel determination of temperatures, gravities or luminosities and interstellar reddenings, and (5) the ability to discover. easily objects with peculiar abundances. This gain of information is still more significant since the accuracy of a metallicity determination from photometry and from high dispersion spectroscopy seems to be of about the same order.

However, this does not mean that photometry can everywhere replace spectroscopy. The photometric method gives a number of quantities (color indices, their differences, reddening-free

D. S. Hayes et al. (eds.), Calibration of Fundamental Stellar Quantities, 285–301.
© *1985 by the IAU.*

parameters, etc.) which are sensitive to abundances. However, these quantities cannot be used without calibration and in this respect high dispersion spectroscopy plays the most important role. In addition, spectroscopy is indispensable for verification of objects with peculiar abundances detected photometrically.

Until recently photometry could not work independently since for the calibration of photometric abundance indicators spectroscopic abundances were necessary. The situation has changed since the appearance of reliable model stellar atmospheres which permit the theoretical calculation of synthetic spectra of stars. These synthetic spectra, convolved with the response functions, give photometric quantities for different temperatures, gravities, and abundances. Now both methods of abundance determination, spectroscopic and photometric, are more and more based on the synthetic spectra. However, the synthetic spectra techniques give reliable results so far only in comparatively narrow spectral regions (tens of Å) for certain temperatures and luminosities. Consequently, this method of calibration is most successful for narrow-band photometry.

The abundance criteria of photometric systems with wide and intermediate band widths are based, as a rule, on the differential blocking effect caused by numerous lines of different elements. Consequently, these systems are able to determine the common effects of hundreds and thousands of lines of different elements. At the same time, narrow-band photometry having half-widths of the order of tens of Å can be used to measure the strengths of strong spectral lines and bands and after corresponding calibration they can give abundances of individual elements. In this respect narrow-band photometry is similar to the equivalent width determination or to the fitting of some absorption feature with its synthetic profile.

In the following sections the abilities of different photometric systems for abundance determinations and their calibration results will be reviewed.

2. WIDE-BAND PHOTOMETRIC SYSTEMS

The (U-B, B-V) diagram is suitable for metallicity determination in two regions. One of them includes metallic-line stars of spectral types A and F and another includes F, G and K subdwarfs and metal deficient giants. With respect to the zero-age main-sequence line the Am stars show ultraviolet deficiencies δ(U-B) of up to 0.15 mag. This deficiency is roughly proportional to metal abundances (Abt 1961; Jaschek and Jaschek 1962). Unfortunately, it is practically impossible to use δ(U-B) as a measure of metallicity due to its strong dependence on luminosity (Eggen and Sandage 1964; Straižys and Kavaliauskaitė 1967) and interstellar reddening. The increase of metallicity, of luminosity and of interstellar reddening shifts the star in the same direction and there is no way to separate them using only photometric UBV data.

The same effects also occur in the region of late-type dwarfs and giants. However, most of the known late-type subdwarfs are relatively close to the Sun, and their interstellar reddening is small. If we neglect the possible small differences in luminosities, the ultraviolet excesses of subdwarfs are a function of metallicity. Fig. 1 shows the position of metal deficient dwarfs of spectral types F, G, and K in the (U-B, B-V) diagram. The ultraviolet excess δ(U-B) for F and G subdwarfs has been calibrated against spectroscopic [Fe/H] by many authors (see Straižys 1982, p.128). One of the latest calibrations is given by Carney (1979). Buser and Kurucz (1978) calibrated the (U-B, B-V) diagram with respect to metallicity by using synthetic colors calculated using model atmospheres by Kurucz (1979). However, the theoretical colors deviate considerably from the observed ones.

The calibration of δ(U-B) in terms of metallicity for K-type dwarfs and subdwarfs is, so far, not possible as there are no reliable spectroscopic values of [Fe/H] with which to calibrate it. At the same time line blanketed model atmospheres or theoretical energy distribution for K-type dwarfs with different metallicities remain unpublished (Eriksson et al. 1979, Bessell and Wickramasinghe 1979, and Thevenin and Foy 1983).

The separation of late subdwarfs can be increased considerably by replacing (B-V) by (R-I) in the two color diagram (Fig. 2). In this diagram the blanketing vectors are almost vertical since the (R-I) color index defines the temperature of G and K stars almost independently of metal abundance. Other indices including red and infrared magnitudes such as (V-R), (V-K), (V-L) and others of the Johnson system, or (R-I) and (G-I) of the Stebbins and Whitford six color system may be also used as temperature criteria.

Smith and Steinlin (1964) were the first who recommended the use of the RGU photographic system for the separation of subdwarfs. The blanketing vectors in the (U-G, G-R) diagram are much steeper than in the (U-B, B-V) diagram and, consequently, the separation of subdwarfs is larger. The diagram was calibrated with respect to metallicity using synthetic colors from model atmospheres by Kurucz (Buser 1979) and spectroscopic [Fe/H] determinations (Trefzger 1981).

Interstellar reddening and the presence of giants reduce the usefulness of the UBV and RGU systems for the detection of subdwarfs and for their metallicity determination.

Metal deficient giants are situated above the solar composition giants both in the (U-B, B-V) and (U-B, R-I) plots, but the ultraviolet excesses, δ(U-B), are much larger in the latter diagram

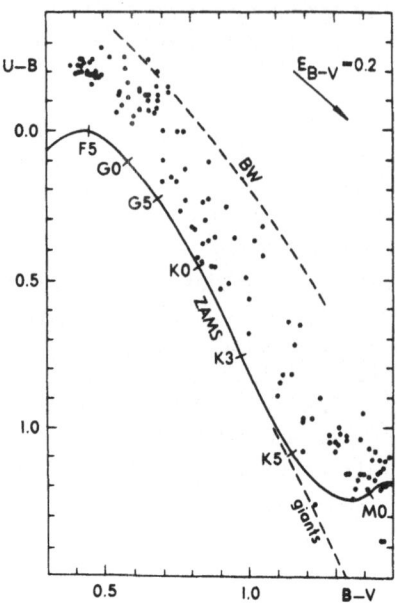

Fig. 1. The (U–B, B–V) diagram for subdwarfs. The smooth line
represents the zero-age main sequence; the broken line
represents the sequence of solar composition giants; and the
line marked BW is a locus of the line-free colors of subdwarf
model atmospheres from Bessell and Wickramasinghe (1979). The
arrow in the upper right corner denotes the reddening line for
$E_{B-V} = 0.2$ mag.

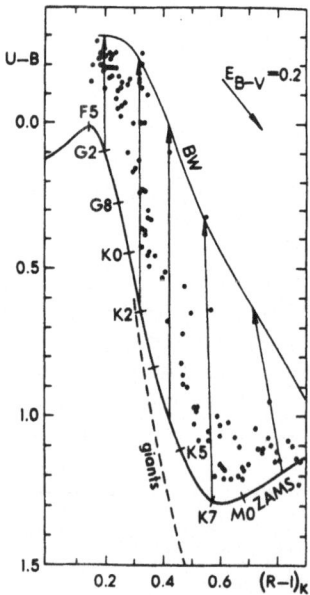

Fig. 2. The (U–B, $[R-I]_K$) diagram for subdwarfs. The designations are
the same as in Fig. 1. The arrow lines pointing upwards
represent the deblanketing vectors.

(Fig. 3). Extreme metal deficient giants such as HD122563 and HD165195 have δ(U-B) $\approx$ 0.2 mag. after correction for interstellar reddening in the (U-B, B-V) plot. At the same time these excesses are as large as 0.7 mag. in the (U-B, R-I) diagram. Model atmosphere analysis of the (U-B, B-V) diagram shows a considerable gravity effect (see Böhm-Vitense 1973; Böhm-Vitense and Szkody 1974; Bell and Gustafsson 1978). Probably, a similar effect is taking place in the (U-B, R-I) diagram. At the same time theoretical colors cannot be used to calibrate these diagrams since most models show a considerable ultraviolet excess relative to stars. This is probably caused by "missing opacity", which increases with metal abundance. Despite these difficulties the (U-B), $(R-I)_K$ diagram has been used by Eggen for metallicity estimates for thousands of G and K giants (for references see Straižys 1982 p.173). The (U-B) excess of the giant branch at $(B-V)_0$=1.0 mag. remains one of the principal metallicity criteria for globular clusters (see Sandage and Hartwick 1977). However, the calibration of δ(U-B) using the field and globular cluster giants studied with high dispersion gives contradictory results and leads to the well known vagueness of the metallicity scales (see the discussion in Straižys 1982).

In order to overcome the difficulty with luminosity effects, Canterna (1976), in collaboration with G. Wallerstein, proposed the Washington photometric system consisting of the four wide bandpasses C, M, T_1 and T_2, with mean wavelengths at 3910, 5085, 6330, and 8050 Å and half-widths of the order of 1000 Å. Later on the system was supplemented by the V magnitude of the UBV system. Canterna and Harris (1979) showed that the $(M-T_1, T_1-T_2)$ relation is almost the same for solar chemical composition stars of different luminosity classes with the reddening line nearly parallel to this relation. They supposed that the "green excess" $\delta(M-T_1)$ is a reddening-free, surface gravity free metallicity parameter. It was calibrated in metallicities both with stars known [Fe/H] from high dispersion spectroscopy and by model atmosphere colors using model atmospheres calculated by Böhm-Vitense and Szkody (1974). The main shortcoming of the method seemed to be a very small abundance effect: the "green excess" $\delta(M-T_1)$ did not exceed 0.15 mag. for stars with [Fe/H] $\approx$-2.5. Recently Straižys and Kurilienė (1983) have shown that the luminosity sequences and the reddening lines in the $(M-T_1, T_1-T_2)$ diagram do not coincide exactly and this limits the usefulness of the system in the determination of metallicity.

3. MEDIUM-BAND PHOTOMETRIC SYSTEMS

3.1 The uvbyβ systems

Much work has been done on the metallicity calibration of one of the most widely used photometric systems: the Strömgren uvby system supplemented by the Hβ index. For the determination of the

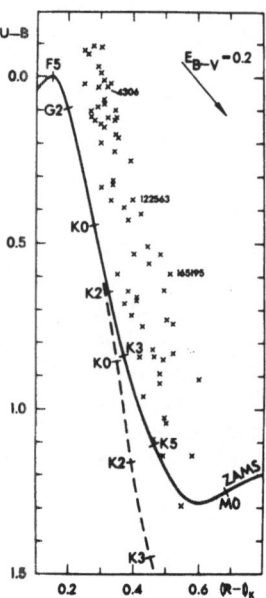

Fig. 3. The (U-B, [R-I]$_R$) diagram for dereddened late-type metal-deficient giants with [Fe/H]<-0.6. The designations are the same as in Fig. 1. The HD numbers of three extreme metal-deficient giants are given. Shown by the courtesy of A. Bartkevičius.

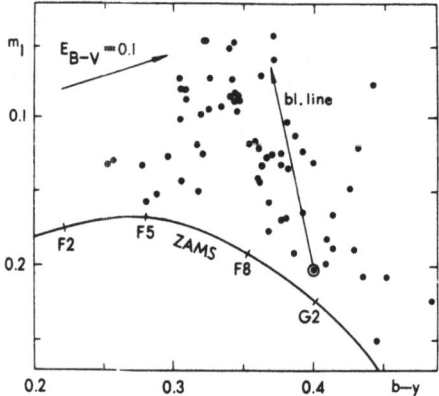

Fig. 4. The (m$_1$, b-y) diagram for F and early G subdwarfs. The upward arrow represents the blanketing vector for the Sun.

metallicity of Am stars and subdwarfs the $(m_1,b-y)$ or (m_1,β) diagrams are used. Sometimes the color difference $m_1 = (v-b) - (b-y)$ is replaced by the reddening-free parameter $[m_1]=Q_{vby}=(v-b)-0.82(b-y)$. In these diagrams A and F-type stars of luminosities V-III form one sequence. Consequently, the deviation δm_1 or $\delta[m_1]$ from the Hyades line is nearly independent of surface gravity and is a measure of metallicity only.

It is known that most of the Am stars usually have larger values of m_1 than the Hyades standard relation at the same effective temperature. The deviation δm_1 is 0.07 mag. in extreme cases. The (m_1,β) diagram was calibrated in terms of iron abundance by Rydgren and Smith (1974), using blocking corrections in different filters determined from high dispersion spectra. The resulting set of isolines of equal metallicity was found to be in reasonable agreement with the iron abundance of 17 normal and Am stars. On the other hand, about 30% of the stars classified as Am spectroscopically are located either near the zero-age main sequence or show negative δm_1 (see Fig. 48 in Straižys (1977). A careful investigation of their photometric and spectral properties is necessary.

Fig. 4 shows the F and early G subdwarfs plotted in the $(m_1, b-y)$ diagram (Straižys 1977). The violet excesses δm_1 are 0.14 mag. for extreme subdwarfs, i.e. half of $\delta(U-B)$ in the $(U-B, B-V)$ diagram. The δm_1 excesses have been calibrated by a number of authors. Strömgren (1964), Crawford (1975) and Crawford and Perry (1976) used [Fe/H] values of stars determined from high-dispersion spectra. Nissen (1970, 1981), Gustafsson and Nissen (1972) and Nissen and Gustafsson (1978) used the metallicities determined from a model atmosphere analysis of narrow band photoelectric observations of two groups of metallic lines. It was shown that differential metal abundances can be determined from δm_1, with an accuracy of ± 0.10 mag. for F0-G2 dwarfs and mild subdwarfs. However, the calibration extends only down to [Fe/H] = -0.6. Relyea and Kurucz (1978) used Kurucz (1979) model atmosphere fluxes for calibration of δm_1. They concluded that there is an excellent agreement between the theoretical colors and the observations for stars with $T_{eff} >8500$ K. However, for the stars cooler than A5 the colors are not in good agreement, so the theoretical calibration of photometric metallicity criteria is unreliable.

The $(m_1,b-y)$ diagram can be used also to determine the matallicity of late-type giants (Fig. 5). Bell and Gustafsson (1978) calculated the indices of the Strömgren system for a grid of model atmospheres. Their results were analyzed by Gustafsson and Bell (1979) and Bond (1980) who used them to calibrate the $(m_1, b-y)$ diagram in temperatures and metallicities. Bond also presented an empirical calibration of δm_1 with respect to [Fe/H] using spectroscopic data and photometric data corrected for interstellar reddening. The rms scatter of the [Fe/H] determinations about the resulting relation is ± 0.15 dex. He concluded that the calibration seems to reproduce [Fe/H] values within the quoted errors of the spectroscopic determinations. Eggen (1977, 1978, 1983) has used a modified version of the uvby system for metallicity determinations of late-type giants and supergiants. He calibrated reddening-free $\Delta[M_1]$ excesses using spectroscopic [Fe/H] values.

The Strömgren system has been used also to determine the metallicity of RR Lyr stars. The parameter m_1 was shown to be constant with phase during the variation. This means that $\delta[m_1]$ is a better representation of metallicity for this type of star than the ΔS parameter which shows a variation with phase. The δm_1 excesses correlate with other metallicity indicators such as ΔS, $\delta(U-B)$ and spectroscopic [Fe/H] (Epstein and Epstein, 1973; Butler 1975).

3.2 The Geneva system

For metallicity estimates in the Geneva system the photometric parameter $m_2 = (B_1 - B_2) - 0.457 (B_2 - V_1)$, which measures the violet blanketing, is most useful. This parameter is close to the reddening-free parameter $Q_{B_1 B_2 V_1}$. Metallic line stars in the (m_2, B_2-V_1) diagram are not well separated from the normal A-stars, except for the coolest ones (Hauck and Curchod 1980). This means that, in addition to line blocking, the photometric parameter m_2 is affected by some other physical characteristics, probably by the projected rotational velocity and binarity. Nicolet and Cramer (1983) suggested for Am stars the second-order photometric parameters l, m, and n which are linear functions of different Q-parameters of the Geneva system. For the interpretation of the (l,m) and (l,n) diagrams, Kurucz models have been used. The authors conclude that the photometric detection of the Am phenomenon must never be considered definite before spectroscopic confirmation. In the case of F and G subdwarfs the situation is much better. The empirical calibration of δm_2 excess against [Fe/H] was given by Hauck (1968, 1973, 1978). Later on North and Hauck (1979) analyzed the $(m_2, B_2 - V_1)$ diagram using Kurucz (1979) model atmosphere fluxes. However, the unsatisfactory agreement of theoretical and empirical indices prevented the use of synthetic colors for calibration purposes. The authors found considerable differences between the theory and observations and suggested that the model atmospheres are not sufficiently blanketed.

In the case of the late-type giants in the Geneva system there is no two-color or (Q, Q) diagram in which abundance effects would be not influenced by luminosity. To overcome this difficulty Grenon (1978, 1981) used a set of second order photometric diagrams exhibiting the effects of gravity and metallicity at given temperature intervals. These second order quantities ε actually are differential color excesses at constant $(V_1 - G)$ index which measures temperature. The

(ε, ε) diagrams were calibrated in terms of M_V and [Fe/H], the latter being taken from spectroscopic determinations in the field and in open clusters. The main shortcoming of the method is the dependence of the ε parameters on interstellar reddening.

Color indices in the Geneva system were calculated for model stellar atmospheres of red giants by Bell and Gustafsson (1978). However, the presence of systematic differences between the theoretical and observed values in the ultraviolet (Gustafsson and Bell 1979) prevents the use of the synthetic spectra for calibration.

3.3 The DDO system

This system was proposed for the classification of G and K stars in terms of temperature, gravity and metallicity. The C_{38-42} index measuring line blocking in the region of $\lambda 3800$ Å is being used for metallicity determination. The δC_{38-42} excess at C_{45-48} = const. shows a good correlation with spectroscopic [Fe/H] (Osborn, 1973a,b). In this respect the system does not differ much from the other medium band systems. A specific property of the system is the ability to measure the strength of the violet CN band shortward of 4216 Å. The deviation of the C_{41-42} index from the standard sequence in the (C_{41-42}, C_{42-45}) diagram indicates anomalous CN strength and can be identified with composition differences after the elimination of the surface gravity effect. The CN anomaly $\delta(CN)$, calibrated by Janes (1975), has been used as a general metallicity parameter for giants of disk population. The linear correlation between $\delta(CN)$ and [Fe/H] later was revised by McClure (1979). For metal-poor giants in the field and in globular clusters the violet CN band is weak and is measured with lower accuracy. In the metal-rich clusters it varies considerably from star to star. It seems that in the halo population there is no one-to-one correlation of $\delta(CN)$ with [Fe/H]. For stars with different CN strength the δC_{38-42} excess is also not suitable because it is affected by ultraviolet CN bands of different strength. As an alternative for the ranking of globular clusters in metallicities Hesser et al. (1977) have used the (C_{45-48}, C_{42-45}) diagram. The δC_{45-48} excess was calibrated in [Fe/H] by Janes (1979). However, the range of variation of this excess is much smaller than of δC_{38-42} and $\delta(CN)$.

Bell and Gustafsson (1978) calculated synthetic spectra and DDO indices for model atmospheres of giants with different metal abundances. The same was done by Bell et al. (1978, 1979), Dickens et al. (1979) and Bell and Dickens (1980) for model atmospheres having various carbon and nitrogen abundances.

All of the photometric abundance criteria in the DDO system can work only when interstellar reddening is absent. When it is present

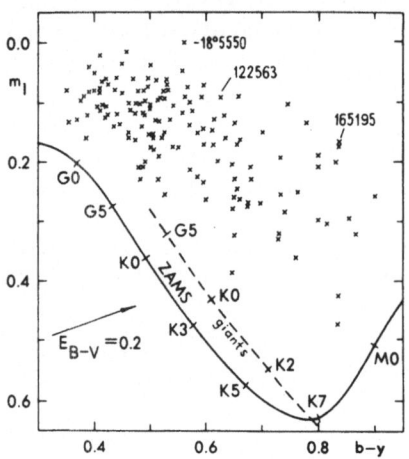

Fig. 5.   The (m_1, b-y) diagram for G and K metal-deficient giants with
[Fe/H]$<$-0.6 shown by the courtesy of A. Bartkevičius.

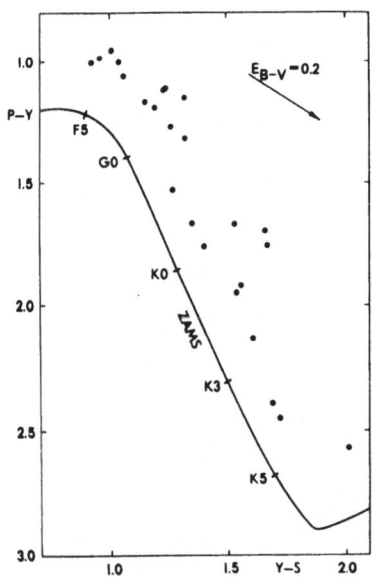

Fig. 6.   The (P-Y, Y-S) diagram of the Vilnius photometric system for
extreme subdwarfs.

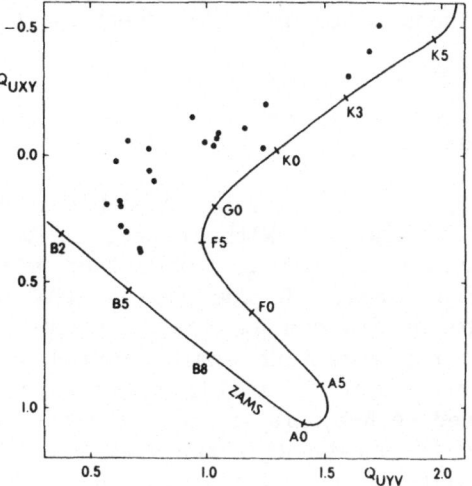

Fig. 7. The reddening-free (Q_{UXY}, Q_{UYV}) diagram of the Vilnius photometric system for extreme subdwarfs.

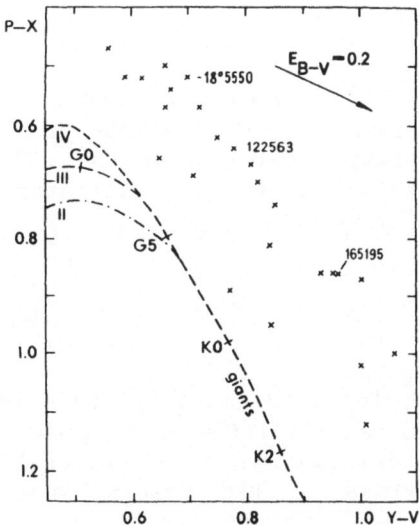

Fig. 8. The (P-X, Y-V) diagram of the Vilnius photometric system for dereddened late-type metal-deficient giants.

the DDO indices must be corrected. The interstellar reddening is usually found using methods based upon the (B-V) indices of stars.

3.4 The Walraven system

Lub and Pel (1977) have recently reviewed the properties of the five-color Walraven VBLUW system. With respect to Am stars the system probably gives no new information in comparison with other systems (Wiertz and van Genderen 1983). In the case of the F and G subdwarfs the most useful two color diagram is the (V-B, B-L) diagram or (L-B, B-V) if we accept the magnitude scale instead of the log I scale used by the Leiden astronomers. In this diagram the gravity effects below FO are small and $\delta(B-L)$ is a metallicity discriminant when interstellar reddening is excluded. Lub and Pel (1977) claim that their photometry is more than twice as sensitive to metallicity differences as is the Strömgren uvby system for F and G stars. In their paper theoretical colors for the Kurucz model atmospheres are used to discuss the possibility of three-dimensional classification in terms of temperature, gravity and metallicity.

The Walraven system has been widely used for the photometric investigation of Cepheids and RR Lyrae type stars. The reddening-free Q_{LBV} parameter was used for the metallicity determination of RR Lyrae stars (Lub 1977, 1979).

3.5 The Arizona system

This 13-color medium-band system was used by Johnson and Mitchell (1968) and Schuster (1976, 1979) for the analysis of F and G subdwarfs. For metallicity determination the (37-45, 45-63) diagram was proposed. The $\delta(37-45)$ excess was calibrated by Schuster (1979) using spectroscopic [Fe/H] values.

3.6 The Vilnius system

This system was developed to classify stars in terms of spectral types, luminosities, and metallicities where interstellar reddening is present and where no information from spectral classification is available. The properties of the system have been described in English by Straižys and Sviderskienė (1972) and Straižys (1973, 1975, 1979) and in Russian by Straižys (1977). The metallicity of unreddened or dereddened F and G subdwarfs can be determined in the surface gravity-free (P-Y, Y-S) diagram (Fig. 6) calibrated in [Fe/H] by Bartkevičius and Straižys (1970a) and Bartkevičius and Sperauskas (1983). The ultraviolet excess $\delta(P-Y)$ at (Y-S) = const for HD 19445 is 0.36 mag., i.e. larger than $\delta(B-L)$ of the Walraven system, δm_1 of the Strömgren system, or δm_2 of the Geneva system. The metallicity and temperature of the reddened F and G subdwarfs can be determined in the reddening-free (Q_{UXY}, Q_{UYV}) diagram (Fig. 7) calibrated by Bartkevičius and Straižys (1970b). The δQ_{UYV} excesses for extreme F and G subdwarfs are $\approx$ 0.4 mag. The separation of subdwarfs from the remaining stars is single-valued, i.e. no other type of single star appears in the subdwarf region of the Q_{UXY}, Q_{UYV} diagram.

Recently it was shown (Straižys et al. 1984) that the same diagrams can be used for metallicity determination of late G and K subdwarfs. These stars are collected from the lists by Eggen (1969) and Bessell and Wickramasinghe (1979) and are plotted in Figs. 6 and 7. Unfortunately, the calibration of these diagrams in terms of metallicity for G and K subdwarfs has not yet been done due to the absence of reliable spectroscopic abundances and model atmospheres.

The metallicity of unreddened or dereddened G and K subgiants and giants can be determined in the same (P-Y, Y-S) diagram as calibrated by Bartkevičius and Sperauskas (1983) or in the (P-X, Y-V) diagram (Fig. 8) suggested by Bartkevičius and Straižys (1970c) and calibrated by Straižys and Bartkevičius (1982) and Bartkevičius and Sperauskas (1983). These diagrams are surface gravity-free for stars of spectral types G5-K2 and of luminosities IV-III-II. The value of $\delta(P-X)$ at (Y-V) = const is 0.33 mag. for the metal-deficient giant BD-18°5550, which has [Fe/H] = -2.8. For the metallicity determination of reddened late-type giants we have no (Q, Q) diagram which would be surface gravity-free. However, Straižys and Bartkevičius (1982) proposed a method based on two diagrams: (P-X, Y-V) and Q_{PYV}, P-X). Both diagrams are surface gravity-free and can be used to determine temperature, metallicity, and color excess. To my knowledge, it is the only method for determining [Fe/H] for reddened late-type giants and subgiants.

All our calibrations in metallicity are based on spectroscopic abundances transformed to a common scale by Bartkevičius (1980, 1984). Our attempts to use the synthetic colors of the Vilnius system calculated using model atmospheres by Bell (1977) showed a certain amount of discrepancy and we avoided using them for calibrations.

Two years ago the joining of the Vilnius and Geneva systems into one seven-color system was suggested (Straižys, Jodinskienė and Hauck 1982; North, Hauck and Straižys 1982). The Vilgen system has mean wavelengths of 350, 374, 402, 468, 516, 550 and 656 nm. This new system has somewhat broader response curves than the original Vilnius system but it maintains all the useful properties of both original systems. In particular, the system allows one to distinguish F and G subdwarfs and G and K metal-deficient giants.

3.7 Narrow-band photometry

In this review only the wide- and medium-band photometric systems currently used for metallicity determinations are considered. Dozens of narrow-band photometric systems measuring the intensities of different atomic lines, their blends and molecular bands are also in use. Even the listing of these systems would take too much space in this paper. On the other hand, narrow-band photometry is not very different from equivalent width measurements by spectroscopic techniques and there is no difference in principle between the two methods. Narrow-band photometry usually determines not the overall metal abundance but only the abundances of individual elements. In this respect it is also closely related to spectroscopic methods. The calibration of measurements of spectral lines and bands is now almost completely based on synthetic spectra calculations. This kind of application of model atmospheres is quite effective since synthetic spectra show the best agreement with observations when comparatively narrow spectral intervals are considered.

4. CONCLUSION

This review shows that astronomers now have a large arsenal of photometric systems and methods for abundance determinations for stars. However, before using them we have to relate their photometric and physical parameters, i.e. to calibrate them. For this we still need model atmospheres giving precise synthetic spectra coinciding with energy distributions of real stars for wide temperature, gravity and abundance ranges.

REFERENCES

Abt, H. A. 1961, Astrophys. J. Suppl., 6 37.
Bartkevičius, A. 1980, A Catalogue of Metal Deficient F-M Stars. Part I. Stars Classified Spectroscopically, Bull. Vilnius Obs. No. 51.
Bartkevičius, A. 1984, Metal Deficient Giants in the Galactic Field. Catalogue and Some Physical Parameters, Bull. Vilnius Obs. No. 66.
Bartkevičius, A. and Sperauskas, J. 1983, Bull. Vilnius Obs. No. 63, 3.
Bartkevičius, A. and Straižys, V. 1970a, Bull. Vilnius Obs. No. 28, 33.
Bartkevičius, A. and Straižys, V. 1970b, Bull. Vilnius Obs. No. 30, 3.
Bartkevičius, A. and Straižys, V. 1970c, Bull. Vilnius Obs. No. 30, 16.
Bell, R. A. 1977, Personal communication.
Bell, R. A. and Dickens, R. J. 1980, Astrophys. J., 242, 657.
Bell, R. A., Dickens, R. J. and Gustafsson, B. 1978, in Astronomical papers dedicated to Bengt Strömgren, A. Reiz and T. Andersen,

eds., (Copenhagen Univ. Obs.) p.249.

Bell, R. A., Dickens, R. J. and Gustafsson, B. 1979, Astrophys. J., 229, 604.

Bell, R. A. and Gustafsson, B. 1978, Astron. Astrophys. Suppl., 34, 229.

Bessell, M. S. and Wickramasinghe, D. T. 1979, Astrophys. J., 227, 232.

Böhm-Vitense, E. 1973, Astron. Astrophys., 24, 447.

Böhm-Vitense, E. and Szkody, P. 1974, Astrophys. J., 193, 607.

Bond, H. E. 1980, Astrophys. J. Suppl., 44, 517.

Buser, R. 1979, in Problems of Calibration of Multicolor Photometric Systems, A. G. Davis Philip ed., (Dudley Obs. Report No.14) p.3.

Buser, R. and Kurucz, R. L. 1978, Astron. Astrophys., 70, 555.

Butler, D. 1975, Astrophys. J., 200, 68.

Canterna, R. 1976, Astron. J., 81, 228.

Canterna, R. and Harris, H. C. 1979, Problems of Calibration of Multicolor Photometric Systems, A. G. Davis Philip, ed. (Dudley Obs. Report No. 14) p.199.

Carney, B. W. 1979, Astrophys. J., 233, 211.

Crawford, D. L. 1975, Astron. J., 80, 955.

Crawford, D. L. and Perry, C. L. 1976, Pub. Astron. Soc. Pacific, 88, 454.

Dickens, R. J., Bell, R. A. and Gustafsson, B. 1979, Astrophys. J., 232, 428.

Eggen, O. J. 1969, Astrophys. J. Suppl., 19, 31.

Eggen, O. J. 1977, Astrophys. J., 215, 812.

Eggen, O. J. 1978, Astrophys. J., 221, 881.

Eggen, O. J. 1983, Astron. J., 88, 386.

Eggen, O. J. and Sandage, A. R. 1964, Astrophys. J., 140, 130.

Epstein, I. and Epstein, A.E.A., 1973, Astron. J., 78, 83.

Eriksson, K., Bell, R. A., Gustafsson, B. and Nordlund, A. 1979, Trans. IAU 17A, part 2, (Reidel, Dordrecht), p.200.

Grenon, M. 1978, Publ. Obs. Geneve, ser. B, fasc. 5.

Grenon, M. 1981, IAU Colloq. No. 68, Astrophysical Parameters for Globular Clusters, A. G. Davis Philip and D. S. Hayes, eds., (L. Davis Press, Schenectady).

Gustafsson, B. and Bell, R. A. 1979, Astron. Astrophys., 74, 313.

Gustafsson, B. and Nissen, P. E. 1972, Astron. Astrophys., 19, 261.

Hauck, B. 1973, IAU Symp. No. 54, Problems of Calibration of Absolute Magnitudes and Temperature of Stars, B. Hauck and B. E. Westerlund eds. (Reidel, Dordrecht) p.117.

Hauck, B. 1968, Publ. Obs. Geneve, ser. A, fasc. 75.

Hauck, B. 1978, Astron. Astrophys., 63, 273.

Hauck, B. and Curchod, A. 1980, Astron. Astrophys., 92, 289.

Hesser, J. E., Hartwick, F.D.A., and McClure, R. D. 1977, Astrophys. J. Suppl., 33, 471.

Janes, K. A. 1975, Astrophys. J. Suppl., 29, 161.

Janes, K. A. 1979, Problems of Calibration of Multicolor Photometric Systems, A. G. Davis Philip ed., (Dudley Obs. Report No. 14) p.103.

Jaschek, M. and Jaschek, C. 1962, in Symposium on Stellar Evolution,
    (La Plata Astron. Obs.) p.137.
Johnson, H. L. and Mitchell, R. I. 1968, Astrophys. J., 153, 213.
Kurucz, R. L. 1979, Astrophys. J. Suppl., 40, 1.
Lub, J. 1977, The RR Lyrae Population of the Solar Neighborhood,
    Leiden, Diss. Thesis.
Lub, J. 1979, Astron. J., 84, 383.
Lub, J. and Pel, J. W. 1977, Astron. Astrophys., 54, 137.
McClure, R. D. 1979, Problems of Calibration of Multicolor Photometric
    Systems, A. G. Davis Philip ed., (Dudley Obs. Report No. 14) p.83.
Nicolet, B. and Cramer, N. 1983, Astron. Astrophys., 117, 248.
Nissen, P. E. 1970, Astron. Astrophys., 6, 138.
Nissen, P. E. 1981, Astron. Astrophys., 97, 145.
Nissen, P. E. and Gustafsson, B. 1978, in Astronomical Papers
    Dedicated to Bengt Strömgren, A. Reiz and T. Andersen, eds.,
    (Copenhagen Univ. Obs.) p.43.
North, P. and Hauck, B. 1979, in Problems of Calibration of Multicolor
    Photometric Systems, A. G. Davis Philip ed., (Dudley Obs. Report
    No. 14) p.183.
North, P., Hauck, B. and Straižys, V. 1982, Astron. Astrophys.,
    108, 373.
Osborn, W. 1973a, in IAU Symp. No. 50, Spectral Classification and
    Multicolor Photometry, Ch. Fehrenbach and B. E. Westerlund eds.,
    (Reidel, Dordrecht) p.176.
Osborn, W. 1973b, Astrophys. J., 186, 725.
Relyea, L. J. and Kurucz, R. L. 1978, Astrophys. J., 37, 45.
Rydgren, A. E. and Smith, M. A. 1974, Astrophys. J., 193, 125.
Sandage, A. and Hartwick, F.D.A. 1977, Astron. J., 82, 459.
Schuster, W. J. 1976, Rev. Mex. Astron. Astrofis, 1, 327.
Schuster, W. J. 1979, Rev. Mex. Astron. Astrofis, 4, 233, 301, 307.
Smith, L. L. and Steinlin, U. W. 1964, Z. f. Astrophys., 58, 253.
Straižys, V. 1973, Bull. Vilnius Obs. No. 36, 3.
Straižys, V. 1975, Multicolor Photometry and the Theoretical HR
    Diagram, A. G. Davis Philip and D. S. Hayes eds., in (Dudley Obs.
    Report No. 9) p.65.
Straižys, V. 1977, Multicolor Stellar Photometry, (Mokslas Publishers,
    Vilnius).
Straižys, V. 1979, Problems of Calibration of Multicolor Photometric
    Systems, A. G. Davis Philip ed., (Dudley Obs. Report No. 14)
    p.215.
Straižys, V. 1982, Metal-Deficient Stars, (Mokslas Publishers,
    Vilnius).
Straižys, V. and Bartkevičius, A. 1982, Bull. Vilnius Obs. No. 61, 22.
Straižys, V., Jodinskienė, E., and Hauck, B. 1982, Bull. Vilnius Obs.,
    No. 60, 50.
Straižys, V. and Kavaliauskaitė, G. 1967, Bull. Vilnius Obs., No. 20,
    3.
Straižys, V. and Kurilienė, G. 1983, Bull. Vilnius Obs. No. 62, 39.
Straižys, V. and Sviderskienė, Z. 1972, Astron. Astrophys. 17, 312.
Straižys, V. Zdanavičius, K., Tautvaisienė, G. and Černis, K. 1984,
    Astrophys. Space Sci., 104, 219

Strömgren, B. 1964, Astrophys. Norvegica, 9, 333.
Thevenin, F. and Foy, R. 1983, Astron. Astrophys., 122, 261.
Trefzger, Ch. F. 1981, Astron. Astrophys., 95, 184.
Wiertz, M.J.J. and van Genderen, A. M. 1983, Astron.
    Astrophys., 121, 35.

DISCUSSION

GUSTAFSSON: It was mentioned that there exists a grid of models for late
G and K dwarfs, calculated by Eriksson et al., which is unpublished.
This grid is available to anybody interested, but should be used at the
user's own risk since we have not yet systematically explored how the
predicted fluxes and colors compare with observations.

FUNDAMENTAL PARAMETERS AND MODELS OF STELLAR ATMOSPHERES

Bengt Gustafsson

Stockholm Observatory

Uffe Graae-Jørgensen

Uppsala Astronomical Observatory

ABSTRACT. The use of photometric and spectroscopic criteria, calibrated by model-atmosphere calculations, for determining effective temperatures, surface gravities and chemical compositions of stars is illustrated and commented on. The accuracy that can be obtained today in such calibrations is discussed, as well as possible ways of improving this accuracy further for different types of stars.

1. INTRODUCTION

In a well-known quotation from the sixties Underhill describes the problem of "fitting a series of beautiful internally-consistent models to honest-to-goodness real stars that are up there" as a "horrible" one (cited by Pecker, 1965). After 20 years of progress in the study of stellar atmospheres, does this problem remain "horrible" and, if so, in which respect?

Many circumstances are important to a consideration of this question. These are related to the remarkable improvements for observational as well as theoretical studies of stellar atmospheres.

The new observational opportunities - the opening up of the vacuum ultraviolet window, in particular by the International Ultraviolet Explorer; the new or improved high-resolution spectrographs, equipped with sensitive multi-element detectors with linear and well defined responses, now used for studies in the visual and near infrared; and the extended or quite new opportunities for observations in the infrared, from the ground or higher up, at high or lower resolution - all are of utmost importance.

Another very significant circumstance is our rapidly increasing ability to carry out large-scale model calculations, and the invention of adequate numerical methods for solving the relevant equations. Thus, the formation of spectral lines and continua may now be modeled without relying on the assumption of LTE, even for relatively complex atoms; the effects of departures from plane-parallel stratification in spheri-

303

D. S. Hayes et al. (eds.), Calibration of Fundamental Stellar Quantities, 303–329.
© *1985 by the IAU.*

cally extended or inhomogeneous atmospheres may be investigated; the blanketing from millions of atomic and molecular spectral lines may be considered; and even convection and its effects on the structure and the radiation field may be modeled in a relatively self-consistent way. In fact, the technological development in this field is so rapid that we need truly daring minds that are willing to start formulating problems that used to be "absolutely hopeless", if we are to take full advantage of this drastic development.

A third circumstance of great importance is our new ability to experimentally measure and accurately calculate a great amount of the physical data (notably cross sections) which are necessary ingredients in models of stellar atmospheres and in quantitative analyses of stellar spectra.

In view of all this progress couldn't one argue that the problem characterized by Anne Underhill is no longer horrible, but even - solved? If we glance through the literature, we shall find that for most stars, errors in the effective temperature estimates are now often stated to be less than 4%, and errors in surface accelerations of gravity and in chemical abundances, less than 0.15 dex. Isn't this enough? No. First, because modern observational techniques and the basic physical data should admit a significantly higher accuracy - our understanding of the physical processes and of the structure of the stellar atmospheres, i.e., our model-building is often nowadays the limiting factor. An improvement in this understanding would then make further progress possible in many fields of astronomy where stellar atmospheres are studied more as probes into the structure or chemical history of our Galaxy, than as physical systems. Secondly, many recent studies have revealed that stars are complicated systems - some of them may even be more complicated than the Sun, which is a shocking thought for any student of solar physics. The models are based on, and presently have to be based on a number of simplifying assumptions. The same assumptions therefore underlie the determination of fundamental parameters using these models, and also (in most cases) the error estimates for these parameters. Thus, we have to live with an uncomfortable feeling that even if our fitting procedure of a stellar spectrum to a calculated one looks satisfactory and self-consistent, the fundamental parameters of the real stars may be astonishingly different from those of the model. It is a major task for research in this field to try to investigate to what extent this sceptical attitude is relevant.

There is also a third important reason why one should not be satisfied with the seemingly rather small errors in the determination of fundamental parameters. This reason will be discussed in Sec. IV, below.

In the present review we shall concentrate on the determination of effective temperatures, surface gravities and chemical abundances by using model atmospheres. Thus, we shall not dwell on the determination of micro - or macro turbulence parameters or magnetic field strengths or other (probably) secondary parameters, nor shall we comment on the determination of angular momentum, which may be a primary parameter for the star itself and certainly is of significant importance for the structure of its atmosphere.

In Sec. II some general comments will be made, while Sec. III reviews the situation for stars of different spectral types. This review will mainly concentrate on "normal stars", although the reader should observe the interesting development that presently is taking place in the study of Wolf-Rayet stars (e.g. DeLoore and Willis 1982, Smith and Willis 1982, Underhill 1983), of hot subdwarfs (Heber et al. 1984, Schönberner and Drilling 1984) and planetary nebula nuclei (Méndez et al. 1983), of early main-sequence stars with peculiar abundances (Cowley and Adelman 1983, Renson 1981) and, not least, of white dwarfs (Sion et al. 1982, Koester et al. 1982, Wegner and Yackovich 1983).

2. SOME GENERAL REMARKS

When estimating the <u>effective temperature</u> of a star from properties of its emitted radiation one uses the continuous flux distribution or temperature-sensitive spectral lines. In principle, these features are not direct measures of the total emissivity per stellar surface area, as is the effective temperature, but rather they measure the temperature in the flux-forming layers. However, in many applications such as in abundance analysis, this characteristic temperature is the relevant quantity to use for estimating the number densities of the atomic or molecular state under consideration. In this case "T_{eff}" is more to be regarded as a label of the relevant model to combine with the star of interest than a measure of the surface flux.

An obvious way to derive T_{eff} is to compare flux gradient measures, obtained from scans or colors, with the corresponding calculated quantities. This procedure is non-trivial in practise, as several other contributions to this symposium will indicate. In addition to the problems with the absolute calibration of the observational system there are theoretical difficulties in calculating the blocking effect from spectral lines in the band-passes observed. The problem is aggravated, for most spectral types, by the fact that the line-blocking gets heavier as the wavelength gets shorter, while, on the other hand, the temperature sensitivity of the (quasi-) continuous flux curve is greater at shorter wavelengths. Thus, for any given problem, a deliberate compromise has to be made as regards which wavelengths to choose - a compromise where practical circumstances, like how easy it is to achieve an absolute calibration of the system proposed, are also of obvious significance.

An important alternative to the flux gradient method is "the integrated flux method" of Blackwell and Shallis (1977), Blackwell et al. (1979). Although first applied to G-K giants it has recently been used extensively for stars of early types (cf. Underhill 1982, and references cited therein, Remie and Lamers 1982) and for very late giants (Tsuji 1981a and b). These applications are natural since the easily measurable gradients of the flux distributions of O and early B stars are not very temperature sensitive, while the visual spectrum of M stars and carbon stars is too heavily blocked by molecular lines to admit any accurate continuum difinition. Following the "multiplicative" formulation of the method of Tobin (1983) we may write

$$\sigma \, T_{eff}^{4} \cong \int_{\lambda_1}^{\lambda_2} f_\lambda \, d\lambda \cdot \frac{\int_0^\omega F_\lambda(T_{mod}) \, d\lambda}{\int_{\lambda_1}^{\lambda_2} F_\lambda(T_{mod}) \, d\lambda} \cdot \frac{F_{\lambda'}(T_{mod})}{f_{\lambda'}} \cdot$$

Here, f_λ is the observed flux corrected for extinction and (in the first
factor) integrated between the wavelengths λ_1 and λ_2, which should span
a considerable wavelength interval. F_λ is the calculated physical flux
from a model atmosphere with effective temperature T_{mod} and λ' a relatively long wavelength. Such a choice of λ' is claimed to be of importance since the line-blocking there is generally smaller and the interstellar reddening is less. Of major significance is that $F_{\lambda'}$ is little
sensitive to the temperature structure of the model. Thus, we arrive
at an iterative procedure, where in iteration i $T^{(i)}(mod) = T_{eff}(i-1)$. It
is convergent (although one should be careful in checking that full
convergence has really been obtained, cf. Remie and Lamers 1982 and
Tobin 1983). The final result is generally not very model dependent, at
least as long as the wavelength interval (λ_1, λ_2) contains most of the
radiation emitted.

The use of temperature sensitive spectral lines for determining
effective temperatures - such as measurements of the ionization of hydrogen through the Balmer lines for F stars, the excitation equilibrium of
neutral iron for G-K giants or TiO bands for late K dwarfs-often requires
a good grasp of the line-forming process, including broadening mechanisms,
and reliable atomic or molecular data. If the lines to be used are
formed at different depths in the atmosphere, which is often the case,
e.g. when ionization equilibria are to be used, detailed knowledge
about the temperature structure may also be necessary.

A broad review of effective-temperature determinations has been
provided by Böhm-Vitense (1981a).

Spectroscopic and photometric determinations of surface gravities
are based on the different pressure sensitivity of different sources of
opacity in neighboring spectral regions. Well-known examples are the
Stark-broadened wings of the Balmer lines in early-type stars (i.e., of
the opacity in the wings relative to the continuum), the different
pressure sensitivity of the absorption on the opposite sides of the
Balmer discontinuity for F-type stars, the pressure sensitivity of
ionization equilibria, the strength of damping wings of strong metal
lines relative to corresponding weak lines in red giants and subgiants,
and the strengths of molecular lines (such as those of MgH relative to
MgI lines) in late-type stars. As will be illustrated below, these
different methods have quite different properties as regards their sensitivity to the uncertainties in the model atmospheres.

As to the determination of chemical abundances there are strong
general arguments for preferring weak spectral lines, situated along the
linear part of the curve-of-growth if they are measurable. This is
the case not merely because the strengths of these unsaturated lines are

independent of line broadening and various shortcomings in the theory thereof, but also because these lines are formed at great depths where the departures from LTE often are smaller than at more shallow atmospheric depths. Moreover, the strength of the weak lines is more insensitive to model uncertainties (cf., e.g., the discussion concerning abundances of metal-poor stars by Gustafsson, 1983). However, for fainter objects one often has to rely on theoretical or semi-empirical calibrations of relatively strong lines or complex spectral features. Examples of such measures are given by Carbon et al. (1982), Cohen (1982) and Thévenin and Foy (1983), and such measures are also represented in most photospheric systems.

Sometimes there may even be no measurable lines available at all for crucial elements; this is the case for e.g. the electron donors in the atmospheres of cool He-rich white dwarfs. In such cases, at the very best, quite indirect methods are available, and it may be that the final determination of fundamental parameters is not unique (Wehrse 1984).

The determination of chemical abundances for stars was discussed at an ESO workshop (Nissen and Kjär, 1980) and at the IAU General Assembly in Patras 1982 (see Gustafsson, 1983, and several other papers in the same volume of the P.A.S.P.).

The presentation above has been artificially divided into the discussion of determination of various parameters, one at a time. In practice, however, any observable quantity responds to changes of several model parameters and sometimes even does so in a remarkably non-linear way. Neglect of these effects may lead to significant errors in derived parameters for stars, as quite recently demonstrated by discussions of several unnecessarily diverging sets of results.

In the following we shall review the present situation as regards model atmospheres and current methods for estimating fundamental parameters when using these models for stars of different types. The discussion of this broad topic must be very schematic, partly because of the lack of space, partly because of our lack of detailed experience from work on early-type stars. We shall try, however, to put the main emphasis on the development during the past three years.

3. SOME MORE DETAILED REMARKS ON STARS OF DIFFERENT TYPES

3.1. O-type stars

An important breakthrough for stellar-atmosphere modelling in general was the development of the complete linearization method by Auer and Mihalas in 1969. They also used their method to construct models of hot stellar atmospheres with a detailed non-LTE treatment of the H and He atoms, demonstrating the great improvement when compared with observations (cf., e.g., Auer and Mihalas, 1972).

In recent years, Kudritzki and collaborators, basically using the methods of Auer and Mihalas, have continued and developed this line of study in an impressive series of papers with detailed modelling of various types of hot stars (cf. Kudritzki, 1980, and references cited therein for a discussion of the methods of analysis). Thus, there are series of analyses published on massive O stars (cf. Simon et al., 1983, for further references), on subluminous O stars (cf. Gruschinske et al., 1983, for further references) and of central stars in planetary nebulae (cf. Mendez et al., 1983, and references cited therein), as well as several analyses of more special objects.

Another very important contribution to the study of early-type stars in general is the grid of model atmospheres by Kurucz (1979). Although in LTE, these models are "blanketed", i.e. the effects of absorption lines have been taken into consideration by means of very extensive lists of data for atomic transitions, and by the so-called opacity-distribution-function (ODF) method. In the models of Kudritzki et al. only blanketing from hydrogen and helium is considered, and thus these two approaches are complementary.

For the formation of visual continua the LTE models seem valid as long as $T_{eff} < 30,000$ K (Kudritzki 1979). However, the structure of the outer layers of the models, the cores of the H and He lines and the formation of lines from many other species and some metal continua in the UV are also effected at much lower temperatures. For a $T_{eff} = 25000$ K main sequence star, the temperature differences in the surface layers may amount to several thousands of K between a non-LTE hydrogen-line blanketed model and a more fully blanketed LTE model, the latter being the cooler one. It is probable that real stars have temperature structures somewhere between these extremes, and it seems reasonable to assume that these differences could show up as lack of agreement between observed and calculated fluxes and spectra.

The non-LTE models with parameters chosen on the basis of spectral lines (see below) generally show good agreement with observed fluxes (cf., e.g. Kudritzki et al. 1983), except for the shortest wavelengths where the neglect of line-blocking becomes obvious in the comparison. For some stars, e.g. the OB subdwarf HD 149382 analysed by Baschek et al. (1982), the cores of the Balmer lines are reported to be deeper than the calculated ones, which is ascribed to the neglected blanketing. The Kurucz models are fairly successful in reproducing the observed fluxes of O stars (cf. Underhill 1982 and references cited therein). One exception is the infrared L and M bands where the Kurucz models are too faint as was pointed out by Castor and Simon (1983). The latter suggest that the flux deficiency of the models may be due to the LTE assumption - the temperature in the atmospheric surface layers would be expected to increase if the lines from the ions were treated in non-LTE, which would increase the infrared flux, although it is not clear that the observed excess of $O^{m}1$ would result.

Castelli et al. (1980) and Remie and Lamers (1982) found that in the Kurucz models an extra UV blocking was needed in the ultraviolet (for $\lambda < 2000$ Å) in order to account for the observed fluxes of late O and early B supergiants and giants. They estimate that at least some fraction of the extra blocking needed (an increase by a factor of 1.5 to 2, according to Remie and Lamers, 1982) can be provided by increasing the microturbulence from 4 km/s to 10 km/s, which seems reasonable. Moreover, the opacity distribution functions of Kurucz are incomplete with respect to lines of higher metal ions.

The effective temperatures of O stars have been estimated from model atmospheres by using the integrated flux method (cf. Underhill et al., 1979, Underhill, 1982, Remie and Lamers, 1982, Heber et al., 1984), from flux gradients (e.g., Schönberner and Drilling, 1984) and from ionization equilibria of He (and H), calculated in non-LTE (Kudritzki and collaborators, cited above). The latter determine He abundance and the gravity (essentially from the Balmer-line profiles) simultaneously with T_{eff} stressing the importance of doing so in their analysis of ζ Pup (Kudritzki et al. 1983). This latter paper demonstrates that the previously claimed systematic differences between effective temperatures of O stars, derived from ionization equilibria, and those obtained from fluxes or angular diameters, may to a great extent be caused by inappropriate choices of surface gravity, small errors in reddening and underestimated errors in the integrated-flux method. Back-scattering of stellar photons by the far UV lines in the stellar wind, or other mechanisms that could heat the outer photospheric layers of O stars have earlier been advocated as explanations for the discrepancy between "photometric" and "spectroscopic" temperatures of O stars. The lack of correlation found by Underhill (1982) between spectral type and effective temperature, determined with the integrated-flux method, may demonstrate the importance of such phenomena.

Typical errors in T_{eff} and log g determinations for massive O stars, as quoted by Kudritzki and collaborators, are ± 2500 K and 0.15 dex, respectively. These uncertainties are intrinsic and basically a reflection of the uncertainties in the procedure whereby calculated H and He lines are fitted to observed ones. Systematic errors in the model atmospheres, such as the effects of the neglect of blanketing from ions, or of back-scattering from winds, are harder to estimate, but seem to be capable of leading to systematic errors of at least the same order of magnitude (cf. Husfeld et al. 1983). The effects of spherical extension and departures from hydrostatic equilibrium are generally thought to be small for the stars close to the main sequence or below it.

Relative errors in the effective temperatures, when derived with the integrated-flux method, are estimated to be within 5% (Underhill 1982, Remie and Lamers, 1982) but found to be considerably greater by Tobin (1983), especially for the hottest stars.

For the central stars of planetary nebulae the errors in the effective temperatures are greater. This is because the HeI lines are masked by nebular emission or because the stars are too hot to show these lines.

The relative errors in the He abundances derived by Kudritzki and collaborators for stars with a roughly solar helium abundance are typically given to be 10-20%, estimates again based on the internal errors in the fitting procedure. For helium in helium-poor stars, and for less abundant elements in general, the accuracy is less. E.g., in the detailed LTE analysis of the subdwarf HD 149382 Baschek et al. (1982) give what they suggest as rather optimistic error estimates, which, nevertheless, range from 0.2 dex (He) to 0.6 dex for the various chemical elements. Much of this error is ascribed to the uncertainty in the microturbulence parameter, which Baschek et al. consider to be the greatest source of error in their model analysis.

3.2. B-type stars

There are several studies showing that blanketed LTE models in the grid of Kurucz (1979) are able to reproduce the visual fluxes and colors of B stars in an accurate and consistent way. Recent examples of such studies are the series of papers by Adelman and collaborators (cf. Adelman and Pyper 1983 and papers cited therein). In this work spectrophotometry was used to test the consistency between effective temperatures obtained for B, A and F stars, using low-resolution scans in direct comparison with models, or using measured and calculated colors like the uvby system (synthesized by Relyea and Kurucz, 1978 and calculated directly from the scans). The internal consistency between these different temperatures was good, as well as the agreement with direct measurements of angular diameters and fluxes; the empirical temperature scale of Code et al. (1976) was found to agree with the more theoretical ones within approximately 5% in T_{eff}. Another example is the agreement between two-color diagrams of the Geneva photometry with corresponding model predictions (cf. Cramer , 1982) and of the theoretical calibration of B-V and U-B versus T_{eff} of Buser and Kurucz (1978) with other calibrations (cf. Cramer, 1984). A comparison with predicted fluxes from Kurucz's models in the vacuum ultraviolet with observations of late B stars showed some discrepencies which were, however, considered by Malagnini et al. (1983) to be "mostly instrumental effects".

Lamers et al. (1983) find that for the B1 supergiant P Cyg, a considerable amount of extra blocking is needed, as compared with the model predictions, and they argue that this cannot solely be due to the stellar wind but must be caused by mainly photospheric lines. Here, an enhanced microturbulence might be reasonable to invoke. In contrast to this, Castelli et al. (1980) found the observed UV blocking in stars later than B5 to be smaller that predicted by the Kurucz models. However, from comparisons between detailed synthetic spectra and high-resolution IUE spectra of early B stars around 1200 Å Kamp (1982) concludes that the Kurucz and Peytremann (1975) line list is seriously incomplete for hot stars.

There are other grids of models, less completely blanketed than those of Kurucz, but still in use for B stars, such as the non-LTE hydrogen-line blanketed models of Mihalas (1972b). In particular, these models reproduce notably the cores of the Balmer lines quite well (cf. the results for Hα, given by Heasley and Wolff, 1983). Yet, in some cases an additional surface cooling, which would be provided by ionic blanketing, can be traced in the line cores, cf., e.g., Heber et al. (1984). For Hα the wings are in better agreement with observations if calculated from the Kurucz models. The difficulties in reproducing the red He I lines λ 5876 and λ 6678 in statistical-equilibrium calculations (Heasley et al. 1982, using Mihalas's grid) might also be diminished if blanketing were taken into acount more completely.

Effective temperatures for B stars have been determined by model atmospheres using the integrated-flux method, the Balmer discontinuity (e.g., measured by the reddening-corrected uvby index c_1) or the ionization equilibria, such as the SiII/SiIII/SiIV ratios. With unblanketed, or only hydrogen-line-blanketed, models temperatures about 1000-2000 K hotter result as compared with the Kurucz models, which is consistent with the first order difference between the temperature structures from the backwarming effect. Nowadays, the Kurucz models are generally prefered.

Underhill et al. (1979) derived effective temperatures using the integrated-flux method and the grid of Kurucz for a great number of B stars. These temperatures seem to agree rather well with those derived from the uvby photometry and the calibration of Relyea and Kurucz (1978). The calibration of Philip and Newell (1975), however, deviates systematically from this (Malagnini et al. 1983). For the early B stars a higher effective temperature was generally found from the Si lines than from the use of other methods. Recently, however, Erhorn et al. (1984) have argued that the consistent use of blanketed model atmospheres may make this difference vanish. In their study LTE was adopted; the non-LTE study of Si in B stars by Kamp (1976,78) suggests that this is acceptable for calculating ratios of equivalent widths and hence for estimating effective temperatures. We note in passing that the ionization equilibria of Si, Al II/ Al III, SII/SIII and CII/CIII in line blanketed LTE models lead to temperature determinations for the extreme He star BD +10°2179 which agree well with what is obtained from the flux distribution (Heber, 1983).

In conclusion, errors less than about 1000 K should be realistic at determinations of T_{eff} from model atmospheres for B stars, with known reddening. Significant systematic errors, however, may occur for the earliest B stars where the blanketing tends to be seriously underestimated by the Kurucz models, which could lead to overestimates of T_{eff}. For B-type supergiants, additional errors may occur due to uncertainties in the value of the microturbulence parameter.

The surface gravities of B stars are usually estimated from the Balmer line profiles. Recent examples of such measurements, calibrated by model-atmosphere calculations, are the studies of Hα in B stars by Heasley and Wolff (1983), the use of Hγ profiles by Adelman (1984) and of Hε in a study of halo OB stars by Keenan, Dufton and McKeith (1982).

Errors in the resulting log g values are generally estimated to be 0.1 - 0.2 dex when T_{eff} is given. However, a systematic study (using different hydrogen lines) of possible errors in these determinations would be valuable, not only in view of the inconsistency of 0.25 dex traced by Lennon and Dufton (1983) between gravity estimates from the Hε profiles and those from Hγ and β indices for stars in NGC 6231.

The problem of determining accurate abundances for B stars is a complicated one, due to the relatively small number of suitable spectral lines, the uncertain microturbulence parameter and departures from LTE. Studies of such departures - for He I (Auer and Mihalas 1973a, Heasley and Wolff 1982), for Be II (references given in Boesgaard et al. 1982), for OI (Baschek et al. 1977), for Ne I (Auer and Mihalas, 1973b), for Mg II (Mihalas 1972a),CaII (Mihalas 1973) and for SiII,III and IV (Kamp 1976,1978) have demonstrated that the effects may well be considerable. However, not all odd spectral line strengths in B stars, such as CaII λ4267 , are easily explained as the result of departures from LTE (cf. the non-LTE calculations of Lennon, 1983).

Errors of ± 0.2 dex or less are often quoted in abundance analyses of B stars. The systematically non-solar results obtained for some elements (e.g., the underabundance of carbon) cast some doubts on our understanding of the systematic errors involved. Some of the stellar-solar differences may, however, be due to errors in the oscillator strengths, since one often cannot use the same lines for B stars and the Sun. One should try to bridge the gap by a two-step differential analysis, with an early F-type dwarf as an intermediate point, as has been suggested by, e.g., Adelman (1984).

3.3. A-type stars

For A stars, as well, the comparisons between observed fluxes and those of the LTE models of Kurucz (1979) show satisfactory over-all agreement, in the visual and with certain exceptions also in the ultraviolet (Böhm-Vitense 1981b and 1982, Kurucz 1980, Malagnini et al. 1982, Adelman et al. 1980, Adelman and Pyper 1983). However, over-ionization of carbon and silicon was found in the non-LTE calculations of Snijders (1977a and b) and this leads to observable effects for the UV continua of late B and early A stars at $\lambda < 1200$ Å. The temperature sensitive intensity jump at 1600 Å has been studied by Böhm-Vitense (1982) and used by her for deriving effective temperatures for A and F stars. Since there are some discrepancies between calculated and observed fluxes in the wavelength region between 1700 and 1800 Å, she chose the flux ratio F_λ (1900 Å)/ F_λ (1420 Å) as a suitable temperature measure. She found, however, that the calculated flux at 1900 Å had to be lowered by 0.1 dex for $T_{eff} < 9000$ K in order to agree with temperatures derived from the F_V/F_λ (1900 Å) ratio, when F_V is the flux in the visual V band.

It seems possible to determine the effective temperature of an early A dwarf from a well observed absolute flux distribution with an accuracy of about $\pm$ 200 K, as is illustrated for the Vega and Sirius cases (cf. Dreiling and Bell, 1980 and Bell and Dreiling, 1982, Selby et al. 1983) and about the same accuracy seems achievable for later type A stars, e.g. using the semi-empirical calibration of 1600 Å jump (Böhm-Vitense, 1982). For the giants and especially the supergiants the fit to the Kurucz models is less satisfactory and the uncertainties in T_{eff} are considerably greater. Note, however, the quite consistent T_{eff} and log g results from different methods that Desikachary and Hearnshaw (1982) obtain for Canopus (F0 Ib-II) on the basis of Kurucz models and near UV - visual - infrared observations. The gravities are generally derived from Balmer-line profiles, and accuracies of about ± 0.2 dex may be expected (Dreiling and Bell, 1980, Adelman et al., 1984).

The abundance determinations, even for the brightest early A stars, are difficult and uncertain, as a result of the small number of suitable spectral lines. E.g., iron in Vega appears underabundant by 0.5 ± 0.3 dex (Dreiling and Bell, 1980), which may at least partly be a non-LTE effect. The errors in the CNO element abundances of Vega and Sirius are considered somewhat smaller ($\pm$ 0.2 dex or less, Lambert et al. 1982). Much of the interest in abundance determinations for A stars has, for natural reasons, been directed towards the study of Ap and Am stars. Although the wealth of sharp metal lines in these spectra simplifies an

abundance analysis, there are several complicating circumstances which
should be borne in mind. One is the fact that the metal-line blocking
in the ultraviolet is different (cf. Böhm-Vilense 1981b) and this should
be considered in the model atmospheres. Muthsam (1979) has calculated
a grid of line-blanketed models for Ap stars and used these for modelling
specific stars (Muthsam and Stepién 1980, and references listed therein).
Other groups have used the over-all metal-enhanced models of Kurucz
(1979).

Other interesting complications are the observed inhomogeneous distribution of the elements across the surface of the star, and the expected
non-uniform distribution with atmospheric depth (cf. Alecian, 1982).

3.4. F-type stars and solar-type dwarfs

It was pointed out by Böhm-Vitense (1970) that theoretical U-B, B-V
colors for F dwarfs did not agree with the observed two-color diagram.
Kurucz's (1979) models are also not quite successful in this respect
(Buser and Kurucz 1978, Böhm-Vitense 1981b), nor do they reproduce the
observed uvby colors with the precision desired for many applications
(Relyea and Kurucz, 1978). The use of a modified mixing-length theory
of convection diminishes some of these discrepancies (Lester et al. 1982,
Buser and Kurucz 1985) but does not solve all problems. Scanner observations (Böhm-Vitense, 1978) tend to show a flux deficiency around 4100 Å
for stars with B-V > 0.25, as compared with radiative equilibrium models.
It was suggested by Nelson (1980) that this phenomenon is caused by
temperature (and thus opacity) inhomogeneities, generated by convective
overshoot. According to Nelson's estimate the existence of inhomogeneities should reduce the flux around 0.4 μ by $0^m.1$ or so, but the effect
was expected to be smaller for the late F stars. This estimate is,
however, very schematic and more detailed simulations of convection in
F stars are needed before any firm conclusions may be drawn.

The problems for the early F type star models may, or may not, be
related to those of the Sun. There is still some uncertainty as regards
which solar photospheric model is the most realistic one, although many
investigations suggest that the semi-empirical Holweger-Müller (HM)
(1974) model is most successful in reproducing continuum fluxes and
lines (Lambert 1978, Holweger 1979, Gehren 1981). Sauval et al. (1984)
find a very impressive consistence between observed and calculated pure
rotation lines of OH when the Holweger-Müller model is used. However,
Rutten and Kostik (1982) have argued that the success of the HM model
would not be so impressive for the iron lines, if departures from LTE
had been taken into consideration properly, and these authors instead
favor models with greater temperature gradients.

The theoretical solar models, although not very far from the
Holweger-Müller model, do not reproduce the observed solar fluxes in
the blue and ultra-violet very well. E.g., the Kurucz (1979) model is
too blue (B-V = 0.60), as is the modified one after the introduction of
the alternative mixing-length formulation (Lester et al. 1982), and they
seem to have too little blocking in the ultraviolet, which is certainly
also the case for the model with solar parameters in the grid of Bell et
al. (1976), cf. Gustafsson and Bell (1979). On the other hand, at least

the latter model is too cool by about 150 K - if the extra blocking required in the uv is introduced this temperature discrepancy vanishes due to the back-warming effect.

Gehren (1981) found that the effective temperature calibration of B-V and of the Strömgren b-y color, based on Kurucz's models did not agree with that from Hα and Hβ profiles. The same is true for the b-y colors calculated from models by Bell and Gustafsson (cited by Nissen and Gustafsson, 1978). Gehren suggests an explanation of this discrepancy in terms of missing line opacity in the calculations, although other possibilities are also mentioned.

Magain (1983)showed that the assumption that the discrepancy between the observed and calculated solar flux in the violet and near ultraviolet is due to a veil of very weak neutral metal absorption lines of rather high excitation would reproduce the observed temperature sensitivity, and the metal abundance sensitivity for the blue-violet-ultraviolet colors of the UBV and Geneva systems. These results were based on blanketed theoretical model atmospheres for solar-type stars, calculated with the program described by Gustafsson et al. (1975). The situation is, however, still unclear and definite conclusions must await detailed simulations of convective overshoot (such as those Nordlund 1978,1982,1984 has made for the Sun) for stars of different T_{eff}, $\log g$ and $[M/H]$,and further studies of veiling of weak spectral lines.

The Balmer discontinuity index of the uvby system, c_1, is an excellent gravity measure for F-type stars, but in view of the problems discussed above one should use theoretical calibrations of it with some care, and prefer more empirical calibrations (such as that of Crawford 1975, and Nissen and Gustafsson 1978). In particular, the metal-abundance sensitivity of the index is still not well known. Similar remarks are valid for the calibration of the abundance criterion m_1 (cf. Olsen, 1982) for which the empirical calibration of Nissen (1981) should be preferred.

Abundance determination is also affected by the problems discussed above; e.g. an uncertainty in effective temperature for a star of 200 K may well correspond to a metal-abundance error of 0.1 - 0.2 dex. A complicating circumstance is the problem with the solar colors; in any differential abundance analysis relative to the Sun where the effective temperature of the star is estimated from a color index, one needs to know the corresponding solar color. There still seems to be an uncertainty of at least ±0.03 in $(B-V)_\odot$ and similar uncertainties in other colors, corresponding to about ±75 K in T_{eff} for solar-type stars.

Another problem in abundance studies, for F- and G-type stars, is the non-LTE effects. The available statistical-equilibrium calculations for atoms and molecules in late-type stars in general are as yet quite uncertain, due to uncertainties in ultraviolet fluxes (of vital importance for photoionization) and in collision cross-sections and possibly too over-simplified model atoms. The recent calculations of Saxner (1984) suggest that iron is significantly overionized in the F star atmospheres, and that this effect is metal-abundance sensitive, since the more metal-poor stars let more photoionizing uv flux through. The resulting iron abundances determined from weak lines are predicted to be underestimated by up to a factor of two for the hottest Intermediate

Population II dwarfs. The second order effects on the model atmospheres are also found to be important.

One should finally keep in mind that the fundamental model assumption as regards plane-parallel stratification may cause errors in abundances of at least 0.1 dex (cf. Hermsen 1982).

3.5. G-K giants

The models in widest use for analyses of yellow-orange giants are the blanketed models of Bell et al. (1976), cf. also Gustafsson et al. (1975). They include blanketing from CN, C_2, CO and other molecules, as well as atomic lines. These models have been confronted with observations in various ways. E.g., the fluxes and colors were compared with observed spectra, scans and colors by Gustafsson and Bell (1979) and a good agreement was found except for the ultraviolet blocking in the models which was too small, leading to too bright uv magnitudes by up to $0^m.4$. This discrepancy was tentatively ascribed to a veil of very weak metal lines, not included in the line list of the models, the inclusion of which would not change the structure of the models significantly. With the recent "hotter" temperature scales for red giants (see below), a similar veil also needs to be invoked in the violet spectral region (cf. Frisk et al. 1982) and its structural effects become somewhat more important. Another discrepancy appeared in the CN band color indices, but that is because solar [N/Fe] ratios we're adopted in most models (cf. Gustafsson and Bell 1979, Lambert and Ries 1981 and Kjaergaard et al. 1982).

Attempts to test the models by comparison between predicted and observed profiles of strong lines also resulted in a reasonable agreement but with not fully conclusive results as regards deviations from the predicted structure of the upper layers (cf. Gustafsson 1980 for references).

The effective temperatures of red giants have been estimated by the integrated flux method, from colors and from studies of excitation equilibria. In addition one should mention the lunar-occultation diameters of Ridgway et al. (1980, 1982a) where the use of model atmospheres is confined to corrections for limb-darkening. We note in passing that the observed limb-darkening as a function of wavelength of α Tau (K5 III) by Ridgway et al. (1982b) gives some checks on the model atmosphere calculations and, within the yet considerable error margin agrees with the predicted one.

The temperature scales based on colors and model atmospheres have recently been revised upwards by almost 200 K, relative to earlier ones such as that of Bell and Gustafsson (1978, cf. also Kjaergaard et al. 1982). This is partly due to the choice of a greater solar color index and the recent use of more well-defined photometric systems like that of Wing (1971) calibrated for G-K giants by Wing et al. (1985), and that of Frisk et al. (1982) and Frisk and Bell (1985). A consequence of this is that the predicted blue color indices are too small and the assumption of a previously unconsidered veil of weak spectral lines in the uv must be extended into longer wavelengths. The temperature for Arcturus of Frisk et al. (1982) is in very good agreement with that of Blackwell and Shallis (1977), derived with the infrared flux method.

High effective temperatures were also derived by Lambert and Ries (1981) from excitation equilibria of neutral iron - however, in view of probable departures from LTE (see below) this scale must be regarded as rather uncertain.

The gravity determinations for G-K giants is still a matter of some controversy (cf. the discussion of Arcturus by Trimble and Bell 1981). The methods available include the use of ionization equilibria, of the wings of pressure-broadened spectral lines and of lines of pressure-sensitive molecules, such as MgH; in fact a fourth method, based on the different pressure sensitivity of the H-opacity relative to that of the HI scattering is exploited in certain photometric systems, see Gustafsson and Bell (1978). These methods are subject to uncertainties of different importance, and they have been intercompared for the test case of Arcturus in a recent study by Bell et al. (1984). The resulting logarithmic gravities range from 1.4 to 1.8 (cgs units), and it is concluded that the method of Blackwell and Willis (1977), based on pressure-broadened lines, should be preferred, especially for stars with greater gravities than Arcturus, i.e., for most G and K giants and subgiants.

The problems of determining abundances for late-type giants have recently been reviewed several times (Gustafsson 1980,83, Taylor 1983), and here we shall only consider them briefly. The possible errors in T_{eff} and $\log g$ may cause considerable problems, especially for estimating abundances of elements such as nitrogen, where the determination is dependent on the molecular equilibrium (cf. Kjaergaard et al. 1982). The model structure uncertainties may be important, especially for strong lines and low excitation lines (Gustafsson 1983). Departures from LTE have been traced empirically in some cases (cf., e.g., Ruland et al. 1980, Brown, Tomkin and Lambert 1983, Steenbock 1983, possibly Kovacs 1983 and Luck and Bond 1983). Unfortunately, one could hardly assume that any elemental abundance for any giant is much better determined than to 0.2 dex, although the internal consistency between different weak lines as measured with modern equipment sometimes suggests a considerably smaller error. The systematic errors may in some cases be significantly greater and, worst of all, it is hard to say when.

In a series of papers Luck has analysed G and K supergiants (cf. Luck 1982a and references cited therein), using models calculated with the program of Gustafsson et al. (1975). He derives effective temperatures and gravities from Fe excitation and ionization equilibria, respectively, and finds a temperature scale which is about 200 K hotter at a given B-V than that of Bell and Gustafsson (1978). It seems natural to ascribe this discrepancy to the veil discussed above, which was not considered in the model fluxes, and to the zero point of colors of the model calibration, even though departures from the Boltzmann excitation equilibrium may also be of importance. Luck also finds his gravity determinations to agree well with estimates from evolutionary tracks, and estimates errors in $\log g$ of, at the most, ±0.3 dex. He gives typical errors in the resulting abundances of ±0.2 - 0.3 dex, but admits that there may be additional systematic errors.

3.6. M giants and supergiants

At least two grids of model atmospheres for M giants and supergiants have been published - that of Tsuji (1978), where blanketing from a number of molecules including TiO and H_2O was taken into account using some rather crude approximations, such as the so-called Voigt-analog - Elsässer band model, and the grid of Johnson et al. (1980) where the more detailed Opacity-Sampling method was used. The models of the latter grid seem to agree quite well with those of Tsuji and also, for temperatures above 4000 K where TiO is not very important, with those of Bell et al (1976). Tsuji (1978) showed that his models agreed well with observed spectral distributions of M giants, and he applied the integrated-flux method to establish an effective-temperature scale for M giants. He estimated that the temperatures are accurate to within ±150 (4%,Tsuji 1981a). This scale agrees nicely with that of Ridgway et al. (1980), based on lunar occultation data. Piccirillo et al. (1981) also found a good agreement between the Ridgway et al. scale and that from the models of Johnson et al. (1980), when used for calibrating the Wing eight-color photometry. The temperature of Betelguese (M2 Ib), derived with the infrared flux method by Tsuji (1981a), T_{eff} = 3800 K, agrees well with that estimated from the excitation equilibria of CO (Lambert et al. 1984) but is significantly greater than the interferometric value of Balega et al.(1982).

The problem of determining the surface gravity for M stars spectroscopically is a very difficult one; due to the lack of unblended ion lines, ionization equilibria cannot be used, while the continuum drawing problems make damping wings of pressure broadened lines difficult to trace. The profound effect of the molecular-line blanketing from TiO was nicely demonstrated by Piccirillo et al. (1981). Attempts to use hydride lines and lines from the corresponding neutral metal atom could be attempted but will require very good temperature estimates.

The placement of the continuum is also a major problem for abundance determinations in M stars from lines in the visual and near infrared. In their analysis of Li and Al in M giants and supergiants Luck and Lambert (1982) estimate that these difficulties lead to an error of 0.2 dex in the Li abundance determination. They find that non-LTE effects should be small and estimate their total errors at ±0.3 dex. Luck (1982b) finds uncertainties of ±0.4 dex in his chemical analysis of two M stars, again referring to the continuum location problems as the major source of error. The analysis of Lambert et al. (1984) of CNO abundances in Betelguese is more satisfactory from this point of view, since it is based on high-resolution spectra beyond 1.5 microns, where the blending problems are less severe. Also, this analysis uses the vibration-rotation bands of CO, NH and OH (in addition to the electronic CN red system) where the departures from LTE are thought to be of less importance.

Departures from LTE in early M stars have been suggested as occuring in the ionization of Ca by Ramsey (1977, 1981). However, it seems reasonable to assume that the empirical support for this may instead be effects of difficulties in locating the continuum (Luck 1982b).

Evidence for inhomogeneities in Betelguese was reported by Goldberg et al. (1982) and by Hayes (1980, 1984)and further discussed by

Goldberg (1984). It is not known what errors these inhomogeneities
introduce into the abundance estimates. The effects of spherical extension should be of importance for M giants and supergiants, as demonstrated by Watanabe and Kodaira (1978, 1979) and Schmid-Burgk et al.
(1981). In fact, the formation of molecules amplifies these extension
effects further, such that they are greater than first order estimates
tend to show, and also depend on the chemical composition in a complicated way. However, the extension d ($\propto$ ($g_{surface}R$)$^{-1}$) can in principle be determined from stellar spectra, and this opens up the possibility of determining the stellar radius, in addition to surface gravity and effective temperature. Thus, stellar masses for single stars
should in principle be possible to determine spectroscopically for extended stars. The practical possibilities for doing this have been
discussed by Scholz and Wehrse (1982).

3.7. Carbon stars

For the hotter carbon stars, the R stars, two grids of models with
detailed blanketing have been calculated - one by Olander (1981) using
the program of Gustafsson et al. (1975) and one by Johnson and O'Brien
(cited by Dominy, 1984). The models of the two grids have been intercompared by Yorka (1981), who found them to agree well, which is of interest since the methods for handling the heavy blanketing are different
(based on the ODF and OS technique, respectively), and since the molecular line lists are based on different compilations of data. A valuable
review on model atmospheres for late-type stars has been published by
Carbon (1979).

There is, as yet, no systematic comparison between predicted fluxes
of these models and observations. However, comparisons like that of
Dominy (1984, his Fig. 1) suggest a reasonable agreement between computed and calculated fluxes. Also, there is no fundamental theoretical
calibration of T_{eff} versus colors yet available - in his analysis
Dominy (1984) had to rely on the corresponding relation for the G-K
stars. It should be noted that the version of the integrated flux method
used by Tsuji (1981a and b) for cooler stars, where the flux in the L
band at 3.5 μm is compared with the integrated flux, cannot be used, due
to the characteristic excess of flux of R stars in the L band. In all,
errors of several hundreds of K are to be expected in the current estimates of temperatures for R stars.

As regards the gravities the situation is very difficult. The heavy
blending and continuum location problems make measured equivalent widths
of the few ionic lines highly uncertain and profiles of pressure-broadened
lines difficult to measure. The molecular equilibria are hard to use
since the chemical composition is unknown and has to be determined from
the same data.

The uncertainties in model parameters and in the continuum location
make the results from abundance analysis of R stars uncertain. In his
pioneering study Dominy (1984) suggests errors in the interval 0.3-0.6
dex, or even more for the N abundances when estimated from the CN lines,
since the propagation of errors in C and O abundances and the gravity
uncertainty makes the situation particularily difficult.

For the cooler carbon stars (of spectral type N) grids of models have been calculated by Querci, Querci and Tsuji (1974), Querci and Querci (1975), by Johnson (1982) and by Eriksson et al. (1984b). Querci and Querci (1976) have compared spectra and colors of their models with observations, with rather inconclusive results, partly because the role of grain opacities is not well known.

A worrying discrepancy between the carbon-star models and the real stars was the fact that the hydrogen quadrupole lines at 2.1 μm were pre- dicted to be strong by the models but not seen in the spectra (Goorvitch et al. 1980). Although H_2 lines were later detected in N-type FTS spectra (Johnson et al. 1983), they were found to be considerably weaker than predicted from the models. A similar problem exists for the H_2 lines in M giants (Tsuji 1983).

The models of Querci et al. and of Johnson do not contain any ab- sorption from polyatomic molecules. Eriksson et al. (1984a) included preliminary opacity estimates by Jørgensen (1982) for HCN and later C_2H_2 and found drastic effects - the pressure at a given optical depth for stars with T_{eff}=2500 K could decrease by two orders of magnitude as a result of the new opacity. Therefore, the H_2 density decreased consi- derably and we conclude that the formation of polyatomic molecules may be an explanation for the weak H_2 lines in carbon star spectra. This conclusion should, however, be taken as preliminary until current ab initio calculations of HCN, C_3 and C_2H_2 have been pursued (Jørgensen et al. 1984). There is the alternative possibility that the outer photo- spheres of N stars are heated by non-thermal processes, which would diminish both the polyatomic opacities and the H_2 number densities (cf. Johnson et al. 1983 and Goebel et al. 1983). Another possibility would be that the carbon stars may be hydrogen poor, which would explain the apparently missing "H^- peak" around 1.6 μm in their spectra (Alexander et al. 1984).

The temperature scale for carbon stars established by Tsuji (1981b), using the integrated-flux method and the Querci et al. grid, must for obvious reasons be considered as preliminary. However, we have recently derived effective temperatures for N stars by using infrared colors, compared with theoretical ones,for models with C_2H_2 and HCN included. We find an agreement within 200 K with Tsuji's scale, which is satis- factory. Tsuji's scale is also consistent with the few angular diameter measurements available for carbon stars.

A major problem for N stars is the derivation of surface gravities from spectra. It is not clear whether this problem can be solved at all with any acceptable errors. It will at least require very reliable T-P structures (which requires good estimates of polyatomic opacities) and very good synthetic spectra around the spectral features to be used. Determinations of CNO abundances, over-all metal abundances and abundance ratios are nevertheless within reach for these stars, when based on lines in the infrared. However, one should note that an acceptable abundance determination requires an iterative procedure when determining the funda- mental parameters, since the structure of the model atmosphere is very dependent on the chemical composition as a consequence of the heavy blocking. The resulting errors are expected to be considerable.

A recent example of an analysis of an SC-type star is the study of UY Cen by Catchpole (1982). Here, the resulting typical errors in the abundances are estimated to ±0.5 dex. The problems of the pure S stars are somewhat special (and not yet fully explored) due to the delicate molecular equilibrium and effects of blanketing from other molecules than those of importance in M and C stars.

For the cooler carbon stars the formation of dust grains may cause major problems, since the calculation of their concentration and their heavy opacity probably cannot be made within the framework of a steady state model (cf. Woodrow and Auman 1982).

3.8. Late-type dwarfs

For the K dwarf atmospheres, usually scaled solar models or model atmospheres from the metal-line-blanketed grid of Peytremann (1970) or from that of Eriksson et al. (1980) are used. (The latter grid, which is consistent with that of Bell et al. 1976, and thus also includes the blanketing from molecular lines, is unpublished but available on request.) Steenbock and Holweger (1981) report indications from the Mg b-line profiles of ε Eri and 40 Eri A (K2 V and K1 V, respectively) that theoretical models seem to be too cool in the upper photosphere. These authors find that scaled HM solar models give better fits.

Effective temperatures for the K dwarfs are usually obtained from the near-infrared colors, but no well-established temperature scale based on the model-atmospheres for these stars has yet been published. The gravities may be derived from bolometric fluxes (parallaxes are often well-known) and temperatures, if the mass is estimated from the luminosity. Alternatively, ionization equilibria, pressure-broadened lines or molecular lines (e.g., from hydrides) may be used for this purpose. Attempts to compare the methods and their resulting gravities have as yet not been fully conclusive (cf. Perrin 1983); earlier reports on significant inconsistencies between the ionization equilibrium gravities and those derived from parallaxes are not valid (Perrin et al. 1975, Hearnshaw 1976b), except for possibly for metal-deficient dwarfs (Hearnshaw 1976a).

Typical errors claimed in abundance determinations relative to the Sun for K dwarfs seem to be ± 0.2 dex, which may be somewhat optimistic in view of the uncertainties in models and in the temperature scale, at least for the late K dwarfs. In their spectra it may also be difficult to locate a true continuum.

For M dwarfs, Mould (1976a) in his pioneering work calculated a grid of model atmospheres with blanketing from atoms and molecules (in particular from H_2O and TiO) taken into account, using a rough ODF method and a smeared line model for the molecular opacities. He applied these models in further studies of M dwarfs (Mould 1976b and references cited therein, see also Hartmann and Anderson 1976) and, to our knowledge, this work still belongs to the most advanced photospheric analyses for M dwarfs. The full arsenal of detailed blanketed models and synthetic spectra should be used in future work for analysing the high-quality spectroscopic material that may be obtained today for these stars.

4. CONCLUSIONS

The superficial review in the previous section may have given the impression that, although there are still some questions of debate left, and some corners in the HR diagram which have not been fully explored with the methods of detailed quantitative spectral analysis, the general success of physical modelling is remarkable. This is partly an illusion, however. It is interesting that most of the cases under intensive debate in this field of research, such as the effective temperature scale of O stars, the temperature scale of F stars, the nature of the so-called super-metal-rich stars (Taylor 1983, and references cited therein), the gravity of Arcturus, the metal abundance of the metal-rich globular clusters (Gustafsson 1982, Bessell 1983, Cohen 1983, D'Odorico et al. 1984), the effective temperature of Betelgeuse, etc., have one characteristic in common: several different methods have been attempted and they tend to give different results. In many other cases, where the errors in the fundamental parameters are thought to be rather small, different or independent methods have not yet been tried. Admittedly, many or most of the conflicting results in the cases listed above may be due to trivial errors in the observations or in the analysis. But some of the discrepancies may reveal more fundamental errors in the models or even in our basic ideas of how stellar atmospheres are structured. This possibility, and the good reasons discussed above in Sec. 1 for improving the parameter determinations further, makes it worthwhile to proceed in attempts to fundamentally improve the models. A third important reason for doing so is the lack of theoretical self-consistency in contemporary model atmospheres.

The future modelling of stellar atmospheres requires more profound theoretical studies of various aspects and detailed numerical simulations of different processes in the atmospheres. This includes investigations of convection and its consequences, of mass flows, of non-thermal heating, of effects of inhomogeneities, of departures from LTE and of grain-formation. It is also important to consider the interplay between these different non-classical aspects of stellar atmospheres since they affect each other in several important ways. Theoretical efforts must be combined with systematic observational studies of direct and indirect diagnostics of the phenomena, e.g., of spectral line shifts and asymmetries, of spectra at different wavelengths in order to disclose the somewhat symbiotic character of spectra from inhomogeneous atmospheres, of suspected departures from LTE in spectral lines, and of thermal emission from dust as compared with temperature diagnostics for the outer photosphere.

The more fundamental studies should be combined with systematic attempts to develop more empirical model atmospheres, at least for a number of "standard stars", based on obtainable and suitable data such as the observations of vibration-rotation lines of different strengths and of different molecules, wing profiles of strong spectral lines and continuum fluxes. This should be supplemented with large scale synthetic-spectrum calculations and comparisons in detail with high signal-to-noise spectra, covering extensive spectral regions for these stars, in order to improve the calibration of low-resolution criteria for fainter stars.

It does not seem possible to achieve any significant improvement in the accuracy of determinations of stellar parameters from model atmosphere comparisons that one can obtain today unless at least part of the program sketched above is realized. This program will require very ambitious collaborative efforts between theorists and observers; and also the use of different methods and an interest in studying why these different methods sometimes give conflicting results. Conflicts should not be swept under the rug, nor should they force us into definitive positions; scientific progress is a dialectic process. In return we shall learn much more about "the stars that are up there".

ACKNOWLEDGEMENTS

Rolf Kudritzki and Rainer Wehrse are thanked for helpful correspondence, Ilse-Beth Byström for typing the manuscript, David Minugh for commenting on it, and the editors of the present proceedings for their patient waiting for it.

REFERENCES

Adelman, S.J. 1984, *Monthly Notices Roy. Astron. Soc.* 206, 637.

Adelman, S.J., Pyper, D.M., White, R.E. 1980, *Astrophys. J. Suppl. Ser.* 43, 491.

Adelman, S.J., Pyper, D.M. 1983, *Astrophys. J.* 266, 732.

Adelman, S.J., Young, J.M., Baldwin, H.E. 1984, *Monthly Notices Roy. Astron. Soc.* 206, 649.

Alecian, G. 1982, *Astron. Astrophys.* 107, 61.

Alexander, D.R., Johnson, H.R., Bower, C.D., Lemke, D.A., Luttermoser, D.G., Petrakis, J.P., Reinhart, M.D., Welch, K.A. 1984, preprint.

Auer, L.H., Mihalas, D. 1969, *Astrophys. J.* 158, 641.

Auer, L.H., Mihalas, D. 1972, *Astrophys. J. Suppl. Ser.* 24, 193.

Auer, L.H., Mihalas, D. 1973a, *Astrophys. J. Suppl. Ser.* 25, 443.

Auer, L.H., Mihalas, D. 1973b, *Astrophys. J.* 184, 151.

Balega, Y., Blazit, A., Bonneau, D., Koechlin, L., Foy, R., Labeyrie, A. 1982, *Astron. Astrophys.* 115, 253.

Baschek, B., Scholz, M., Sedlmayer, E. 1977, *Astron. Astrophys.* 55, 375.

Baschek, B., Kudritzki, R.P., Scholz, M., Simon, K.P. 1982, *Astron. Astrophys.* 108, 387.

Bell, R.A., Eriksson, K., Gustafsson, B., Nordlund, A. 1976, *Astron. Astrophys. Suppl. Ser.* 34, 229.

Bell, R.A., Gustafsson, B. 1978, *Astron. Astrophys. Suppl. Ser.* 34, 229.

Bell, R.A., Dreiling, L.A. 1982, *Publ. Astron. Soc. Pacific* 94, 50.

Bell, R.A., Edvardsson, B., Gustafsson, B. 1984, submitted to *Monthly Notices Roy. Astron. Soc.*

Bessell, M.S. 1983, *Publ. Astron. Soc. Pacific* 95, 94.

Blackwell, D.E., Shallis, M.J. 1977, *Monthly Notices Roy. Astron. Soc.* 180, 177.

Blackwell, D.E., Shallis, M.J., Selby, M.J. 1979, *Monthly Notices Roy. Astron. Soc.* 188, 847.

Blackwell, D.E., Willis, R.B. 1977, *Monthly Notices Roy. Astron. Soc.* 180, 169.

Boesgaard, A.M., Heacox, W.D., Wolff, S.C., Borsenberger, J., Praderie, F. 1982, *Astrophys. J.* 259, 723.

Böhm-Vitense, E. 1970, *Astron. Astrophys.* 8, 283.

Böhm-Vitense, E. 1978, *Astrophys. J.* 223, 509.

Böhm-Vitense, E. 1981a, *Ann. Rev. Astron. Astrophys.* 19, 295.

Böhm-Vitense, E. 1981b, *Astrophys. J.* 244, 938.

Böhm-Vitense, E. 1982, *Astrophys. J.* 255, 191.

Brown, J.K., Tomkin, J., Lambert, D.L. 1983, Astrophys. J. Letters 265, L93.

Buser, R., Kurucz, R.L. 1978, *Astron. Astrophys.* 70, 555.

Buser, R., Kurucz, R.L. 1985, *IAU Symp.* 111 (this volume), p. 513.

Carbon, D.F. 1979, *Ann. Rev. Astron. Astrophys.* 17, 515.

Carbon, D.F., Langer, G., Butler, D., Kraft, R., Suntzeff, N., Kemper, E., Trefzger, C., Romanishin, W. 1982, *Astrophys. J. Suppl. Ser.* 49, 207.

Castelli, F., Lamers, H.J.G.L.M., Llorente de Andrès, F., Müller, E.A. 1980, *Astron. Astrophys.* 91, 32.

Castor, J.I., Simon, T. 1983, *Astrophys. J.* 265, 304.

Catchpole, R.M. 1982, *Monthly Notices Roy. Astron. Soc.* 199, 1.

Code, A.D., Davis, J., Bless, R.E., Hanbury Brown, R. 1976, *Astrophys. J.* 203, 417.

Cohen, J. 1982, *Astrophys. J.* 258, 143.; Cohen, J.G. 1983, ibid 270, 654.

Cowley, C.R., Adelman, S.J. 1983, *Quarterly J. Roy. Astron. Soc.* 24, 393.

Cramer, N. 1982, *Astron. Astrophys.* 112, 330.

Cramer, N. 1984, *Astron. Astrophys.* 132, 283.

Crawford, D. 1975, *Astron. J.* 80, 955.

DeLoore, C.W.H., Willis, A.J. (eds.) 1982, *IAU Symp. No.* 99, Wolf-Rayet Stars: Observations, Physics, Evolution, D. Reidel Publ. Company, Dordrecht, Holland.

Desikachary, K., Hearnshaw, J.B. 1982, *Monthly Notices Roy. Astron. Soc.* 201, 707.

D'Odorico, S., Gratton, R.G., Ponz, D. 1984, *Astron. Astrophys.*, in press.

Dominy, J.F. 1984, *Astrophys. J. Suppl. Ser.* 55, 27.

Dreiling, L.A., Bell, R.A. 1980, *Astrophys. J.* 241, 737.

Erhorn, G., Groote, D., Kaufmann, J.P. 1984, *Astron. Astrophys.* 131, 390.

Eriksson, K., Bell, R.A., Gustafsson, B., Nordlund, Å. 1980, unpubl.

Eriksson, K., Gustafsson, B., Jørgensen, U.G., Nordlund, Å. 1984a, *Astron. Astrophys.* 132, 37.

Eriksson, K., Gustafsson, B., Nordlund, Å, 1984b, in preparation.

Frisk, U., Bell, R.A., Gustafsson, B., Nordh, L.H., Olofsson, S.G. 1982, *Monthly Notices Roy. Astron. Soc.* 199, 471.

Frisk, U., Bell, R.A. 1985, *IAU Symp.* 111 (this volume), p. 543.

Gehren, T. 1981, *Astron. Astrophys.* 100, 97.

Goebel, J.H., Bregman, J.D., Cooper, D.M., Goorvitch, D., Langhoff, S.R., Witteborn, F.C. 1983, *Astrophys. J.* 276, 190.

Goldberg, L., Hege, E.K., Hubbard, E.N., Strittmatter, P.A., Cocke, W.J. 1982, in *Second Cambridge Workshop on Cool Stars, Stellar Systems and the Sun*, eds. M.S. Giampapa and L. Golub, Smithsonian Spec. Report No. 392, Vol I. p. 131.

Goldberg, L. 1984, in *Lecture Notes in Physics* 193, *Cool Stars, Stellar Systems and the Sun*, eds. S.L. Baliunas and L. Hartmann, Springer-Verlag, p. 333.

Goorvitch, D., Goebel, J.H., Augason, G.C. 1980, *Astrophys. J.* 240, 588.

Gruschinske, J., Hamann, W.-R., Kudritzki, R.P., Simon, K.P., Kaufmann, J.P. 1983, *Astron. Astrophys.* 121, 85.

Gustafsson, B. 1980, in *ESO Workshop on Methods of Abundance Determination for Stars*, eds. P.E. Nissen, K. Kjär, p. 31.

Gustafsson, B. 1982, in *Highlights in Astronomy* 6, ed. R.M. West, D. Reidel Publ. Co., Dordrecht, Holland, p. 101.

Gustafsson, B. 1983, *Publ. Astron. Soc. Pacific* 95, 101.

Gustafsson, B., Bell, R.A., Eriksson, K., Nordlund, Å. 1975, *Astron. Astrophys.* 42, 407.

Gustafsson, B., Bell, R.A. 1979, *Astron. Astrophys.* 74, 313.

Hartmann, L., Anderson, C.M. 1976, *Wisconsin Astrophys.* 40.

Hayes, D.P. 1980, *Astrophys. J. Letters* 241, L165.

Hayes, D.P. 1984, in *Lecture Notes in Physics* 193, *Cool Stars, Stellar Systems and the Sun*, eds. S.L. Baliunas and L. Hartmann, Springer-Verlag, p. 342.

Hearnshaw, J.B. 1976a, *Astron. Astrophys.* 51, 71.

Hearnshaw, J.B. 1976b, *Astron. Astrophys.* 51, 85.

Heasley, J.N., Wolff, S.C., Timothy, J.G. 1982, *Astrophys. J.* 262, 663.

Heasley, J.N., Wolff, S.C. 1983, *Astrophys. J.* 269, 634.

Heber, U. 1983, *Astron. Astrophys.* 118, 39.

Heber, U., Hunger, K., Jonas, G., Kudritzki, R.P. 1984, *Astron. Astrophys.* 130, 119.

Hermsen, W. 1982, *Astron. Astrophys.* 111, 233,

Holweger, H. 1979, in *Les Elements et leur Isotopes dans l'Universe*, Proc. 22nd Liège International Astrophys. Symp., p. 117.

Husfeld, E., Kudritzki, R.P., Simon, K.P. 1983, *Mitt. Astron. Gesellschaft* 60, 306.

Johnson, H.R. 1982, *Astrophys. J.* 260, 254.

Johnson, H.R., Bernat, A.P., Krupp, B.M. 1980, *Astrophys. J. Suppl. Ser.* 42, 501.

Johnson, H.R., Goebel, J.H., Goorvitch, D., Ridgway, S.T. 1983, *Astrophys. J. Letters* 276, L63.

Jørgensen, U.G. 1982, Thesis, University of Copenhagen (in Danish).

Jørgensen, U.G., Almlöf, J., Larsson, M., Siegbahn, P.E.M. 1984, submitted to *J. Chem. Phys.*

Kamp, L.W. 1976, *NASA TR R-455*.

Kamp, L.W. 1978, *Astrophys. J. Suppl. Ser.* 36, 143.

Kamp, L.W. 1982, *Astrophys. J. Suppl. Ser.* 48, 415.

Kapranidis, S. 1983, *Astrophys. J.* 275, 342.

Keenan, F.P., Dufton, P.L., McKeith, C.D. 1982, *Monthly Notices Roy. Astron. Soc.* 200, 673.

Kjærgaard, P., Gustafsson, B., Walker, G.A.H., Hultqvist, L. 1982, *Astron. Astrophys.* 115, 145.

Koester, D., Weidemann, V., Zeidler,-K.T., E.-M. 1982, *Astron.*
Astrophys. 116, 147.

Kovács, N. 1983, *Astron. Astrophys.* 120, 21.

Kudritzki, R.P. 1979, in *Les Elements et leur Isotopes dans l'Universe*,
Proc. 22nd Liège International Astrophys. Symp., p. 295.

Kudritzki, R.P. 1980, *Astron. Astrophys.* 85, 174.

Kudritzki, R.P., Simon, K.P., Hamann, W.-R. 1983, *Astron. Astrophys.*
118, 245.

Kurucz, R.L. 1979, *Astrophys. J. Suppl. Ser.* 40, 1.

Kurucz, R.L. 1980, *Smithsonian Astrophys. Obs. Spec, Report No.*387.

Kurucz, R.L., Peytremann, E. 1975, *Smithsonian Astrophys. Obs. Spec.*
*Report No.*362.

Lambert, D.L. 1978, *Monthly Notices Roy. Astron. Soc.* 182, 243.

Lambert, D.L., Ries, L.M. 1981, *Astrophys. J.* 248, 228.

Lambert, D.L., Roby, S.W., Bell, R.A. 1982, *Astrophys. J.* 254, 663.

Lambert, D.L., Brown, J.A., Hinkle, K.H., Johnsson, H.R. 1984,
Astrophys. J. (in press).

Lamers,H.J.G.L.M.,deGroot,M.,Casatella, A. 1983, *Astron. Astrophys.*
128, 299.

Lennon, D.J. 1983, *Monthly Notices Roy. Astron. Soc.* 205, 829.

Lennon, D.J., Dufton, P.L. 1983, *Monthly Notices Roy. Astron. Soc.*
203, 443.

Lester, J.B., Lane, M.C., Kurucz, R.L. 1982, *Astrophys. J.* 260, 272.

Luck, R.E. 1982a, *Astrophys. J.* 256, 177.

Luck, R.E. 1982b, *Astrophys. J.* 263, 215.

Luck, R.E., Lambert, D.L. 1982, *Astrophys. J.* 256, 189.

Luck, R.E., Bond, H.E. 1983, *Astrophys. J. Letters* 271, L75.

Magain, P. 1983, *Astron. Astrophys.* 122, 225.

Malagnini, M.L., Faraggiana, R., Morossi, C., Crivellari, L. 1982,
Astron. Astrophys. 114, 170.

Malagnini, M.L., Faraggiana, R., Morossi, C. 1983, *Astron. Astrophys.*
128, 375.

Méndez, R.H., Kudritzki, R.P., Simon, K.P. 1983, in *Planetary Nebulae*,
IAU Symp. 103, D.R. Flower (ed.), D. Reidel Publ. Company,
Dordrecht, Holland, p. 343.

Mihalas, D. 1972a, *Astrophys. J.* 177, 115.

Mihalas, D. 1972b, *NCAR, Technical Note STR*-16.

Mihalas, D. 1973, *Astrophys. J.* 179, 209.

Mould, J.R. 1976a, *Astrophys. J.* 210, 402.

Mould, J.R. 1976b, *Astron. Astrophys.* 48, 443.

Muthsam, H. 1979, *Astron. Astrophys.* 73, 159,

Muthsam, H., Stepién, K. 1980, *Astron. Astrophys.* 86, 240.

Nelson, G.D. 1980, *Astrophys. J.* 238, 659.

Nissen, P.E. 1981, *Astron. Astrophys.* 97, 145.

Nissen, P.E., Kjär, K. (eds.) 1980, *ESO Workshop on Methods of*
Abundance Determination for Stars.

Nissen, P.E., Gustafsson, B. 1978, in *Astronomical Papers Dedicated*
to Bengt Strömgren, eds. A. Reiz and T. Andersen, Copenhagen
1978, p. 43.

Nordlund, Å. 1978, ibid., p. 95.

Nordlund, Å, 1982, *Astron. Astrophys.* 170,1.

Nordlund, Å. 1984, in *Small Scale Dynamical Processes in Quiet Stellar Atmospheres*, ed. S. Keilz, *Sacramento Peak Publ.* (in press).

Olander, N. 1981, *Uppsala Astron. Obs. Report No.* 21.

Olsen, E.H. 1982, *Astron. Astrophys.* 110, 215.

Pecker, J.C. 1965, *Ann. Rev. Astron. Astrophys.* 3, 135.

Perrin, M.-N. 1983, *Astron. Astrophys.* 128, 347.

Perrin, M.-N., Cayrel de Strobel, G., Cayrel, R. 1975, *Astron. Astrophys.* 39, 97.

Peytremann, E. 1970, Thesis, Obs. Genève.

Philip, A.G.D., Newell, B. 1975, in *Multicolor Photometry and the Theoretical HR Diagram*, eds. A.G.D. Philip and D.S. Hayes, *Dudley Obs. Report No.* 9, p. 161.

Piccirillo, J., Bernat, A.P., Johnson, H.R. 1981, *Astrophys. J.* 246, 246.

Querci, F., Querci, M., Tsuji, T. 1974, *Astron. Astrophys.* 31, 265.

Querci, F., Querci, M. 1975, *Astron. Astrophys.* 39, 113.

Querci, M., Querci, F. 1976, *Astron. Astrophys.* 49, 443.

Ramsey, L.W. 1977, *Astrophys. J.* 215, 827.

Ramsey, L.W. 1981, *Astrophys. J.* 245, 984,

Relyea, L.J., Kurucz, R.L. 1978, *Astrophys. J. Suppl. Ser.* 37, 45.

Remie, H., Lamers, H.J.G.L.M. 1982, *Astron. Astrophys.* 105, 85.

Renson, P.(ed.) 1981, *Upper Main Sequence Chemically Peculiar Stars, 23rd Liège Int. Astrophys. Symp. Liège*.

Ridgway, S.T., Joyce, R.R., White, N.M., Wing, R.F. 1980, *Astrophys. J.* 235, 126.

Ridgway, S.T., Jacoby, G.H., Joyce, R.R., Siegel, M.J., Wells, D.C. 1982a, *Astron. J.* 87, 808.

Ridgway, S.T., Jacoby, G.H., Joyce, R.R., Siegel, M.J., Wells, D.C. 1982b, *Astron. J.* 87, 1044.

Ruland, F., Holweger, H., Griffin, R. and R., Biel, D. 1980, *Astron. Astrophys.* 92, 70.

Rutten, R.J., Kostik, R.L. 1982, *Astron. Astrophys.* 115, 104.

Sauval, A.J., Grevesse, N., Brault, J.W., Stokes, G.M., Zander, R. 1984, *Astrophys. J.* (in press).

Saxner, M. 1984, Thesis, Uppsala University, to be subm. to *Astron. Astrophys.*

Schmidt-Burgk, J., Scholz, M., Wehrse, R. 1981, *Monthly Notices Roy. Astron. Soc.* 194, 383.

Scholz, M., Wehrse, R. 1982, *Monthly Notices Roy. Astron. Soc.* 200, 41.

Schönberner, D., Drilling, J.S. 1984, *Astrophys. J.* 278, 702.

Searle, L., Zinn, R. 1978, *Astrophys. J.* 225, 357.

Selby, M.J., Mountain, C.M., Blackwell, D.E., Petford, A.D., Leggett, S.K. 1983, *Monthly Notices Roy. Astron. Soc.* 203, 795.

Simon, K.P., Jonas, G., Kudritzki, R.P., Rahe, J. 1983, *Astron. Astrophys.* 125, 34.

Sion, E.M., Guinan, E.F., Wesemael, F. 1982, *Astrophys. J.* 255, 232.

Smith, L.J., Willis, A.J. 1982, *Monthly Notices Roy. Astron. Soc.* 201, 451.

Snijders, M.A.J. 1977a, *Astron. Astrophys.* 60, 377.

Snijders, M.A.J. 1977b, *Astrophys. J. Letters* 214, L35.

Steenbock, W. 1983, *Astron. Astrophys.* 126, 325.

Steenbock, W., Holweger, H. 1981, *Astron. Astrophys.* 99, 192.

Thévenin, F., Foy, R. 1983, *Astron. Astrophys.* 122, 261.
Tobin, W. 1983, *Astron. Astrophys.* 125, 168.
Trimble, V., Bell, R.A. 1981, *Quarterly J. Roy. Astron. Soc.* 22, 361.
Tsuji, T. 1978, *Astron. Astrophys.* 62, 29.
Tsuji, T. 1981a, *Astron. Astrophys.* 99, 48.
Tsuji, T. 1981b, *J. Astrophys. Astron.* 2, 95.
Tsuji, T. 1983, *Astron. Astrophys.* 122, 314.
Underhill, A.B. 1982, *Astrophys. J.* 263, 741.
Underhill, A.B. 1983, *Astrophys. J.* 266, 718.
Underhill, A.B., Divan, L., Prévot-Burnichon, M.-L., Doazan, V. 1979, *Monthly Notices Roy. Astron. Soc.* 189, 601.
Watanabe, T., Kodaira, K. 1978, *Publ. Astron. Soc. Japan* 30, 21.
Watanabe, T., Kodaira, K. 1979, *Publ. Astron. Soc. Japan* 31, 61.
Wegner, G., Yackovich, F.H. 1983, *Astrophys. J.* 275, 240.
Wehrse, R. 1984, private communication.
Wing, R.F. 1971, in *Proc. Conf. on Late-Type Stars,* eds. G.W. Lockwood and H.M. Dyck, *Kitt Peak Natl. Obs. Contr. No.* 554, p. 145.
Wing, R.F., Eriksson, K., Gustafsson, B. 1985, *IAU Symp.* 111 (this volume), p. 571.
Woodrow, J.E.J., Auman, J.R. 1982, *Astrophys. J.* 257, 247.
Yorka, S.B. 1981, Ph.D. Thesis, Ohio State University.

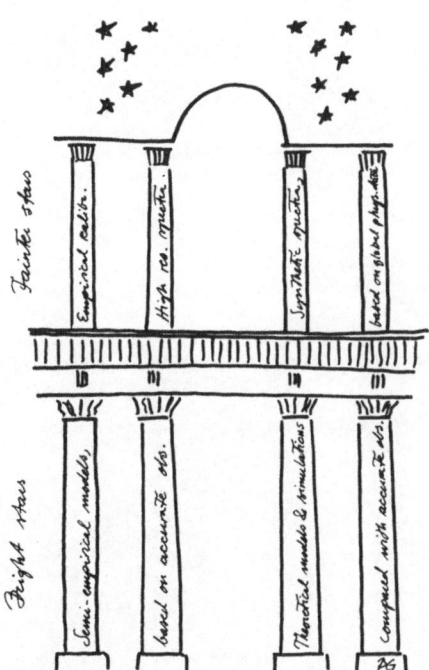

Steps towards improved stellar fundamental parameters from model atmosphere analysis – a subjective interpretation of the architecture of the great meeting room at Villa Olmo, Como.

DISCUSSION

GUSTAFSSON: I would like you to look upwards and consider the nice paint-
ing by Dominico Pozzi on the ceiling of this beautiful room. The paint-
ing could be called: "Il dilemma delle calibrazioni di astrofisiche",
and if you translate that, with some Underhillian sharpening, you may
call it the "horrible problem of astrophysical calibration". You see
three groups of people and also, with some difficulty you may find stars
and two stellar constellations. There are two groups of people in the
sky, the photometrists and the spectroscopists, and they seem to be
interested, to some extent, in the stars. They are not debating very
much, however. And then, on the ground, we have the theorists. They
are in intensive debate, but they don't even look at the stars. I think
that the point the painter wants to make is that, in order to improve,
we have to collaborate more. More interaction, more discussions, even
more conflicts. After all, scientific progress is a dialectic process.

ROSSI: Concerning the empirical temperature calibration for late-type stars, I have to make a warning about the fact that when analyzing the 2.2 micron sky survey and the AFCR2 catalogues, you see IR excesses for K-type stars. I wish to make a remark about the atmospheres of C stars. I am very interested in your models for C stars. In these stars grain formation in the atmosphere is very likely and one has to take it into account.

GUSTAFSSON: I wonder if IR excesses for "normal" K giants are important even in the near IR? Grain formation in carbon star atmospheres is really problematic; at least for effective temperatures < 2500 K. The models predict temperatures cold enough for graphite and SiC grains to form in the outer layers. Probably, this extra important opacity cannot be treated within the framework of static models and possibly not even with models with stationary flows.

PETERS: I was delighted to hear your cautionary remarks concerning the use of the microturbulence parameter in the analysis of stellar spectra. For the early B stars, the meaning of this parameter is not entirely clear. It could be a real microturbulence, a non-LTE coverup, a reflection of an incorrect T-tau relationship in the atmosphere, etc. A curious result still prevails that the deduced microturbulence parameter for early B stars increases with increasing effective temperature and decreasing log g.

Temperatures for early B stars deduced from the continuous energy distribution observed in the region longward of the near ultraviolet are indeed strongly dependent on the assumed reddening correction because we are observing only the Rayleigh–Jeans part of the distribution. The situation is improved if we use flux data in the far UV to the Lyman limit where these stars radiate most of their energy. Recent Voyager UVS continuum observations of such stars from 900 – 1700 Å, which are being analyzed by Ron Polidan and myself, promise to be especially valuable for obtaining the temperatures of B stars and I refer you to our paper in this symposium volume for further details.

STANDARD VALUES AND INFORMATION IN DATA BANKS

C. Jaschek

Stellar Data Center, Strasbourg

ABSTRACT. This paper is an attempt to provide the background information
necessary to find data outside of one's specialized field. The most important computer readable catalogs pertaining to the different stellar
parameters are reviewed.

The purpose of my talk is to inform you of the ways in which you
may obtain stellar data outside of your specialized field. The fact
that specialists and non-specialists require different sources can best
be illustrated by the example of a photometrist who needs to know the
color of a star. As a specialist, he certainly would not want to use
an average color index, but would want to derive his own "best" averages,
taking the necessary elements from an observational compilation catalog
(OCC). On the other hand, a non-specialist, such as myself, would prefer a critical compilation catalog (CCC) for which a respected photometrist has sifted, judged, weighed, and averaged the observation, so
that my average or "best" values will be based upon an expert's recommendation. For instance, the La Plata catalog of MK types is an
observational compilation catalog (OCC) -it only compiles observations
without adding any additional information. Alternatively, the UBV catalog by Nicolet (1978) is a critical compilation catalog (CCC) -it gives
averages but provides no individual values. Even better would be a
catalog providing both individual data and averages, such as the uvbyβ
catalog by Hauck and Mermilliod (1980) (2057). (Readers wishing more
information on different catalog types will find a detailed discussion
in Jaschek (1984a)).
    In this paper, I will list and briefly discuss the different cata-
logs available for the various kinds of stellar data. Since the number
of existing catalogs is very large -for astrometry alone, Sevarlic et
al. (1978) have listed 2,087 for the 70's- my discussion is limited to
only the most recent compilation catalogs in each field and among these,
priority is given to those readable by machine. This is hardly a very
serious limitation if we consider that almost all compilation catalogs
are now available in this form.
    I suggest that we begin by describing the various ways of ob-

D. S. Hayes et al. (eds.), Calibration of Fundamental Stellar Quantities, 331–342.
© *1985 by the IAU.*

taining data for a given star. Let us consider the following; you have
a current designation of a star and want to know:
a) its observational parameters, and
b) the papers in which the star has been discussed.
You may contact, either by letter or telex, a data center such as the
CDS, or contact it through a computer network (TRANSPAC, TYMNET). You
may consult the microfiches of the "Catalog of Stellar Identifications
and Supplementary Data" by Ochsenbein et al. (1981) (4009), which provides cross identifications and data for 435,000 stars. It should be
clear that if you query the CDS, you will get the state of art at present, rather than for 1980 (575,000 objects). For a recent description
of the data base, see Ochsenbein (1984).

     With regard to star bibliographies, the CDS now has 5.5 x 10^5 ref-
erences to 1.1 x 10^5 stars. Non-stellar objects outside the solar system were included since 1983, and are now available upon request. For
Vega alone, there are references to 536 paper titles.

     After having obtained the main observational parameters and ref-
erences to the literature on the star, you may proceed to more specialized items, such as spectroscopic orbits, masses, and diameters. I
shall discuss some of these items in order of the different parameters,
although the order I use is somewhat arbitrary. Beside each catalog
title, you will find the author's name, the date, and a second number
in parenthesis. This reference number will enable you to obtain a tape
version of the catalog from any of the major data centers, namely:
- USA : NASA-ADC, GSFC, Code 680, Greenbelt, Maryland 20771
- USSR : Astronomical Council of the USSR Academy of Sciences, SADC,
           Pyatnitskaya ul. 48, Moscow 109017
- Japan : Kanazawa Institute of Technology, 7-1, Ogigaoka Nonoichimachi,
           Ishikawa 921
- German Democratic Republic : Zentralinstitut für Astrophysik,
           Rosaluxemburgstr. 17, 1502 Postdam
- France: Centre de Données Stellaires, Observatoire Astronomique,
           11, rue de l'Université, Strasbourg
All of these data centers work in close cooperation; they exchange cata-
logs so that you may use the most the convenient center and they use the
same numbering system. The numbers of objects in the catalogs which
are quoted have been rounded off.

POSITIONS AND PROPER MOTIONS

     The most widely-used catalog is the SAO (Smithsonian Astronomical
Observatory) by Haramundanis (1966) (1001). It provides positions and
proper motions for 259,000 stars in the whole sky. However, it should
be noted that the SAO is not complete down to a given magnitude since
in crowded areas some stars were omitted.

     For the northern sky, the AGK3 by Heckmann and Dieckvoss (1975)
provides data for 193,000 stars north of S = -2°. It comes in three
versions (1002, 1061, and 1969), each differing in detail. For a discussion of these catalogs, see Fresneau (1981). Nothing similar exists
for the southern sky.

Positions for about 3.6×10^5 stars are known to an arc second, but most of them are brighter than magnitude ten. Proper motions are known for about the same number of stars (Ochsenbein et al. 1981), but in the AGK2-3 only one-third have proper motions larger than three times its mean error (Fresneau 1978). For stars with very large proper motions one can use Luyten (1955) (1054), which provides 4,500 stars with annual proper motions $>0\rlap{.}''5/y$.

A general file on kinematic data of HD and HDE stars may be found in Mennessier et al. (1978). This catalog provides what are considered to be the best available data and contains 178,000 stars, of which 8×10^4 possess good proper motions.

PARALLAXES

For trigonometric parallaxes, one may use the Jenkins catalog (1963) (1081) which contains data for 7,300 stars. However, only a fraction of them have accuracies better than 10% or so. These are "nearby stars" compiled by Gliese and Jahreiss (1979) (5001, 5035). Gliese (1983) reports that about 700 stars have this level of accuracy ($\Delta M < \pm 0\rlap{.}^m3$). A new catalog by van Altena is forthcoming.

RADIAL VELOCITIES

The standard catalog in this field is Wilson (1953) (3021) which gives average radial velocities for 15,100 stars. Evans (1967) (3047) provides additional average radial velocities for 7,800 stars; however his survey is incomplete and I recommend that users also employ the OCC by Abt and Biggs (1972) (3004), which provides bibliographic references up to 1970, and Barbier (1975) (3038), which updates the literature as far as 1974. A summary of the data in these four publications may be found in Ochsenbein et al. (1981) (5026).

Barbier has announced in a private communication that a new edition of the bibliographic catalog is forthcoming. However, no catalog of average radial velocities is available. Radial velocities are known for about 27,000 stars. Standard stars are listed by Batten (1983).

PHOTOMETRY

The largest body of non-photoelectric measures is the 3.6×10^5 stars reduced to a homogeneous system by Ochsenbein (1974) (5026). These values sould be used whenever photoelectric values are unavailable.

Photoelectric measures are known for about 1.02×10^5 stars (Mermilliod, 1984). The forthcoming "General Catalog of Photometric Data" by Hauck will provide references to the system in which each of these stars was observed. At present, only a first version of this catalog by Magnenat (1976) (2039) is available. It provides data for 3.2×10^4 stars.

For stars observed in given systems, Egret and Philip (1979) (5016) have provided response functions for 33 photometric systems as well as a list of 800 stars used as standards in the seven most widely-used systems.

I shall now comment on catalogs providing information on objects observed in different photometric systems. Table I summarizes the situation to date.

TABLE I : Photometric catalogs

System	Author		Catalog	Number of stars	Notes
UBV	Mermilliod & Nicolet	(1977)	2035	53,000	measures
	Mermilliod	(1983)	2089	26,000	update of meas.
	Nicolet	(1978)	2051	59,000	averages
uvbyβ	Hauck and Mermilliod	(1980)	2057	20,000	measures and averages
Geneva	Rufener	(1981)	2072	14,600	averages
U_cBV	Nicolet	(1975)	2027	7,200	measures and averages
Vilnius	North	(1980)	2058	1,900	averages
DDO	McClure & Forrester	(1981)	2080	2,200	averages
RI	Jasniewicz	(1982)	2075	5,700	measures and averages
UBVRI..N	Morel & Magnenat	(1978)	2007	4,500	measures and averages

Infrared observations ($\lambda > 1\mu$) are reported in the "Catalog of Infrared Observations" by Gezari et al. (1982) (6020), which lists sources for 55,000 observations of about 3×10^4 objects. There are a number of source catalogs for the ultraviolet region of the spectrum – the most important to date is the "TD1 Four-Band Photometry Catalog" by Thompson et al. (1978) (2059) giving data for 31,000 stars. Other catalogs include the "Celescope Four-Band Catalog" by Davis et al. (1973) (2006) giving data for 5,700 stars, the "ANS Photometry Catalog" (six bands) by van Duinen et al. (1975) (2060) giving data for 3,600 stars and several more dealing with a smaller number of objects.

SPECTROSCOPY

For unidimensional spectral types, the most important collection of data in this field has been embodied in the "Henry Draper Catalog" by Cannon and Pickering (1924) (3001) covering 225,000 stars.

Two dimensional types (MK) have been compiled by Jaschek et al. (1964) (3018) for 21,000 stars, and by Kennedy (1983) (3078) for 36,500 additional stars.

Houk is currently reclassifying all HD stars by declination zones; the first volumes of this work are already available; vol. I (-90° to -53°), Houk and Cowley (1975) (3031); vol. II (-53° to -40°), Houk (1978) (3051); and vol. III (-40° to -26°), Houk (1982) (3080), providing types for 97,000 stars.

MK standard stars have been published by Morgan et al. (1978) and by Keenan (1983) for stars later than G0.

Mercedes Jaschek (1978) (3042) has produced a catalog of selected spectral types basing her "best" spectral types on the literature for 31,000 stars.

Spectral classifications for 1,900 stars based upon the ultraviolet region have been published by Cucchiaro et al. (1979) (3053) (TD1 experiment); and by Heck et al. (1984) (3083) for 230 stars, (IUE experiment).

Stars with spectral peculiarities are listed by Egret and Jaschek (1984) (3081); the catalog lists about 25,000 stars belonging to 53 different groups.

Stellar abundancies for 700 stars in the form of $[Fe/H]$ ratios are given in a bibliographic catalog by G. Cayrel et al. (1981) (3061). A new edition is forthcoming.

For further details on spectroscopic computer readable catalogs, see Jaschek (1984b).

SPECTROPHOTOMETRY

The two standard catalogs in this field are Breger (1976) (3048), which gives data for 940 stars, and Ardeberg and Virdefors (1980) (3069), which gives data for 380 stars. Glushneva's important catalog (1982), which gives data for 700 stars, does not exist on tape. A list of standards has been published by Glushneva (1983) and additional standards by Taylor (1981).

There are three important spectrophotometric catalogs for the ultraviolet region, Jamar (1976) (3039), giving data for 1,800 stars, and its complement, Macau-Hercot et al. (1978) (2086), giving data for 400 stars. The third is the IUE flux catalog by Heck et al. (1984) (3083), giving data for 230 mostly normal stars. There are a number of other catalogs providing data for smaller number of stars.

STELLAR ROTATION

The latest catalog is from Uesugi and Fukuda (1982) (3063). It provides average values for 6,500 stars. Standards are provided by Slettebak (1975).

DIAMETERS

Fracassini et al. (1981) (2061) have produced a bibliographic catalog of the existing determinations for 4,300 stars.

MASSES

Popper has edited a critical compilation catalog (1980), but it is not available on tape. For additional information see the section on "Binaries".

POLARIZATION

The only catalog available on this subject is Mathewson et al. (1978) (2034). It provides data for 7,500 stars but no update is available. A list of standards has been provided by Breger and Hsu (1981).

MAGNETIC FIELDS

The only catalog on this subject is Didelon (1983), which lists magnetic field measurements for 800 stars.

BINARIES

Worley's "Index Catalog" (1983) (1107) deals with visual binaries giving data for 70,000 objects. The catalog of individual measures is however not yet published. A new catalog of visual binary orbits became recently available (Worley and Heintz, 1983) (5039). The orbit catalog of Batten et al. (1978) (3016), dealing with spectroscopic binaries, provides almost 1,000 orbits. This catalog has been updated by Pedoussaut et al. (1979) (3065). Physical parameters derived from these data are given by Kraicheva (1980) (3057) for over 900 binaries. Eclipsing binaries are included in the catalog of orbits of close binaries by Svechnikov and Bessonova (1984) (3082) which gives data for about 300 pairs.

VARIABLE STARS

For variable stars, one may use the third edition of the "General Catalog of Variable Stars" by Kukarkin et al. (1971) (2011), providing data for 23,000 stars, and its supplement (suspected variables) by Kukarkin (1981) (2079), giving data for 15,000 additional stars.

A recent bibliography of variable stars by Huth and Wenzel (1981) (6022) provides 290,000 references for 28,000 variables.

CLUSTER DATA

The standard catalog of open clusters is Lyngå (1983) (7041), providing data for 1,200 objects. Globular clusters are dealt with in the bibliographic catalog of Ruprecht et al. (1983)(7044)(140 objects). Mermilliod (1976) (2029) takes on the more difficult problem of data for stars in open clusters, and provides UBV values and MK classifications for 8,900 and 2,300 stars, respectively, in 223 clusters. Another important catalog is Mermilliod's (1979) (3055) catalog of radial velocities in clusters, giving data for 800 stars in 78 open clusters. A new update is in preparation.

There is no general catalog to date which will enable us to separate open clusters into field clusters. However, Humphreys, et al. (1984) (7046) have taken a step in this direction by compiling a catalog which provides 2,300 objects in associations.

In conclusion, I would like to thank all those astronomers who have devoted their time to catalog making. All of us should regard this worthwhile occupation as a very significant contribution to astronomy, because without it, we would face the impossible task of trying to find information in the haystack of astronomical literature – a literature so dense that for 1982 alone, it was of the order of 300 books and 17,000 papers.

REFERENCES

Abt, H.A. and Biggs, E.S.   1972, *Bibl. of Stellar Radial Velocities*,
    Latham Corp., New York

Ardeberg, A. and Virdefors, B.   1980, *Astron. Astrophys. Suppl.* 40, 307

Barbier, M.   1975, 'CDS microfiche 3038' (*Bibl. Cat. Radial Velocities*)

Batten, A.H., Fletcher, J.M., Mann, P.J.   1978, *Publ. Dom. Astrophys.*
    Obs. 15, 121

Batten, A.   1983, *Bull. Inform. CDS* 24, 3

Breger, M.   1976, *Astrophys. J. Suppl.* 32, 7

Breger, M. and Hsu, J.C.   1981, *Bull. Inform. CDS* 23, 51

Cannon, A.J. and Pickering, S.   1924, *Harvard Annals* 91 to 100

Cayrel, G., Bentolila, C., Hauck, B. and Lovy, D.   1981, *Bull. Inform.*
    CDS 20, 112

Cucchiaro, A., Jaschek, M. and Jaschek, C.   1979, *Bull. Inform. CDS*
    17, 93

Davis, R.J., Deutschman, W.A., Haramundanis, K.L.   1973, *S.A.O. Special*
    Report 350

Didelon, P.   1983, *Astron. Astrophys.* 53, 119

Egret, D. and Philip, A.G.D.   1979, *Dudley Obs. Report* 14

Egret, D. and Jaschek, M.   1984, *Catalog of Stellar Groups*, Strasbourg

Evans, D.S.   1967, *IAU Symp.* 30, 57, eds. A.H. Batten and J.F. Heard,
    Acad. Press

Fracassini, M., Pasinetti, L.E., Manzolini, F.   1981, *Astron. Astrophys.*
    Suppl. 44, 155

Fresneau, A.   1978, *Bull. Inform. CDS* 15, 67

Fresneau, A.   1981, *Bull. Inform. CDS* 20, 110

Gezari, D., Schmitz, M., Mead. J.M.   1982,   NASA TM 83819

Gliese, W. and Jahreiss, H.   1979, *Astron. Astrophys. Suppl.* 38, 423

Gliese, W.   1983, *IAU Coll.* 76, 5, eds. A.G.D. Philip and A.R. Upgren

Glushneva, I.N.   1982, *Spectrophotometry of Bright Stars*,   Nauka,
    Moscow

Glushneva, I.N.   1983, *Bull. Inform. CDS* 24, 7

Haramundanis, K.L.   1966,   *SAO Catalog*, Smithsonian Inst., Washington,
    D.C.

Hauck, B. and Mermilliod, M.   1980, *Astron. Astrophys. Suppl.* 40, 1

Heck, A., Egret, D., Jaschek, M. and Jaschek, C.   1984, *IUE Low-Dis-*
    persion Spectra Reference Atlas. Part I. Normal Stars, ESA SP-1052

Heckmann, O. and Dieckvoss, W.   1975, AGK3, Hamburg

Houk, N. and Cowley, A.P.   1975, *Michigan Obs. Publ.*, vol. I

Houk, N.   1978, *Michigan Obs. Publ.*, vol. II

Houk, N.   1982, *Michigan Obs. Publ.*, vol. III

Humphreys, R.M., McElroy, D.B. and Ahigo, K.   1984, *Catalogue of Stars*
    in Stellar Associations and Young Clusters, University of Minnesota

Huth, H. and Wenzel, W.   1981, *Bull. Inform. CDS* 20, 105

Jamar, C., Macau-Hercot, D., Monfils, A., Thompson, G.I., Houziaux, L.,
    Wilson, R.   1976, *ESA Sp. Report* 27

Jaschek, C., Conde, H., de Sierra, A.C.   1964, *Publ. La Plata Ser.*
    Astr. 28, 2

Jaschek, M.   1978, *Bull. Inform. CDS* 15, 121

Jaschek, M.   1984a, *Quarterly J. Roy. Astron. Soc.* (*in press*)

Jaschek, C. 1984b, in *The MK Process and Stellar Classification*, R. F. Garrison, ed., David Dunlap Obs., Toronto, p. 94.

Jasniewicz, G. 1982, *Astron. Astrophys. Suppl.* 49, 99

Jenkins, L.F. 1963, *General Catalog of Trigonometric Stellar Parallaxes and Supplement*, Yale University Observatory

Keenan, P.C. 1983, *Bull. Inform. CDS* 24, 19

Kennedy, P.M. 1983, *MK Classification Catalogue*, Mt. Stromlo and Siding Spring Observ., Australia

Kraicheva, Z., Popova, E., Tutukov, A., Yungelson, L. 1980, *Bull. Inform. CDS* 19, 71

Kukarkin, B.V., Kholopov, P.N., Efremov, Yu.N., Kukarkina, N.P., Kurochkin, N.E., Medvedeva, G.I., Perova, N.B., Fedorovich, V.P., Frolov, M.S. 1971, *General Catalogue of Variable Stars*, Moscow State University

------------------ 1981, *Supplement to the General Catalogue of Variable Stars*, Moscow State University

Luyten, W. 1955, *A Catalog of 1849 Stars with Proper Motions Exceeding 0."5 Annually*, Lund Press, Minnesota

Lynga, G. 1983, *Catalog of Open Cluster Data*, Lund Observatory

Macau-Hercot, D., Jamar, C., Monfils, A., Thompson, G.I., Houziaux, L., and Wilson, R. 1978, *ESA Sp. Report* 28

McClure, R.D. and Forrester, W.T. 1981, *Publ. Dom. Astrophys. Obs.* 15, 439

Magnenat, P. 1976, *Bull. Inform. CDS* 11, 17

Mathewson, D.S., Ford, V.I., Klare, G., Neckel, Th., and Krautter, J. 1978, *Bull. Inform. CDS* 14, 115

Mennessier, M.O., Gomez, A., Crézé, M. and Morin, D. 1978, *Bull. Inform. CDS* 15, 83

Mermilliod, J.C. 1976, *Astron. Astrophys. Suppl.* 24, 159

Mermilliod, J.C. and Nicolet, B. 1977, *Astron. Astrophys. Suppl.* 29, 259

Mermilliod, J.C. 1979, *Bull. Inform. CDS* 16, 2

Mermilliod, J.C. 1983, *Bull. Inform. CDS* 25, 79

Mermilliod, J.C. 1984, *Bull. Inform. CDS* 26, 3

Morel, M. and Magnenat, P. 1978, *Astron. Astrophys. Suppl.* 34, 477

Morgan, W.W., Abt, H.A., and Tapscott, J.W. 1978, *Revised MK Spectral Atlas for Stars Earlier than the Sun*, Yerkes and Kitt Peak Obs.

Nicolet, B. 1975, *Astron. Astrophys. Suppl.* 22, 239

Nicolet, B. 1978, *Astron. Astrophys. Suppl.* 34, 1

North, P. 1980, *Astron. Astrophys. Suppl.* 41, 395

Ochsenbein, F. 1974, *Astron. Astrophys. Suppl.* 15, 215

Ochsenbein, F., Bischoff, M. and Egret, D. 1981, *Astron. Astrophys. Suppl.* 43, 259

Ochsenbein, F. 1984, *Bull. Inform. CDS* 26, 75

Pedoussaut, A., Bourdoncle, P. and Capdeville, A. 1979, *Bull. Inform. CDS* 17, 21

Popper, D.M. 1980, *Ann. Rev. Astron. Astrophys.* 18, 115

Rufener, F. 1981, *Astron. Astrophys. Suppl.* 45, 207

Ruprecht, J., Balazs, B., and White, R.E. 1983, *Soviet Astronomy* 27, 358

Sevarlic, B.M., Teleki, G. and Szadeczky-Kardoss, G. 1978, *Publ.*

Beograd 7
Slettebak, A.   1975, *Astrophys. J. Suppl.* <u>29</u>, 137
Svechnikov, M.A. and Bessonova, L.A.   1984, *Bull. Inform. CDS* <u>26</u>, 99
Taylor, B.J.   1981, *Astrophys. J. Suppl.* <u>54</u>, 259
Thompson, G.I., Nandy, K., Jamar, C., Monfils, A., Houziaux, L.,
    Carnochan, D.J., and Wilson, R.   1978, *Catalog of Stellar Ultra-
    violet Fluxes*, Science Research Council 1978
Uesugi, A. and Fukuda, I.   1982, Revised Catalog of Stellar Rotational
    Velocities, Dept. of Astron., Kyoto Univ.
van Duinen, R.J., Aalders, J.W., Wesselius, P.R., Wildeman, K.J., Wu,
    C.C., Luinge, W. and Snel, D.   1975, *Astron. Astrophys.* <u>39</u>, 159
Wilson, R.F.   1953, *General Catalog of Stellar Radial Velocities*,
    Carnegie Inst. Washington, Publ. 601
Worley, C.E. and Heintz, W.D.   1983, *Publ. Naval Obs. Washington*
    2nd ser., vol. <u>24</u>, 7
Worley, C.E.   1984, *Index Catalogue of Visual Double Stars*, Washington
    D.C.

DISCUSSION

GLUSHNEVA: With respect to your reference to the Sternberg Institute spectrophotometric catalogue published by Glushneva, et al., the catalogue now includes 867 stars and 7 standards connected directly and carefully with the main spectrophotometric standard Alpha Lyrae.

HAUCK: You have forgotten one catalogue: the Catalogue of Catalogues of the Information Bulletin of the CDS at Strasbourg.

JASCHEK: Yes, indeed!

GRIFFIN: I should like to make two comments. First, I would like to encourage you to keep the Bibliographical Star Index up to date. At present there is a gap of five or six years since the most recent published Index. As a journal editor, some time ago I accepted a paper sent from the CDS containing an analysis of the 1983 Index. I would have preferred to see the Index itself rather than the analysis of it. Please understand that this comment is by way of encouragement and not criticism - we are all very grateful for the existence of the BSI.

JASCHEK: We have a backlog of the microfiches for the BSI, but the BSI itself is up to date. Bibliographical references become available immediately after incorporation, which is one about week after the journal arrives at the library.

We welcome queries of the CDS by TELEX, letter or computer link.

GRIFFIN: My other point concerns MK types. I noticed that you did not refer to Buscombe's catalogues, and those catalogues, which have no references, are for many purposes unsatisfactory. Is there any way of retrieving MK classifications made in the last ten years?

JASCHEK: Yes, the 1983 edition of the Kennedy catalogue is available from the data centers. It covers the MK literature up to 1982.

GLIESE: Thank you for your very clear description of the situation concerning radial velocity catalogues. As the compiler of a new catalogue of nearby stars I urgently need radial velocity data not necessary with an accuracy of 0.1 km/sec but with 1 km/sec, which is quite sufficient for combination with tangential velocities. How do I find the best data available today, that is, with one value for each star for which different measurements are published? What procedure is recommended?

JASCHEK: This is a task beyond those of any data center; it must be asked of the specialists of velocities. In the case of the CDS I have always refused to get involved in cataloging if no specialist is doing it; this is the only way to guarantee the quality of the catalogues we distribute.

BATTEN: I should mention the concern of a group at Marseille with the production of a new catalogue of radial velocities and the definition of a "best" value of radial velocity.

JASCHEK: Yes, a catalogue is in preparation, but I do not know when it will be ready.

NORDSTRÖM: With respect to Gliese's question about how to find radial velocities for the nearby stars, I would like to mention that a large catalogue, Radial Velocities of 790 Bright Stars (Andersen, et al. 1984), has just been submitted to the Astron. and Astrophys. Suppl. Series. This catalogue contains accurate CORAVEL velocities for all late-type stars in the Bright Star Catalog without previously published velocities. Preliminary values for some of these stars were included in the last edition of the Bright Star Catalogue.

WALKER: While I certainly applaud the accumulation of data into catalogues I think it can lead to problems. The rotational velocity catalogue of Uesugi and Fukuda attempts to put rotational velocities from a number of sources into a single system using linear transformations, but, in my view, many of the transformations are incomplete and as a result the original quality of the data has been lost. It is important, in my view, that the appropriate commissions of the IAU have input when such catalogues are being prepared.

JASCHEK: We accept what the specialists provide us with, but we encourage other specialists to provide us with different answers for the same parameters.

LYNGÅ: In my catalogue I have tried to select references rather than taking mean values. For example, the [Fe/H] for a cluster is a single determination with its reference. It must be in the vein of the purpose of this symposium that we have data, the calibration of which are known. I don't recommend the other method, that of taking mean values. Particularly, I hope this will guide the selection of radial velocity data.

JASCHEK: Again, it is entirely up to the specialists to determine what they will put into the catalogue!

SCARFE: One matter that is somewhat outside the scope of the specialists is that of cross references between catalogues. Does the CDS provide this kind of information?

JASCHEK: Yes, indeed. The "Catalogue of Stellar Identifications" provides cross identifications for a half million stars and carries up to 40 different designations for a given star. Automatically, when you ask the CDS for data concerning a given star, you also get all the cross-identifications.

ROUNTREE: Do you have any advice for the compilers of catalogues as to the best identification numbers to use for their stars?

JASCHEK: Yes, indeed. 1) Provide always <u>two</u> identifiers; one can be the position, at a specified epoch. 2) Explain the catalogue abbreviations, using the "first dictionary" by Fernandez, Loret and Spite (A & A Suppl. 1983). 3) If you list newly discovered objects, do not call them 1,2,...n but introduce a catalogue abbreviation for future use. Call them CJSO1 (Carlos Jaschek, Strasbourg Observatory first object), because otherwise different people will refer in different ways to them, causing unnecessary confusion. 4) If you publish charts, provide the scale, orientation and magnitude of the stars. The leading astronomical journals have recently agreed to act along these lines.

MILLWARD: On a number of the microfiches available from the CDS the objects are listed by CSI number. Now, while there is an HD to CSI cross reference available on fiche, the CSI numbers are not in any order in the catalogues. This makes it impossible to find objects from the fiche at our own institutions without writing to the Data Center.

JASCHEK: We will look into that.

CALIBRATION OF FUNDAMENTAL STELLAR QUANTITIES: SUMMARY AND CONCLUSIONS

Gösta Lynga

Lund Observatory

WHAT THE SYMPOSIUM WAS ABOUT

At a time when so much of the prestigious astronomical research is devoted to new and highly speculative phenomena, it has been rewarding to see and hear the cautions and accurate approach to the measuring and interpreting of stellar quantities. It has been shown by many speakers, perhaps clearest of all by our participant of honor Professor Popper, how insufficient attention to calibration problems can lead to erroneous results.

We have listened to many important discussions of the details of the determining of different quantities. I am proposing not to repeat all this but rather to concentrate on those aspects that deal with the calibration of the data.

It may at this time be appropriate to analyse what we have meant by our title, the four major words of which have been interpreted differently by different participants:

Calibration   relating of quantities inside a system to other quantities outside that system
or
the process of keeping a system autonomous

Fundamental   a primary characteristic of the star itself
or
a basic parameter known to the observer but not to the star

Stellar   referring to all stars
or
referring to all stars except the Sun

Quantities   numbers
or
numbers, classifications, descriptions

D. S. Hayes et al. (eds.), Calibration of Fundamental Stellar Quantities, 343–349.
© 1985 by the IAU.

STANDARD STARS AND STANDARD SYSTEM

The symposium was given an ideal start by the chairman of the
scientific organizing committee who told us what requirements a standard star should fulfill and that a reference star need not necessarily be a standard star. The obvious requirement, that a standard
star is constant in the relevant characteristic, is not always easy
to check but all the same quite fundamental. Batten also discussed
the characteristics of a system of standards and pointed out that
the standards should be suitable for the testing of the experimental
set-up. He also pointed out that it is unfortunately necessary to
have different standard stars for different quantities. However,
Batten concluded, we may in the future not need standard stars at
all; hopefully we will be able to measure the quantities directly.

Further analysis of standard star systems was made by Garrison,
particularly with application to the MK system but in its philosophical
approach of far-reaching consequences. In that review as well as in
Keenan's review it came out quite clearly that the MK system provides
an autonomous description of stellar spectra classified without regard
to other observational data. Precisely this property of independence
makes it useful in the study of stars by confrontation with other data.
The construction of the system, given by the standards in Keenan's
paper, is a model of calibration. It is also interesting to see how
this method of spectrum description has evolved to contain more information without changing its basic layout.

One more point made by Garrison is worth repeating and that is
that the large telescopes now available will probably contribute very
little to the calibration of fundamental data, merely because of the
short, incidental runs that are normally granted. The important work of
establishing standard systems has to be carried out elsewhere and one
can only hope that large telescope observations are satisfactorily tied
in with such systems.

Carlos Jaschek described the enormous quantities of information
that are available from the data centers of the world. The degree of
collaboration between them is gratifying. Great credit is due to
Jaschek and to the Centre des Données Stellaires, CDS, where so much of
the compilation has taken place. It has always been his and his collaborators aim to make this material calibrated, standardised and
unambiguous. Yet, the task is a very difficult one and it is only in
certain sectors that they have succeeded so far. I think the astronomical community should wish the CDS the best of luck but also be
willing to help in the endeavor to make the data handled by these data
banks homogeneous and trustworthy. The danger of a very efficient
procedure spreading second-rate data is obvious. Jaschek's review paper
showed us the difference in concept between the observational compilation and critical compilation catalogues; the realisation of this
difference is an important step and we should all appreciate that the
large data bank of CDS is handled in such a responsible way.

Several papers dealt with radial velocity standards, notably the
one by Scarfe who investigated the constancy of the IAU standards and
the one by Walker et al. who studied the precision attainable for

measurement of radial velocity variations particularly for observations with Reticon arrays.

The issuing of a microfiche containing information on standard stars was proposed by Davis Philip in a poster paper and was subsequently discussed a number of times during the meeting. The plan adopted was that a prototype should be made by Davis Philip in collaboration with the CDS and circulated among the participants for comments. The microfiche would eventually become an enclosure of the present volume.

Problems regarding calibrations for Space Telescope (ST) were discussed by Bohlin who invited comments about suitable standard sources and standard stars.

In the poster session the Griffins presented a tracing of a high-dispersion (10Å/mm) spectrum of a B7 star with line identifications. More such useful displays are planned for other spectral types.

EVER MORE ACCURATE MEASUREMENTS

Several review papers dealt with questions of positional astronomy. Corbin described how transit circles need to be accurately calibrated and how the FK5 is envisaged to go faint and to avoid some of the systematic errors of FK4. The relating of the system to faint objects, particularly counterparts of radio sources, was also discussed, as was the link with the Hipparcos and ST observations.

Upgren showed how the accuracy of trigonometric parallaxes has increased dramatically during the last decade. The main reason is the use of modern measuring machines, particularly the PDS machines but also Starscan at US Naval Observatory. This is so much the more important as it allows the ever more precise stellar angular diameters to be translated into linear diameters, one of fundamental stellar quantities. As was pointed out by Hanbury Brown, the trigonometric parallaxes are less accurate than the angular diameters for the 10 stars where both quantities are known. In the field of parallax measurements much hope is attached to the Hipparcos satellite which will measure stars down to 12th magnitude and will be complete to V = 8. In a poster paper Gliese and Jahreiss gave a luminosity calibration of the lower main sequence on the basis of available trigonometric parallaxes. The importance of making calibrations valid for constant volume was discussed several times during the symposium. In any case, it should always be clear whether a calibration is based on material selected according to apparent magnitude or from a volume of space.

During the last twenty years, the greatest leap towards understanding the fundamental stellar properties has been the determination of angular diameters. Three basic ways have been used:

1. The intensity interferometry, a technique which has yielded 32 diameters for early type stars.
2. Occultation observations, particularly lunar occultations, which have given information on stellar diameters for late type giants.

3.   Speckle interferometry, which has yielded diameters of a number of
     late type stars; this technique came into use at the time when the
     Narrabri interferometer already had finished its diameter
     program.

     The future development may depend on the amplitude interferometer
planned by the Sydney University group. The prototype is nearing
completion and if this is successful we might look forward to the
building of a 1 km instrument capable of observing stars down to V > 7.
Davis outlined this and he also pointed out that both occultation
methods and speckle interferometry are inherently limited in comparison
with the Michelson interferometer. Resolutions down to $4 \cdot 10^{-5}$ arcsec
are expected for OB stars and $2 \cdot 10^{-4}$ arcsec for other stars.
     In the discussion following this paper, Evans described the
achievements and the capacity of the lunar occultation method, particularly useful for late type stars. Evans also pointed out that when
observing a lunar occultation of a binary from different sites, the
different geometric aspects give added information for determination of
the separation and position angle.
     Statistical methods of obtaining parallaxes using the principle of
maximum likelihood were presented by Heck in a poster. Results for
absolute magnitude calibrations were given.

QUANTITIES KNOWN TO THE STARS

     Masses   and   luminosities   are   fundamental   quantities,   the
determinations of which have been the subject of several papers at
this symposium. McAlister reviewed the interferometric studies of
binary orbits, particularly methods of speckle interferometry. The
addition of data by interferometric studies of binaries, particularly
at the high-mass end of the mass-luminosity relation, have made a
large impact. Although speckle observations concern very close binary
pairs there is a certain overlap with orbits from visual binaries.
Comparisons that can be made show no systematic differences.
     On the question of a standard orbit for calibration purposes,
McAlister discussed Capella's orbit but even this could not fill a
requirement of 1% accuracy. An alternative would be a set of 21 standard stars with well-known separations and position angles which, if
observed frequently could yield a set of standard orbits. For the
future of high resolution studies McAlister pointed out the great
potentials of long baseline interferometry.
     The arduous task of determining masses from astrometric observa-
tions of binaries was described by Heintz. Although the new catalogue
contains 850 orbits, only about 40 stars have good masses since high
precision data are required for several different quantities: mass
ratios, parallaxes, orbits, inclinations etcetera. A good mass calibration of the lower main sequence could, however, be made with these
data.
     Popper gave a lucid review of practices in evaluating eclipsing
binary observations. The determination of masses, radii and luminosi-

ties from spectroscopic and photometric observations requires careful attention to correct practices. He showed with several examples how insufficient inspection of the observed spectra or non-standard photometry can lead to erroneous results, the presence of which in the literature can be of disadvantage. Computer usage is of particular value in studying photometric light-curves where different model fitting parameters can be tried out. A similar situation does not exist in the spectroscopic analysis which is more straight-forward. While the matter of determining good masses, radii and luminosites for eclipsing binaries is mainly a question of high standard of work, for the observing quantities the calibration is of highest importance.

In one poster paper Bell et al. studied the surface gravity spectroscopically for Arcturus and were able to deduce a mass - the one method available for mass determination of single stars.

In his review over the measurement of stellar rotation Slettebak described the principles and achievements of the three different methods:

1. Modulation of light by rotation when the stellar surface has an uneven brightness.
2. Distortions in radial velocity curves of eclipsing binaries.
3. Effects on spectrum line profiles.

These methods are all old, suggested before this century, and have yielded a lot of data. The third method is now also used in Fourier analysis of line profile and in CORAVEL observations, where the width of the minimum gives information on rotation. It appears that the Fourier methods make the most use of the line profile information but that their application requires a good understanding of the light distribution over the stellar disk.

PHOTOMETRY, SPECTROSCOPY AND THE UNDERSTANDING OF SUCH DATA

The review by Rufener gave a good insight into the problems of photometric calibration. Particularly the problems of changing detector technology were discussed. Rufener made a distinction between the stars that are used for comparison purposes in the reduction of the observations and the standard stars. A standard system should contain stars in all areas of the HR diagram and be used to understand the final result of each stellar observation.

While Rufener's concern was to give an internally consistent photometric system, the paper by Hauck tried to relate photometric data from a number of systems to determinations of effective temperature. The calibrations of B-V and B2-V1 from the UBV and Geneva systems in terms of effective temperature were given for unreddened stars. A catalogue of 104 stars were proposed as a standard for T_{eff} calibrations.

Several important contributions concerned the calibration of the uvby system. Three different papers by Ardeberg and Lindgren, by Antonello and by Nelles et al., all showed endeavors to push the uvby system towards later spectral types. Manfroid has investigated the effect on the uvby system of rectangular spectral response curves, such as will be achieved by a dispersing photometer with slits but without filters. He suggests the use of two different uvby systems.

Regarding the absolute calibration of photometry we heard a most impressive account of the state of affairs by Hayes. He first concluded that despite a number of other indications, Vega is constant and good for use as a primary standard. The calibration of Vega against black-body sources on the Earth has been carried out by several groups and the results agree in the visual region to within about $0^m.01$, in the infrared somewhat less well. The absolute flux calibration is made with an accuracy of about 1½ percent. Comparisons between Vega and the Sun are very difficult but similar results have been achieved by several groups. The fact that the B-V value for the Sun thus determined does not agree with the B-V for stars with the spectral class (G2V) for the Sun could be indicative of a peculiarity in the Sun's color, but the low precision in the B-V determinations make such a conclusion premature. It would be of value to study the Sun's color also in other photometric systems. One of the high precision data series on solar radiation referred to by Hayes, was the measurement by Neckel and Labs which was presented as a poster at this symposium.

Giusa Cayrel gave an inspired talk about high-dispersion spectro-scopy, particularly for the Hyades stars where the dispersion between [Fe/H] values is now much lower. It is quite clear that Reticon obser-vations have a tremendous potential, but at the same time this is obviously a detector that as yet needs to be perfected. The long discussion about what S/N really means when such a detector is discussed would have been better among detector specialists. We congratulate Cayrel on her results and wish her better and better oscillator strengths to work with.

The usefulness of M67 as a standard of reference was pointed out by Janes. This open cluster is very well studied and its composition and age are very close to those of the Sun. Physically it is more similar than the Hyades to the Sun and its distance can be determined directly.

Straižys discussed the different photometric systems in terms of metallicity indices. For several of the systems reddening effects can not easily be removed while other systems are so designed as to estab-lish a valid metallicity scale even in the presence of reddening effects. However, Straižys pointed out the lack of model atmosphere data, general enough to provide calibration of metallicity indices.

Gustafsson showed us, on the other hand, that the state of the art in model atmosphere calculation is high - that a lot of the discrepan-cies that were present earlier now have been removed through better programs which take more lines into account, through a better under-standing of the geometry of the atmosphere and through other, more reasonable parameters. Some effects are less clearly understood. Recent studies of microturbulence and convection need to be introduced. As a whole the situation is improving fast with better observations, also in UV and IR, with better computational programs and with new physical data for input to the models.

Several poster papers gave results from model atmosphere work in terms of expected photometric data. Bessell and Scholz had calculated colors from models for M giants and supergiants with different abundances. Buser and Kurucz attempted to calibrate the UBV system in terms of physical parameters.

Code told us about the problems with calibration of UV observations and also about the usefulness of UV data; the UV colors discriminate better than the optical between stellar temperature classes and they measure the interstellar extinction better. Also, the addition of UV colors should help in the study of several astrophysical quantities, temperature, surface gravity, metallicity. Analysis of composite spectra is another area of future space research.

HOPES FOR THE FUTURE

It seems quite clear that vast improvements in the quantity and quality of available data can be expected for the next ten-year period for:

    trigonometric parallaxes
    angular diameters
    radial velocities
    high dispersion spectroscopy
    model atmospheres

Let me end by a quotation from D.M. Popper: "The heart of the matter is care and judgment."

CONTRIBUTED PAPERS

THE MICROFICHE OF STANDARD STARS

A. G. Davis Philip

Van Vleck Observatory and Union College

Daniel Egret

Stellar Data Center, Strasbourg

We are preparing a microfiche, at the Stellar Data Center in Strasbourg, concerning standard stars in various systems which will be included with the proceedings of this meeting. In the first part of the microfiche we will present a list of stars which are standards in spectroscopic, photometric systems or for which fundamental determinations of temperature, gravity, radius or mass have been made or for which spectrophotometry has been done, fluxes measured or [Fe/H] determined. If a star is a standard in one of these systems an X will be placed in the column opposite its name.

Part two of the microfiche will contain the lists of the standard stars in each system, with V magnitude, positions (1950) and the parameters of each system listed. Part three of the microfiche will contain lists of the standard stars in each system, sorted by the parameters of the system so that on can pick out stars with specified values of the parameters.

The systems originally picked for coverage are:

Spectroscopic Standards

      1. MK Types
      2. Radial Velocities
      3. V Sin i
      4. Equivalent Widths

Photometric

      1. UBV
      2. Four-Color
      3. Geneva
      4. DDO
      5. H_α, H_β, H_γ
      6. Infrared

D. S. Hayes et al. (eds.), Calibration of Fundamental Stellar Quantities, 353–356.
© *1985 by the IAU.*

                    7.   Ultraviolet
                    8.   Polarization
                    9.   Vilnius
                   10.   Walraven

Fundamental Parameters

                    1.   Temperatures
                    2.   Gravity
                    3.   Radius
                    4.   Mass

Other

                    1.   Flux Measurements
                    2.   Spectrophotometry
                    3.   [Fe/H]
                    4.   Magnetic Stars

Appendix

                    1.   Solar Analogs
                    2.   Binaries with well studied properties
                    3.   Clusters with well derived distances

     Many of these lists are already on file at the Stellar Data
Center, other lists have been requested during the last few months.
An effort will be made during the meeting to complete the lists and
to take suggestions concerning the data to be displayed in the
microfiche. Please contact either author if you wish to make a
comment. Reidel has agreed to include the microfiche in a pocket on
the back cover of the proceedings volume.

DISCUSSION

PHILIP: The original plan was to have a preliminary version of the microfiche ready in time for this meeting. This has not been possible, but the Stellar Data Center is working on preparing the microfiche of standard stars. When a preliminary version is ready, copies will be mailed to all participants of the symposium so that they will have a chance to comment on the organization of the microfiche and the data that it contains.

The photometric standard star data are pretty well in hand. Infrared photometry does not yet have suitable standards. We have sources for spectroscopic standard data (radial velocities - Batten, Vsini - Slettebak, equivalent widths - Cayrel). Concerning the listing of stars with fundamental parameters determined we should list "stars with well determined values" of mass, temperature, gravity etc. The review article by Popper can be used for most of these, supplemented by the list by Davis of well observed angular diameters. Other quantities with which we are concerned are flux measurements and [Fe/H]. Dr. Cayrel has promised a list of the best observed stars for abundance.

BESSELL: The Cayrel catalogue has no "censorship"?

PHILIP: Only the very best observed stars are being selected. In the an appendix to the microfiche we will list solar analogs, binaries with sell-studied properties and clusters with good distances.

JASCHEK: The CDS is willing to produce the microfiche and to add it to the proceedings of this symposium. We would like to get from the astronomers present explicit instructions as to what lists of standards to use, in order to complete work on the microfiche as soon as possible. This is not an attempt to force anyone to use any set of standard values, it is only to help astronomers locate standard data.

PHILIP: I agree. This information is important to us.

HAYES: You have mentioned catalogues of spectrophotometry; Breger, or Ardeberg and Virdefors. You should use the lists of secondary standards by Taylor and by Glushneva.

JASCHEK: Yes.

GARRISON: It would help if you would classify the lists as primary standards, secondary standards, etc. instead of generally listing the stars as standard stars.

JASCHEK: I emphasize that the lists of standard stars must be prepared by specialists.

MISSANA: For the spectra of fundamental stars I would express the hope that in the future accurate wavelengths, central intensities and equivalent widths will be available.

BESSELL: Concerning IR standards, the AAT–CTIO values of Koorneef are available.

HAYES: I said that because the major systems are very filter dependent and they differ from observatory to observatory. There are several of them and certainly the Kitt Peak system is a major one. But the filter transmissions and list of stars has not been published. The others, as you say, are available and have been published. Koorneef's is a compiled system and attempts to reproduce the Johnson system, so it is different from the others. I have been encouraging my colleagues to prepare the data for publication.

BOHLIN: How about including UV spectrophotometric standards in the microfiche of standards? I suggest the use of the stars which have been well observed during the first year of IUE operations. The data obtained during this time period was used to define the IUE calibrations, so that the question of IUE sensitivity changes is not relevant. These data cover the range 1160 to 3250 Å. The appropriate stars are:

HD 60753	B3
HD 93521	O9 Vp
BD +28° 4211	Op
BD +75° 235	sdO
BD +33° 2642	B2 IV

Extension of fluxes to shorter wavelengths might be provided by Polidan for the hotter of these sources. Errors in these fluxes should not exceed about 10% for the wavelengths longward of 1250 Å.

STANDARD ASTRONOMICAL SOURCES FOR THE SPACE TELESCOPE

Ralph C. Bohlin

Space Telescope Science Institute

ABSTRACT.   The Space Telescope (ST) will require many types of
standard sources for a diverse range of calibrations to be performed
after launch.   The scientific instruments are sensitive to a wide
range of wavelengths from 1050 to 11,000Å and encompass a broad range
of measurement capabilities including astrometry, photometry, imaging,
polarimetry, and spectroscopy.   To verify proper operations of each
instrument and to provide quantitative calibrations, a diverse range
of standard sources and fields are required.   In order to select
targets that satisfy the requirements of the Instrument Definition
Teams and the long term responsibilities of the Science Institute, six
groups containing a total of 25 astronomers are defining the
calibration targets to be observed after launch.   The six categories
of ST standard sources are:

1)   Ultraviolet Spectrophotometric
2)   Ground Based Spectrophotometric and Photometric
3)   Wavelength
4)   Astrometric
5)   Polarimetric
6)   Spatially Flat Field

The data in these categories will be collected from the literature or
through new observing programs as appropriate.   These six reports of
the working groups outline the calibrations and proposed targets for
all of the scientific instruments on ST.   The collected data on each
set of standard sources should be published in the refereed
literature.

1.   CALIBRATION TARGETS

The background material for the choice of calibration targets comes
from the individual instrument teams, which have the responsibility
for calibrating their scientific instrument during the first six months
after launch. At the end of the six-month period, calibration becomes
the responsibility of the ST ScI.   Therefore, cooperation and joint

D. S. Hayes et al. (eds.), Calibration of Fundamental Stellar Quantities, 357–360.

planning of calibration efforts is imperative for all concerned parties. Through an intercomparison of all of the team plans, the centralization of the calibration planning at the ST ScI leads to the identification of omissions and the elimination of redundant targets that appear in the calibration plans developed by each team acting alone.

Table I contains the list of calibration target categories and the scientists who are contributing. The first name listed is the scientist at the ST ScI who is leading the efforts to obtain a consensus on the requirements for each instrument and to identify a minimal set of targets that satisfy the complete matrix of requirements. Presently, the preparation of the combined target lists is about 50% complete. After distribution of these proposed target lists, some changes are expected as new ideas come into the ST ScI from the astronomers in the ST project and from the general community of potential ST users.

2. RESULTS

Currently, data has been collected only for the category of UV spectrophotometry. Using the IUE observatory, the spectra shown in Figure 1 were obtained in December of 1983. The goal of the IUE program is to obtain spectrophotometry of hot stars in low dispersion from the bright limit of third mag to the faint limit of ~15 mag with a spacing of about one mag. This ST standard star program has additional observing time scheduled for 1984 at the ESA and the NASA stations.

Notes to Table I.

FGS - Fine Guidance Sensor. This instrument is being developed by the Perkin Elmer Corporation for the primary purpose of fine pointing the Space Telescope. The FGS can also be used for astrometry, and these scientific uses are being directed by an astrometry team led by W. Jefferys at the University of Texas.

FOC - The Faint Object Camera has been developed by the European Space Agency. P. Jakobsen is the Project Scientist, replacing F. Macchetto, who is now the science team leader.

FOS - Faint Object Spectrograph. The principal investigator is R. Harms from the University of California at San Diego and the Applied Research Corporation.

HRS - High Resolution Spectrograph. The principal investigator is J. Brandt of the Goddard Space Flight Center.

HSP - High Speed Photometer. The principal investigator is R. Bless of the University of Wisconsin.

WFPC - Wide Field and Planetary Camera. The principal investigator is J. Westphal of the California Institute of Technology.

TABLE I

ST CALIBRATIONS REQUIRING ASTRONOMICAL TARGETS

Category of Standard Source	Affected Instrument	Study Team Members
1. UV Spectrophotometric	FOC, FOS, HRS, HSP, WFPC	R.C. Bohlin, J.C. Blades, A.V. Holm, B.D. Savage, C.C. Wu
2. Ground Based Spectrophoto-metric and Photometric	FGS, FOC, FOS, HSP, WFPC	J. Koornneef, W.A. Baum, R.C. Bohlin, J.F. Dolan, J.B. Oke, D.A. Turnshek
3. Wavelength	FOC, FOS, HRS, WFPC	H.C. Ford, L.M. Hobbs, D.G. York
4. Astrometric	All	A. Fresneau, P.D. Hemenway, B.G. Marsden, P.K. Seidelmann, W.F. van Altena
5. Polarimetric	FOC, FOS, HSP, WFPC	H.S. Stockman, J.R. Angel, A.D. Code, J.F. Dolan, O.L. Lupie, R.L. White
6. Spatially Flat Fields	FOC, FOS, HRS, HSP, WFPC	R.E. Griffiths, P.D. Feldman, J.E. Gunn

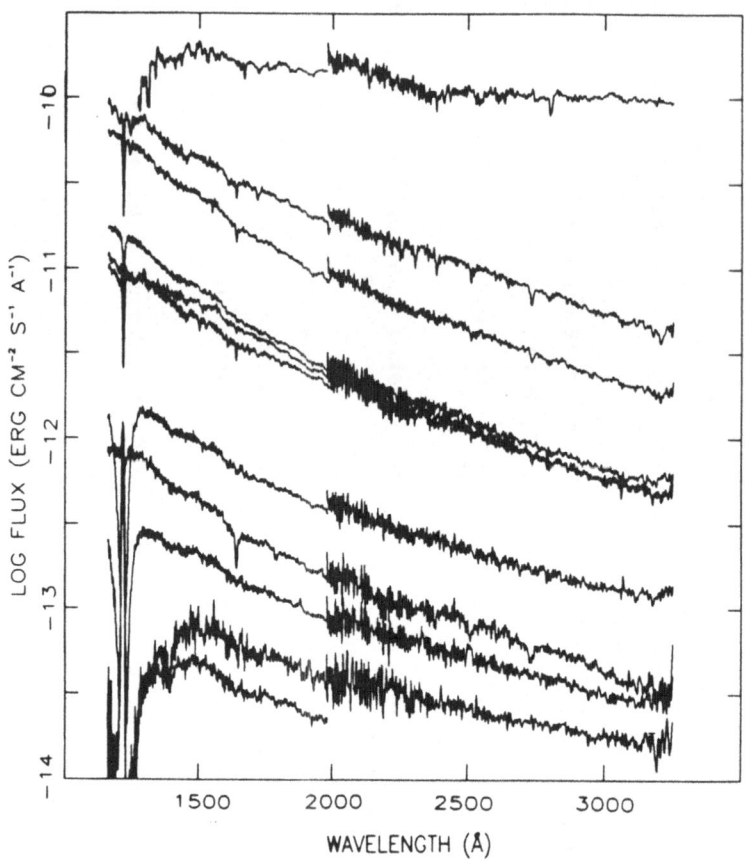

Fig. 1. Preliminary UV spectrophotometry of ST standard stars from the IUE spacecraft. The fainter stars were obtained at the Vilspa site using ESA time, while the brighter targets were observed at GSFC on NASA time.

DISCUSSION

JASCHEK: What are the wavelengths you are concerned with? Do you have the list of wavelengths or the identification of the wavelengths in the stellar spectra?

BOHLIN: We are seeking stars with well determined wavelengths. The accuracy we need is about 1 km/sec. The HRS observes at wavelengths below 3200 Å so we need a hot star, such as a B-star. Alternatively, we may do better with interstellar lines.

THE OPEN CLUSTER M67 AS A FUNDAMENTAL STANDARD OF REFERENCE FOR STELLAR PROPERTIES

Kenneth Janes

Boston University

ABSTRACT. With the possible exception of the Hyades, M67 is the best-studied star cluster. Accurate photoelectric photometry exists well down onto the main sequence and proper motion studies have isolated cluster stars from the field. From photometry and spectroscopy, its composition is determined to be almost exactly the same as the Sun, with an age about one-half billion years less. This similarity to the Sun permits an accurate determination of distance to M67 independently of other distance measurements. Using the Sun as a reference, the distance modulus of M67 is found to be 9.48 mag. An extensive analysis of possible sources of error leads to an uncertainty (standard error) of ± 0.15 mag., with the chief sources of error being the color index of the Sun, the composition of the cluster, and its age. The M67 distance uncertainty compares favorably with Hanson's (1975) Hyades modulus of 3.29 ± 0.08.

1. INTRODUCTION

Because of the close resemblance in age and composition between M67 stars and the Sun, the M67 distance modulus can be found using the Sun as a reference, with no dependence on trigonometric distances. Its accuracy depends on the extent to which M67 and the Sun are matched and on the accuracy of the photometry of both M67 and the Sun.

A key reason why some confidence can be placed on the distance determination of M67 lies in the fact that there have been a number of photometric studies of the cluster (see Janes and Smith, 1984, for a review of the photometry). Schild (1983) and Janes and Smith (1984) find agreement among the various systems to better than 0.01 mag. (mean deviation) in both the V and B-V indices. For this analysis, Racine's (1971) photometry is used, since it includes much of the earlier photometry.

The absolute visual magnitude of the Sun is M_v = 4.83 (Allen, 1973) and its color index, B-V = 0.65 (Vandenberg and Bridges, 1984). A mean line through the M67 main sequence between m_v = 13.5 and m_v = 15.5 yields m_v = 14.53 ± 0.05 at the solar B-V. After correcting for a

361

D. S. Hayes et al. (eds.), Calibration of Fundamental Stellar Quantities, 361–364.

reddening of E(B–V) = 0.056 (Janes and Smith, 1984) and assuming
A_v = 3.2E(B–V), then the distance modulus of M67 is $(m-M)_0$ = 9.52.
Janes and Demarque (1983) found M67 to be about 4.1 billion years old.
Assuming an age of 4.6 billion years for the Sun (Ostic, et al, 1963),
an evolutionary correction of 0.04 mag. is required, bringing the M67
distance modulus to 9.48.

2. ANALYSIS

How good is this value? The estimated error derives mainly from the
uncertainty in the color index of the Sun, the composition of M67 and
the age of M67. The consequences of these and other sources of error
are summarized in Table I and in the following paragraphs.

Table I – Sources of Error in M67 Distance Modulus

Parameter	Value	Adopted Error	Contributions to Error in (m–M)
$(B-V)_0$	0.65 mag	±0.015	±0.075
M67 composition	solar	±20%	±0.075
M67 Age	4×10^9 yrs	$\pm 10^9$	±0.08
Solar Age	4.6×10^9 yrs	$\pm 2 \times 10^8$	±0.02
M_{v0}	4.83 mag	±0.03	±0.03
E(B–V)	0.056 mag	±0.005	±0.025
m.s. at $(B-V)_0$	14.53 mag	±0.05	±0.05
$R = A_v/E(B-V)$	3.2	±0.2	±0.01
$(m-M)_{M67}$	9.48	----	±0.15

The B–V color index of the Sun has been disputed for some time;
based on an extensive review, Vandenberg and Bridges (1984) derived a
value of 0.65. Since the value is almost certainly greater than 0.63
and less than 0.67, an uncertainty of ±0.015 is adopted here. The slope
of the M67 main sequence is about 5 so an uncertainty of ±0.015 in B–V
translates into ±0.075 in V.

Based on the discussion in Janes and Smith (1984) and a survey of
recent literature, M67 must have a metallicity very close to that of the
Sun and a little less than the Hyades metallicity. The uncertainty in
this value is of the order of 20% (0.08 dex). From the Vandenberg and
Bridges (1984) theoretical main sequences, the effect of metallicity on
M_{Bol} at the solar T_{eff} is 0.91 mag/dex.

In the context of current models (Janes and Demarque, 1983), M67
has an age of 4 billion years, but although the internal error is rather
small, residual uncertainties in the models suggest a possible age
uncertainty as large as one billion years. An analysis of the
Vandenberg (1983) isochrones shows that at the solar effective
temperature, the main sequence increases in luminosity by 0.08 mag/10^9

years in the period between 3 and 5 billion years of age.

Other factors necessary to consider are the uncertainties in the solar luminosity and age, the interstellar reddening to M67, the finite width of the (observed) M67 main sequence and the uncertainty in the ratio of total-to-selective absorption, R. The adopted uncertainties for these parameters are shown in Table I.

The helium abundance is an unknown quantity. Vandenberg and Bridges (1984) find that it does not strongly affect the luminosity of the main sequence at a given color, and it is likely, in view of the age and metallicity of the cluster, to be close to the solar value. Consequently, the helium question is ignored.

The various sources of error discussed in the preceeding paragraphs are essentially independent of one another, so a straightforward calculation of the r.m.s. error in the M67 distance modulus gives ±0.15 mag. This compares favorably with Hanson's (1975) Hyades modulus of 3.29 ± 0.08.

3. SUMMARY

The apparent magnitude of the Hyades main sequence at the solar B-V is $m_v = 8.15$. If the age of the Hyades is about 0.8 billion years, an evolutionary correction of -0.20 is necessary to compare the Hyades to the M67. The Hyades metallicity [Fe/H] = 0.15, so an additional correction of +0.14 is required. These figures imply a distance modulus of M67 that is larger than that of the Hyades by 6.26; or using the M67 value of (m-M) = 9.48, a value for the Hyades of (m-M) = 3.22 results. For comparison, Hanson (1975) found (m-M) = 3.29 for the Hyades. While the possibility of compensating errors exists, the self-consistent set of results presented here strongly suggest that the stellar evolution models of Vandenberg (1983) are reliable and furthermore that M67 is close to the Sun in composition and 4 billion years old.

4. REFERENCES

Allen, C.W. 1973, Astrophysical Quantities (London: Athlone Press).
Hanson, R.B. 1975, Astron. J., 80, 379.
Janes, K.A. and Demarque, P. 1983, Astrophys. J., 264, 206.
Janes, K.A. and Smith, G.H. 1984, Astron. J., 89, 487.
Ostic, R.G., Russell, R.D. and Reynolds, P.H. 1963, Nature, 199, 1150.
Racine, R. 1971, Astrophys. J., 168, 393.
Schild, R.E. 1983, Pub. Astron. Soc. Pacific, 95, 1021.
Vandenberg, D.A. 1983, Astrophys. J. Suppl., 51, 29.
Vandenberg, D.A. and Bridges, T.J. 1984, Astrophys. J., 278, 679.

DISCUSSION

POPPER: What kind of analysis leads to your confidence that the chemical abundance of M 67 is close to that of the Sun? Is there an analysis of high-dispersion spectra of main sequence stars or only of the brighter giants?

JANES: Most of the work that has been done is low-resolution work (photometry and spectroscopy). There are, however, a few high-resolution studies, both of giants and the upper main sequence.

CAYREL: I agree with Dr. Janes that the [Fe/H] value of M 67 is equal to that of the Sun. Very recently one of my students, Dominique Proust, analyzed detailed spectra of two yellow giants of M 67 and found that their Fe-abundance was equal to that of the Sun.

GARRISON: Some years ago I did MK classification of the main sequence of NGC 752. What I found was that the dwarfs were identical to Hyades dwarfs of the same spectral type. This was confirmed by Crawford using Strömgren photometry. Most previous photometry and high-dispersion analyses had been done on giants, hinting at a possible difference in composition between giants and dwarfs. Do you see this effect for M 67?

JANES: In my survey of the literature relating to the chemical composition of M 67 I included studies both of giants and dwarfs in the cluster, as well as studies based on photometry and spectroscopy at various spectral resolutions. For the most part, they are in good agreement with one another. Although I did not check on NGC 752 in this survey, I seem to recall an impression that it might be a little metal-rich. Finally, there is no confirmation of the Spinrad and Taylor super-metal-rich designation of M 67.

GARRISON: Morgan classified M 67 stars from the same spectra used by Spinrad and Taylor and found them to be normal, in contrast to the results of Spinrad and Taylor who found them metal-rich. This is published in the Astrophys. J. about 12-15 years ago. Spinrad's results were due to use, or, misuse, of out-of-focus standards with respect to the cluster stars.

WHITE DWARF CANDIDATES FOR TRIGONOMETRICAL PARALLAX DETERMINATIONS

T. J. Moffett

Dept. of Physics, Purdue University

Thomas G. Barnes and David S. Evans

McDonald Obs. and Dept. of Astronomy, Univ. of Texas

ABSTRACT. The visual surface brightness relation is applied to the determination of parallaxes of white dwarfs on the assumption, borne out by previous studies of white dwarfs of known parallax, that these show only a small range of linear diameters.

In a previous paper (Moffett et al. 1978) the visual surface-brightness relation (Barnes et al. 1978) was shown to be applicable to white dwarfs by observation of white dwarfs of known parallax. It was also demonstrated that the photometric system of Eggen and Greenstein could be transformed reliably to that required for the application of the visual surface-brightness relation. In this way, values of the angular diameter ϕ' (in arc ms) can be deduced from the visual surface brightness using

$$\log \phi' = 8.4414 - 0.2V - 2F_v,$$

where $\qquad F_v = \log Te + 0.1 C,$

C being the bolometric correction, tabulated in terms of (V-R) in Barnes et al. (1978).

These circumstances enable us to deduce angular diameter values for the white dwarfs in the lists of Eggen and Eggen and Greenstein. The stars are identified by their EG numbers in the comprehensive list given by Greenstein (1976) which gives other names and coordinates. The list is in order of right ascension, not EG numbers.

The paper of Moffett et al. strongly suggests that white dwarf radii do not scatter much about a mean lying between one and two hundredths of that of the Sun. Note how close the values are to those deduced from the theory of degenerate matter. The adoption of a particular linear radius for a star of given angular diameter determines its parallax. Table I gives the trigonometrical parallaxes deduced for a number of white dwarfs by this method, on a range of assumptions for their linear radii. It will be seen that the following may have parallaxes in excess of 0.1 arc seconds: -- EG Nos 21, 27, 309 and 320 (none of which may be white dwarfs) and EG Nos 111 and 188 (of which the latter

D. S. Hayes et al. (eds.), Calibration of Fundamental Stellar Quantities, 365–368.
© *1985 by the IAU.*

may be composite) and 382, 388. On firmer ground the following can be expected to have parallaxes in excess of 0.03 arc seconds: —— EG Nos: 76, 91, 103, 178, 184, 248, 302, 307, 312, 319, 321 (possibly not a white dwarf), 322, 323, 327, 329, 335, 336, 342, 344, 349, 350, 353, 359, 364, 368, 370, 374, 380, 381, 391, or a total of four very close candidates and 30 within some 30 parsecs of the Sun.

REFERENCES

Barnes, T. G., Evans, D. S. and Moffett, T. J. 1978, Mon. Not. R. Astron. Soc., 183, 285.
Greenstein, J. L. 1976, Astron. J., 81, 323.
Moffett, T. J., Barnes, T. G. and Evans, D. S. 1978, Astron. J., 83, 820.

TABLE I

PREDICTED PARALLAXES FOR WHITE DWARFS

EG	Sp	F_v	$\log \phi'$	p R=.011$R_\odot$ (0".001)	p R=.013$R_\odot$ (0".001)	p R=.015$R_\odot$ (0".001)	Notes
3	DB	4.071	-2.764	17	14	12	
21	sdM	(3.417)	-1.478	325	275	238	3
27	sdM	(3.471)	-1.604	243	206	178	3
31	DA	4.129	-2.632	23	19	17	
63	DB	4.060	-2.810	15	13	11	
72	DA	3.802	-2.544	28	24	20	
76	DA	4.098	-2.362	42	36	31	1
77	DB	3.971	-2.688	20	17	15	
80	DA	4.145	-2.842	14	12	10	
86	DO	(4.295)	-3.074	8	7	6	
91	DB	4.015	-2.420	37	31	27	4
103	DA	4.060	-2.356	43	36	31	
128	DA	3.916	-2.510	30	25	22	
133	DB	4.060	-2.592	25	21	18	
136	DAe	3.920	-2.472	33	28	24	4
154	DA-F	3.905	-2.608	24	20	18	
178	DA-F	(3.781)	-2.424	37	31	27	
184	DA	4.129	-2.310	48	40	35	1
185	DA	4.098	-2.544	28	24	20	
193	DB	4.084	-2.662	21	18	16	
198	DA	3.994	-2.722	19	16	13	2
248	DC2p	3.802	-2.068	83	71	61	
267	DC	(4.377)	-3.218	6	5	4	
272	DB	3.912	-2.572	26	22	19	
297	DA	4.071	-2.750	17	15	13	
302	DA-F	3.836	-2.374	41	35	30	
303	DA	4.198	-2.896	12	11	9	
304	DAwk	(4.216)	-2.908	12	10	9	

TABLE I   cont'd

EG	Sp	F_v	$\log \phi'$	p R=.011R $(0\overset{''}{.}001)^\oplus$	p R=.013R $(0\overset{''}{.}001)^\oplus$	p R=.015R $(0\overset{''}{.}001)^\oplus$	Notes
305	DB	4.060	−2.548	28	23	20	
307	DC	3.844	−2.244	56	47	40	
308	DA	4.198	−2.576	26	22	19	
309	sdB	3.916	−1.860	134	114	99	1,3
310	DA	4.198	−2.844	14	12	10	
311	DA	4.036	−2.616	24	20	17	
312	DA	3.900	−2.328	46	39	34	
314	DA	4.129	−2.572	26	22	19	
315	DB	4.036	−2.620	23	20	17	
317	DA	3.889	−2.536	28	24	21	
318	DA	3.961	−2.686	20	17	15	
319	DA	3.864	−2.224	58	49	43	
320	RSL	(3.664)	−1.774	164	139	121	3
321	DC-K	(3.639)	−2.112	75	64	55	3
322	DC	3.910	−2.400	39	33	29	
323	DA	3.840	−2.388	40	34	29	
325	DC	3.941	−2.636	23	19	17	2
326	DA	4.145	**−3.140**	7	6	5	
327	DC	3.818	−2.436	36	30	26	
328	DAs	3.886	−2.632	23	19	17	
329	DC	3.860	−2.298	49	42	36	
330	DC-F	(3.775)	−2.532	29	24	21	
331	DA	4.060	−2.712	19	16	14	
332	DA	4.129	−2.856	14	11	10	
333	DXp	4.071	−2.666	21	18	15	
335	DC	3.920	−2.426	37	31	27	
336	DA	3.924	−2.032	91	77	67	
341	DA	3.910	−2.514	30	25	22	
342	DA-F	(3.750)	−2.392	40	33	29	
343	DA	3.947	−2.586	25	21	19	2
344	DC	3.808	−2.316	47	40	35	
347	DA	4.162	−3.094	8	7	6	
348	DAp	4.060	−2.808	15	13	11	
349	DC	(3.611)	−2.186	64	54	47	
350	DA	3.823	−2.146	70	59	51	
351	DA	3.902	−2.752	17	15	13	
352	DA	4.015	−2.668	21	18	15	
353	DC-G	(3.652)	−2.148	69	59	51	
354	DA	4.005	−2.790	16	13	12	
355	DA	3.941	−2.659	21	18	16	
356	DC	(3.787)	−2.558	27	23	20	
357	DA	4.036	−2.932	11	10	8	
358	DF	(3.775)	−2.460	34	29	25	
359	DB+M	3.848	−2.058	85	72	63	4
361	DC	3.902	−2.762	17	14	12	

TABLE I   cont'd

EG	Sp	F_v	$\log \phi'$	p R=.011R$_\odot$ (0".001)	p R=.013R$_\odot$ (0".001)	p R=.015R$_\odot$ (0".001)	Notes
362	DF–C	3.814	−2.596	25	21	18	
363	DA+K	3.961	−2.712	19	16	14	
364	DA	3.941	−2.392	40	33	29	
365	DA	3.875	−2.480	32	27	24	
366	DC	3.879	−2.682	20	17	15	
367	DC–F	3.910	−2.688	20	17	15	
368	DA	4.015	−2.030	91	77	67	1
370	DA	4.036	−2.294	50	42	36	
373	DA–F	3.836	−2.518	30	25	22	
374	DXp	3.802	−2.278	51	43	38	
376	DA	4.113	−3.036	9	8	7	
378	DA	4.275	−2.574	26	22	19	1
379	DC	3.947	−2.842	14	12	10	
380	DAwk	(3.698)	−2.368	42	35	31	2
381	DA–F	3.802	−2.064	84	71	62	
382	DK–G	(3.664)	−1.870	132	111	97	
383	DB	3.961	−2.718	19	16	14	
384	DB	4.010	−2.684	20	17	15	
386	DB	3.994	−2.470	33	28	24	
387	DB	3.924	−2.572	26	22	19	
388	DC+M	(3.475)	−1.644	222	188	163	4
389	DBe	3.889	−2.542	28	24	21	4
391	DG	3.814	−2.404	39	33	28	
392	DC	3.848	−2.544	28	24	20	
393	DB	4.024	−2.572	26	22	19	
394	DA	3.941	−2.476	33	28	24	
395	DA	3.977	−2.804	15	13	11	
396	DA	3.920	−2.578	26	22	19	
397	DA	(4.216)	−2.668	21	17	15	
398	DC	3.907	−2.524	29	25	21	
399	DO?	(4.234)	−2.648	22	19	16	
400	DA	4.071	−2.938	11	9	8	
401	DA	4.005	−2.628	23	19	17	
402	DA	4.113	−2.828	15	12	11	
403	DA	4.084	−2.908	12	10	9	
404	DA	3.941	−2.692	20	17	15	
405	DA	3.941	−2.668	21	18	15	

NOTES TO TABLE I

1 – Our photometry used.
2 – Greenstein's (1976) photometry uncertain.
3 – Possibly not a white dwarf.
4 – Composite.

THE ABSOLUTE MAGNITUDES OF THE NEARBY STARS: CALIBRATIONS OF MEAN LUMINOSITY RELATIONS

W. Gliese and H. Jahreiss

Astronomisches Rechen-Institut, Heidelberg

The nearby stars are most favored for determining precise absolute magnitudes and for calibrating spectral-type, luminosity relations and color, luminosity relations. To demonstrate the main problems we discuss the $(M_V,B-V)$ relation since $(B-V)$ data are available for most of our candidates. At Yale Observatory van Altena is compiling a new catalog of trigonometric parallaxes. We feel deeply indebted to van Altena for making available to us a preliminary version of this catalog.

Based on these new data Figs. 1-3 show $(M_V,B-V)$ diagrams of the stars nearer than 25 pc. The stars with the best-determined absolute magnitudes (s.e. $\leq 0^m.20$ and $\leq 0^m.30$) are plotted in Figs. 1 and 2 resp.; in Fig. 3 all the stars with $\sigma_\pi/\pi \geq 0.14$ are given. In addition we have plotted the Zero-Age Main Sequence in the upper part of the diagrams; for $B-V > +0.40$ the mean main sequence, as derived from the nearby stars, data available in 1982 was marked.

Fig. 3 points out that a sample of stars with large accidental errors shifts the mean main sequence to lower luminosities; such effects could be eliminated by statistical corrections ("Lutz-Kelker corrections") if the observational material fulfils certain conditions. However, if the number of stars available for calibration is sufficiently large it is recommended that the sample of objects be restricted to those with accurately determined absolute magnitudes as shown in Figs. 1 and 2 where the Lutz-Kelker correction is smaller than a tenth of a mag.

In Fig. 2 the mean curve fits the data fairly well whereas the best determined stars (Fig. 1) in the region $+8^m < M_V < +10^m.5$ are significantly brighter. Obviously, this deviation is not caused by a change in the trigonometric parallax system. For 42 stars in this region with s.e. of $M_V < 0^m.3$ the mean difference $\langle M_V \rangle_{new} - \langle M_V \rangle_{old}$ is only $-0^m.02 \pm 0^m.02$. Also an incompletely applied Lutz-Kelker correction cannot account for it. However there is a slight dependence of the mean absolute magnitudes on the tangential velocities. A least squares analysis yields $\Delta M_V = +0^m.003(\pm0^m.001)$ per km s^{-1}. In the 1982 sample the portion of stars with large tangential velocities was relatively high, producing a shift to lower luminosities.

At M_V of about $+10^m$ these effects are strengthened by a phenomenon

369

D. S. Hayes et al. (eds.), Calibration of Fundamental Stellar Quantities, 369–372.
© *1985 by the IAU.*

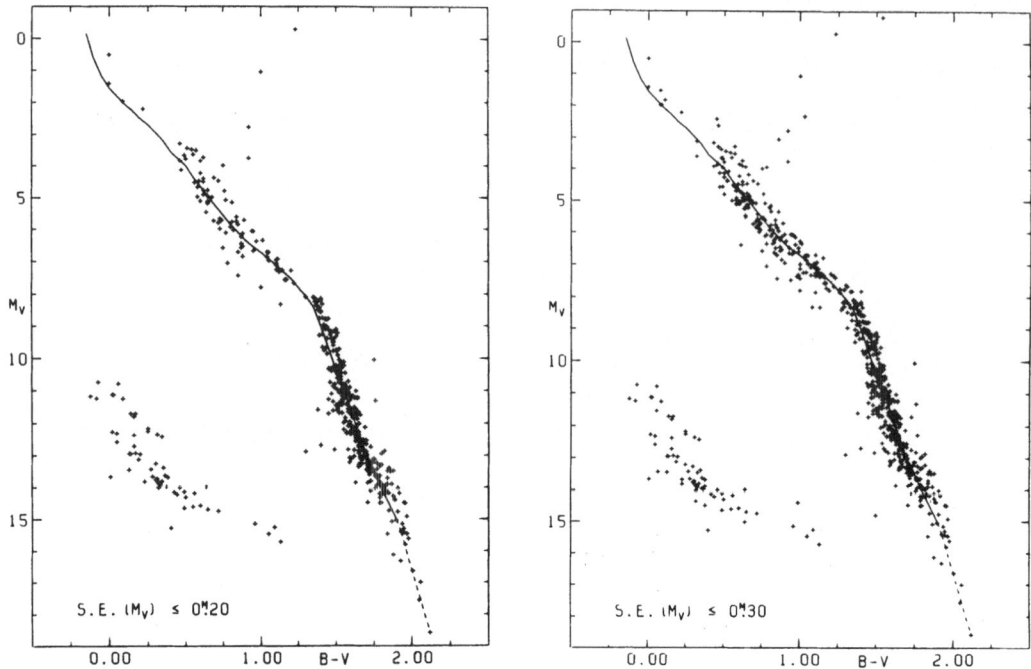

Fig.1 CM – Diagram of the stars within 25 pc; s.e.$(M_V) \leq 0^m.20$.

Fig.2 CM – Diagram of the stars within 25 pc; s.e.$(M_V) \leq 0^m.30$.

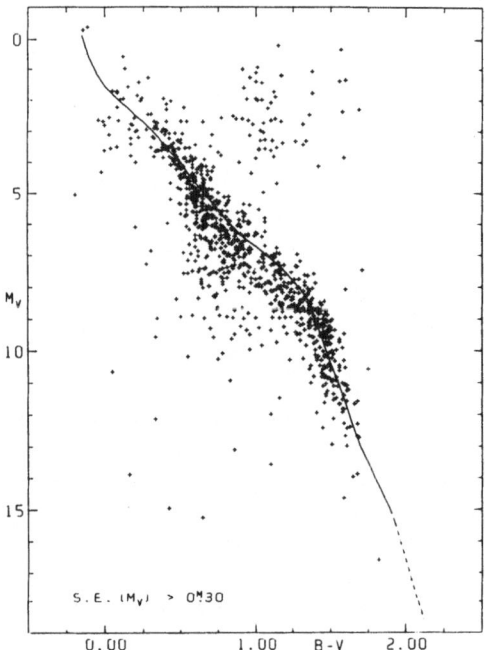

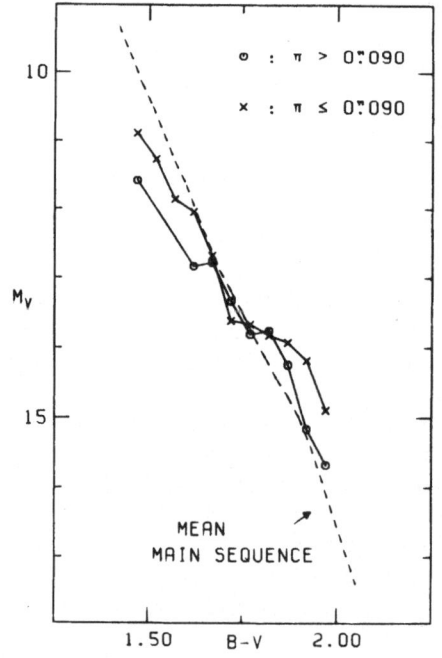

Fig.3 CM – Diagram of the stars within 25 pc; s.e.$(M_V) > 0^m.30$.

Fig.4 $\langle M_V \rangle$ for faint stars with π_t published 1969 to 1978.

represented in Fig. 4. This diagram gives mean values of M_V in (B-V)-intervals of $0^m.05$ for faint stars with precise trigonometric parallaxes published from 1969 to 1978 ("GJ stars"). This sample has not only a faint limit in apparent magnitude but also a bright one at about $V = 10^m$. A bright limit implies that in a certain color range the absolutely brighter dwarfs were not observed and the derived mean main sequence proves to be too faint. On the other hand, near the faint limit the observed mean main sequence will be too bright ("Malmquist bias").

These effects vary with the distance modulus M-m. The nearest stars in such a series define a mean main sequence somewhat fainter than that of the more distant objects. Our sample was subdivided into very nearby stars ($\pi \geq 0".090$) and those with parallaxes between $0".040$ and $0".090$. Unfortunately the numbers of objects per interval are very small (1 to 13) but even so we see that the expected deviations are observable. In the region $+1.45 < B-V < +1.60$ such stars of lower luminosities may have shifted the mean curve in the 1982 derivation which, however, is no longer confirmed by the new parallax data.

We may conclude that a mean main sequence, even if derived from the best determined trigonometric parallaxes, should be used only with caution for different samples of dwarf stars. A mean curve depends on the limits (whether in magnitude or in volume) of the observational series used for calibration; furthermore the luminosities are affected by the ages which are correlated with the velocities and/or the chemical compositions of the stars.

REFERENCES

Calibrations of absolute magnitudes for nearby stars used up to now
    are published in:
Gliese, W. 1971, Veröffentl. Astron. Rechen-Institut Heidelberg, No. 24.
Gliese, W. 1982, Astron. Astrophys. Suppl., 47, 471.
Jahreiss, H. and Gliese, W. 1983, in Statistical Methods in Astronomy,
    ed. E. J. Rolfe, (ESA Spec. Publ., SP-201), p. 201.

The theoretical aspects of the calibration of absolute magnitudes from
    trigonometric parallaxes are described by:
Lutz, T. E. 1983, in IAU Colloquium No. 76, The Nearby Stars and the
    Stellar Luminosity Function, ed. A. G. D. Philip and A. R. Upgren,
    (L. Davis Press, Schenectady), p. 41.

DISCUSSION

MILLWARD: In your Figure 2, it looks as if the upper envelope of your main sequence at (B-V)= 0.3 - 0.7 may be composed of undetected binaries. They all lie about half a magnitude above the main locus of stars.

GLIESE: You are right. In this region the majority of our stars lie above the curve given. Maybe some of these objects are binaries but for (B-V) < 0.4 the ZAMS is plotted and it is connected to our curve at 0.4. Obviously, in this region evolutionary effects may occur among our field stars.

JASCHEK: Have you compared with open cluster results?

GLIESE: Not yet!

LYNGÅ: How much does your new calibration differ from the earlier calibrations made for similar, apparent magnitude-limited, samples?

GLIESE: In 1971 I used all good M_v's of nearby stars available at that time. The material was a mixture of various observational series which had different magnitude limits. In 1982 I used Eggen's data for stars with large proper motions. From stars with a transverse velocities < 80 km/sec. I derived a mean main sequence which deviated from the 1971 sequence in the range $8 < M_v < 13$ by 0.2 to 0,3 mag. But from $M_v = 4$ to 8 the differences were insignificant (< 0.1 mag.)

FIFTEEN YEARS OF STATISTICAL PARALLAXES BY THE PRINCIPLE OF MAXIMUM LIKELIHOOD

A. Heck

Astronomical Observatory, Strasbourg

ABSTRACT. This paper presents a brief review of the various applications carried out with the statistical-parallax algorithm based on the principle of maximum likelihood.

1. INTRODUCTION

The algorithm of statistical parallaxes based on the principle of maximum likelihood and developed successively by Rigal (1958), Jung (1968, 1970) and Heck (1975a) has **proven** to be a reliable method for determining the absolute luminosity and kinematics of a number of stellar groups under a minimum of assumptions. For comparative methodological reviews, please refer to Heck (1976 & 1978b).

Let us just recall here that, under its present generalized form, the RJH algorithm permits the calibration of a relation between individual stellar absolute magnitudes (M_i) of a given sample and a number of observable quantities (Q_{ij}) of the type

$$M_i = \sum_{j=1}^{n} a_j \, Q_{ij}.$$

Obviously, if $n = 1$ and $Q_{i1} = 1$ for all i, a_1 represents the sample mean absolute magnitude. The mean solar motion and the corresponding velocity ellipsoid are simultaneously determined. Moreover, numerical simulations provide the precision of each estimate. Stars with extreme kinematical properties (i.a. high-velocity stars) are eliminated from the sample by tests on the space velocities.

The method requires for each star the knowledge of the following data: position (not critical), proper motion components and errors, radial velocity and error, apparent magnitude, and the observable quantities Q_{ij} which can be of various kinds (period, photometric indices or colors, metallicity indices, ...).

Although the algorithm can deal with raw or already corrected apparent magnitudes, it includes optionally a statistical correction for the interstellar reddening based on Parenago(1940)'s law.

D. S. Hayes et al. (eds.), Calibration of Fundamental Stellar Quantities, 373–375.
© *1985 by the IAU.*

This paper presents a very brief review of the applications carried out with the algorithm and which have led to a consistent set of absolute magnitudes (see Heck, 1980) for the various stellar groups considered up to now. Space is lacking in these proceedings to provide detailed synthetic tables, but we intend to do it in a forthcoming paper.

2. APPLICATIONS OF THE RJH ALGORITHM

The following stellar groups have been calibrated in luminosity with the RJH algorithm (roughly in chronological order):

- RR Lyrae stars (Heck, 1972, 1973a&b, 1975b; Lakaye, 1977; Heck & Lakaye, 1978),
- Mira variables (Foy et al., 1975),
- Strömgren's "intermediate" and "late" groups (Heck 1975c, 1977),
- late-type stars (Grenier et al., 1976a&b, 1977),
- F-type stars (Heck, 1978a),
- Hg-Mn stars (Jaschek et al., 1980),
- Am stars (Gomez et al., 1981),
- Ap stars (Grenier et al., 1981b),
- δ Delphini stars and δ Scuti stars (Grenier et al., 1981a),
- G5-M3 giants (Egret et al., 1982),
- F-G-K-M stars (Mikami & Heck, 1982),
- B5-F5 dwarf and giant stars (Grenier et al., 1984).

More calibrations are considered (see i.a. Heck et al., 1981) and will be carried out every time the number of stars of a given type with all the necessary measurements will be large enough to obtain statistically significant results.

3. REMARKS

Another algorithm based on the maximum-likelihood principle has been described by Clube & Jones (1971) and Clube & Dawe (1980a&b). Its differences from the RJH method have been discussed in Clube & Jones (1974) and Heck & Jung (1975). A comparative study of the performance of both methods has been carried out by running simulations on a RR Lyrae star sample (Jones et al., 1980). To the best of our knowledge, the CJD algorithm has only been applied to RR Lyrae stars.

REFERENCES

Clube, S.V.M., Dawe, J.A. 1980a, Monthly Not. Roy. Astron. Soc. 190, 575
Clube, S.V.M., Dawe, J.A. 1980b, Monthly Not. Roy. Astron. Soc. 190, 591
Clube, S.V.M., Jones, D.H.P. 1971, Monthly Not. Roy. Astron. Soc. 151, 231
Clube, S.V.M., Jones, D.H.P. 1974, Astron. Astrophys. 33, 153
Egret, D., Keenan, P.C., Heck, A. 1982, Astron. Astrophys. 106, 115
Foy, R., Heck, A., Mennessier, M.O. 1975, Astron. Astrophys. 43, 175
Gomez, A.E., Grenier, S., Jaschek, M., Jaschek, C., Heck, A. 1981, Astron. Astrophys. 93, 155

Grenier, S., da Silva, L., Heck, A. 1976a, in Abundance Effects in Classification, ed. B. Hauck & P.C. Keenan, (D. Reidel Publ. Co., Dordrecht), p. 215

Grenier, S., Gomez, A.E., Jaschek, C., Jaschek, M., Heck, A. 1981a, in Upper Main Sequence CP Stars, 23rd Liège Astrophys. Coll., Liège, p. 491

Grenier, S., Gomez, A.E., Jaschek, C., Jaschek, M., Heck, A. 1984, in preparation.

Grenier, S., Jaschek, M., Gomez, A.E., Jaschek, C., Heck, A. 1981b, Astron. Astrophys. 100, 24

Grenier, S., Heck, A., Jung, J. 1976b, Astron. Astrophys. 52, 385

Grenier, S., Heck, A., Jung, J. 1977, Astron. Astrophys. Suppl. 27, 267

Heck, A. 1972, Astron. Astrophys. 21, 231

Heck, A. 1973a, Astron. Astrophys. 24, 313

Heck, A. 1973b, in Problems of Calibration of Absolute Magnitudes and Temperature of Stars, ed. B. Hauck & B. Westerlund, (D. Reidel Publ. Co., Dordrecht), p. 21

Heck, A. 1975a, Ph. D. Thesis, Univ. Liège

Heck, A. 1975b, Astron. Astrophys. 42, 131

Heck, A. 1975c, Astron. Astrophys. 43, 111

Heck, A. 1976, Quelques méthodes de détermination de la magnitude absolue, ed. Centre de Données Stellaires, Strasbourg

Heck, A. 1977, Astron. Astrophys. 56, 235

Heck, A. 1978a, Astron. Astrophys. 66, 335

Heck, A. 1978b, Vistas in Astron. 22, 221

Heck, A. 1980, Astron. Astrophys. 82, 370

Heck, A., Gomez, A.E., Grenier, S., Jaschek, C., Jaschek, M. 1981, Inf. Bull. Strasbourg Stellar Data Center 20, 34

Heck, A., Jung, J. 1975, Astron. Astrophys. 40, 323

Heck, A., Lakaye, J.M. 1978, Monthly Not. Roy. Astron. Soc. 184, 17

Jaschek, M., Jaschek, C., Grenier, S., Gomez, A.E., Heck, A. 1980, Astron. Astrophys. 81, 142

Jones, D.H.P., Heck, A., Dawe, J.A., Clube, S.V.M. 1980, Astron. Astrophys. 89, 225

Jung, J. 1968, Bull. Astron. Paris, Sér. III, 3, 461

Jung, J. 1970, Astron. Astrophys. 4, 53

Lakaye, J.M. 1977, Master Thesis, Liège

Mikami, T., Heck, A. 1982, Publ. Astron. Soc. Japan 34, 529

Parenago, P.P. 1940, Astron. Zh. 17, 3

Rigal, J.L. 1958, Bull. Astron. Paris 22, 171

AN EMPIRICAL Hγ - LUMINOSITY CALIBRATION

Christopher G. Millward and Gordon A. H. Walker

University of British Columbia

ABSTRACT. High signal-to-noise Reticon spectra for 87 members of 8 open clusters and associations together with 37 stars having reliable parallaxes (early A-type stars with reliable trigonometric parallaxes, eclipsing binaries, and visual binaries) have been used to calibrate the W(Hγ)-Mv relation for spectral types O to early A of luminosity classes III-V. The new calibration has a mean probable dispersion of $\pm$0.28 mag. The distance modulus of the Pleiades is 5.54 $\pm$ 0.06 mag, which is in excellent agreement with other, recent determinations, as are the distance moduli for all the calibrating clusters. The use of visual-binary parallaxes implies a Hyades distance modulus of about 3.0 which is significantly smaller than the Hanson (1980) value of 3.30 mag. Although no spectral-type corrections are necessary, stellar evolution probably affects the construction of the new calibration and special care should be taken when determining distance moduli from slightly evolved cluster sequences or for individual stars. Systematic departures from the calibration may be present for stars with Vsin i $\geq$ 220-250 km/sec. Significant residuals are found between our values of W(Hγ) and those of Petrie in the range 1-13 Å equivalent width, which are due in part to systematic errors in Petrie's W(Hγ) measures. Our distance modulus of 11.11 mag for NGC 2244 is in excellent agreement with the photometric distance. The new calibration is compared to other early type star calibrations for main sequence stars. It is 1.2 mag brighter than Petrie's (1965) Hγ calibration at spectral type O6 and 0.7 mag brighter at A3. For types B1 and earlier the new calibration averages 0.4 mag brighter than the Balona and Crampton (1974) Hγ calibration. There is generally good agreement with the Blaauw (1963) MK calibration although the latter is 0.4 mag brighter at spectral type B0. The Crawford (1978) Hβ calibration is up to 0.5 mag brighter for the earlier spectral types and 0.4 mag fainter for later types. More complete discussions of the Hγ-luminosity calibration are available in Millward and Walker (1984, 1985).

D. S. Hayes et al. (eds.), Calibration of Fundamental Stellar Quantities, 377–380.

REFERENCES

Balona, L. A. and Crampton, D. 1974, Mon. Not. Roy. Astron. Soc.,
    166, 203.
Blaauw, A. 1963, in Basic Astronomical Data, Vol. III of Stars and
    Stellar Systems, ed. K. Aa. Strand (U. of Chicago, Chicago), p. 383.
Crawford, D. L. 1978, Astron. J., 83, 48.
Hanson, R. B. 1980, in IAU Symposium No. 85, Star Clusters, ed. J. E.
    Hesser (Reidel, Dordrecht),p. 71.
Millward, C. G. and Walker, G. A. H. 1984, in The MK Process and
    Stellar Classification, ed. R. F. Garrison (David Dunlap Obs.,
    Toronto), p. 261.
Millward, C. G. and Walker, G. A. H. 1985, Astrophys. J. Suppl.,
    57, 63.
Petrie, R. M. 1965, Pub. Dominion Astrophys. Obs., 12, 317.

DISCUSSION

ROUNTREE: What is the systematic difference between your new H-gamma calibration and some other well known absolute-magnitude calibrations for early-type stars - for example the H-gamma calibrations of Petrie and Guetter, the H-beta calibration of Crawford and the more general calibration of Blaauw?

MILLWARD: Our calibration is up to 1.2 magnitudes brighter than the Petrie H-gamma calibration for the O stars. It is up to 0.4 magnitudes fainter than Crawford's H-beta calibration for spectral types earlier than B0 V and averages 0.3 magnitudes brighter for later spectral types. The same type of systematic difference between various H-gamma calibrations and Crawford's H-beta calibration has been noted by others. Shobbrook has recently indicated that Crawford's H-beta calibration may be too bright for the earlier spectral types. This may explain some of the residuals. There is generally good agreement with the Blaauw-MK calibration except at type B0 V where we are about 0.5 magnitudes fainter. This gives support to the claims of Walborn and Turner that the Blaauw calibration is too bright for types O9 - B2 V.

Finally, our new H-gamma calibration averages 0.4 magnitudes brighter than the Balona and Crampton H-gamma calibration for types earlier than about B2. A number of comparisons are shown in the poster paper.

LYNGÅ: Does emission in B stars worry you? For H-beta measurements this is quite serious.

MILLWARD: Emission is less severe at H gamma than at H beta and with our high signal-to-noise Reticon spectra any features observed in the H-gamma line can be assumed to be real, whether it is a blend in the line wings or emission in the line core. It is a very straightforward matter to correct for any emission that is present unless it is so severe that it is filling in most of the line.

FRACASSINI: I would remind Dr. Millward and Dr. Walker that Dr. Martin of the Observatory of Marseille published similar results in 1959 (Compt. Rend. Acad. Science, Paris, 248, 1776) and in 1964 (thesis presented at the Faculty of Sciences of the University of Marseille). The equivalent widths were determined by means of the method of the Swedish astronomer Ohman.

PARTHASARATHY: Rotation, emission in the line and duplicity may influence the M_v-H-gamma calibration.

MILLWARD: We have investigated the effect of rotation on the calibration and find that the systematic effect, if present, is only important for Vsini > 220 - 250 km/sec. Even though the lines are broad and shallow,

any emission present can be detected easily and corrected for due to our
high signal-to-noise spectra. Binarity can be a problem with the H
gamma calibration as it affects both the apparent magnitude and the H
gamma index. However, we have made corrections to known and suspected
binaries in the calibration.

THE WILSON-BAPPU EFFECT FOR Mg II k-LINE EMISSION WIDTHS

M. Parthasarathy

Indian Institute of Astrophysics

ABSTRACT. The existence of a tight linear correlation between the stellar absolute magnitude M_v and the Mg II k-line emission width log $W_{Mg\ II\ k}$ (kms^{-1}) is confirmed using IUE high-resolution (0.2A) data for 100 late-type stars. A least-squares fit to the data gives the relation:

$$M_v = 37.80 - 16.06 \log W_{Mg\ II\ k}(kms^{-1}).$$

1. INTRODUCTION

Several investigators in the recent past have shown that the Mg II k lines are good indicators of chromospheric phenomena in late-type stars. The width and strength of these lines were used to calibrate and correlate with various stellar quantities (Kondo et al. 1976, McClintock et al. 1975, Weiler and Oegerle 1979, Böhm-Vitense 1982, Linsky et al. 1978, Stencel et al. 1980). Weiler and Oegerle (1979), from a Copernicus survey of Mg II emission in 49 late-type stars, found a good correlation between the stellar absolute magnitude and the Mg II k-line width analogous to the Wilson-Bappu effect for Ca II k emission in late-type stars. During the past few years the IUE has produced high-resolution (0.2Å) long-wavelength (2000Å $<\lambda<$ 3300Å) spectra of a large number of late-type stars. The IUE data are less noisy than the Copernicus data.

In this paper the relationship between M_v and log W for the Mg II k line is reported using IUE data for 100 late type stars.

2. DATA

The Mg II k-emission full-line widths were measured near the base of the line in IUE spectra of several late-type stars. Stencel et al. (1980) gave Mg II k-line emission widths for 50 late-type stars from IUE spectra. Garcia-Alegre et al. (1981) gave Mg II k-line emission

D. S. Hayes et al. (eds.), Calibration of Fundamental Stellar Quantities, 381–383.
© *1985 by the IAU.*

widths for 15 F and G main-sequence stars. The accurate width determination is complicated by the fact that most of the central reversals in Mg II h and k emission lines are due to interstellar absorption or, possibly in some cases, to circumstellar absorption (Böhm-Vitense 1981). Böhm-Vitense (1982) has pointed out that a few peculiar late-type giants have anomalous Mg II h and k line widths. Stars with anomalous Mg II k line widths are not included in the present analysis.

3. M_v - Mg II k EMISSION WIDTH RELATION

In Fig. 1 absolute magnitudes are plotted for 100 stars against $\log W_{Mg\ II\ k}$ (kms^{-1}). M_v values based on trigonometric parallaxes or spectroscopic parallaxes were adopted. A least-squares fit to the data gives the relation:

$$M_v = 37.80 - 16.06 \log W_{Mg\ II\ k}(kms^{-1}),$$

with a correlation coefficient which is nearly unity. The relation found in this work confirms the existence of a tight linear correlation between the logarithm of the Mg II k line width and absolute visual magnitude. The high-resolution Mg II h and k line data is now available for several hundred late-type stars from the IUE data bank. If we make use of all this data a more accurate calibration can be derived. In the near future with the Hubble Space Telescope it will be possible to obtain Mg II k line profiles of faint late type stars in galactic clusters, globular clusters and external galaxies. The M_v - $\log W_{Mg\ II\ k}$ relation can be an important tool for calibrating the galactic and extragalactic distance scale.

ACKNOWLEDGEMENTS

I am thankful to the staff of IUE, RDAF and NSSDC. I am grateful to Prof. Harlan J. Smith for his encouragement and help during my stay in Austin.

REFERENCES

Böhm-Vitense, E. 1981, Astrophys. J., 244, 504.
_____ . 1982, Astrophys. J., 252, 628.
Garcia-Alegre, M. C., Ponz, J. D. and Vazquez, M. 1981, Astron. Astrophys., 96, 17.
Kondo, Y., Morgan, T. H. and Modisette, J. L. 1976, Astrophys. J., 209, 489.
Linsky, J. L. et al. 1978, Nature., 275, 389.
McClintock, W., Henry, R. C., Moos, H. W. and Linsky, J. L. 1975, Astrophys. J., 202, 733.
Stencel, R. E., Mullan, D. J., Linsky, J. L., Basri, G. S. and Worden, S. P. 1980, Astrophys. J. Suppl., 44, 383.
Weiler, E. J. and Oegerle, W. R. 1979, Astrophys. J. Suppl., 39, 537.

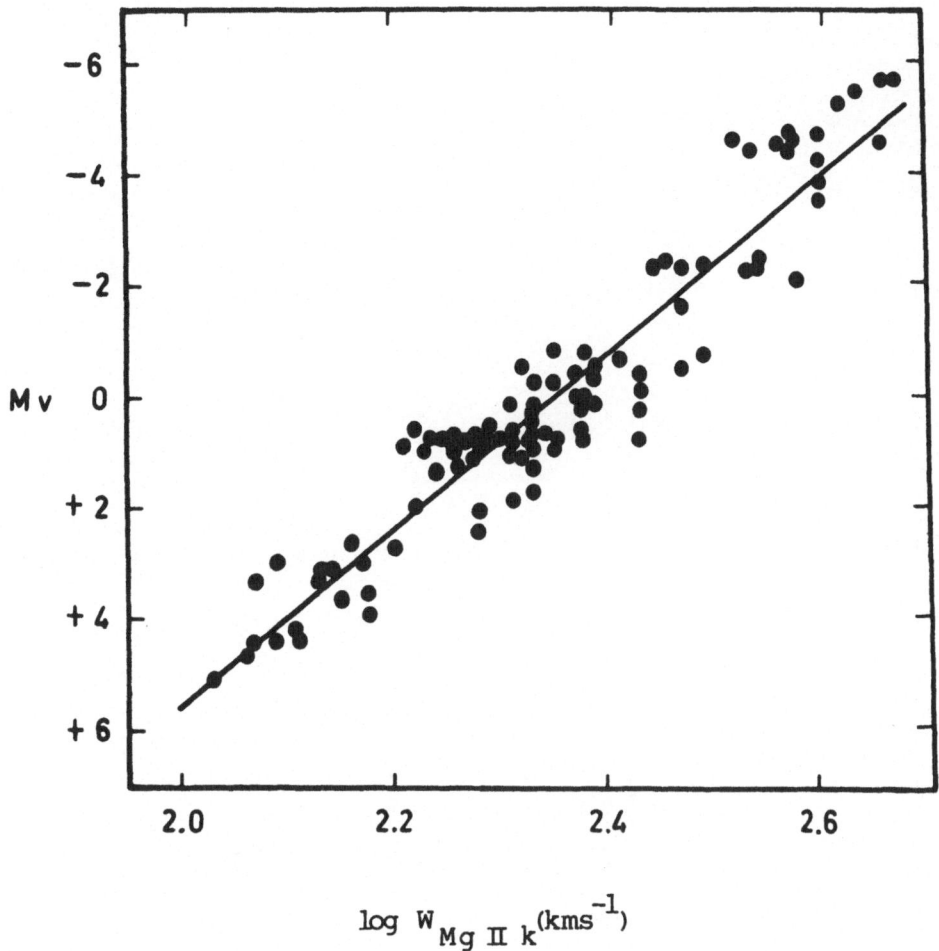

$$\log \ W_{Mg\ II\ k}^{(kms^{-1})}$$

Fig. 1. Absolute magnitude M_V plotted against the logarithm of the emission full width at the base of the Mg II k line, in units of kilometers per second.

DISCUSSION

BESSELL: How were the luminosities determined for the stars used in your calibration?

PARTHASARATHY: I used trig parallaxes or spectroscopic parallaxes.

JASCHEK: I wanted to ask the same question. The scatter in your diagram results from the use of spectroscopic parallaxes, because they are not as good as the trig parallaxes.

PARTHASARATHY: Yes, but we do not have trig parallaxes for many of these stars.

THE INFLUENCE OF AN OUTFLOWING GASEOUS STREAM ON THE DETERMINATION OF MASSES OF SEMIDETACHED BINARY SYSTEMS

D. Chochol

Astronomical Institute of the Slovak Academy of Sciences

A. Vittone

Capodimonte Astronomical Observatory, Naples

ABSTRACT. A general model of gaseous streams in semidetached systems is proposed from the study of the eclipsing binary symbiotic stars CI Cyg and V 1329 Cyg. The influence of gaseous streams on the determination of the masses of semidetached systems is shown.

1. INTRODUCTION

The interpretation of the observational properties of semi-detached binary systems is very difficult. The main problem is connected with the exact location of the line formation regions. The aim of this paper is to show that the lines used for the determination of the radial velocity curve are strongly affected by the gaseous streams forming an excretion disk around the system. The symbiotic stars reveal the best physical laboratories for a detailed study of the accretion and excretion processes including the exact locations of the line forming regions. In some systems it is possible to estimate the influence of gaseous streams on the radial-velocity curve and to obtain reliable values for the masses of the components. Moreover a model of gaseous streams in the system, which appears as a general model of semidetached binaries, can be made.

2. THE MODEL OF GASEOUS STREAMS

A model was made of gaseous streams in semidetached binaries from the detailed study of the two eclipsing binary symbiotic stars CI Cyg (Chochol et al. 1984) and V 1329 Cyg (Chochol and Vittone 1984). These symbiotic stars are highly interacting binaries. The loser is a cold M giant and the gainer, surrounded by an accretion disk, is a B9 main sequence star and a white dwarf in CI Cyg and V 1329 Cyg, respectively. Due to the impact of the accretion disk into the inflowing stream, which penetrates the disk, a turbulent region is

385

D. S. Hayes et al. (eds.), Calibration of Fundamental Stellar Quantities, 385–388.
© 1985 by the IAU.

formed. The turbulent region is responsible for flickerings, flares and eruptions. The outflowing stream from a turbulent region into the outer excretion disk, which surrounds both components was found. While the outburst is caused by the atmospheric expansion of an accreting star due to the mass transfer burst or due to the thermonuclear runaway in a hydrogen/helium rich envelope of the accreting white dwarf, the eruption is caused by the outflow of hot matter from the inner part of the accretion disk after the deep penetration of disk by the inflowing stream. Both cases can be distinguished by observations. The excretion is more important than the accretion. The mass and angular momentum escape from the system through L_2.

The influence of the outflowing stream from the turbulent region was studied in V 1329 Cyg by a careful analysis of the line splitting present in the spectrum. The radial velocity curve (dashed line in Fig. 1) published by Iijima et al. (1981) gave $f(M) = 23\ M_\odot$. From the assumption that the mass of accreting white dwarf is $1\ M_\odot$, a mass of $25\ M_\odot$ was derived for the cold component. The new radial velocity curve (full line in Fig. 1) reflects only the orbital motion of the accreting star. A more reliable $f(M) = 5.9\ M_\odot$ and a mass $7.6\ M_\odot$ for the cold component was derived.

3. THE LAW OF INTERACTING BINARIES AND ITS APPLICATIONS

The main characteristic of the outflowing gaseous stream in the semidetached systems can be formulated as a law of semidetached binaries: The outflowing gaseous stream preserves the characteristics of the accretion disk and the turbulent region in the outer excretion disk. The law is valid both for absorption and emission line objects. The hot turbulent region is responsible for the formation of emission lines. The hot matter from the turbulent region is carried away by the outflowing stream into the outer excretion disk. For this reason we observe the emission lines during the primary eclipse in some interacting binaries.

The proposed model appears as a general model of semidetached binaries and can be applied to Algol binaries, W Ser stars, ζ Aur stars, Be stars, novae, dwarf novae and symbiotic stars. The differences in various kinds of binaries depend on the nature of the loser and gainer, characteristics of the accretion disk, orbital period, cooling rate of the matter in the outer excretion disk, etc.

An application of this model to ε Aur (FO Ia + ?) can explain the behavior of this very interesting eclipsing binary. The radial velocity curve of ε Aur is influenced by the gaseous streams to such an extent that it cannot be used for a reliable estimate of the masses. The mass function is overestimated. The main characteristics of the system can be obtained from the length of the eclipses and the assumption that the cold component fills up the Roche lobe:

M_F = 13 $M_\odot$ (assumed), M_c = 0.143 $M_\odot$, R_F = 280 $R_\odot$, R_c = 480 $R_\odot$ and distance A = 4570 $R_\odot$. The cold component is a protostar, which can not evolve into a normal star owing to a large mass loss through the accretion and excretion disk.

If the asymmetric line profiles observed in Cyg X-1 are interpreted in the framework of this general model, the masses in the system may be overestimated. As a consequence, the presence of a black hole in this system is doubtful.

REFERENCES

Chochol, D., Vittone, A., Milano, L., and Rusconi, L. 1984, <u>Astron</u>. <u>Astrophys</u>., (in press).
Chochol, D. and Vittone, A. 1984, <u>Astron</u>. <u>Astrophys</u>., (submitted).
Iijima, T., Mammano, A., and Margoni, R. 1981, <u>Astrophys</u>. <u>Space</u> <u>Sci</u>., <u>75</u>, 237.

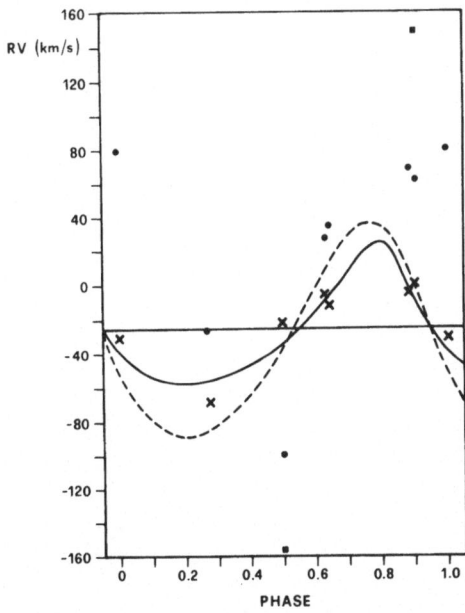

Fig. 1. Radial velocity curves of V 1329 Cyg. The full points and squares indicate the radial velocities of gaseous streams.

DISCUSSION

BIANCHI: Gas streams in binary systems with mass exchange should produce distortion of the line profiles and theoretical models exist which predict the influence of accretion wakes on profiles. Can the problem addressed in your paper be solved by very high-resolution observations of the line profile at different phases?

CHOCHOL: Yes, the problem can be solved by a detailed study of line splitting using high-resolution spectra.

THE ABSOLUTE DIMENSIONS AND MASSES OF IQ PERSEI

Claud H. Lacy

Department of Physics, University of Arkansas

ABSTRACT. High signal-to-noise ratio spectrometric observations of the large light-ratio eclipsing binary IQ Per (B7 + A2) have been obtained with the coudé Reticon spectrograph of the 2.7 m reflector at McDonald Observatory. Absorption lines of the secondary are seen at the 4481 Å MgII line and the 4549 Å TiII + FeII blend. Radial velocities of both components have been measured by cross-correlation techniques and spectroscopic orbits have been computed. The fitted orbits have an eccentricity (0.075 $\pm$ 0.007) that is consistent with the photometric orbit of Hall, Gertken and Burke (1970). Young's (1975) "provisional" estimate of K_2 is about 14% too small. Absolute dimensions and masses have been computed from Hall et al.'s (1970) photometric orbit and the new spectroscopic orbit. Additionally, V-R photometric observations obtained by M. Frueh at McDonald Obs. have been analysed by Popper with light curve synthesis techniques (WINK and EBOP). The relative radii of the new photometric orbits differ by less than 2% from the previous orbits, and the other orbital elements also show excellent agreement. The absolute dimensions and masses are (3.51 $\pm$ 0.04 $\mathcal{M}_\odot$, 2.46 $\pm$ 0.04 $R_\odot$) for the primary and (1.73 $\pm$ 0.02 $\mathcal{M}_\odot$, 1.50 $\pm$ 0.03 $R_\odot$) for the secondary. Both stars are near the zero-age main sequence. The value of ω has changed significantly between the epoch of Hall, Gertken, and Burke's (1970) observations and my own due to apsidal motion. The apsidal motion period is estimated to be in the interval $90 \leq \tau \leq 180$ yr.

REFERENCES

Hall, D. S., Gertken, R. H. and Burke, E. W. Jr. 1970, Publ. Astron. Soc. Pacific, 82, 1077.
Young, A. 1975, Publ. Astron. Soc. Pacific, 87, 717.

D. S. Hayes et al. (eds.), Calibration of Fundamental Stellar Quantities, 389.

UN DIAGRAMME FONDAMENTAL LUMINOSITE-TYPE SPECTRAL POUR LES ETOILES DOUBLES

P. COUTEAU

Observatoire de Nice

ABSTRACT. In the Hertzsprung-Russell diagram, which is not well known for the binary stars, we can use the unit-mass brightness instead of the brightness itself. For each component this parameter can be obtained using the photometry and the orbital elements of the system. It is only through the orbital constant that the parallax comes in the calculation. This new diagram can be applied to any component of binary stars which orbit and class are known. By using a mass luminosity relationship one can define the zero-age main sequence on the diagram. The evolved stars can thus easily be noticed on the diagram; the mean age of the solar vicinity binary systems should be about 10^{10} years.

Les paramètres stellaires fondamentaux que permet d'obtenir l'observation des étoiles doubles sont essentiellement les magnitudes absolues et les masses. La connaissance exacte de ces paramètres est liée à la précision sur les mesures de parallaxe. Actuellement, comme on sait, les parallaxes sont mal connues dès que leur dimension est inférieure à $0\overset{..}{,}05$, ce qui limite à quelques dizaines le nombre d'étoiles dont les magnitudes et les masses sont à la fois bien connues.

Plusieurs auteurs, Russell et Moore (1938), Baize et Romani (1946), Eggen (1967) ont calculé des relations empiriques masse-luminosité à l'aide d'étoiles doubles d'orbites et de parallaxes connues. Ces relations, très simples, appliquées aux binaires donnent des parallaxes dynamiques, puis des masses et des magnitudes absolues.

On arrive ainsi à établir le diagramme de Hertzsprung-Russell $(M_V, Type)$ et le diagramme masse-luminosité. Ce sont deux diagrammes fondamentaux de l'astrophysique, ils situent les étoiles en puissance rayonnée, en masse et en température effective. Mais le diagramme de Hertzsprung-Russell construit à partir des parallaxes dynamiques n'est plus fondamental, puisqu'établi à partir d'une loi empirique.

Nous proposons ici un diagramme similaire, applicable à toute étoile qui fait partie d'un système d'orbite connue, mais établi uniquement sur des valeurs observées, sans hypothèse et qui a, ainsi, un caractère fondamental. C'est le diagramme luminosité par unité de masse en fonction de l'indice (B-V).

391

D. S. Hayes et al. (eds.), Calibration of Fundamental Stellar Quantities, 391–396.

Définissons ce que nous appelons la luminosité par unité de masse. Considérons une binaire d'orbite connue de composante A et B. Soient μA et μB les masses en Soleils, mA et mB les magnitudes apparentes dans un système photométrique quelconque et MA, MB les magnitudes absolues dans le même système photométrique. On connaît le demi-grand axe apparent a", la période en années P. On sait que :

$$\Sigma\mu = \mu A + \mu B = a''^3/\pi^3 P^2 = \alpha^3/\pi^3$$

π étant la parallaxe, ou en faisant intervenir le rapport des masses

$$R = \mu B/\Sigma\mu \qquad \mu A = (1-R)\alpha^3/\pi^3 \qquad (1)$$

D'autre part, la loi de Pogson s'écrit :

$$MA = mA + 5 + 5 \log\pi \qquad (2)$$

ce qui donne :

$$MA = mA + 5 + 5 \log \alpha + (5/3)\log(1-R) - (5/3) \log\mu A \qquad (3)$$

C'est une relation masse-luminosité dynamique pour un couple donné, elle a été mise en évidence bien des fois, surtout par P.J. Morel et A. Baglin (1983). Cette relation peut s'écrire aussi :

$$M_x = MA + (5/3) \log \mu A \qquad (4)$$

M_x est la magnitude absolue de la composante A par unité de masse, elle s'écrit :

$$M_x = mA + 5 + 5 \log \alpha + (5/3)\log(1-R)$$

Ce paramètre est connu en fonction de la magnitude apparente, de la constante orbitale α et du rapport des masses R. Sa connaissance ne fait appel à aucune hypothèse, sauf, parfois, dans la détermination de R, dont l'imprécision possible n'a que peu d'influence sur la valeur de M_x.
    Les observateurs d'étoiles doubles ont l'habitude de considérer les magnitudes visuelles, ou photovisuelles, V. C'est ce système que nous utiliserons par la suite et nous écrivons :

$$M_{Vx} = V + 5 + 5\log\alpha + (5/3)\log(1-R) = M_V + (5/3)\log \mu A \qquad (4bis)$$

L'observation donne également les types spectraux et, d'une manière plus précise, quoique plus rare, les différences (B-V) entre les magnitudes photographiques et photovisuelles. Le diagramme M_{Vx}, (B-V) des points représentatifs de chaque étoile est obtenue sans hypothèse. Son intérêt est de nécessiter ni la connaissance de la parallaxe, ni celle de la masse, seuls sont nécessaires les éléments de l'orbite, connus pour près d'un millier de couples, et la mesure photométrique (B-V), ou à défaut la connaissance du type spectral.

Ceci étant, on peut obtenir d'autre part, M_{Vx} par la relation masse-luminosité et comparer les valeurs obtenues. On trouve ainsi une mesure réelle de l'écart à la relation masse-luminosité. Nous allons préciser cela.

Les relations masse-luminosité se ressemblent toutes. Nous considérerons celle de Baize-Romani (1946) :

$$\text{Log}\mu = - k(M-M_{\odot}) \tag{5}$$

k est un facteur voisin de 0,1117 et $M_{\odot}$ est la magnitude absolue bolométrique du Soleil. Il convient de ramener (5) au système photovisuel. Les liaisons entre les types spectraux, les (B-V), les M_V, les corrections bolométriques B.C., et les magnitudes bolométriques M_b sont dispersées dans la littérature. Les accords entre les auteurs sont acquis, mais relativement récents. Nous avons donc constitué le tableau I, en puisant à différentes sources sûres qui font autorité.

Tableau I

1	2	3	4	5	6	7	8
Sp	(B-V)	M_V	B.C.	M_b	log μ	log μ	M_{Vx}
B7V	-0,12		-1,04				
B8V	- ,09	0,71	-0,85	-0,14	0,548	0,523	1,623
B9V	- ,06	1,04	-0,66	+0,38	,490	,474	1,857
A0V	,00	1,55	-0,40	1,15	,404	,400	2,223
A1V	,03	1,70	- ,32	1,38	,379	,379	2,332
A2V	,06	1,85	- ,25	1,60	,354	,358	2,440
A3V	,09	2,00	- ,20	1,80	,338	,338	2,563
A5V	,15	2,25	- ,15	2,10	,298	,304	2,747
A7V	,20	2,50	- ,12	2,38	,267	,271	2,945
F0V	,33	3,10	- ,08	3,02	,196	,195	3,427
F2V	,38	3,41	- ,06	3,35	,159	,157	3,675
F5V	,45	3,88	- ,04	3,84	,104	,102	4,053
F6V	,47	4,01	- ,04	3,97	,089	,087	4,158
F7V	,50	4,20	- ,04	4,16	,068	,066	4,313
F8V	,53	4,38	- ,05	4,33	,049	,046	4,462
G0V	,60	4,80	- ,06	4,74	,003	,002	4,805
G2V	,64	5,04	- ,07	4,97	-,022	-,022	5,003
G5V	,68	5,28	- ,10	5,18	-,046	-,046	5,203
G8V	,72	5,50	- ,15	5,35	-,065	-,067	5,392
K0V	,81	5,94	- ,19	5,75	-,110	-,108	5,757
K2V	,92	6,40	- ,25	6,15	-,154	-,148	6,143
K3V	,98	6,70	- ,35	6,35	-,177	-,173	6,405
K5V	1,18	7,56	- ,71	6,85	-,232	-,238	7,173
K7V	1,38		-1,02				

Colonne 1.-   Type spectral, série principale.

Colonne 2.-   Indice (B-V) donné par D.L.Harris III (1963).

Colonne 3.-   Magnitude absolue visuelle donnée par A.Blaauw(1963).

Colonne 4.-   Correction bolométrique par D.L.Harris III (1963).

Colonne 5.-   Magnitude absolue bolométrique $M_b=M_V+B.C.$

Colonne 6.-   log de la masse en Soleils selon Baize-Romani (1946).

Colonne 7.-   log de la masse selon (6), voir ci-dessous.

Colonne 8.-   Magnitude absolue visuelle par unité de masse $M_{Vx}=M_V+(5/3)\log\mu$

Ce tableau donne une courbe (M_V, log μ). On considère la parabole qui passe au mieux par trois points dont log μ = 0 et M_V = 4,81 (magnitude absolue photovisuelle du Soleil), et on obtient les relations :

$$\log \mu = 0,0059\ M_V^2 - 0,16\ M_V + 0,634 \qquad (6)$$

ou

$$\log \mu_A^2 - 1,794 \log \mu_A + 1 - 0,208\ M_{VxA} = 0 \qquad (7)$$

La colonne 7 donne les valeurs log μ en fonction de M_V d'après l'équation (6), valeurs très proches de celles de la colonne 6. La relation (6) est la transformée de la relation de Baize-Romani dans le système photovisuel.

La relation (7) donne la masse de la composante A (ou B) en fonction de M_{VxA} donnée par l'observation. Une seule des deux racines est valable, la plus petite. On obtient ensuite M_V et la parallaxe par (4) et (3).

Remarquons que l'équation (7) est une relation entre la masse et la luminosité, celle-ci étant ramenée à l'unité de masse par (5).

On peut illustrer tout cela par le diagramme M_{Vx}, (B-V). Dans ce diagramme, les étoiles se placent selon des valeurs observées, donc sans hypothèse. On trace la courbe M_{Vx}, (B-V) du tableau I. Cette courbe est le lieu des étoiles d'âge zéro selon la relation de Baize-Romani. Les autres relations sont très proches.

Nous avons porté cent cinquante points. Chaque point représente l'étoile principale d'un couple d'orbite bien connue. Ces orbites ont été choisies soit dans le catalogue Finsen-Worley (1970), soit dans les fichiers du Centre d'étoiles doubles de l'Observatoire de Nice. Les (B-V) ont été extraits du catalogue de Blanco et al. (1970). Les (B-V) indiqués dans ce catalogue concernent généralement le couple AB. Dans ce cas, ou bien les deux étoiles sont de même éclatet leur (B-V) est, en principe, le même, ou bien sont d'éclats différents et le (B-V) concerne la primaire.

Pour un (B-V) donné, la plupart de ces points se placent au dessus de la courbe d'âge zéro, représentant la série principale. Ceci traduit une évolution nette de l'astre. Cette évolution démarre à partir de (B-V) = 0,6, correspondant au type GOV. Le domaine des naines rouges se situe bien sur la courbe théorique, alors que les types chauds, surtout à partir de FOV, montrent souvent un excès de luminosité. Le point de démarrage indique que les étoiles de type solaire commencent à quitter la série principale. L'âge moyen des binaires du voisinage solaire serait de dix milliards d'années, âge comparable à celui de la Galaxie. Le diagramme met en évidence quelques sous-naines, moins de dix pour cent du total.

Ce diagramme ne concerne que les étoiles principales des systèmes, les compagnons se placeraient plus près de la courbe d'âge zéro, les données photométriques précises font défaut.

Une étude détaillée portant sur les 350 binaires d'orbite et de type assez bien connus permettrait de comparer plus finement la composition chimique et l'évolution des bianires de notre voisinage avec les amas ouverts proches, les Hyades, les Pléiades et Praesepe. Les modèles théoriques de la séquence principale d'âge zéro étudiés tout récemment entre autres par VandenBerg et Bridges (1984) pourraient s'appliquer aux étoiles doubles et permettraient peut-être de voir si les binaires forment une famille homogène.

BIBLIOGRAPHIE

Baize, P., Romani, L. 1946, Ann. d'Astrophys. t 9, n° 1-2.
Blaauw, A. 1963, Basic Astronomical Data, ed. K. Aa. Strand (U. of Chicago, Chicago), p. 383.
Blanco, V.M., Demers, S., Douglass, G.G., Fitzgerald, M.P. 1970, Photometric Catalogue, Publ. U.S.N.O. Sec. Ser. XXI.
Eggen, O.J., 1967, Ann. Rev. Astron. Astrophys., 5, 105.
Finsen, W.S., Worley, C.E. 1970, Third Catalogue of Orbits of Visual Binary Stars, R.O.J.C., 7, n° 129.
Harris III, D.L. 1963, Basic Astronomical Data, ed. K. Aa. Strand (U. of Chicago, Chicago), p. 269.
Johnson, H.L. 1963, Basic Astronomical Data, ed. K. Aa. Strand (U. of Chicago, Chicago), p. 205.
Morel, P.J., Baglin, A. 1983, Cptes Rendus Journées de Strasbourg, 5ème réunion.
Russell, H.N., Moore, C.E. 1938, Astrophys. J., 87, 389.
VandenBerg, D.A., Bridges, T.J. 1984, Astrophys. J., 278, 679.

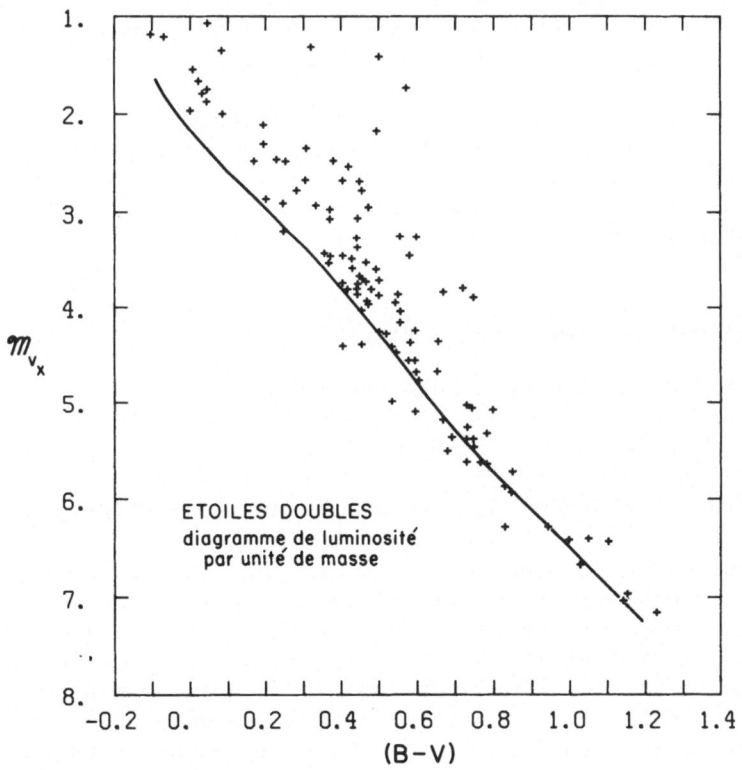

Fig. 1.   Etoiles Doubles.

A MODIFIED HR DIAGRAM FOR VISUAL BINARY STARS

SUMMARY. The HR diagram is poorly known for visual binaries because of uncertaintiy in the parallaxes. However, when the orbits of these stars are known, one can plot a modified diagram using observed quantities and draw the same conclusions.

By combining Kepler's laws with Pogson's scale the formula

$$M_A = m_A + 5 + 5 \log \alpha - \frac{5}{3} \log (\mu_A + \mu_B)$$

can be derived relating the main component's absolute (M_A) and apparent (m_A) magnitudes. $\alpha = ap^{-2/3}$ is the orbital parameter and μ_A, μ_B are the masses of the two components in solar units. It is convenient to introduce M_{*A}, which represents the luminosity of component A per unit mass and is defined by

$$M_{*A} = M_A + \frac{5}{3} \log \mu_A \quad .$$

This magnitude is of particular interest because it only involves observed quantities. It does not appear to have been used previously in the way suggested here. A modified HR diagram can be drawn by plotting M_* against (B-V) for stars with known color index which are members of binary systems whose orbits have been computed.

In order to plot the ZAMS curve on this diagram it is necessary to know the mass μ_A. For this we adopt the mass-luminosity relationship of Baize and Romani (1946) and transform it to the photovisual system by making use of a table which relates spectral types, (B-V), M_V, B.C., and M_{bol}. This table is completed by including M_{*V} and μ_A which are related as follows:

$$M_{*V} = 4.81 \log \mu_A^2 - 8.625 \log \mu_A + 4.81 \quad .$$

In terms of the visual magnitude this becomes:

$$M_{*V} = 0.0098 \, M_V^2 + 0.733 \, M_V + 1.057,$$

where both of these relationships are valid for $- 0.1 \leq (B-V) \leq 1.2$.

As an illustration of the method we present a diagram containing data points which correspond to the main sequence components of 164 visual binaries with measured orbits. The early-type stars lie on the ZAMS, but as stars become hotter (from G0V onwards) they begin to diverge from the curve, thus indicating a mean age of 10^{10} years for binaries in the neighborhood of the Sun. The modified HR diagram enables one to conclude whether binaries form a family and to compare it with the population of local open clusters.

VISUAL BINARIES AND THE CALIBRATION OF THE MAIN SEQUENCE

Karl D. Rakos

Institute for Astronomy Vienna

ABSTRACT. 147 of the best known visual binary stars have been used for an independent calibration of the main sequence in the interval $-0\overset{m}{.}15 < B-V < 0\overset{m}{.}80$. The $(M_v, B-V)$ diagram is found to be equal to the composite open cluster diagram for the age group between 8.35 and 8.43 x 10^8 years - very hot stars excepted. The cosmic scatter of the main sequence is discussed. It appears that only a small number of visual binaries in our sample are simple binary systems.

1. INTRODUCTION

In the past there were a number of calibrations of the absolute magnitudes of the main sequence stars. The use of visual binaries provides an independent method for calibration. Usually the components of a binary system have different spectral types, but the difference in their apparent magnitudes mostly reflects the difference in their luminosities. In this way the absolute magnitudes of primaries can be found from the luminosities of the secondaries and vice versa. Starting with the best known calibration of the lower part of the main sequence, the shape of the sequence can be improved independently of the parallax of the binary system used. The systematic error along the main sequence will be minimized.

2. CALCULATION PROCEDURE

However, the best known binary systems were not used for the investigations until now. The components are too close together and therefore not suitable for classical photoelectric photometry. In particular the complete lack of B-V colors for components closer than 5 arc sec prevents the use of this group of stars - usually provided with reasonable orbits and distances - for the calibration. For the past few years it has been possible, by means of the area scanner technique, to measure the difference in brightness for binaries as

D. S. Hayes et al. (eds.), Calibration of Fundamental Stellar Quantities, 397–400.
© *1985 by the IAU.*

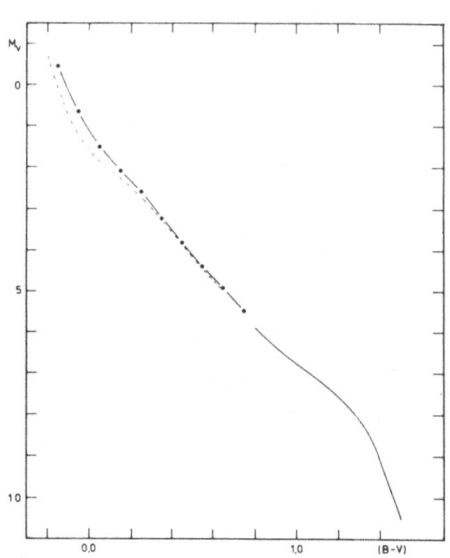

THE MAIN SEQUENCE

B–V	M_V
$-0^{m}_{.}15$	$-0^{m}_{.}44$
−0.05	0.66
0.05	1.51
0.15	2.10
0.25	2.58
0.35	3.23
0.45	3.82
0.55	4.40
0.65	4.92
0.75	5.50

Fig. 1. (See text).

close as 0.5 sec of arc with a high degree of accuracy. Observations of this type for a total of 215 double stars, published by Rakos et al. (1982), have been considered. An additional 28 visual binaries with larger separations measured by Lutz (1971) at KPNO and 153 binaries observed by Hurly and Warner (1982) at Cape Town have been added for analysis. In general the standard deviation for a typical B–V value does not exceed 0.05 magnitudes. The same is true for the difference of the V magnitudes of the components. The accuracy is more than an order of magnitude higher than the cosmic scatter in the main sequence. The quality of the calibration therefore depends on the good statistical sample used in the analysis and not on the accuracy of the individual observations. 147 systems with components on the main sequence have been selected for analysis.

For this reason we start from the main sequence for red stars found by Gliese (1982) from his catalogue of nearby stars. Because the main sequence lifetimes of these late type stars are greater than the age of the galaxy, this sample is free from evolutionary effects. In the HR diagram, see Figure 1, this sequence is a solid line for colors between $1^{m}_{.}60 > B{-}V > 0^{m}_{.}80$. We start with the assumption that the absolute magnitude of a secondary component of a binary system is exactly the tabular value for that B–V. The difference in the magnitudes of the components is then subtracted from this absolute magnitude yielding the absolute magnitude of the primary component. This procedure was applied to those binaries with secondaries in the interval $0^{m}_{.}80 < B{-}V < 1^{m}_{.}60$. The first components of these binaries define a continuation of the main sequence toward higher temperatures. We continued the procedure using the secondaries with higher and higher temperatures.   The dots in Fig. 1, connected by the solid line, are this new mean sequence. As a check the ZAMS calibration of Mermilliod (1981) was used, and is shown as the broken line in Fig. 1.   The complete new mean main sequence is also given in the Table. Comparing the results with the comparative studies of young clusters by Mermilliod (1981b), we find the composite diagram of an age group between 8.35 and 8.43x10^8 years will be most similar to our sequence, very hot stars excepted.

3. COSMIC SCATTER OF THE MAIN SEQUENCE

Double stars are also very suitable for deriving the intrinsic scatter in the luminosities along the main sequence. Several binary systems are composed of components of the same colors. These stars provide direct information on the cosmic scatter. The modern theory of star formation predicts that multiplicity should be the natural configuration in which stars are found. The observed binary frequency along the main sequence is one of the interesting parameters useful in obtaining the luminosity distribution function for multiple star systems.   This function is very poorly known, and the results of several investigations – due to extreme selection effects – are still controversial, but it is one of the most important touchstones for the theory of star formation.

In our sample 46 binaries have components of the same B-V colors, that is, the difference in the colors is less than or equal to 0^m03. From the distribution of ΔV as a function of B-V for these stars it appears that the scatter is homogenous for all values of B-V between A0 and K7.   The color versus brightness difference, Fig. 2, and the histogram, Fig. 3, show three different groups of stars. Stars with $\Delta V > 0^m80$ are hot and evolved from the main sequence. The group around $\Delta V = 0^m70$ are visual binaries and one of the binary components has a close companion of the type A, B-C, all three components are of equal colors.   In the third group the stars with $\Delta V < 0^m1$ are simple binaries and the deviation in the brightness is introduced by the observing errors, or of intrinsic origin.   For $\Delta V > 0.1$ with the peak about 0^m15 we have triple systems but the contribution of the small companion to the color of the unresolved system does not exceed 0^m03.   This third group has the same structure of the histogram for hot and cool stars and is not influenced by the evolutionary effects. It seems that we are able to attribute the observed scatter to the presence of a large number of triple systems among the so called binary stars. Only 6 of the total sample of 46 are probably simple binary systems.

REFERENCES

Gliese, W. 1982, Astron. Astrophys. Suppl., 47, 471.
Hurly, P. R. and Warner, B. 1982, "Area Scanner Observations of Close
    Visual Double Stars II: Results for 153 Southern Stars", preprint.

Lutz, T. E. 1971, Publ. Astron. Soc. Pacific, 83, 488.
Mermilliod, J. C. 1981a, Astron. Astrophys., 97, 235.
Mermilliod, J. C. 1981b, Astron. Astrophys., 93, 136.
Rakos, K. D., Albrecht, R., Jenkner, H., Kreidl, T., Michalke, R.,
    Oberlerchner, D., Santos, E., Schermann, A., Schnell, A. and
    Weiss, W. 1982, Astron. Astrophys. Suppl., 47, 221.

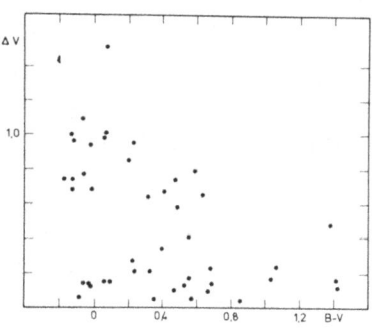

Fig.   2. (see text)

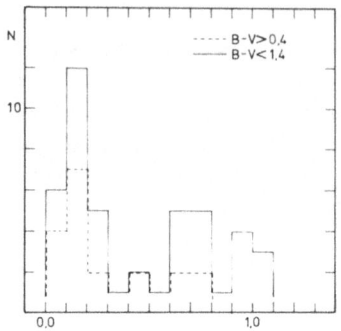

Fig.   3. (see text)

DISCUSSION

GARRISON: I have a question and a comment. The comment is to emphasize
the importance of what Corbally has called the "gold zone", that is,
binaries with separations of 1 - 5 seconds. These stars are true
physical systems, since they are so close, yet they are not so close
that there are interactions. Therefore, they are an ideal group for
calibration of fundamental quantities.

        The question concerns the overlap between your work and that of
Hurly. There are only a few stars in the overlap, if I remember
correctly, but there is a larger than expected difference in the
magnitudes and colors. Can you give us your opinion on the reason for
the difference?

RAKOS: The difference can be at least partly attributed to the
instrumental effect of the area scanner used by Hurly. When scanning
the stars by means of a slit aperture, attention should be paid to the
telescope entrance pupil; that is, it must be projected on the same
place on the photocathode of the photomultiplier during the scanning
cycles. This is possible only by moving the slit aperture in the focal
plane of the telescope. Hurly's area scanner violates this principle.
I would like to continue my observations in the future. At the moment I
am looking for a suitable telescope and the necessary financial
support.

ROTATIONAL PROPERTIES AND SYNCHRONIZATION IN EARLY-TYPE CLOSE BINARIES

G. Giuricin, F. Mardirossian, M. Mezzetti

Astronomical Observatory of Trieste

We have rediscussed the synchronism between rotation and revolution in close binaries by an inspection of the published projected rotational velocities V sin i of about 250 early-type (from 0 to F5) eclipsing and double-lined spectroscopic binaries. Corrections of the V sin i - values (which are mainly taken from the catalog of Uesugi and Fukuda, 1982) for the aspect effect is straightforward for the eclipsing binaries with analyzed light-curves; in the other cases we have estimated the value of the orbital inclination angle i from the primary's minimum mass $M_1 \sin^3 i$ on the assumption that its mass follows Straižys and Kurilienė's (1981) mass-spectrum relations for different luminosity classes. For the components of non-eclipsing binaries, for which the absolute radii are not directly known, we have adopted values of the absolute radii in accordance with Straižys and Kurilienė's (1981) radius-spectrum relations for different luminosity classes. By using our estimates of the radii, for each component we have evaluated the synchronized velocity V_k (corresponding to the mean orbital angular velocity) and the pseudosynchronized velocity V_e, which corresponds to a synchronization with the instantaneous orbital angular velocity at periastron of an eccentric orbit; in close binaries with appreciably eccentric orbits synchronization is attained with $V/V_k > 1$ and it is probably quickly reached at periastron (Hut 1981).

We have examined the degree of synchronism (expressed by the ratios Log V/V_k and Log V/V_e) as a function of the stellar fractional radius r. (Whenever it is not directly known from light-curve analyses, we have estimated r from the above-mentioned estimates of the absolute radii and from evaluation of the binary separation, which results - via Kepler's law - from the total mass of the system). In accordance with the view that tidal forces are stronger for large r, we have found that the fraction of stars showing pronounced deviation from synchronism ($V/V_k > 2$) or pseudosynchronism ($V/V_e > 2$) tends to increase as we go down to smaller r. Synchronism greatly prevails at r > 0.15 (90% or 92% have $V/V_k < 2$ or $V/V_e < 2$, respectively), whereas supersynchronous and superpseudosynchronous rotators become

D. S. Hayes et al. (eds.), Calibration of Fundamental Stellar Quantities, 401–402.

essentially the rule for r<0.05 only (where their percentage is 98% and 78% respectively), i.e. for wider binaries than generally held (Levato 1976). In fact, the fraction of synchronous and especially pseudosynchronous rotators remains relatively high (49% and 73%, respectively) in the range 0.05<r<0.15.

The degree of synchronism appears to be on average stronger in late A-and early F-type stars than in earlier-type stars. This can be due to the fact that the envelopes of late A-type and F-type stars are partially convective so that tidal forces are more efficient therein than in fully radiative envelopes. But, in contrast with earlier contentions (Levato 1976), the degree of synchronism does not appear to be stronger in "evolved" (subgiant, giant, supergiant) components than in "unevolved" (dwarf) ones.

Remarkably, the relatively high fraction of synchronized rotators with r<0.10 appears to be incompatible with current theoretical views on tidal interaction in early-type close binaries (Zahn 1977), even if stellar models including a plausible amount of overshooting are considered.

REFERENCES

Hut, P. 1981, Astron. Astrophys., 99, 126.
Levato, H. 1976, Astrophys. J., 203, 680.
Straižys, V. and Kurilienė, G. 1981, Astrophys. Space Sci., 80, 353.
Uesugi, A. and Fukuda, I. 1982, Revised Catalogue of Stellar Rotation
    Velocities, (Kyoto University, Kyoto).
Zahn, J.-P. 1977, Astron. Astrophys., 57, 383.

DISCUSSION

SCARFE: For the eccentric binaries, did you use the angular velocity at periastron to compare with the rotational angular velocity?

GIURICIN: Yes.

UV STANDARD STARS BY THE VARIABLE PROCRUSTEAN BED APPROACH

A. Heck

Astronomical Observatory, Strasbourg

D. Egret

Stellar Data Center, Strasbourg

Ph. Nobelis and J. C. Turlot

Center for Statistical Studies, Strasbourg

ABSTRACT.   Statistical algorithms have been appplied to a set of stellar low-dispersion IUE spectra in order to derive objective spectral classification schemes in the UV.   This paper describes how standard stars can be selected, particularly in the context of the VPB approach. Comparisons to existing schemes proved quite satisfactory.

1. INTRODUCTION

The preparation of the IUE Low-Dispersion Spectra Reference Atlas (Heck et al. 1984a) put at our disposal a large set of representative IUE low-dispersion stellar spectra. With this sample, we investigated the possibilities of constructing an objective clasification scheme in the UV (Egret and Heck 1983; Egret et al. 1984a,b; Heck et al. 1984b) which could be confronted with existing schemes, i.e., with the sequences published in the Atlas and established by a classical morphological approach (Heck et al. 1984a; Jaschek and Jaschek, 1984).   We also explored how standard stars could be selected statistically.

2. THE VPB APPROACH

The VPB approach is essentially based on an information condensation at the level of the spectral-continuum shape and an objective selection of the most discriminant spectral features. It is described in detail in Heck et al. (1984b) and has been applied to our original set of 384 IUE spectra. By spectrum we understand here a table of 410 absolute flux values covering the whole IUE range (1150-

403

D. S. Hayes et al. (eds.), Calibration of Fundamental Stellar Quantities, 403–406.

3200 Å) and binned by 5 Å steps. We must point out here that the VPB approach can deal with both reddened and unreddened spectra.

3. UV SPECTRAL GROUPS AND STANDARDS

In this application, an exploratory analysis has been carried out on the sample of reddened and unreddened normal stars with the SPAD package (Lebart and Morineau, 1982; Lebart et al., 1982) consisting of a principal component analysis (PCA) establishing the number of significant dimensions in the problem and then a cluster analysis aggregating the stars in groups around moving centers.

Thirty groups have been requested of the algorithm and aggregated on the basis of the mutual distances of the stars in the multivariate space defined by the first twenty PCA factors. The groups obtained are quite homogeneous and their distribution is illustrated in the planes of the first two PCA factors (Fig. 1a) and of the first and third PCA factors (Fig. 1b).

The first factor given by the PCA is strongly correlated with the effective temperature, while the second one has a functional relationship with it, and can be considered as parasitic. The third factor discriminates the luminosity among hot stars, but not among cool ones (likely due to the under-representation of these stars in the original sample). Other factors reveal no clear tendencies, apart from individual stellar influences.

Each group on the graphs has been named according to the stars closest to its barycenter where standard stars have also to be looked for. A comparison with the standards selected for the Atlas proved to be quite satisfactory, in the sense that Atlas standards lie close to the barycenters of the groups they should naturally belong to.

It is difficult to quantify the homogeneity in the groups, but the one achieved in this application is the best so far, and this, for a sample containing also reddened stars. In the earlier types, the groups rarely cover more than two subtypes. The discrimination is not as good in the later types for the reason stated earlier. The discrimination between the luminosity classes is of the same quality.

Requesting more groups of the cluster analysis might be a way to obtain a better discrimination both in spectral type and luminosity class, but we must keep in mind the limitations imposed by the size of the original sample if we want to derive statistically significant results. As to the peculiar stars, they are still a problem, but we think of dealing with them by an appropriate comparative weighting of the different variables.

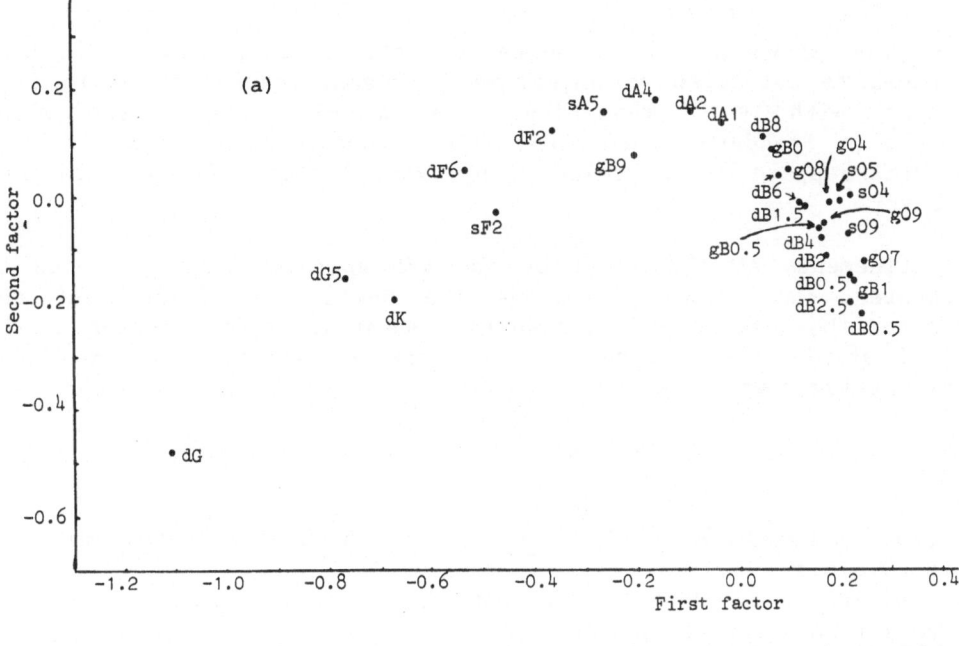

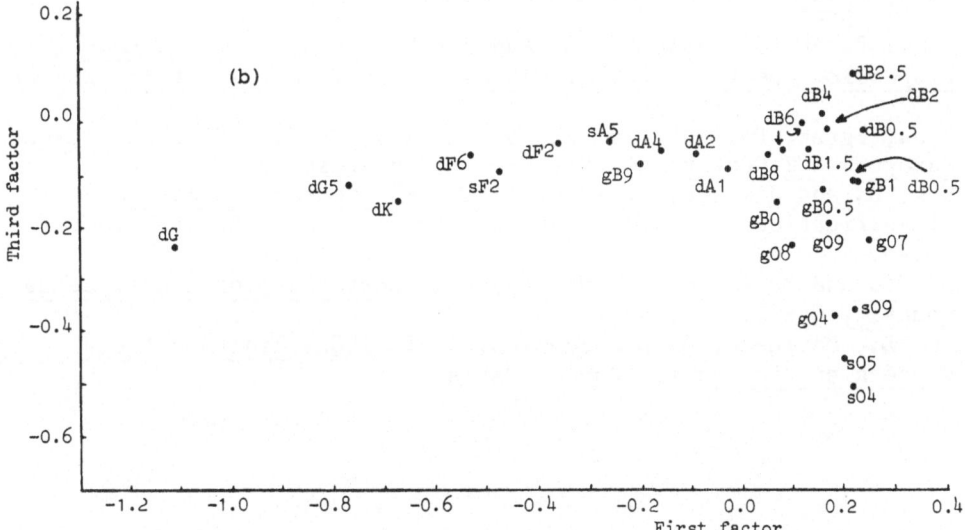

Fig. 1. Distribution of the group barycenters in the planes of the first two PCA factors (a) and of the first and third PCA factors (b).

4. CONCLUSION AND FINAL REMARKS

We have shown that it is quite feasible to use purely statistical approaches to establish an objective UV classification frame in full agreement with that resulting from a classical morphological approach. Standard stars for the groups constructed in the multivariate space can be selected by examining the distances to each barycenter.

Refinements at the level of the VPB approach have still to be implemented, which should improve the overall performance of the method and the quality of its results. An increase of the size of the original sample should also have a positive influence on the final discriminating power.

REFERENCES

Egret, D., and Heck, A. 1983, in Statistical Methods in Astronomy, ESA SP-201, p.59.

Egret, D., Heck, A., Nobelis, Ph. and Turlot, J. C. 1984a, Inf. Bull. Stellar Data Center, 26, 61.

Egret, D., Heck, A., Nobelis, Ph. and Turlot, J. C. 1984b, in Future of Ultraviolet Astronomy Based on Six Years of IUE Research, NASA CP, in press.

Heck, A., Egret, D., Jaschek, M. and Jaschek, C. 1984a, IUE Low-Dispersion Spectra Reference Atlas - Part I. Normal Stars, ESA SP-1052.

Heck, A., Egret, D., Nobelis, Ph. and Turlot, J. C. 1984b, in Fourth European IUE Conference, ESA SP-218, in press.

Jaschek, M. and Jaschek, C. 1984, in The MK Process and Stellar Classification, ed. R. F. Garrison (David Dunlap Obs.: Toronto), p.290.

Lebart, L. and Morineau, A. 1982, Système portable pour l'analyse de données, C.E.S.I.A., Paris.

Lebart, L., Morineau, A. and Fénelon, J. P. 1982, Traitement des données statistiques, (Dunod, Paris).

MKJ AND MSS CLASSIFICATION OF SOLAR-TYPE STARS WITHIN 100 PARSECS OF THE SUN: PRELIMINARY RESULTS

M. Fracassini, L. E. Pasinetti and M. Borella

Physics Department, University of Milan

A. Pasinetti

CERN, Geneva

ABSTRACT. A study of the distribution of spectral types of Solar Type Stars (STS) in the revised MKJ and MSS classifications is made on 3919 F8-K3 HD spectral-type stars brighter than m_v=10. By means of the solar color indices U-B and B-V 697 STS were selected. The spectral types G3V and G5V have the highest percentages in MSS and MKJ, respectively, confirming statistically the results published by Keenan and Pitts (1980) and by Hardorp (1982). The distribution of the color indices U-B and B-V in the revised G2V spectral type shows that these are good selection criteria for STS and are in the range $0.06 \leq U-B \leq 0.10$ and $0.58 \leq B-V \leq 0.65$.

1. INTRODUCTION

The search for solar spectral analogs, the stars that match the solar line spectrum as well its energy distribution, is clearly important for several astronomical domains. In particular, for problems of standards calibration, it is essential to know solar analogs in as many parts of the sky as possible, where the brightest G2 stars are contained.

The HR diagrams derived from the Michigan Spectral Catalogue show that the 4700 HD stars within 100 pcs have spectral types between F3 and G8 with a maximum at G0, and the majority of the 184 HD stars within 25 pcs are early G dwarfs (Houck and Fesen 1978).

An extensive survey of the HR and color-luminosity diagrams of nearly 500 stars in the solar neighborhood (within 32 pcs) shows also systematic effects on luminosity due to the errors in trigonometric parallax measurements but also, in a smaller part, to the spectral classification (Gliese 1978).

Moreover, the color and metallicity of the Sun relative to other stars remains controversial and there is evidence that the color and spectrum of the Sun are more similar to G3 and G4 than G2 (Barry 1978). Since 1978, Hardorp has performed extensive and careful investigations of solar spectral analogs. One of them in particular (Paper V; Hardorp 1982) has been very useful to the present study, which is a statistical study of solar-type stars within 100 pcs of the Sun by means of

407

D. S. Hayes et al. (eds.), Calibration of Fundamental Stellar Quantities, 407–410.
© *1985 by the IAU.*

color and spectral type.

2. DATA AND RESULTS

A list of 3919 stars, between F8 and K3 HD spectral types and magnitudes $m_v \leqq 10$, was kindly provided by the Centre de Données Stellaires of Strasbourg. This list contains equatorial coordinates (1950), U-B and B-V color indices, revised spectral types by Jaschek et al. (MKJ, 1964) and Houk and Cowley (MSS 1975, 1978), and a complete and updated bibliography for each star. Taking into account the solar color indices U-B= 0.20 and B-V=0.66 (Hardorp 1982) we have selected 697 STS which are divided into the following three groups:
1) Solar Type stars (ST, 48), with U-B = 0.20 ±0.02 and B-V = 0.66 ±0.01, which are reported in Table I;
2) Solar Analog stars (SA, 311), with $0.07 \leqq U-B \leqq 0.28$ and $0.62 \leqq B-V \leqq 0.71$, and the CN 3850 Å, Fe 3740 Å indices relative to the Sun (Hardorp 1982) in the range $\Delta CN, \Delta Fe = \pm 0.05$;
3) Solar Candidate stars (SC, 338) within the limits of the list (Table I) by Hardorp (1982) $0.02 \leqq U-B \leqq 0.38$ and $0.41 \leqq B-V \leqq 0.80$.

The distributions of the revised MKJ and MSS spectral types give the following maximum percentages:
1) ST : MKJ = G5V (46%), MSS = G3V (47%), MKJ+MSS = G5V (44%);
2) SA : MKJ = G5V (38%), MSS = G3V (45%), MKJ+MSS = G5V (34%);
3) SC : MKJ = G0V (45%), MSS = G0V (35%), MKJ+MSS = G0V (42%);

The whole group of the STS (arbitrarily weighted 3xST, 2xSA, 1xSC) give: MKJ = G5V (30%), MSS = G3V (36%), MKJ+MSS = G5V (28%). A secondary maximum in the spectral type G2V is shown for all groups in the MKJ and MKJ and MSS revision and in SC for the MSS revision. On the other hand, a secondary maximum in the spectral type G5V is shown for groups ST, SA and STS in the MSS revision.

The distribution of color indices for the spectral type G2V in the revised MKJ and MSS classifications gives:
MKJ : U-B (90) = 0.06 (9%), 0.12 (8%), $0.02 \leqq U-B \leqq 0.12$ (61%), $\overline{U-B}$ = 0.10;
     B-V(110) = 0.60(13%), 0.66(11%), $0.60 \leqq B-V \leqq 0.68$ (83%), $\overline{B-V}$ = 0.64;
MSS : U-B (22) = 0.07(13%), 0.19(13%), $0.02 \leqq U-B \leqq 0.12$ (55%), $\overline{U-B}$ = 0.11;
     B-V (50) = 0.60(20%), 0.64(16%), $0.59 \leqq B-V \leqq 0.66$ (64%), $\overline{B-V}$ = 0.62;
MKJ $\cap$ MSS : $0.06 \leqq U-B \leqq 0.10$, $0.58 \leqq B-V \leqq 0.65$.

3. CONCLUSIONS

These results confirm statistically those published by Keenan and Pitts (1980) and Hardorp (1982). The common agreement, as regards the revised MKJ and MSS classifications, among the solar groups emphasizes that the color indices U-B and B-V are good selection criteria for the STS and points out also the spectral homogeneity of these stars within 100 pcs of the Sun.

The range $0.58 \leqq B-V \leqq 0.65$, found for the G2V stars in common between the revised MKJ and MSS classifications, may be considered at present as the range of color precision for the spectral classification of ST stars

(Taylor 1984).

Finally, it would be useful to consider also the dynamical homogeneity of these solar groups. A study of the color-luminosity arrays of stellar groups within 20 pcs from the Sun was made by Woolley and Eggen (1958) showing that the Sun could belong to groups similar to Hyades and Praesepe. In this regard, a further investigation has been proposed (and accepted) for the ESA Astrometry Satellite Mission HIPPARCOS.

REFERENCES

Barry, Don C. 1978, IAU Symp. N° 80, The HR Diagram, A.G.D. Philip and D.S. Hayes eds., (D. Reidel, Dordrecht), p. 89.

Gliese, W. 1978, IAU Symp. N° 80, The HR Diagram, A.G.D. Philip and D.S. Hayes eds., (D. Reidel, Dordrecht), p. 79.

Hardorp, J. 1982, Astron. Astrophys. 105, 120.

Houk, N., Cowley, A.R. 1975, 1978, Michigan Spectral Catalogues I, II, Michigan Observatory, Ann Arbor.

Houk, N., Fesen, R. 1978, IAU Symp. N° 80, The HR Diagram, A.G.D. Philip D.S. Hayes eds., (D. Reidel, Dordrecht), p. 91.

Jaschek, C., Conde, H., De Sierra, A.C. 1964, Publ. Astron. Univ. Nac. La Plata, Ser. Astron. 28, N° 2.

Keenan, P.C., Pitts, R.E. 1980, Astrophys. J. Suppl. 42, 541.

Taylor, B.J. 1984, Astrophys. J. Suppl. 54, 167.

Woolley, R.v.d.R., Eggen, O.J. 1958, Mon. Not. Roy. Astron. Soc. 118, 57.

TABLE I. Solar-Type Stars

HD	U-B	B-V	m$_v$	MSS	MKJ	HD	U-B	B-V	m$_v$	MSS	MKJ
11112	0.22	0.65	7.13	G3V	G4V	101563	0.19	0.66	6.44		GO5
16141	.21	.66	6.83		G5IV	104212	.21	.66	8.38	G3V	
16417	.22	.66	5.79		G5IV	105585	.22	.67	8.49		G5V
27149	.22	.67	7.51		G5V	110314	.22	.65	8.31		G2V
27685	.22	.67	7.84		G4V	111398	.19	.66	7.07		G5V
29461*	.20	.66	7.96		G5V	112257	.18	.66	7.84		G2V
32963	.21	.66	7.58		G5IV	112502	.22	.66	9.86		G3V
34052	.20	.67	8.83		G2V	112753	.18	.65	7.97		G1V
44594	.20	.66	6.60	G3V	G4V	114174	.18	.67	6.80		G5IV
45289	.20	.67	6.65	G5V	G5V	132256	.22	.66	7.31		G2V
50787	.20	.65	8.33	G3V		141690	.19	.65	8.66		GOIV
64184	.19	.67	7.52	G5V		146233	.18	.65	5.48		GOV
66653	.18	.67	7.55	G5V		150680	.20	.66	2.81		GOIV
67738	.20	.65	8.46	G3IV		153344	.21	.67	7.06		G5IV
70081	.21	.66	8.91	G3V		159656	.19	.65	7.16	G2V	G4V
71334	.18	.67	7.82		G4V	165271	.21	.66	7.66	G5V	G5IV
74497	.20	.66	7.81	G3V	G3V	183877	.21	.67	7.15		G5IV
76440	.18	.65	8.44	G5V		186408	.18	.65	5.87		G2V
76943	.19	.65	5.97		G5V	186427*	.20	.66	6.20		G5V
78429	.20	.66	7.30	G5V		191069	.21	.67	8.11		G5V
78418	.20	.65	6.00		G5IV	206301	.20	.65	5.18		G2IV
90520	.21	.65	7.52	GOV	G3V	206828	.19	.65	8.45		G2V
93215	.19	.67	8.05		G5V	215274	.20	.67	8.03		G5V
98562	.21	.66	8.78		G2V	217014	.21	.67	5.49		G4V

MSS = Revised Spectral Types by Houk and Cowley (1975, 1978)
MKJ = Revised Spectral Types by Jaschek et al. (1964)
HD = Solar Analogs (* Closest Analogs) in the list of Hardorp (1982)

DISCUSSION

SCARFE: I notice that the (U-B) values for the 48 "solar" stars is much redder than the range for G2 V stars in the last part of the paper. This discrepancy is larger than for (B-V). Why was this value of (U-B) chosen for the "solar" group?

FRACASSINI: We adopted values of (U-B) or (B-V) from Hardorp (1982) for the "Solar Type" stars.

GARRISON: As I pointed out in my paper yesterday, it is extremely dangerous to take averages of inhomogeneous data to make any inference about the standard stars, including the Sun. Fracassini et al. were only interested in a sample of stars for future dynamical studies and this list serves their purposes, but I wish they would not try to make any inferences about either the color or the spectral type of the Sun. While the Houck (MSS) catalogue is homogeneous, the rest of the data they have used in their study is so inhomogeneous as to be completely inconclusive for determination of the color of the Sun. The exercise is, however, interesting, just so long as it is not misinterpreted or overinterpreted.

CODE: The (U-B) index of the Sun is in as good agreement with the (U-B) of solar analog stars as the (B-V) index. These comparisons go back as early as the work of Stebbins and Kron which still appear valid.

UV SPECTRAL CLASSIFICATION OF B STARS

Janet Rountree

University of Denver

George Sonneborn

Computer Sciences Corporation

Robert J. Panek

Raytheon Company

Previous studies of ultraviolet spectral classification have been insufficient to establish a comprehensive classification system for ultraviolet spectra of early-type stars because of inadequate spectral resolution. We have initiated a new study of ultraviolet spectral classification of B stars using high-dispersion IUE archival data. High-dispersion SWP spectra of MK standards and other B stars are retrieved from the IUE archives and numerically degraded to a uniform resolution of 0.25 or 0.50 Å. The spectra (in the form of plots or photowrites) are then visually examined with the aim of setting up a two-dimensional classification matrix. We follow the method used to create the MK classification system for visual spectra. The purpose of this work is to examine the applicability of the MK system (and in particular, the set of standard stars) in the ultraviolet, and to establish classification criteria in this spectral region.

The spectra displayed in the figure on the following two pages represent a sequence of MK-standard main sequence B stars, processed according to the methods to be used in this study. The spectra have been rebinned to 0.25 Å and normalized (not dereddened). Some prominent features, which may or may not prove to be useful as classification criteria, include Si II, 1260 Å; Si III, 1295-1300 Å; C II, 1335 Å; Si IV, 1393 and 1402 Å; and C IV, 1548 and 1550 Å. It should be emphasized that these MK standards are shown for illustrative purposes only -- they will not necessarily be standards for ultraviolet classification. We estimate that about 100 B stars from the IUE archives will be used in setting up the ultraviolet classification system. We intend to publish an atlas of IUE spectra of B stars with ultraviolet spectral types and classification criteria, so that other observers and users of archival data may visually place their objects within the grid of standard stars.

D. S. Hayes et al. (eds.), Calibration of Fundamental Stellar Quantities, 411–413.
© *1985 by the IAU.*

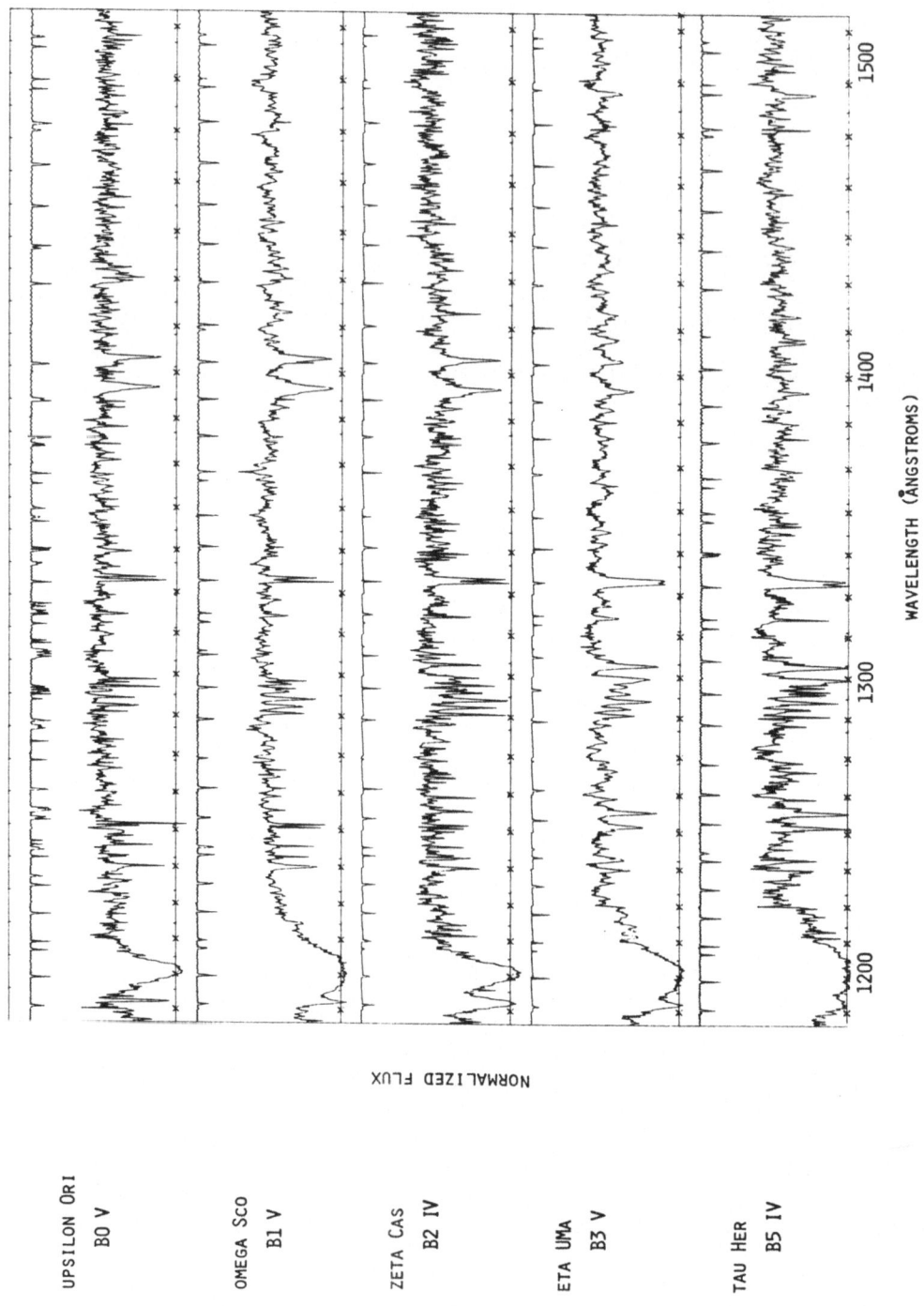

UPSILON ORI
B0 V

OMEGA SCO
B1 V

ZETA CAS
B2 IV

ETA UMA
B3 V

TAU HER
B5 IV

Fig. 1a.   Spectra of MK Standard Main Sequence B Stars.   (1200-1500 Å).

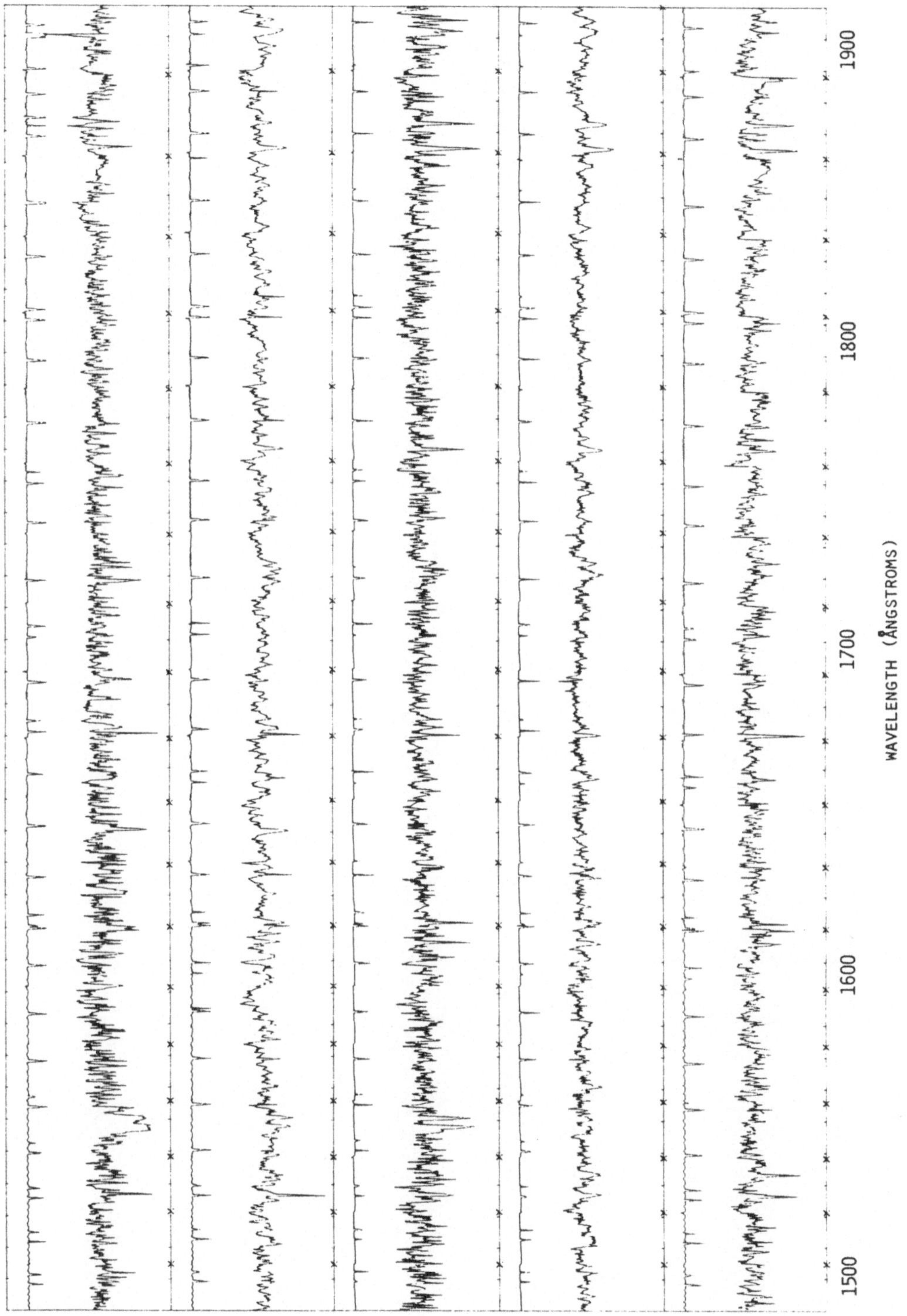

WAVELENGTH (ÅNGSTROMS)

Fig. 1b. Spectra of MK Standard Main Sequence B Stars. (1500-1900 Å).

SPECTRAL ANOMALIES IN THE HYADES AND PLEIADES AND IN FIELD STARS WITH ACTIVE CHROMOSPHERES

James A. Rose

Institute for Astronomy, University of Hawaii

ABSTRACT. Widened photographic image-tube spectra at 50 Å mm^{-1} dispersion have been obtained for a large number of late-type field dwarfs and giants with well-determined atmospheric parameters and for 35 Hyades dwarfs and 31 Pleiades dwarfs. The spectra have a resolution of 2.5 Å and cover the wavelength range $\lambda\lambda 3400-4500$ Å. A new quantitative three-dimensional spectral classification system is derived for late-type stars and is calibrated using the sample of field dwarfs and giants with known atmospheric parameters. Diagnostic indices are defined by comparing the counts in the bottoms of two neighboring absorption lines or by comparing the counts in two neighboring pseudocontinuum peaks.

It is found that the Hyades and Pleiades dwarfs and field dwarfs with strong Ca II H and K emission reversals exhibit well-defined anomalies in their diagnostic indices when compared with normal field dwarfs. These results are summarized in Rose (1984). In LaBonte and Rose (1984), it is shown, from spectra of solar magnetic plages, that all spectral and photometric peculiarities of the Hyades discovered to date, including the original "Hyades anomaly" (Crawford 1969; Strömgren et al. 1982), are manifestations of plage activity.

REFERENCES

Crawford, D. L. 1969, in Theory and Observation of Normal Stellar Atmospheres, O. Gingerich, ed. (MIT Press, Cambridge), p. 72.

LaBonte, B. J. and Rose, J. A.: 1984, in preparation for Pub. Astron. Soc. Pacific.

Rose, J. A.: 1984, Astron. J., 89, 1238.

Strömgren, B., Olsen, E. H. and Gustafsson, B.: 1982, Pub. Astron. Soc. Pacific 94, 5.

D. S. Hayes et al. (eds.), Calibration of Fundamental Stellar Quantities, 415–416.

DISCUSSION

GARRISON: In classification, we no longer rely on Fe/H ratios, as you have for all three of your indices. May I suggest that you add 4250/4254/4260 to your criteria since it is relatively free of abundance effects and is sensitive to temperature? It should be within your resolution possibilities.

ROSE: The Hyades and Pleiades show anomalies in a number of spectral indices that are not Fe/H ratios. Only two diagrams were exhibited in the poster paper; the rest are discussed in Rose (1984 Astron. J., 89, 1238). The 4250/4254/4260 lines that you refer to would be interesting to try; however they are near the resolution limit of my spectra and do not appear to be sensitive over the entire spectral range that I have considered.

GUSTAFSSON: How are your indices affected by stellar rotation? Have you looked for such effects, empirically and/or theoretically using synthetic spectra?

ROSE: I have artificially broadened the stellar spectra to simulate the effects of stellar rotation and of degradation in resolution of the observing equipment. For a Gaussian broadening with FWHM < 100 km/sec the effect is negligible. Only in extreme cases such as FK Comae (Vsini 100 km/sec) will stellar rotation affect the indices.

A NEW ANALYSIS OF THE ABUNDANCE STANDARDS ι HERCULIS AND τ SCORPII

Geraldine J. Peters

University of Southern California

Ronald S. Polidan

LPL, University of Arizona

ABSTRACT. Recent continuum and line data in the far ultraviolet for
ι Her (B3IV) and τ Sco (BOV) have been combined with published observations in the near uv and visible spectral regions and interpreted
with the aid of the Kurucz (1979) line-blanketed model atmospheres.
For ι Her and τ Sco, respectively, interpolated models of (T_{eff}/log g)
17500±500K/3.75±0.15, and 31500±1500K/4.3±0.2 fit the observed continua
from 900 - 6500Å and the profiles of Hγ and Hδ. Adopting these models,
solar abundances are suggested for both stars.

1. INTRODUCTION

Model atmospheres have successfully been employed to obtain effective temperatures, surface gravities, and elemental abundances for B
stars now for about twenty years (cf. Underhill and Doazan 1982 for a
review of techniques and results). Heretofore, however, except for a
very few limited investigations only ground-based spectroscopic data
have been analyzed using earlier blanketed or even unblanketed model
atmospheres. Utilizing recently acquired continuum observations from
the Voyager UVS, published high resolution line data from the Copernicus
satellite, and line blanketed model atmospheres of Kurucz (1979), we
have completed new abundance analyses for the sharp-lined B stars ι Her
(B3IV) and τ Sco (BOV) and summarize some of the results in this paper.
These efforts represent the initial steps of a more extensive
recalibration of the temperature scale and abundances for B stars.

2. ANALYSIS OF THE CONTINUUM

The far uv continua (900 - 1700Å) of ι Her and τ Sco have been observed with the Voyager ultraviolet spectrometers (UVS) with a spectral
resolution of about 20Å. We have combined these calibrated flux measurements with published near uv and visible data to provide uninterrupted
continuous spectra from 900 - 6500Å. These observations are compared
with our best fits to the Kurucz model continua in Figures 1 and 2.

D. S. Hayes et al. (eds.), Calibration of Fundamental Stellar Quantities, 417–421.
© *1985 by the IAU.*

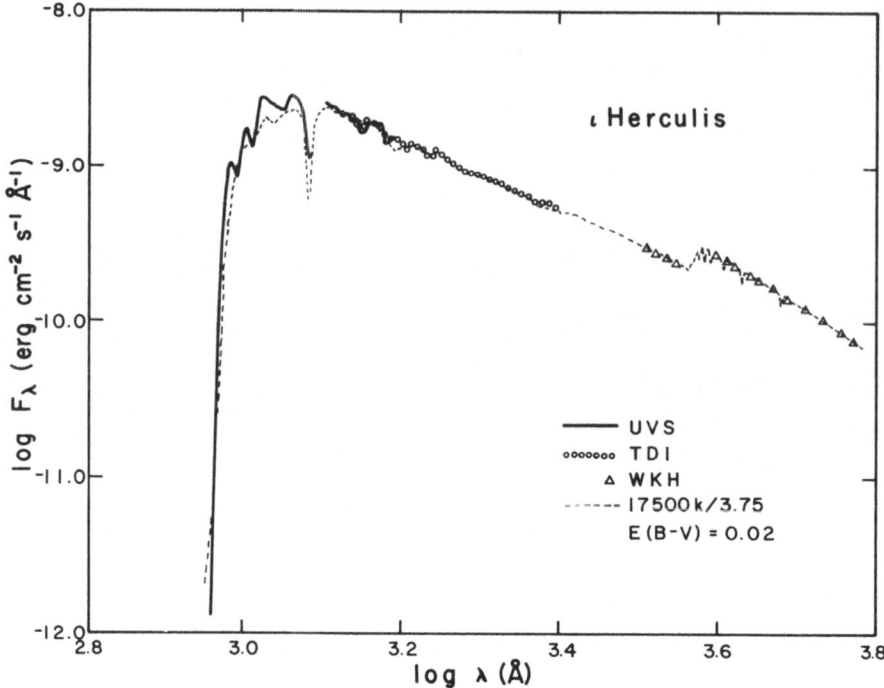

Fig. 1 – Combined <u>Voyager</u>, TD1, and ground-based (Breger 1976) flux data
for ι Her compared with Kurucz model which produced the best fit.

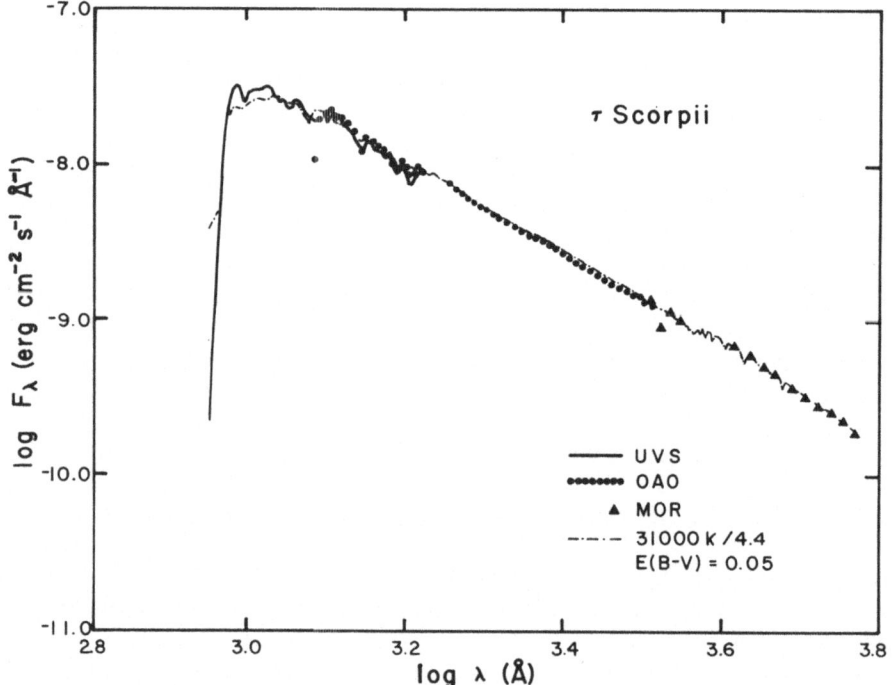

Fig. 2 – Same type of display as above but for τ Sco.

For ι Her and τ Sco, respectively, we obtain values of T_{eff}, log g of 17500±500K, > 3.5, and 31000±2000K, 4.4±0.2. The Voyager observations place tight constraints on the adopted model atmospheres since for ι Her and τ Sco, respectively, about 45% and 70% of their total flux is emitted in the spectral region covered by the Voyager instrumentation. The continuum shortward of the flux maximum is fit as well as the Rayleigh-Jeans tail of the distribution. To correct the observations for a small amount of interstellar reddening, we employed the extinction curve (extrapolated to 900Å) published by Savage and Mathis (1979). Uncertainties in the reddening produce only small errors in T_{eff} (<1000K at 30000K, <500K at 17000K) since the flux level and distribution shortward of 1200Å is so strongly dependent on temperature. The far uv data suggest that log g $\geq$ 4.2 for τ Sco as the model continua show a striking increase in the flux level <1200Å with increasing gravity for a star near 30000K. This effect is not seen at 17000K, however. For both program stars, the observed flux between 900-1100Å is slightly larger than predicted by the models. It is premature to state whether this mismatch might be a result of errors in the calibration or model computations.

3. ANALYSIS OF THE LINE SPECTRA

The technique for analysis and a description of the UCLA spectrum systhesis code which was used are given in Peters (1976). The far uv data from the Copernicus satellite (Upson and Rogerson 1980 and Rogerson and Upson 1977) provide valuable supplements to older ground-based observations (Peters and Aller 1970, Hardorp and Scholz 1970) allowing us to improve the published chemical compositions as well as determine abundances for Ti, V, Cr, and Mn which do not display measurable lines in the visible spectrum. Surface gravities obtained by fitting observed profiles of Hγ and Hδ to theoretical profiles (Kurucz 1979) are in agreement with the values of log g suggested by the energy distributions. The abundance results are summarized in Table I. Details will be discussed elsewhere but, in general, a solar composition is suggested for both stars within the uncertainties of the observations and analysis. Ionization balance is acceptable. In ι Her, there was excellent agreement between the iron abundance suggested by the visible multiplet 118 and uv multiplet 1 (1130Å).

REFERENCES

Breger, M. 1976, Astrophys. J. Suppl., 32, 7.
Kurucz, R. L. 1979, Astrophys. J. Suppl., 40, 1.
Hardorp, J. and Scholz, M. 1970, Astrophys. J. Suppl., 19, 193.
Peters, G. J. 1976, Astrophys. J. Suppl., 30, 551.
Peters, G. J. and Aller, L. H. 1970, Astrophys. J., 159, 525.
Rogerson, J. B., Jr., and Upson, W. L., II 1977, Astrophys. J. Suppl., 35, 37.
Ross, J. E. and Aller, L. H. 1976, Science, 191, 1223.
Savage, B. D., and Mathis, J. S. 1979, Ann. Rev. Astron. Astrophys., 17, 73.

Underhill, A. B., and Doazan, V. 1982, B Stars With and Without Emission
    Lines, NASA SP-456 (NASA, Washington, D. C.).
Upson, W. L., II, and Rogerson, J. B. 1980, Astrophys. J. Suppl.,
    42, 175.

TABLE I

THE CHEMICAL COMPOSITIONS OF ι Her[1] and τ Sco[2]

Ion	Number of Lines	Log N[3]	Number of Lines	Log N[3]	Log $N_\odot$[4]
		ι Her		τ Sco	
He I	9	10.90 ± .24	3	11.16 ± .13	10.8 ± .2
C II	32	8.42 ± .46	14	8.27 ± .46	8.62 ± .12
C III	1[5]	8.91	14	8.36 ± .33	
N II	28	7.89 ± .39	32	8.35 ± .42	7.94 ± .15
N III	$\cdots$	$\cdots$	14	8.38 ± .30	
O I	2[6]	8.99 ± .15	$\cdots$	$\cdots$	8.84 ± .07
O II	49	8.67 ± .44	75	8.69 ± .26	
O III	$\cdots$	$\cdots$	5	8.75 ± .30	
Ne I	10	8.64 ± .26			7.57 ± .12
Ne II	$\cdots$	$\cdots$	9	8.86 ± .37	
Mg II	7	7.34 ± .09	3	7.51 ± .19	7.60 ± .15
Al III	9	6.42 ± .24	$\cdots$	$\cdots$	6.52 ± .12
Si II	12	7.04 ± .47	$\cdots$	$\cdots$	7.65 ± .08
Si III	7	7.41 ± .44	13	7.51 ± .54	
Si IV	2	7.39 ± .06	7	7.34 ± .28	
P II	17	6.37 ± .55	$\cdots$	$\cdots$	5.50 ± .15
P III	3[7]	5.02 ± .60	$\cdots$	$\cdots$	
S II	67	7.17 ± .35	$\cdots$	$\cdots$	7.20 ± .15
S III	5	6.95 ± .36	9	7.09 ± .38	
Ar II	15	6.86 ± .59	3	8.17 ± .44	6.0 ± .2
Ca II	1[8]	6.16	$\cdots$	$\cdots$	6.35 ± .10
Ti III	9[7]	5.37 ± .55	5[7]	5.52 ± .63	5.05 ± .12
V III	14[7]	4.30 ± .83	3[7]	3.96 ± .66	4.02 ± .15
Cr III	9[7]	5.41 ± .63	3[7]	6.31 ± .75	5.71 ± .14
Mn III	8[7]	5.55 ± .40	4[7]	5.23 ± .13	5.42 ± .16
Fe II	15	6.82 ± .41	$\cdots$	$\cdots$	7.50 ± .08
Fe III	13[9]	7.50 ± .31	3	7.35 ± .18	

[1] T_{eff} = 17500K; log g = 3.75; ξ_T = 0 km s^{-1}
[2] T_{eff} = 31500K; log g = 4.3; ξ_T = 5 km s^{-1}
[3] N_H = 12.0
[4] Ross, J.E., and Aller, L.H. (1976)
[5] λ1247.38
[6] $\lambda\lambda$1304, 1306
[7] All FUV lines
[8] K-line
[9] Includes multiplets 118 and 1 uv

DISCUSSION

PETERS: I would like to make a few comments about the UV energy distributions we get from Voyager. Considering B-type MK standards, there is a striking functional dependence of the energy distribution of the star on the effective temperature, or spectral type of the star. (See Fig. 1)   The energy distributions for neighboring subtypes are clearly separated. There are 1 1/2 orders of magnitude in the flux at about 1000 Å between a B0 star and a B7 star. Note also the short-wavelength cutoff of the energy distribution is also very dependent on the spectral type of the star. From either of these dependencies one can get the effective temperature.

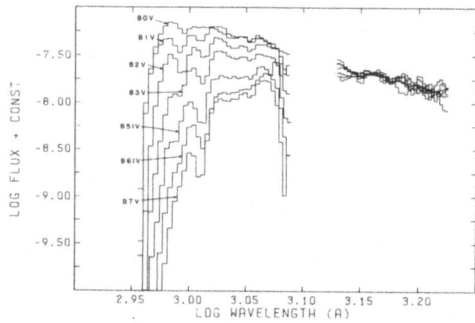

Fig. 1 - The fuv energy distributions (900 - 1700 Å) of selected B-type standard stars observed with the Voyager UVS and normalized from 1350 - 1450 Å.

ELEMENTAL ABUNDANCES OF NORMAL SHARP-LINED B AND A STARS FROM OPTICAL REGION ANALYSES

Saul J. Adelman

Department of Physics, The Citadel

ABSTRACT. Optical-region elemental-abundance analyses were performed for ten sharp-lined main sequence B and A stars. The derived abundances are generally in good agreement with those of the Sun. Multiple high dispersion spectrograms, fully line-blanketed solar composition model atmospheres, optical spectrophotometry, and the most accurate gf values were employed. This study provides initial parameters for studies of these stars in the ultraviolet and a consistent set of values for comparison with abundances of more exotic stars.

1. OPTICAL REGION ANALYSIS TECHNIQUES

Elemental-abundance analyses using optical region data and fully line-blanketed solar composition model atmospheres (Kurucz 1979) have been performed in a consistent manner for ten sharp-lined normal B5 through A2 type stars (see Table I) (Adelman 1984, 1985). The effective temperatures and surface gravities were determined by comparison of model predictions with optical-region spectrophotometry consistent with the Hayes-Latham (1975) calibration of Vega and Hγ profiles, respectively. For each star the equivalent widths were measured on several spectrograms, which were for the most part 4.3 Å/mm IIaO spectrograms obtained with the 2.5-m telescope of Mt. Wilson Observatory. Least-squares relations were derived to convert values derived from other types of spectrograms to this system. Microturbulent velocities were determined by minimizing the dependence of the equivalent widths of Fe I and Fe II lines on the derived abundances. Of order 20 lines is required to determine this quantity properly for each species. Fe I gf values based primarily on those of Blackwell and his collaborators (see, e.g., Blackwell et al. 1982) and Fe II gf values chosen to be as consistent as possible with them result in derived iron abundances which are in good agreement with solar values and in microturbulent velocities between 0.0 and 2.1 km s^{-1}, which are somewhat smaller than those found by previous investigators of similar stars (Baschek and Reimers 1969). When appropriate, corrections for non-LTE effects were made to the LTE abundances.

D. S. Hayes et al. (eds.), Calibration of Fundamental Stellar Quantities, 423–426.
© *1985 by the IAU.*

2. THE RESULTS

Table I contains the abundances for each star, the mean abundances of the normal stars, and solar values, consistent with the adopted gf values, from the literature. The mean abundances are in good agreement with solar values for helium, carbon, neon, magnesium, silicon, sulfur, calcium (from Ca I lines), scandium, titanium, chromium, iron, and nickel. The Ca II K line yields smaller values than the Ca I lines by an average of 0.28 dex. The average aluminum abundance is 0.28 dex less than solar, but the stars with Al I lines present show a trend of values with the hottest stars having the solar value and the coolest stars sub-solar values. The mean vanadium and strontium abundances are, respectively, 0.41 and 0.23 dex greater than solar. This suggests possible gf value scale offsets between my studies and those of the Sun. The mean offset of 0.05 dex for those stars with abundances derived from both Fe I and Fe II lines may be partially due to systematic errors in the Fe II gf values, particularly, of those lines with lower excitation potentials of several electron volts. The differences in the abundances of yttrium and zirconium between θ Leo and the Sun are difficult to interpret as Y II and Zr II lines are definitely present in only one normal star.

3. RELATED PROGRAMS

Elemental abundance analyses of two hot Am stars and three HgMn stars have also been performed in a manner consistent with this study (Adelman, Young, and Baldwin 1984, Adelman 1985). Such studies illustrate the importance of similar normal-star analyses in finding the magnitudes of the anomalous abundances. Ultraviolet region elemental abundance analyses of one hot Am and five normal stars, o Peg, π Cet, 21 Aql, 134 Tau, ν Cap, and θ Leo, using IUE data are being performed in collaboration with Dr. David S. Leckrone, NASA Goddard Space Flight Center. The use of parameters derived from the optical region studies as starting values for the ultraviolet studies is an important technique in analyzing these relatively complex spectra. In addition optical region analyses of several other B, A, and early F stars are in progress.

REFERENCES

Adelman, S. J.   1984, Monthly Notices Roy. Astron. Soc. 206, 637.
Adelman, S. J.   1985, in preparation.
Adelman, S. J., Young, J. M. and Baldwin, H. E.   1984, Monthly
    Notices Roy. Astron. Soc. 206, 649.
Baschek, G. and Reimers, D.   1969, Astron. Astrophys. 2, 240.
Blackwell, D. E., Petford, A. D., Shallis, M. J. and Simmons, G. J.
    1982, Monthly Notices Roy. Astron. Soc. 199, 43.
Hayes, D. S. and Latham, D. W.   1975, Astrophys. J. 197, 593.
Kurcuz, R. L.   1979, Astrophys. J. Suppl. 40, 1.

TABLE I. NORMAL STAR AND SOLAR ABUNDANCES

Species	τ Her log N/H	HR 2154 log N/H	HR 5780 log N/H	π Cet log N/H	21 Aql log N/H	5 Aqr log N/H	14 Cyg log N/H	134 Tau log N/H	ν Cap log N/H	θ Leo log N/H	Stellar Mean log N/H	Stellar Mean n	The Sun log N/H
He I	-1.04	-1.12	-0.98	-1.09	-1.13	-1.12	-0.99	-1.13	-1.26	-1.00	-1.09	10	-1.07
C II	-3.48	-3.55	-3.59	-3.45	-3.61	-2.86	-3.26	-3.43	-3.40	...	-3.40	9	-3.33
N II	...	...	-3.40:	-3.87	-3.64	...	...	...	...	...	...	...	-4.01
O I	-2.71:	-3.17	...	-3.50	-3.27	...	...	-3.31	-3.04	-3.25	-3.26	6	-3.08
Ne I	-3.91	-4.06	...	...	-4.23	...	...	...	...	...	-4.07	3	-4.15
Na I	...	...	...	...	...	...	...	...	-5.12	-4.59	...	...	-5.68
Mg I	...	...	...	...	...	...	...	-4.13	-4.13	-4.42	-4.23	3	-4.38
Mg II	-4.15	-4.28	-4.15	-4.31	-4.34	-4.19	-4.29	-4.39	-4.11	-4.40	-4.26	10	-4.38
Al II	...	...	...	...	...	-5.57	-5.71	-5.88	-5.86	-5.91	-5.79	5	-5.51
Si II	-4.59	-4.52	-4.38	-4.40	-4.39	-4.08	-4.14	-4.59	-4.42	-4.28	-4.38	10	-4.37
Si III	-4.80	-4.48	-4.28	-4.05	-4.05	...	...	...	...	...	-4.40	4	-4.37
S II	-4.67	-4.73	-4.58	-4.72	-4.71	...	...	...	...	-4.68	-4.68	5	-4.77
Ca I	...	...	...	...	...	...	...	...	-5.76	-5.57	-5.66	2	-5.66
Ca II	-6.03	-5.68	-5.78	-5.93	-5.69	-5.87	-6.07	-6.09	-6.17	-6.12	-5.94	10	-5.66
Sc II	...	...	...	...	...	...	...	-9.20	-9.14	-8.98	-9.11	3	-8.96
Ti II	...	-5.92:	-6.87	-7.15	-7.19	-6.99	-6.89	-6.99	-6.82	-6.85	-6.97	8	-7.02
V II	...	...	...	...	...	...	...	-7.26	-7.54	-7.35	-7.38	3	-7.79
Cr I	...	...	...	...	...	...	...	-5.80	-6.02	-6.26	-6.03	3	-5.88
Cr II	...	...	-5.48	-5.72	-6.12	-5.60	-5.74	-5.83	-5.60	-5.82	-5.78	7	-5.88
Mn I	...	...	...	...	...	...	...	...	...	-6.59	...	...	-7.16
Mn II	...	...	...	...	...	...	...	...	...	-6.49	...	...	-7.16
Fe I	-4.60	...	...	...	...	-4.30	-4.16	-4.29	-4.34	-4.25	-4.27	5	-4.37
Fe II	...	-4.79	-4.56	-4.45	-4.46	-4.22	-4.38	-4.50	-4.23	-4.21	-4.38	8	-4.37
Ni I	...	...	...	...	...	...	...	...	...	-5.55	...	...	-6.70
Ni II	...	...	-6.80	-6.82	-6.99	-6.32	-6.39	-6.68	-6.39	-6.19	-6.57	8	-6.70
Sr II	...	...	...	...	...	...	-8.91	-9.20	-8.88	-8.49	-8.87	4	-9.10
Y II	...	...	...	...	...	...	-8.39:	...	...	-9.50	...	...	-9.76
Zr II	...	...	...	...	...	...	...	...	...	-8.57	...	...	-9.44
Ba II	...	...	...	...	...	...	...	...	-9.44	-8.99	...	...	-9.91

DISCUSSION

ADELMAN: The normal stars being analyzed in addition to those in this paper are HR 6559 and Eta Lep. The latter star is classified FO IV. Some of the offset in the average abundance of the ten normal stars and that of the Sun may be due to small systematic errors in the gf values. One way to demonstrate such a hypothesis would be to perform a differential analysis of Eta Lep and the Sun.

HEINTZE: How do your new effective temperatures agree with others?

ADELMAN: They agree rather well with the scale by Code and his collaborators.

JASCHEK: I am happy to see that someone is doing a whole series of stars along the main sequence, and I hope you will publish the line identifications.

ADELMAN: Although I am not publishing complete line identifications with these stars, I am publishing the equivalent widths of all the lines used in the analysis as well as those of other unblended lines for which at present good gf values do not exist. I am planning to include some normal stars in my current program of HgMn stars at Dominion Astrophysical Observatory.

SEGGEWISS: Could you briefly summarize the most important abundance differences between "normal" A and Am stars as deduced from your studies?

ADELMAN: The hot Am stars are 2-3 times solar iron-rich. Most of the major anomalies occur for elements heavier than the iron peak. Adelman, Young and Baldwin (1984 Mon. Not. R. Astron. Soc. 206, 649) give the analysis of these stars.

THE RELATION BETWEEN THE ANOMALOUS CN STRENGTH, δ CN, AND [FE/H] ABUNDANCES

M. Fracassini,[1] L.E. Pasinetti,[1] L. Pastori[2] and R. Pironi[1]

[1]Department of Physics, University of Milan

[2]Astronomical Observatory of Milano-Merate

Following a previous study of stellar abundances (Pasinetti 1980), a relation has been derived between the DDO cyanogen-strength index and values of [Fe/H] reported in the catalog of Cayrel de Strobel, et al. (1980). From this relation we obtain [Fe/H] = 0.08 for the Hyades. This value is slightly lower than that given by Janes (1979), [Fe/H] = 0.13, and is in better agreement with other recent estimates of the Hyades metallicity (Cayrel de Strobel 1980). This improvement is due to the larger quantity of more precise spectroscopic data used in our correlation.

REFERENCES

Cayrel de Strobel, G., Bentolila, C., Hauck, B. and Curchod, A. 1980, Astron. Astrophys. Suppl., 41, 405.
Cayrel de Strobel, G. 1980, in IAU Symposium No. 85, Star Clusters, ed. J. E. Hesser (Reidel, Dordrecht), p. 91.
Janes, K. 1979, Astrophys. J. Suppl., 39, 135.
Pasinetti, L. E. 1980, in Cosmic Abundances of the Elements, ed. M. Hack, Mem. Soc. Astron. It., 51, 71.

D. S. Hayes et al. (eds.), Calibration of Fundamental Stellar Quantities, 427.

ON THE GRAVITY DETERMINATION FROM CARBON STAR SPECTRA

H. Gass

Institute of Theoretical Astrophysics, Heidelberg

ABSTRACT. Spectra of carbon dwarf and giant stars may look very similar, as is demonstrated, e. g., by TT CVn and G77-61. Model calculations show that the bands of C_2, CN and CO cannot be used in the range log g = 0... ...6 for log g determinations, since they are affected by abundance. If, moreover, lines of two ionization stages cannot be observed in any element, the only log g indicator would seem to be the ratio of the MgH band and the Mg I b lines. Simple analytic expressions are derived which demonstrate the relation between spectral appearance, abundances and gravity.

Generally, carbon stars are considered to be giants; the only exception up to now is G77-61 (Dearborn 1983), which has strong C_2 bands in its spectrum but should be a dwarf according to astrometric data (Dahn, Liebert and Hintzen 1977). In order to confirm this spectroscopically, and to determine the effective temperature, T_{eff}, and the element abundances, we perform a fine analysis mainly based on Lick IDS, MMT echelle and Palomar multichannel data, kindly provided by J. Liebert.

We calculate synthetic spectra for T = 4000 K and log g = 0 ....6. It turns out that by adjusting the C-abundance, the spectrum of G77-61 can be reproduced well by models in this whole range of gravity. This can be understood in the following way: Since the optical spectrum of G77-61 is completely dominated by the C_2 band, we approximate the absorption coefficient

$$\kappa = \kappa_o^C \, P_e P_g + \varepsilon_C^2 \kappa_o^L \, P_g^2 \; , \tag{1}$$

where P_g and P_e are the total gas and the electron pressure; ε_C is the fractional carbon abundance; κ_o^C and κ_o^L are constants. In all models the first term of Eqn. (1) reduces to

D. S. Hayes et al. (eds.), Calibration of Fundamental Stellar Quantities, 429–431.
© 1985 by the IAU.

$$\kappa = \varepsilon_C^2 \, \kappa_o^L \, P_g^2 \tag{2}$$

Insertion of this expression into the hydrostatic equation

$$\frac{dP_g}{d\tau} = \frac{gm_H \mu P_g}{\kappa kT} \tag{3}$$

(g=gravity, m_H=mass of the H-atom, μ relative molecular weight, k=Boltzmann constant, T=temperature) gives

$$\frac{dP_g}{d\tau} = \frac{gm_H \mu}{\varepsilon_C^2 \kappa_o^L kTP_g} \tag{4}$$

and by integration

$$P_g^2 = \frac{2gm_H \mu}{\kappa_o^L \varepsilon_C^2 k} \int \frac{d\tau}{T} \quad . \tag{5}$$

The monochromatic τ_λ-scale

$$\tau_\lambda = \int \kappa_\lambda(\tau)/\kappa(\tau) \, d\tau \quad . \tag{6}$$

(κ_λ=monochromatic absorption coefficient) is constructed by combination of Eqs. (2), (5) and (6)

$$\tau_\lambda = \frac{\kappa_o^\lambda}{\kappa_o} \tau \tag{7}$$

where κ_o^λ describes the wavelength dependence of the absorption coefficient.

If identical temperature distribution $T(\tau_\lambda)$ for all carbon stars with the same T_{eff} - but possibly quite different gravities - are assumed, then the flux emerging from the star

$$F_\lambda = 2 \int_0^\infty E_2(\tau_\lambda) \, B_\lambda(T(\tau_\lambda)) \, d\tau_\lambda \tag{8}$$

(E_2 is the second integral exponential function, B_λ=Kirchhoff-Planck-function) is independent of log g for all frequencies. Deviations in the temperature stratifications may be corrected to first order by changes of the abundances.

Similar arguments hold for CN and CO, so that they also cannot be used for the calculation of log g. If, moreover, atomic lines of two ionization stages cannot be observed for any element - as is the case for G77-61 -. the only gravity indicator seems to be the ratio of the strength of the MgH band and the MgI b lines. The latter determines the Mg abundance. In a He-rich atmosphere a higher Mg abundance is needed to get the same line strength as in a normal atmosphere. As a consequence the MgH band is not dependent on the H-abundance. If the intensity of the Mg is kept constant in all models with different gravity, then the strength of the MgH band, which is due to the calculations comparable with the C_2-bands in the case of log g = 5 and vanishes at all if log g = 3.5, is a log g indicator. In G77-61 a very weak MgH feature leads in this way to a gravity of log g = 4.

We think that this is the only way to determine the gravity where independent data on the luminosity - as e. g., for field carbon stars, are not available.

Details of this investigation will be published in a forthcoming paper.

ACKNOWLEDGEMENTS

Thanks are due to Dr. R. Wehrse for the supervision and inspiring discussions. Dr. J. Liebert kindly placed unpublished material at our disposal. This work is supported by the Deutsche Forschungsgemeinschaft (SFB 132).

REFERENCES

Dahn, C. C., Liebert, J. and Hintzen, P. M. 1977, Astrophys. J., 216, 757.
Dearborn, D. S. P. 1983, private communication.

A NOTE ON THE CALIBRATION OF EFFECTIVE TEMPERATURES

J. R. W. Heintze

"Sonnenborgh" Observatory, Utrecht

ABSTRACT. The discrepancy in the derived chemical compositions between B-A1 stars and stars of later spectral types can be removed if another T_{eff}-calibration is applied for A and F type stars than the one usually used.

1. MASS-EFFECTIVE TEMPERATURE RELATIONS

In the lower part of Figure 1 the experimental $(M/M_{\odot}, T_{eff})$ relations according to Habets and Heintze (1981) and Schmidt-Kaler (1982) are given together with the theoretical relation of Straižys and Kurileinė (1981) for stars of 2.3 to 0.9 solar masses. In that mass range the difference between Habets and Heintze and Schmidt-Kaler is considerable, except for the Sun. For masses exceeding five solar masses this difference becomes smaller. The relations of Straižys and Kurilienė, for the main sequence stars with masses larger than that of the Sun, agree rather well with that of Habets and Heintze (1981). This is not true for the gravities of stars of luminosity class III as can be seen from the upper part of Figure 1.

In Figure 1 some observed values are plotted also. Most of the main sequence or nearly main-sequence stars (log g in or close to V area) agree rather nicely with Schmidt-Kaler's relation [V805 Aql and EE Peg (Popper 1981), KM Hya (Andersen and Vaz 1984), RS Cha (Clausen and Nordström 1980), PV Pup (Vaz and Andersen 1984) and V1143 Cyg (Van Hamme and Wilson 1984)].

TY Pyx (Andersen et al. 1981) lies far below the Schmidt-Kaler main-sequence relation, (see section 3). The components of SZ Cen (Grønbech et al. 1977) lie below that relation, too; however, these components are of luminosity class IV at least. Components close to the Straižys and Kurilienė and/or Habets and Heintze relations are YZ Cas (h) (De Landtsheer 1983b and De Landtsheer and Mulder 1983), Vega (Dreiling and Bell 1980), Sirius (Bell and Dreiling 1981), AS Eri (h) (Van Hamme and Wilson 1984) and YZ Cas (c) (De Landtsheer 1983b and De Landtsheer and Mulder 1983). It has to be noticed, that for the hot main-sequence stars, the relations of Straižys and Kurilienė, Schmidt-Kaler and Habets and Heintze agree rather well with each other and with the positions of some well-studied stars/components [Zet Pup (Kudritzki et al. 1983, $M/M_{\odot} = 40$), VV

433

D. S. Hayes et al. (eds.), Calibration of Fundamental Stellar Quantities, 433–437.
© 1985 by the IAU.

Ori (Chambliss 1984, $M/M_{\odot}$ = 10.2, 4.5), V539 Ara (Andersen 1983, $M/M_{\odot}$ = 6.25, 5.33), CV Vel (Clausen and Grønbech 1977, $M/M_{\odot}$ = 6.10, 5.99), Zet Phe (Anderson 1983, $M/M_{\odot}$ = 3.93, 2.55), χ^2 Hya (Clausen and Nordström 1978, $M/M_{\odot}$ = 3.61, 2.64), TV Cas (h) (De Landtsheer 1983a, $M/M_{\odot}$ = 2.8) and RW Tau (h) (Plavec and Dobias 1983, $M/M_{\odot}$ = 2.55)]. U Cep, according to Plavec (1983), fits better than according to Tomkin (1981).

2. THE EFFECTIVE TEMPERATURES OF THE COMPONENTS OF YZ CAS

At the Utrecht Observatory we have tried to determine T_{eff} of the components of YZ Cas as accurately as possible. Lacy (1981) determined a very accurate mass ratio. Narrow-band lightcurves at 472, 672, 782 and 872 nm provided very accurate radii (De Landtsheer 1983b). As a result we determined $\log g_h$ = 3.974 ± 0.003 and $\log g_c$ = 4.295 ± 0.003. A low- dispersion long-wavelength IUE spectrum outside phase 0 could be fitted best with a Kurucz (1979) model of T_{eff} = 10300 K, $\log g$ = 4 and 10x solar abundance. With a high-dispersion short-wavelength IUE spectrum the overabundance of the metals turned out to be about 10x the solar abundance (De Landtsheer and Mulder 1983).

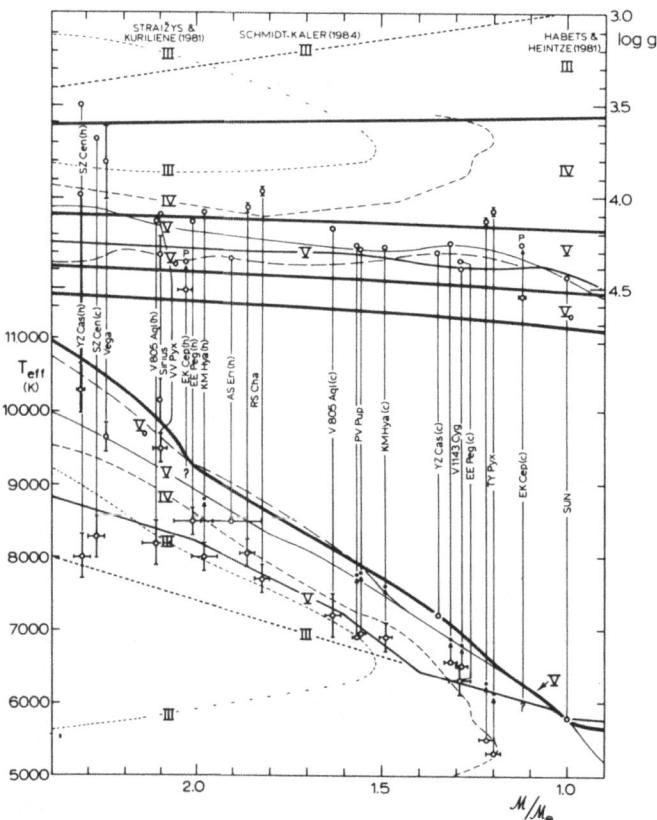

Fig. 1.   Theoretical and empirical ($M/M_{\odot}$, $\log g$) and ($M/M_{\odot}$, T_{eff}) relations with some recent observational results.

The light curves give for the cool component T_{eff} = 7200 K, in perfect agreement with the Habets and Heintze relation.

In the literature the observed (b-y) index of the primary of YZ Cas varies from 0.008 to 0.020 or even 0.036 (for references see De Landtsheer (1983b)). Eggen (1963) finds (B-V) = 0.005, De Landtsheer and Mulder (1983) find E (B-V) = 0.07. For $(B\text{-}V)_0$ = -0.02 the spectral type is B9.5 according to Popper (1980) and A0 according to Schmidt-Kaler (1982). The published spectral types lie between A1 V, A2 IV and A3.7m (see De Landtsheer 1983b for references.).

3. TY PYX

In the (log T_{eff}, log g) diagram, TY Pyx could not be fitted to Hejlesen's evolutionary tracks (Hejlesen 1980) with the correct mass and with (X,Z) = (0.60, 0.02), (0.70, 0.02), (0.70, 0.04) and (0.80, 0.02). Andersen et al. (1981) suggest that the colors of TY Pyx might be reddened by some unknown (presumably circumstellar) mechanism to which the Crawford calibrations do not apply and they propose to withhold final judgement until improved calibrations become available and until the physical phenomena occurring in RS CVn binaries and their causes are better understood.

However, giving the components of TY Pyx the effective temperatures as indicated by the points in Figure 1 (interpolated between the Straižys and Kurilienė relations for luminosity classes IV and V), the positions of these components in the (log T_{eff}, log g) diagram fall on the Hejlesen track with the correct mass and with (X,Z) = (0.70, 0.02). See Figure 2.

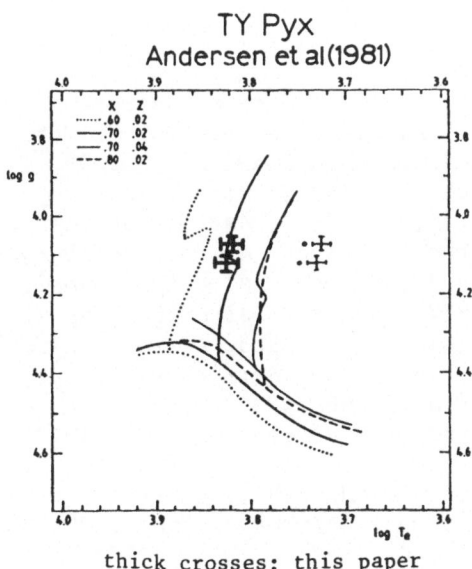

thick crosses: this paper

Fig. 2.   TY Pyx in the (log T_{eff}, log g) diagram.

4. CONCLUSIONS

Andersen et al. (1984a) have pointed out that B1 - A1 stars fit $(X,Z) \approx (0.70, 0.02)$, whereas stars of later spectral types fit $(X,Z) \approx (0.80, 0.02)$. With the same procedure as described in section 3 , the positions of PV Pup (Vaz and Anderson 1984) and KM Hya (Andersen and Vaz 1984) can be brought to the Hejlesen tracks with $(X,Z) \approx (0.70, 0.02)$. In the case of YZ Cas (see section 2) an excessive $E(B-V) = 0.07$ had to be applied whereas the distance is only about 75 pc. A remarkable example of a detached early A-type system, for which Andersen et al. (1984b) found a normal chemical composition, is VV Pyx. The same calibration methods, as used earlier by the Copenhagen group, were applied in this case. See Figure 1.

It would be interesting to find detached systems at high galactic latitudes with masses between 2.2 and 1 solar masses that feature a total eclipse. During the total eclipse energy distributions should be obtained as completely as possible to get reliable temperatures and chemical compositions of the atmospheres. It is hoped that $E(B-V)$ could be found in such cases unambiguously at the same time. In this way the determination of reliable T_{eff} [$M/M_\odot$, log g, abundance] relations should be possible. EK Cep could be a candidate (Tomkin 1983), although its galactic latitude is +12 degrees.

REFERENCES

Andersen, J. 1983, *Astron. Astrophys.* 118, 255

Andersen, J., Clausen, J.V., Jørgensen, H.E., Nördström, B. 1984, in IAU Symposium No. 105: Observational Tests of Stellar Evolution Theory, ed. A. Maeder and A. Renzini (Reidel, Dordrecht), p. 391.

Andersen, J., Clausen, J.V., Nordström, B. 1984, *Astron. Astrophys.* 134, 147

Andersen, J., Clausen, J.V., Nordström, B., Reipurth, B. 1981, *Astron. Astrophys.* 101, 7

Andersen, J. & Vaz, L.P.R. 1984, *Astron. Astrophys.* 130, 102

Bell, R.A. & Dreiling, L.A. 1981, *Astrophys. J.* 248, 1031

Chambliss, C.R. 1984, *Astrophys. Space Science* 99, 163

Clausen, J.V. & Grønbech, B. 1977, *Astron. Astrophys.* 58, 131

Clausen, J.V. & Nordström, B. 1978, *Astron. Astrophys.* 67, 15

Clausen, J.V. & Nordström, B. 1980, *Astron. Astrophys.* 83, 339

De Landtsheer, A.C. 1983a, *Astron. Astrophys. Suppl.* 52, 213

De Landtsheer, A.C. 1983b, *Astron. Astrophys. Suppl.* 53, 161

De Landtsheer, A.C. & Mulder, P. 1983, *Astron. Astrophys.* 127, 247

Dreiling, L.A. & Bell, R.A. 1980, *Astrophys. J.* 241, 736.

Eggen, O.J. 1963, *Astron. J.* 68, 483

Grønbech, B., Gyldenkerne, K., Jørgensen, H.E. 1977, *Astron. Astrophys.* 55, 401

Habets, G.M.H.J. & Heintze, J.R.W. 1981, *Astron. Astrophys. S.:* 46, 193

Hejlesen, P.M. 1980, *Astron. Astrophys. Suppl.* 39, 347

Kudritzki, R.B., Simon, K.P., Hamann, W.R. 1983, *Astron. Astrophys.* 118, 245

Kurucz, R.L. 1979, *Astrophys. J. Suppl.* 40, 1

Lacy, C.H. 1981, *Astrophys. J.* 251, 591
Popper, D.M. 1980, *Ann. Rev. Astron. Astrophys.* 18, 115
Popper, D.M. 1981, *Astrophys. J.* 244, 541
Plavec, M.J. 1983, *Astrophys. J.* 275, 251
Plavec, M.J. & Dobias, J.J. 1983, *Astrophys. J.* 272, 206
Schmidt-Kaler, Th. 1982, in: Landolt-Börnstein, New Series, Vol. 2,
  ed. in chief: K.H. Hellwege, Springer-Verlag, Berlin, p. 31,453,454
Straizys, V. & Kuriliene, V. 1981, *Astrophys. Space Science*, 80, 353
Tomkin, J. 1981, *Astrophys. J.* 244, 546
Tomkin, J. 1983, *Astrophys. J.* 271, 717
Van Hamme, W. & Wilson, R. E. 1984, *Astron. Astrophys.* 141, 1.
Vaz, L.P.R. & Andersen, J. 1984, *Astron. Astrophys.* 132, 219

DISCUSSION

NORDSTRÖM: Concerning TY Pyx and the stars mentioned in paragraph 4 of your poster paper, I would like to make the following comment: To me it seems strange to "correct" the temperatures of specific stars in order to fit the evolutionary tracks of a desired chemical composition when the composition of actual stars is not known to be constant. Moreover, accurate studies show that a given star cannot be expected to have exactly the mean temperature for all stars of the same mass within the main-sequence band, not even for the same log g.

HEINTZE: I am not "correcting" temperatures. I am only saying that I do not believe the published temperatures of some of the stars mentioned in spite of the fact that they were obtained by well calibrated indices. It is quite possible that some still unknown effects influence these indices. As long as two stars with about the same mass (about 2 $M_\odot$) and gravity (about 4.1) differ in T_{eff} as much as 1300 K (about 15%) [V 805 Aql with T_{eff} = 8200 K and VX Pyx with T_{eff} = 9500 K respectively] it is likely that something is wrong. Therefore, for V 805 Aql, just as in the case of TY Pyx, the comparison with theoretical evolutionary tracks is premature.

THE Be II λ 3130 Å REGION IN THE SPECTRA OF VEGA AND SIRIUS

R. F. Griffin and R. E. M. Griffin

Cambridge Observatories and Mount Wilson Observatory

Spectrograms of very high quality have been obtained of Vega and Sirius with the Mount Wilson 100-inch (2.5-m) telescope and coudé spectrograph. Examples of these plates, showing the Be II region in the ultraviolet, are exhibited. The reciprocal dispersion is 0.83 Å/mm (83 nm/m) and the FWHM is about 19 mÅ (1.9 pm). The spectrograms have a trailed width of 3 mm and are on IIa or IIIa emulsions.

No beryllium absorption can be detected in either star. The equivalent width of the Be II λ 3131.06 Å line is 0.7 ± 1.2 mÅ (70 ± 120 fm) in Vega and 0.38 ± 0.41 mÅ (38 ± 41 fm) in Sirius. The 1-σ upper limits of 1.9 and 0.8 mÅ are thought to correspond approximately to [Be/H] abundances of <-11.4 (Vega) and <-11.6 (Sirius).

We are very grateful to the Mount Wilson Observatory for our appointments there as Visiting Associates.

A paper describing our work on the Be II region has been submitted to Astronomy and Astrophysics.

DISCUSSION

MOROSSI: I have a colleague who is looking at the Be region with IUE showing a new possible employment of the International Ultraviolet Explorer.

D. S. Hayes et al. (eds.), Calibration of Fundamental Stellar Quantities, 439.
© *1985 by the IAU.*

MOUNT WILSON SPECTRA OF STANDARD STARS

R. F. Griffin and R. E. M. Griffin

Cambridge Observatories and Mount Wilson Observatory

We show part of a tracing of π Ceti, an example of a uniform series of high-quality tracings of standard stars covering the wavelength region 3850 - 4650 Å and derived from 10 Å/mm photographic spectrograms taken with the coudé spectrograph of the Mount Wilson 100-inch reflector.

DISCUSSION

MISSANA: You should publish lists of wavelengths and not just spectral atlases.

GRIFFIN: That is a separate and substantial task that we cannot guarantee to undertake.

JASCHEK: I want to stress the present importance of the generation of these atlases, and I hope you will get the new ones out as soon as possible. I would like to add a small comment with respect to the spectrum of Vega. My wife, Mercedes, has worked recently on the identification of lines in the infrared region, from 5000 - 10000 Å, and finds a lot of evident traces of Fe II in absorption.

D. S. Hayes et al. (eds.), Calibration of Fundamental Stellar Quantities, 441–442.
© *1985 by the IAU.*

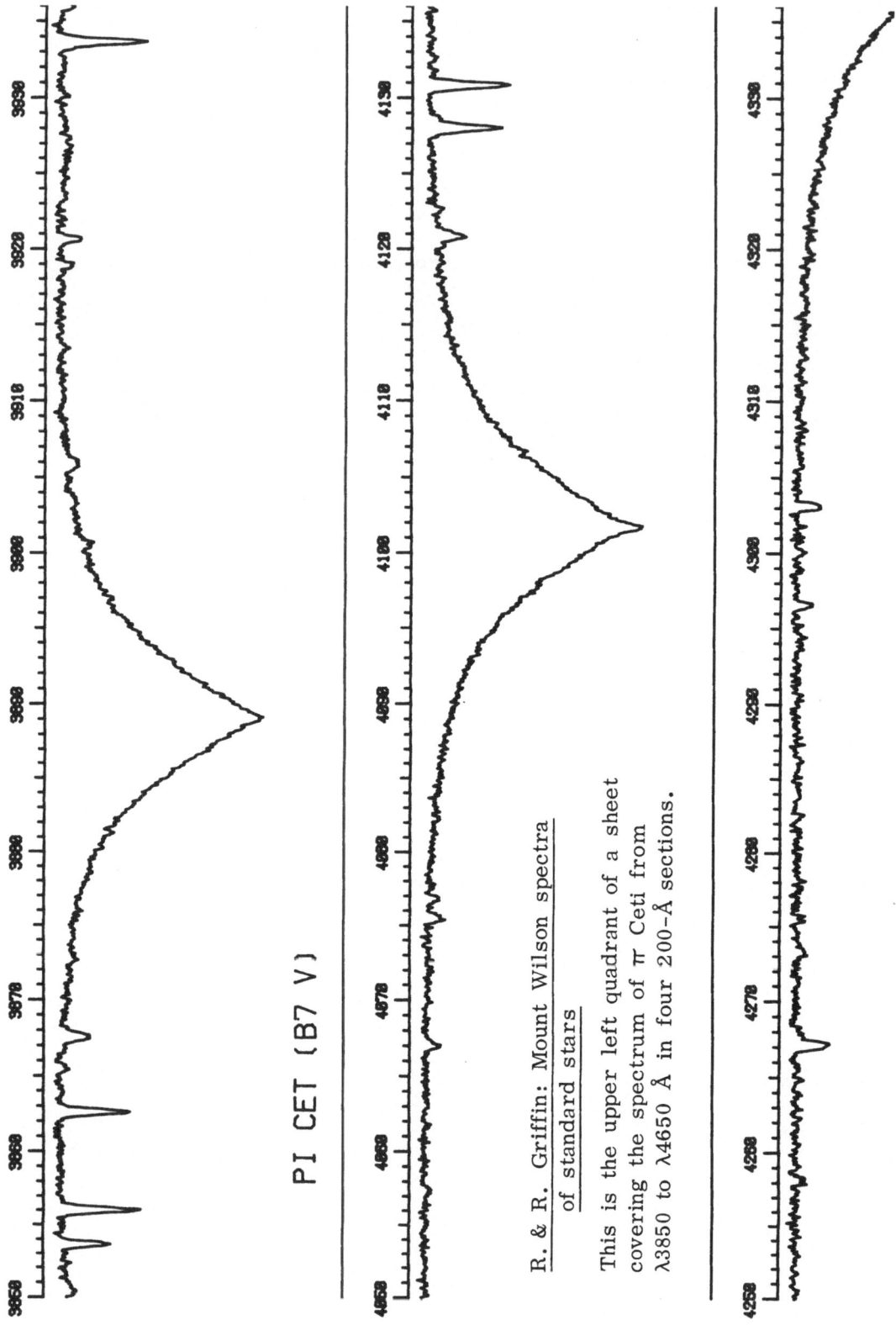

PI CET (B7 V)

R. & R. Griffin: Mount Wilson spectra
of standard stars

This is the upper left quadrant of a sheet
covering the spectrum of π Ceti from
λ3850 to λ4650 Å in four 200-Å sections.

ABOUT THE PHOTOMETRIC CALIBRATION OF IUE HIGH RESOLUTION SPECTRA:
QUANTIFICATION OF THE ORDER OVERLAP FOR THE SWP CAMERA

Luciana Bianchi

Astronomical Observatory of Turin

Ralph Bohlin

Space Telescope Science Institute

ABSTRACT. In order to quantify the errors in the IUE line profiles
caused by the order overlap, we have compared line depths in IUE and
Copernicus spectra. The excess line depth in IUE spectra suggests
that the amount of order overlap is about 32% at 1150Å and decreases
to zero at about 1400Å, for spectra extracted with the recent version
of IUESISPS (the IUE standard extraction software). The transfer of a
spectral feature from one order to the next is below the 5% level.

Based on these results, a correction technique is described.

1. INTRODUCTION

A longstanding problem with the IUE high resolution data is the
determination of the true background level in the region of the
spectral format where the orders are closely spaced. Order overlap is
primarily an artificially—raised background level caused by the
overlapping wings of the order profile in the higher echelle orders.
A secondary effect is the transfer of a spectral feature from one
order to the neighboring ones, due to the long range term of the point
spread function (PSF).

An earlier evaluation of the problem and a correction technique
is given by Bianchi (1980). In order to quantify the amount of order
overlap we have compared line depths in IUE spectra to line depths
observed by Copernicus. The excess line depth, i.e., the order
overlap, in IUE spectra can be expressed as a percent of the local net
continuum level. The net is the most appropriate quantity to scale
the order overlap, because gross and background are affected by the
radiation level and the camera null drift.

D. S. Hayes et al. (eds.), Calibration of Fundamental Stellar Quantities, 443–446.
© *1985 by the IAU.*

Fig.1. The measured order
overlap (in % of the net
flux). Typical error bar
is indicated for one
point.

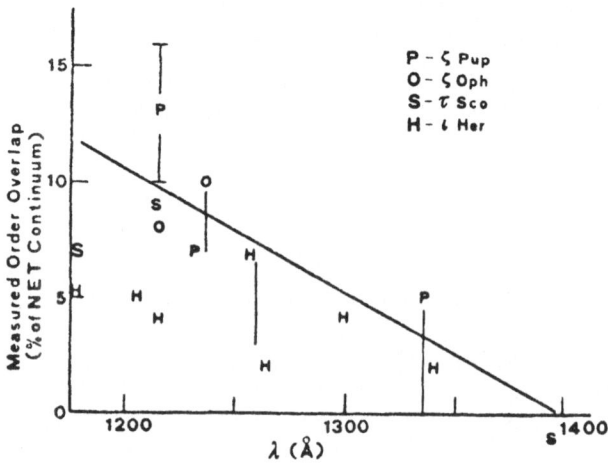

2. COMPARISON OF IUE AND COPERNICUS SPECTRA

The IUE resolution is 0.1Å, intermediate between the Copernicus
U1 and U2 spectra (resolution 0.05 and 0.2Å respectively); therefore
we have chosen absorption lines broad enough so that the U1 and U2
line depths agree within 5%. We further restricted the choice to
lines with central depth between 0 and 35% of the continuum.

We analyzed spectra of stars for which complete U1 and U2 scans
were available. All the IUE spectra were reprocessed with the recent
version of the standard software (Bohlin and Tunrose 1982). The
details of the data and the lines used are given in Bianchi and Bohlin
(1984). The measured order overlap is shown in Fig. 1. The amount of
order overlap is found to be about 32% of the net spectrum at 1150Å,
decreasing to zero at about 1400Å. In the spectra extracted with the
older software, the order overlap is worse by 10%. However, larger
errors could be present in these spectra, since the spectral
registration was less accurate and less stable (see e.g., Thompson and
Bohlin 1982).

3. CORRECTION TECHNIQUES

For spectra extracted with the old software a correction
technique is recommended such as that of Bianchi (1980). For spectra
extracted with the recent software, in which the background and the
spectral registration are more stable, a simple correction routine
based on the results of this paper is outlined here. It should give
results accurate to about 5%, for large aperture spectra (point
sources, in focus). Let's consider only the order m itself and the
two neighboring orders m-1 and m+1. The corrected net N_o can be
expressed in terms of the net n_o on the software tapes as:

$$N_o = n_o + \Delta B_o + (\Delta B_- + \Delta B_+)/2 - \Delta N_- - \Delta N_+ \qquad (1)$$

The subscripts $-$, o and $+$ refer to the orders $m-1$, m and $m+1$, respectively. The corrections ΔB are normalized to the extraction slit height and are due to the fact that the extraction background is too high. The corrections ΔN are the excess contribution to the gross from the wings of the adjacent orders. For the case of a deep line as measured by the correction C in Fig. 1, $\Delta B_o=0$. On the average, the nearby orders have approximately the same net continua n. With these assumptions and some knowledge of the order profile shape, a solution can be obtained. The precise PSF for IUE is not known, but Bianchi (1980) has shown that the core of the profile is gaussian with a longer range component in the wings. These wings produce the elevated background in the short wavelength orders and probably drop off as r^{-2} (r=distance from the peak of the order). Thus, if b is the background contribution due to one order, then this order contributes as an increase of b/4 to the neighboring net and as b/9 to the background on the other side of order m. In summary:

$$\Delta B_o=0, \quad \Delta B_-=\Delta B_+=b+b/9, \quad \Delta N_-=\Delta N_+=b/4, \quad n_-=n_+=n.$$

Eq. 1 becomes:

$$N_o-n_o=b+b/9 - b/4 - b/4 = 11b/18 \qquad (2)$$

Since the difference N_o-n_o is just what has been measured (Fig. 1), if C is the fractional correction in terms of n, then $b=18Cn/11$. The general solution is therefore:

$$N_o=n_o+1.636C\left(n_o\right)+0.5C\left(n_-\right)+0.5C\left(n_+\right) \qquad (3)$$

where $\left(\ \right)$ indicates the appropriately smoothed net spectrum. The appropriate smoothing is 31 points filter done twice, just the same as the background smoothing for the new software, since the correction is essentially for errors in the smooth background used to compute the net on the tape. In the case where the three continua are all equal, then $(N_o-n_o)/(n_o)=2.636C$ (eq. 3), which is used to estimate the maximum order overlap of 32% when C=0.12 at 1150Å.

A FORTRAN program to implement this correction is given in Bianchi and Bohlin (1984). In Fig. 2 we show an example of a corrected IUE line profile (SiIII λ1206.5Å in zeta Pup) compared to the line profile from Copernicus.

REFERENCES

Bianchi, L. 1980, in IUE Data Reduction, ed. W. Weiss et al., (Vienna: Austrian Solar and Space Agency), p.161.
Bianchi, L. and Bohlin, R. 1984, Astron. Astrophys., in press.
Bohlin, R. and Tunrose, B. 1982, NASA IUE newsletter N.18, p.29 and ESA IUE newsletter N.13, p.14.
Thompson, R. and Bohlin, R. 1982, NASA IUE newsletter N.18, p.45.

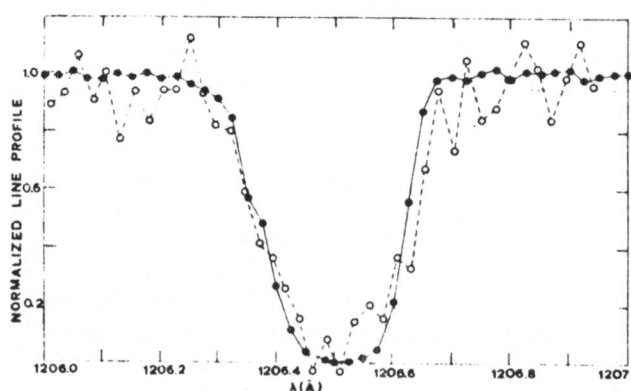

Fig.2. An IUE line profile
corrected by our technique
(open circles) and the U1
scan from <u>Copernicus</u>
(filled circles) with
0.05Å resolution.

DISCUSSION

BIANCHI: I want to point out that an error of 32% in the continuum is an
error of 32% in the equivalent width.

DETERMINATION DES DIAMETRES STELLAIRES PAR OCCULTATIONS. TECHNIQUE DE REDUCTION.

Michel Froeschlé

C. E. R. G. A.

Georges Helmer

Nice Observatory

Claude Meyer

C. E. R. G. A.

ABSTRACT. Lunar occultations provide measurements of stellar angular diameters, leading thus to a determination of their effective temperatures. Basically, the data reduction process relies on two methods: model fitting and deconvolution. "Integrated deconvolution", presented here, is a derived method of deconvolution and its results can compare with the two methods. The purpose is to determine the variation of the uncovered surface of the source throughout the occultation instead of extracting the brightness profile of the source. The main advantage provided is that computing the signal derivative is no longer necessary. In addition to saving a lot of computing time, the method affords a good estimate of the apparent speed, precise dating, and the angular separation of double stars down to 0.002 arcsec. After being tested against the usual methods, "integrated deconvolution" is now currently used. Some results are presented here.

1. INTRODUCTION

Lorsque l'on observe l'occultation d'une étoile par la Lune, on constate au voisinage de l'ombre géométrique la présence de franges de diffraction, parallèles au bord lunaire. L'aspect de ces franges dépend de nombreux paramètres.

Les paramètres physiques tels que l'angle de position de l'occultation, la vitesse apparente de la Lune, le diamètre angulaire de l'étoile ou son éventuelle duplicité, le relief lunaire et la scintillation atmosphérique, ont été étudiés par de nombreux auteurs (Nather et Evans, 1970 ;

D. S. Hayes et al. (eds.), Calibration of Fundamental Stellar Quantities, 447–453.
© *1985 by the IAU.*

Knoechel et von der Heide, 1978).

Les effets des paramètres instrumentaux tels que le diamètre du télescope, l'échantillonnage temporel du phénomène observé ou la bande
passante spectrale des filtres utilisés ont été souvent décrits (Ridgway,
1977).

Toutes les méthodes de réduction modélisent d'une façon ou d'une
autre ces différents paramètres. Il existe essentiellement deux méthodes
de réduction qui permettent de déterminer les instants d'occultations
géométriques, les diamètres apparents d'étoiles et une composante de la
séparation angulaire d'étoiles doubles. Ce sont la méthode par ajustement
des paramètres et la méthode par déconvolution.

1.1. Méthode par ajustement des paramètres

C'est la plus ancienne. Elle consiste à ajuster le signal observé
à un signal théorique. Cette méthode a surtout été développée par Nather
et McCants (1970). Pour chacun des N points observés, on calcule la valeur du modèle et on exprime les résidus sous la forme de N équations
que l'on traite par la méthode des moindres carrés pour obtenir les corrections aux valeurs initiales des paramètres de modélisation. On est
amené à utiliser un processus itératif pour affiner la valeur de ces corrections. Dans le cas d'une étoile double, on recherche la séparation ρ
et l'écart de magnitude Δm par la superposition de deux modèles du type
précédent.

1.2. Méthode par déconvolution

Développée par les radioastronomes (Scheuer, 1962) pour déterminer
le diamètre de radiosources, la méthode consiste à résoudre l'équation
de convolution

$$I(x) = F(x) * O(x) \tag{1}$$

où $I(x)$ est l'intensité observée, $F(x)$ l'intensité correspondant à une
source ponctuelle et $O(x)$ la fonction de distribution de luminosité de
la source.

L'équation (1) peut encore s'écrire, par dérivation,

$$I'(x) = F'(x) * O(x) \tag{2}$$

et on peut en déduire le profil de la source par

$$O(x) = \overline{F'(-x)} * I'(x) \tag{3}$$

en utilisant la propriété $\overline{F'(x)} * F'(-x) = \delta(x)$, ce qui caractérise la
fonction représentative de la diffraction.

Dans le domaine optique, l'équation (3) n'est pas facile à traiter.

Elle utilise la dérivée du signal observé et, compte tenu du bruit, il faut obligatoirement filtrer.

2. METHODE DE DECONVOLUTION INTEGREE

Les difficultés liées à la dérivation du signal observé nous ont conduits à développer une solution qui consiste à intégrer l'équation (3), ce qui nous fait calculer non plus le profil $O(x)$ de la source mais la variation de sa surface pendant l'occultation, c'est à dire à chaque instant l'aire $S(x)$ délimitée par le profil de la source et le bord de l'écran, en l'occurence le bord lunaire (Froeschlé, Meyer, 1983).

Dans le cas monochromatique, on écrira à la longueur d'onde λ_O

$$S_{\lambda_O}(x) = \int_{-\infty}^{x} O_{\lambda_O}(u)\, du$$

ou $\quad S_{\lambda_O}(x) = C_{\lambda_O}(x) * H(x)$

En utilisant l'équation (3), on obtient finalement

$$S_{\lambda_O}(x) = I_{\lambda_O}(x) * \overline{F'_{\lambda_O}(-x)}$$

C'est la convolution du signal observé par la transposée de la dérivée du signal théorique d'une source ponctuelle.

La méthode s'applique aussi bien au cas d'étoiles simples qu'au cas d'étoiles doubles. Les figures (1) et (2) montrent respectivement le signal simulé et la déconvolution intégrée correspondante dans chacun de ces cas.

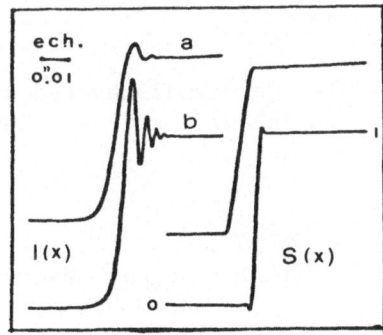 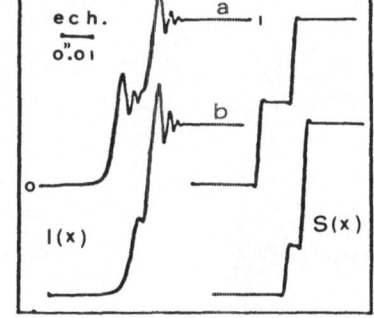

Fig. 1. Simulation étoile simple
(a) diamètre apparent 0".01
(b) diamètre apparent 0".002
I(x) : signal simulé
S(x) : déconvolution

Fig. 2. Simulation étoile double
(a) séparation ρ = 0".01 ; Δm = 0
(b) séparation ρ = 0".005; Δm = 1
I(x):signal simulé
S(x):déconvolution

Le problème est devenu un simple problème géométrique car, si la source a un profil circulaire de rayon r, on peut représenter l'aire par une forme analytique

$$A(x) = \frac{1}{\pi}\left(\frac{x}{r}\left(1 - \frac{x^2}{r^2} \right)^{1/2} + \arcsin\frac{x}{r} + \frac{\pi}{2} \right) \quad \text{pour } -r \leqq x \leqq r$$

A partir d'une valeur initiale r_0 du rayon, on peut linéariser l'équation précédente pour déterminer des corrections Δr à r_0. On calcule Δr par moindres carrés en comparant $S(x)$ obtenu par convolution à $A(x)$ théorique.

3. RESULTATS

Le tableau I donne les diamètres angulaires obtenus lors de six observations effectuées d'une part à l'aide d'une instrumentation prototype de l'observatoire de Nice avec un télescope de 40 cm et, d'autre part, à l'aide d'une instrumentation en cours de développement sur le site du CERGA et qui utilise le télescope de 150 cm de diamètre affecté à la télémétrie laser-Lune.

Tableau I. Diamètres mesurés au CERGA et à l'observatoire de Nice.

SAO	m	Sp	date	diamètre télescope (m)	λ (μm)	bande passante (μm)	diamètre x 10^{-3} arcsec (disque uniforme)
78297	3,2	MO	09.10.82	1.50	0,77	0.05	14.89 ± 1.14
93954	3,6	KO	24.01.83	1.50	0,77	0.04	3.35 ± 0.30
94027	1,1	K5	30.09.80	0.40	0,52	0.04	18.77 ± 0.70
94027	1,1	K5	30.09.80	0.40	0,433	0.05	19.51 ± 0.78
119035	4,2	MO	31.01.83	1.50	0,77	0.04	7.03 ± 0.56
158427	4,3	KO	17.07.83	1.50	0,77	0.05	3.15 ± 0.22

La figure (3) montre, à titre d'exemple, le signal observé au CERGA le 24/01/83 ainsi que la déconvolution intégrée correspondante.

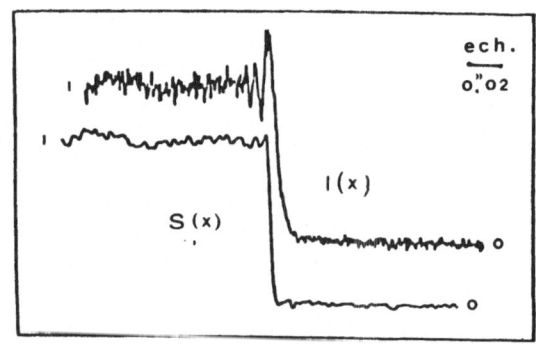

Fig. 3

I(x) : signal observé de SAO 93954

S(x) : déconvolution intégrée

La précision limite accessible par cette méthode est de l'ordre de 0".002. Si le bruit ne dépasse pas 5 % du signal, on peut atteindre les objectifs astrophysiques et astrométriques fixés. La datation précise des instants d'occultations permet de contribuer au rattachement des systèmes de référence géométriques et dynamiques (Froeschlé, Meyer, 1981). La haute résolution angulaire atteinte permet la mesure de la séparation d'étoiles doubles serrées avec un écart de magnitude important et, en association avec des mesures photométriques, la détermination du diamètre apparent des enveloppes stellaires et la mesure des températures effectives (Rigdway, 1980).

Remerciements : Nous remercions l'équipe du laser-Lune du CERGA qui a mis à notre disposition le télescope de 1,50 m, et en particulier J.F. Mangin et J.M. Torre qui ont participé activement aux observations.

REFERENCES

Froeschlé, M. et Meyer, C. 1981, in IAU Colloquium No. 56: Reference Coordinate Systems for Earth Dynamics, ed. E. M. Gaposhkin and B. Kołaczek (Reidel, Dordrecht), p. 317.
Froeschlé, M. èt Meyer, C. 1983, Astron. Astrophys., 121, 319.
Knoechel, G. et von der Heide, K. 1978, Astron. Astrophys., 67, 209.
Nather, R. E. et Evans, D. S. 1970, Astron. J., 75, 575.
Nather, R. E. et McCants, M. M. 1970, Astron. J., 75, 963.
Ridgway, S. T. 1977, Astron. J., 82, 511.
Ridgway, S. T. 1980, Astrophys. J., 235, 126.
Scheuer, P. A. G. 1962, Austr. J. Phys., 15, 333.

DETERMINATION OF STELLAR DIAMETERS BY OCCULTATIONS.   REDUCTION TECHNIQUE

SUMMARY. This paper is concerned with a method of reduction of occultation measurements of stellar diameters which the authors call "integrated deconvolution". They briefly discuss two older methods: "parameter adjustment" and "deconvolution". The former is iterative, while
the latter - devised by radio astronomers - is not easy to apply in the
optical domain, because it is based on the derivative of the observed
signal, which must be filtered to eliminate the noise.

The authors, therefore, devised their new method, which is based
on an integration of equation 3 (see Section 1.2 of the French text)
which enables them to obtain the convolution of the observed signal
with the transform of the derviative of the theoretical signal from a
point source (Section 2). The method can be applied either to single
stars or to double stars. The method reduces the problem to a simple
geometric one since the area of an occulted circular source may be
represented by an analytic formula, and the radius of the source can be
obtained from differential corrections determined by least squares.

Results for six stars are presented in the Table. They were
obtained with an 0.4-m telescope at Nice and the 1.5-m telescope at
CERGA normally used for lunar-laser telemetry. The limiting precision
of the method is about 0".002. Precise timing of occultations permits
the relation of the geometric and dynamic frames of reference, while the
high angular resolution enables close double stars of large magnitude
difference to be measured, or, in association with photometric measurements, the determination of the diameters of stellar envelopes and of
effective temperatures.

DISCUSSION

EVANS: It is not true that all that matters is the sensitivity of the detector and not the size of the telescope. Larger telescopes produce lower seeing noise but in principle for a large telescope a correction must be applied because the event takes place at one time on one edge of the aperture and later at the other.

FROESCHLÉ: The choice of the telescope diameter is a compromise between the wish to average out the atmospheric effects and the wish not to diminish excessively the contrast between the diffraction fringe observed. On the other hand the diameter is taken into account in the data reduction method.

JASCHEK: I would like to encourage the specialists to compose a catalogue of all occultation measurements.

EVANS: I try to keep a catalogue of all photoelectric occultation observations. Please send me data.

JASCHEK: Is it on tape?

EVANS: No.

MEYER: A catalogue already exists in which the diameters of stars, obtained by direct and indirect methods, are compiled. It should be updated, because now the measures are scatted all through the literature.

STANDARD STARS FOR THE COMPARISON OF METHODS OF DIAMETER DETERMINATION

L. Pastori[1], L. E. Pasinetti[2], E. Antonello[1] and G. Malaspina[1]

[1]Astronomical Observatory of Milano-Merate

[2]Department of Physics, University of Milan

Fracassini, et al. (1983) suggested a preliminary list of stars suitable for use as standards for methods of determination of stellar apparent diameters and absolute radii. According to their criteria, only stars with at least three methods of determination were considered; moreover, only if the percentage error of the data was < 10% were they considered as "standards". This simple statement is not completely satisfactory, because of some ambiguous cases. For instance, the interferometric method (code 1), Wesselink's method (6H) and Barnes and Evans' (6G) method are not independent.

In order to supply a useful set of stars for the comparison of methods of diameter determination, more restrictive criteria are proposed: we designate as 'primary standards' those stars whose diameter or radius is determined by at least a) three independent methods and has a percentage error ε less than 10% (the most common error in the stellar dimension determinations), or b) two direct methods (generally considered more reliable than the indirect ones) with $\varepsilon < 5\%$. Table I gives the only nine stars we found in CADARS (Fracassini, et al. 1981) which satisfy the criteria. The columns give, respectively: the identification, the spectral type (Hoffleit, et al. 1982), the averaged values of d" (uniform disk) or $R/R_\odot$ with their standard deviations, the number of values available and a code designating the method (Fracassini, et al. 1981). The small number of primary standards emphasizes the need for many other determinations to cover all spectral types and luminosity classes in the HR diagram.

At present, to extend this set of stars, we designate as 'secondary standards' those stars whose dimensions are determined by at least two independent methods, of which one is a direct method; moreover, the percentage error has to be less than 5%. Table IIa gives 17 secondary standards for measurements of stellar apparent diameters; most of them are giants of late spectral type. Other stars belonging to CADARS and satisfying the above-mentioned criteria were disregarded owing to their peculiarity with respect to the different methods (Pastori et al. 1985).

D. S. Hayes et al. (eds.), Calibration of Fundamental Stellar Quantities, 455–457.
© *1985 by the IAU.*

TABLE I. Primary standards

HD	Name	Sp.Type	d"(arcsec) or R/R$_\odot$	n	Code
14386	o Cet	M7IIIe	0.0571 ± 0.0010	4	1,3B,6G
29139	α Tau	K5III	0.0200 ± 0.0010	5	1,2,6G
48915	α CMa	A1Vm	0.0060 ± 0.0004	6	1,6F,6G,6I
124897	α Boo	K1IIIb	0.0201 ± 0.0010	5	1,3B,6G
148478	α Sco	M1Iab-Ib	0.0409 ± 0.0020	5	2,3B,6G
172167	α Lyr	A0Va	0.0032 ± 0.0002	3	1,6G,6F
48915	α CMa	A1Vm	1.83 ± 0.13	5	4,6F,6H,6I
61421	α CMi	F5IV-V	2.207 ± 0.029	3	4, 6G,6I
213306	δ Cep	F5Ib-G2Ib	41.4 ± 1.58	15	5,6G,6I

TABLE IIa. Secondary standards: apparent diameters

HD	Name	Sp.Type	d"(arcsec)	n	Code
18191	45 Ari	M6III	0.01009 ± 0.00039	4	2,6I
38307	γ Tau	C5II	0.00838 ± 0.00028	3	2,6I
47105	γ Gem	A0IV	0.00130 ± 0.00003	2	1,6F
86663	π Leo	M2IIIab	0.00504 ± 0.00022	3	2,6H
87837	31 Leo	K3.5IIIb	0.00340 ± 0.00014	3	2,6I
102212	ν Vir	M1IIIab	0.00640 ± 0.00022	4	2,6I
102647	β Leo	A3V	0.00129 ± 0.00005	2	1,6F
112142	Ψ Vir	M3III	0.00568 ± 0.00023	3	2,6I
123934		M2III	0.00426 ± 0.00019	3	2,6I
169916	λ Sgr	K1IIIb	0.00430 ± 0.00014	2	2,6I
172816		M4III	0.00845 ± 0.00035	2	2,6I
187642	α Aql	A7V	0.00270 ± 0.00012	4	1,6F
193495	β Cap	F8V+A0	0.00313 ± 0.00007	3	2,6I
196777	Ups Cap	M2III	0.00453 ± 0.00015	4	2,6H,6I
216386	λ Aqr	M2.5IIIa	0.00793 ± 0.00038	4	2,6I
223075	TX Psc	C5II	0.00942 ± 0.00011	10	2,6I
	V774 Sgr		0.00568 ± 0.00004	2	2,6I

     Table IIb gives a preliminary list of secondary standards for
absolute radius determinations. However, these stars should be used
with particular caution as, so far, comparisons of various methods of
radius determination have not been carried through.

TABLE IIb. Secondary standards: absolute radii

HD	Name	R/R$_\odot$	n	Code
33088	TT Aur S	3.493 ± 0.011	7	4,6H
34364	AR Aur G	1.838 ± 0.003	10	4,6H
34364	AR Aur S	1.802 ± 0.039	10	4,6H
40183	β Aur G	2.55 ± 0.11	8	4,6H
44701	IM Mon G	3.86 ± 0.028	2	4,6H
44701	IM Mon S	2.71 ± 0.014	2	4,6H
45412	RT Aur	23.9 ± 0.85	6	5,6I
46052	WW Aur G	1.98 ± 0.071	10	4,6H
46052	WW Aur S	1.938 ± 0.026	10	4,6H
72257	VZ Hya G	1.256 ± 0.009	5	4,6H
72257	VZ Hya S	1.055 ± 0.007	5	4,6H
121909	BH Vir G	1.12 ± 0.014	3	4,6H
121909	BH Vir S	1.06 ± 0.028	3	4,6H
139006	α CrB	2.695 ± 0.007	3	4,6H
156247	U Oph G	3.31 ± 0.042	10	4,6H
156247	U Oph S	3.11 ± 0.025	10	4,6H
156965	TX Her G	1.79 ± 0.072	9	4,6H
170470	V451 Oph G	2.52 ± 0.085	5	4,6H
170470	V451 Oph S	2.019 ± 0.083	5	4,6H
170757	RX Her G	2.348 ± 0.117	11	4,6H
170757	RX Her S	1.925 ± 0.078	10	4,6H
185507	σ Aql S	3.502 ± 0.068	8	4,6H
185912	V1143 Cyg G	1.4 ± 0.057	3	4,6H
185912	V1143 Cyg S	1.265 ± 0.007	3	4,6H
188727	S Sge	56.37 ± 0.52	7	5,6H
205234	EI Cep S	2.39 ± 0.106	3	4,6H
209147	CM Lac G	1.506 ± 0.033	8	4,6H
209147	CM Lac S	1.319 ± 0.058	8	4,6H
216014	AH Cep G	6.262 ± 0.002	7	4,6H
218066	CW Cep S	4.365 ± 0.078	4	4,6H
52.03383	RT And G	1.438 ± 0.045	4	4,6H

REFERENCES

Fracassini, M., Pasinetti, L. E. and Manzolini, F. 1981, Astron. Astrophys. Suppl., 45, 145.

Fracassini, M., Pasinetti, L. E., and Valentini, B. 1983, Inform. Bull. CDS, (Strasbourg), 24, 31.

Hoffleit, D. and Jaschek, C. 1982, Catalogue of Bright Stars (Yale U. Obs., New Haven).

Pastori, L., Pasinetti, L. E. and Antonello, E. 1985, in IAU Symposium No. 111: Calibration of Fundamental Stellar Quantities, ed. D. S. Hayes, L. E. Pasinetti and A. G. Davis Philip (Reidel, Dordrecht), p. 459.

A COMPARISON OF DIRECT AND INDIRECT METHODS OF DETERMINATION OF STELLAR ANGULAR DIAMETERS

L. Pastori[1], L. E. Pasinetti[2] and E. Antonello[1]

[1]Astronomical Observatory of Milano-Merate

[2]Department of Physics, University of Milan

Several direct and indirect methods for the determination of stellar apparent diameters have been developed in the past; they are summarized in Table 1 by Fracassini, et al. (1981) with a code number and references. So far, no detailed comparison of the methods has been carried out, the main difficulty being the extremely small number of common stars. Nevertheless, from the stars listed in CADARS (Fracassini, et al. 1981), we have obtained some correlations in order to test the reliability of some methods and to define the regions of the HR diagram where they can be applied.

We selected all the stars whose apparent diameter (uniform disk) is determined by two or more methods, of which one is the interferometric method. If many values of d'' are available for a given method, a weighted average was obtained; for homogeneity, we considered only values measured in the visual region. When possible, the correlation parameters were computed by the weighted least-squares method; otherwise the classical procedure was adopted. Table I reports: the code of the methods, the number N of common stars, the correlations and the correlation coefficients r.

Only three of the direct methods, namely the interferometric (code 1, which includes both the intensity interferometer and Michelson interferometry), lunar occultations (2) and speckle interferometry

TABLE I. Results of the analysis

Code	N	Correlation	r
1,3B	4	$d(3B) = (1.16\pm0.11)d(1) - 0.0068\pm0.0036$	0.99
1,6F	7	$d(1) = (1.01\pm0.02)d(6F) + 0.000035\pm0.000056$	0.99
1,6H	10	$d(1) = (1.11\pm0.04)d(6H) - 0.000082\pm0.000038$	0.98
1,6G	11	$d(6G) = (1.01\pm0.05)d(1) + 0.0019\pm0.0009$	0.99
1,6I	31	$d(6I) = (0.97\pm0.01)d(1) + 0.00001\pm0.00001$	1.00

459

D. S. Hayes et al. (eds.), Calibration of Fundamental Stellar Quantities, 459–461.
© *1985 by the IAU.*

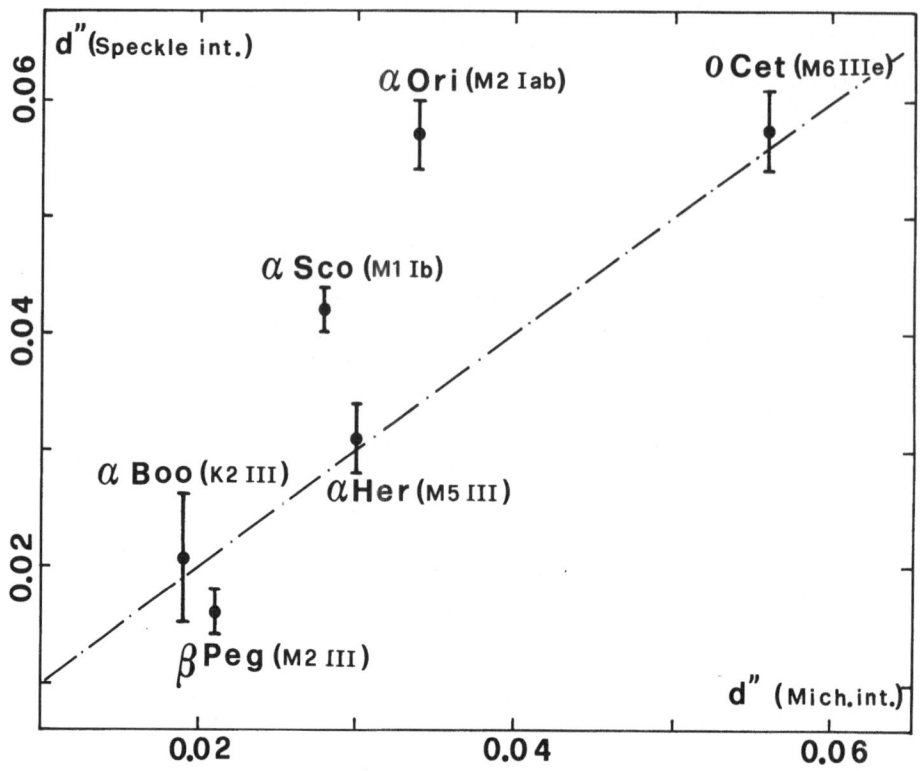

Fig. 1 - Comparison of the apparent diameters d" (arcsec) determined
by Michelson interferometry and speckle interferometry.

(3b), have stars in common. Fig. 1 shows the comparison between (1) and
(3b); the diameters of all six stars were determined by Michelson
interferometry. The values obtained by the two methods are in good
agreement, except for those of α Ori and α Sco. The comparison between
(1) and (2) is not significant, as it involves only three stars. Again,
the diameter of α Sco obtained from (2) differs from the interferometric
value, whereas it is in excellent agreement with (3B). Thus Michelson
interferometry seems to give understimated diameters for supergiant
stars of late spectral types; actually,little can be stated about the
reliability of this technique as its diameters may suffer from very
large errors (Hanbury Brown, 1968).

Disregarding the methods of purely historical interest, we have made
correlations between the interferometric method (mainly the intensity
interferometry) and four indirect methods. Methods (1) and (6F) show
good agreement for the common stars (up to A7), as already verified by
Fracassini et al. (1980). However the applicability field of (6F) is
restricted to dwarf stars belonging to the B5-F5 spectral range. Methods
(1) and (6H) are interdependent (Wesselink, 1969); the comparison is not
quite satisfactory owing to γ Aql (K3II), δ CMa (F8Ia) and κ Ori (BOIa).
As already pointed out by Fracassini et al. (1973) and in more detail
by Barnes et al. (1976) and Pastori et al. (1984), surface gravity effects

bias the Sv -(B-V)$_\wedge$ relation so that Wesselink's calibration, in its
original form, is not suitable for supergiant stars; these stars were not
considered in the computations. The comparison between (1) and (6G)
shows a disagreement for late supergiants (α Ori, α Her, α Sco and
β Peg) whose diameters are measured by the Michelson interferometry
technique. These stars are irregular or semi-regular variables. The 6G
method is based on the comparison between the observed absolute energy
distribution and an energy distribution predicted from a model-atmosphere
computation (Gray 1967). Thus, besides the unknown errors of
Michelson interferometry, the disagreement is probably due to the
difficulty of computing a reliable model atmosphere for variable stars,
especially of late spectral type in which metallic absorption lines
bias the energy distribution; this conclusion is also strengthened
by the late variables β And (M0III), α Cet (M2III) and α Aur (G8III),
which scatter from the straight line. The above-mentioned four supergiants
and α Boo (peculiar spectrum) were ignored in the correlation. The
angular diameters of the comparison between (1) and (6I) were obtained
taking into account the limb-darkening effect; all the spectral types
are earlier than F8. The two methods are in good agreement, however.
This test is not significant as, like the (1)-(6H) comparison, (1)
and (6I) are not independent.

From these comparisons, the following general conclusions may be
drawn: a) Michelson interferometry fails when applied to supergiant
stars of late spectral type, b) for the spectral range (B0-F8)
considered, the intensity interferometer shows its validity whatever
the luminosity class is, c) it is not yet possible to obtain correlations for other direct methods as common stars are not available,
d) the indirect methods are reliable in the most regions of the HR
diagram, however they should be applied with great caution to the
supergiants, especially of late spectral type; in particular, the reliability of the 6G method might be poor when applied to variable stars.

REFERENCES

Barnes, T.G., Evans, D.S., Parson, S.B. 1976 Monthly Notices Roy.
    Astron. Soc. 174, 503
Fracassini, M., Gilardoni, G., Pasinetti, L.E. 1973 Astrophys. Space
    Sci. 22, 141
Fracassini, M., Manzolini, F., Pasinetti, L.E., Ruggenini, M. 1980
    Astrophys. Space Sci. 69,401
Fracassini, M., Pasinetti, L.E., Manzolini, F. 1981 Astron.Astrophys.
    Suppl.   45, 145
Gray, D.F. 1967 Astrophys. J. 149, 317
Hanbury Brown, R. 1968 Ann. Review Astron. Astrophys. 6, 13
Pastori, L., Malaspina, G. 1984 Astron. Astrophys. Suppl., 57, 219.
Wesselink, A.J. 1969 Monthly Notices Roy. Astron. Soc. 144, 297

PROBLEMS CONCERNING PLANCK'S BLACKBODY LAW

Luigi Galgani

Department of Mathematics, University of Milan

For the calibration of radiation detectors, use is currently made of blackbodies, assuming they satisfy Planck's Law. The first problem considered here is then: how well has this law been checked experimentally? Now, it has been pointed out (Crovini and Galgani 1984) that essentially no new experiments have been made after 1921 (Rubens and Michel 1921), when the data were interpreted as fitting the theoretical law within 1%. But, in fact, this work made use of the value 14300 (in suitable units) of the second radiation constant, while the presently adopted value is 14388. When one inserts this value into the calculations, one finds that, indeed, Planck's Law has been checked only to 3%.

Furthermore, some theoretical considerations, based on Arnold's diffusion, lead to the idea that Planck's Law could be just a first approximation to the radiation law, and that one could expect some corrections in the Rayleigh-Jeans region for values of $x=h\nu/kT$ less than 0.2. No experiments on blackbodies are available in that region. Looking, instead, at the data for the Sun, clear deviations are observed, giving a plateau.

REFERENCES

Crovini, L. and Galgani, L. 1984, Lett. Nuovo Cimento, 39, 10.
Rubens, H. and Michel, G. 1921, Z. für Phys., 22, 569.

D. S. Hayes et al. (eds.), Calibration of Fundamental Stellar Quantities, 463.
© *1985 by the IAU.*

EFFECTIVE TEMPERATURES OF STARS WITH "STANDARD" ANGULAR DIAMETERS

I. N. Glushneva

Sternberg State Astronomical Institute, Moscow

ABSTRACT. For 12 stars from the list of stars with "standard" angular diameters (Fracassini et al. 1983), effective temperatures, bolometric corrections, radii and luminosities were determined. These stars are included in the stellar spectrophotometric catalog of the Sternberg Astronomical Institute and three of them were used as spectrophotometric standards. A comparison was made of T_{eff} obtained directly using angular diameters from the list of Fracassini et al. (1983) and by means of joint determination of T_{eff} and Θ (Blackwell and Shallis 1977). For 7 stars the differences in T_{eff} values don't exceed 1-1.5% and the maximum discrepancies are about 6% for BS 2294, 2943 and 4% for the spectrophotometric standard α Aql (BS 7557). Effective temperature values of α Lyr obtained by these two methods are in the agreement within 0.5%.

Fracassini, Pasinetti and Valentini (1983) published a preliminary list of standard stars for the determination of apparent diameters and absolute radii. This list includes stars with reliable values of angular diameters obtained with different methods of determination. There are 22 stars in the list and 12 of them are in the stellar spectrophotometric catalog of the Sternberg State Astronomical Institute (Glushneva et al. 1982). Three stars, α Lyr, γ Ori and α Aql, were used as spectrophotometric standards and T_{eff} and other parameters for these stars were obtained (Glushneva 1983).

It is possible to compare the angular diameters Θ and the effective temperatures T_{eff} obtained by means of approximations (Blackwell and Shallis 1977) with those obtained directly, using the values of Θ from the list by Fracassini, et al. (1983). In both cases, the total flux is obtained from ultraviolet spectrophotometry from TD-1 (Jamar, et al. 1976) and OAO-2 (Code and Meade 1979), data from the stellar spectrophotometric catalog in the region 3200-10800Å (Glushneva, et al. 1982), and infrared photometry by Johnson, et al. (1966).

Values of Θ_{UD} from the list by Fracassini et al. were transformed into Θ_{LD} using the coefficients for these stars from Code et al. (1976).

D. S. Hayes et al. (eds.), Calibration of Fundamental Stellar Quantities, 465–467.
© 1985 by the IAU.

TABLE I

Angular diameters, effective temperatures and total fluxes for stars with "standard" angular diameters.

BS	HD	Θ''	Θ^*	T_{eff},K	T^*_{eff},K	F,erg. $cm^{-2}sec^{-1}$
1543	30652	1.45×10^{-3}	1.70×10^{-3}	6660	6151	1.39×10^{-6}
1713	34085	2.85	2.64	11023	11453	40.1
1790	35468	0.707	0.741	21204	20909	35.1
2294	44743	0.482	0.541	25903	24470	35.1
2421	47105	1.36	1.39	9440	9346	4.93
2943	61421	5.22	5.87	6660	6278	17.9
4534	102647	1.34	1.37	8854	8726	3.64
4662	106625	0.781	0.782	12317	12313	4.70
5056	116658	0.876	0.860	24837	25070	97.7
6556	159561	1.65	1.68	7855	7781	3.46
7001	172167	3.25	3.23	9546	9581	29.4
7557	187642	3.04	2.82	7952	8263	12.4

The results of determinations of Θ and T_{eff} are presented in Table I, where the data obtained using Θ_{UD} from the list by Fracassini et al. are marked by asterisks. For BS 1543 we used $\pi = 0".137$.

The comparison of the values T_{eff} and T^*_{eff} for stars with "standard" angular diameters shows that there are no significant systematic differences among them. The average difference is about 1.5% as a rule. For most of the stars $T_{eff} > T^*_{eff}$, but for two stars the differences are about 6% (α CMi, β CMa) and for α Aql they are 4%. However, for some stars (for example, γ Gem, α Oph, γ Crv) these differences do not exceed 1%. The agreement between T_{eff} and T^*_{eff} for α Lyr is better than 0.5%.

The comparison of effective temperatures obtained by means of the two methods presents an independent confirmation that the accuracy of the method of the joint determination of T_{eff} and Θ (Blackwell and Shallis 1977) may be not worse than 1%.

REFERENCES

Blackwell, D. E. and Shallis, M. J. 1977, Mon. Not. Roy. Astron. Soc., 180, 177.
Code, A. D., Davis, J., Bless, R. C. and Hanbury Brown, R. 1976, Astrophys. J., 203, 417.
Code, A. D. and Meade, M. R. 1979. Astrophys. J. Suppl., 39, 195.
Fracassini, M., Pasinetti, L. E. and Valentini, B. 1983, Inform. Bull. CDS, 24, 31.

Glushneva, I. N., 1982, Spectrophotometry of Bright Stars, (Moscow, Nauka).
Glushneva, I. N. and Ovchinnikov, S. L. 1982, Soviet Astron., 59, 908.
Glushneva. I. N. 1983, Soviet Astron.. 60, 560.
Jamar, C., Macan-Hercot D., Monfils A., Thompson. G. I., Houziaux, L. and Wilson, R. 1976, Ultraviolet Bright-Star Spectrophotometric Catalogue, ESA SR-27.
Johnson, H. L., Mitchell, R. I., Iriarte, B. and Wisniewski, W. Z. 1966, Comm. Lunar and Plan. Lab., No. 63, 4, part 3.
Voloshina I. B., Glushneva. I. N. and Shenavrin, V. I. 1980. Soviet Astron., 57, 1003.

DISCUSSION

GLUSHNEVA: This paper is a part of a rather large work on the determination of effective temperatures. We obtained values of T_{eff} for 73 stars of B - G spectral types and constructed a new scale of effective temperature. The differences between T_{eff} obtained by the two methods for two stars, out of the total number of twelve stars studied, are analyzed in the poster paper. The differences are connected with the accuracy of infrared flux measurements.

FAINT SECONDARY STANDARDS FOR SPECTROPHOTOMETRY AND THE ENERGY DISTRIBUTIONS OF HORIZONTAL-BRANCH A-STARS

D. S. Hayes

Kitt Peak National Observatory,
National Optical Astronomy Observatories[*]

A. G. Davis Philip

Van Vleck Observatory and Union College

The energy distributions of 16 horizontal-branch A-stars and 11 horizontal-branch stars in globular clusters have been measured using the Harvard Scanners at KPNO and CTIO and the Oke multichannel spectrophotometer on the 5-m telescope at Mt. Palomar (Philip and Hayes 1983, Hayes and Philip 1983). Wavelengths between 3400 and 6800 Å were measured and reduced to absolute energy distributions on the system of Hayes and Latham (1975). The internal measuring errors were ± 0.034 mag. per observation for the 15th mag. globular cluster stars and ±0.025 mag. per observation for the 7th to 11th mag. field stars, averaged over all wavelengths. Eleven of the field stars have been observed over nine times each and have low internal measurement errors; these stars plus four globular cluster stars with low internal measurement errors are recommended as secondary standard stars. (See Table I.)

Three secondary standard stars recommended by Breger (1976) ξ^2Cet (HR 718), η Hya, (HR 3454) and 109 Vir (HR 5511) have been used as the standards in the observations of the field stars. BD +17°4708, the primary standard of the four faint secondary standards proposed by Oke and Gunn (1983) has been used as a standard for the observations of the globular cluster stars. We have discussed the internal consistency of our observations of the Breger standards in a general way (Philip and Hayes 1983). Oke and Gunn's energy distribution for BD +17° 4708 is on the system of Hayes and Latham (1975) and we have used observations of five stars observed both at Palomar and at KPNO and CTIO to show the excellent consistency of their calibration with that of the Breger standards over this wavelength range (Hayes and

---

[*]Operated by the Association of Universities for Research in Astronomy, Inc., under contract with the National Science Foundation.

469

D. S. Hayes et al. (eds.), Calibration of Fundamental Stellar Quantities, 469–472.
© *1985 by the IAU.*

Philip 1983).

A new discussion of the secondary standards for spectrophotometry has been published recently by Taylor (1984). Taylor includes the Breger standards plus some new ones. The result of this work is a set of energy distributions for a compete set of secondary standards on the system of Hayes and Latham, and this catalog of secondary standards is being proposed to be a replacement for Breger's. Taylor's discussion appears to be quite convincing but we thought that a detailed comparison of our own observations of five of these stars with Taylor's new energy distributions would be of value.

Our own observations include θ Crt (HR 4468) and 29 Psc (HR 9087) in addition to the three stars mentioned previously. In order by HR number 718, 3454, 4468, 5511 and 9087 we have 23, 53, 34, 37 and 22 scans of each star, with the observations being made at both KPNO and CTIO over the years 1978, 79 and 80. The observations were described in detail elsewhere (Philip and Hayes 1983). We should emphasize that all five standard stars were observed in various combinations depending on the season and the observatory. We attempted to observe at least three standards each night. In order to give the final reduction the greatest possible coherence, we used the minimum number of standards to reduce the data and we chose the three stars listed above. We used Breger's energy distributions for 3400 – 6056 Å and those of Taylor (1979) for 5840 – 6790 Å. We made some minor modifications to Breger's energy distributions (described in Philip and Hayes 1983). We included the wavelength 3704 Å, which is not in Taylor's tables (thus this wavelength is not included in the present comparison).

For the present comparison we have fitted each of our energy distributions to those of Taylor by normalizing them to minimum deviations. Then the residuals were used to calculate an "external" standard deviation per observation, using all wavelengths. This is our measure of the agreement between our values and those of Taylor (or Breger). We have also calculated an "internal" standard deviation which is based on the internal agreement of the scans for each star; it has been averaged over all wavelengths. In order of HR number, the internal standard deviations are ±0.009, 0.016, 0.015, 0.006 and 0.010 mag. The "external" standard deviations are ±0.006, 0.010, 0.007, 0.011 and 0.008 against Taylor's values and ±0.007, 0.016, 0.007, 0.007 and 0.012 against Breger's values. The agreement with Taylor and Breger is excellent. Except for 109 Vir (HR 5511) the agreement is actually better with Taylor than with Breger. This comparison is slightly circular since Taylor used our observations of 29 Psc (HR 9087) as one of the four contributors to his energy distribution for this star. Comparing the agreement with Taylor and Breger is somewhat redundant, since Taylor used much of the same source material as Breger. Yet, Taylor uses new material and his treatment is different. We show the comparison for each star in

Fig. 1. It is clear that the systematic agreement is excellent, except at a few wavelengths, such as 4464 Å, as noted by Taylor. In this regard we should note that the corrections for He and Mg lines have been applied to his values for the points at 4036 and 4464 Å in η Hya (HR 3454) and 29 Psc (HR 9087). Note that the Balmer discontinuity of 29 Psc agrees very well, whereas the Balmer jump measured by Breger was 0.008 mag. smaller than ours.

Our conclusion is that our measurements for these stars are in agreement with the new system of secondary standards proposed by Taylor (1984) over the wavelength range covered. Further, we can extend this conclusion to include our examination of the agreement of Oke and Gunn's (1983) energy distribution of BD +17° 4708 with that of Breger's standards. The energy distribution for BD +17° 4708 is in excellent agreement with the system of Taylor's standards as well. These conclusions then imply that the secondary standards that we recommend from our data are in agreement with the system of Taylor's standards.

Finally, we note that one of the stars used, 109 Vir (HR 5511) is one that has been classified by Taylor as "archival" since he (Taylor 1982) has found possible evidence of episodes of variability. We have examined our own data, including both the scanner data discussed here and extensive Strömgren four-color photometry, plus some other data from the literature (Philip and Hayes 1983, 1984) and found no evidence for variability. Taylor recommends replacing 109 Vir with 108 Vir, in this part of the sky. We have concluded that our use of 109 Vir does not compromise our investigation of HB stars. We note that the observational history of 109 Vir as a spectrophotometric standard is extensive (Philip and Hayes 1984) whereas that of 108 Vir is limited. We recommend that observers who use 109 Vir as a standard do so with caution, and that monitoring it for variability is important. (We are planning such monitoring, ourselves.) We also recommend that those who prefer to use 108 Vir contribute observations of this star which will strengthen its position as a secondary standard.

REFERENCES

Breger, M. 1976, Astrophys. J. Suppl., 32, 1.
Hayes, D. S. and Latham, D. W. 1975, Astrophys. J., 197, 593.
Hayes, D. S. and Philip, A. G. D. 1983, Astrophys. J. Suppl., 53, 759.
Oke, J. B. and Gunn, J. E. 1983, Astrophys. J., 266, 713.
Philip, A. G. D. and Hayes, D. S. 1983, Astrophys. J. Suppl., 53, 751.
Philip, A. G. D. and Hayes, D. S. 1984, Publ. Astron. Soc. Pacific, 96, 546.
Taylor, B. J. 1982, Astron. J. 84, 96.
Taylor, B. J. 1982, Publ. Astron. Soc. Pacific, 94, 663.
Taylor, B. J. 1984, Astrophys, J. Suppl., 54, 259.

Table I.

The Secondary Standards

Star	BD	Type	HD	V mag	b-y mag	c mag	Source	n	UV	Blue	Red
HD 2857	-06 86	FHB	A2	9.98	0.132	1.216	P&H	21	0.026	0.025	0.027
57336	-79 243	FHB?	A0	7.4	0.126	1.210	P&H	17	0.023	0.022	0.020
74721	+13 1981	FHB	B9	8.72	0.028	1.255	P&H	22	0.026	0.033	0.027
83041	-28 7417	FHB?	A0	8.78	0.221	0.733	P&H	11	0.021	0.019	0.024
86986	+15 2156	FHB	A0	7.99	0.088	1.265	P&H	15	0.020	0.020	0.021
107369	-31 9638	Hi Lum	A0	9.58	0.143	1.614	P&H	17	0.029	0.028	0.027
109995	+40 2558	FHB	A0	9.07	0.045	1.285	P&H	9	0.010	0.029	0.008
117880	-17 3883	FHB	A0	8.13	0.055	1.205	P&H	16	0.023	0.018	0.026
130095	-26 10505	FHB	B8	8.13	0.060	1.244	P&H	9	0.025	0.017	0.018
161817	+25 3344	FHB	A0	6.96	0.125	1.207	P&H	10	0.011	0.009	0.013
12 17 24		FHB?		11.99	0.121	1.103	H&P	15	0.019	0.014	0.013
M5 III 17		BHB		15.0	0.11	1.22	H&P	4	0.028	0.022	0.017
M13 SA531		BHB		14.9	0.10	1.17	H&P	4	0.036	0.018	0.019
M13 SA 16		BHB		15.0	0.06	1.34	H&P	4	0.031	0.030	0.026
M92 XII 6		BHB		15.9			H&P	4	0.057	0.065	0.062

P & H: Philip and Hayes, 1983
H & P: Hayes and Philip, 1983

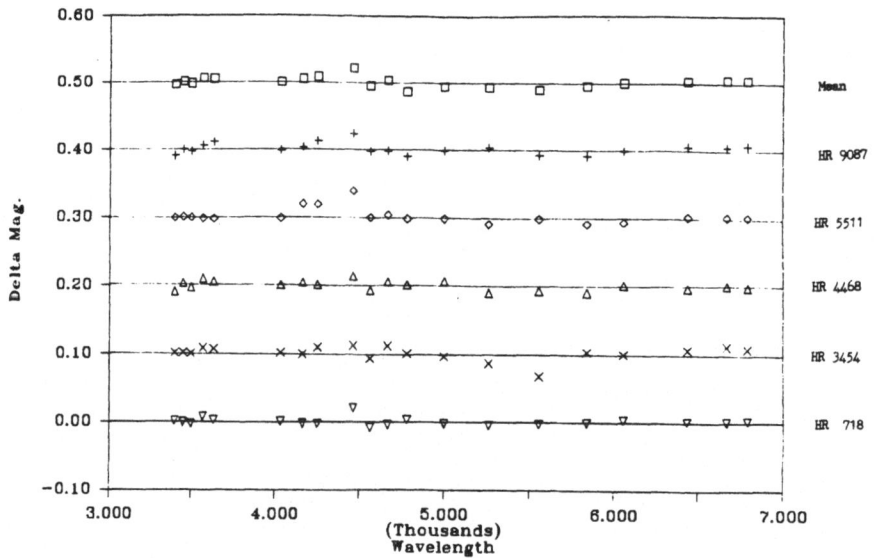

Fig. 1.   Comparisons of scans made by Taylor (1984) and by
Philip and Hayes (1983).

HIGH-PRECISION SOLAR RADIATION DATA FOR $\lambda\lambda 3300$–$12500\overset{\circ}{A}$

Heinz Neckel

Hamburg Observatory

Dietrich Labs

Heidelberg Observatory

Very precise data on the solar radiation has been obtained by merging: (1) the absolute integrals of the disk-center intensity for 20Å-wide spectral bands, observed in the 1960's by Labs and Neckel (1962, 1963, 1967); (2) the ratios of the mean-to-central intensity following from observations of the center-to-limb variation of (a) all 20Å bands below 6600Å (Neckel and Labs 1984), and (b) the intensities at selected continuum wavelengths (Pierce and Slaughter 1977a,b); and (3) the high-resolution Fourier transform spectra (FTS) obtained by J. Brault at Kitt Peak for the center of the disk and for the irradiance.

The main result is (1) a compilation of 10, 20, and 50Å averages for both the intensity at the center of the disk and the disk-averaged radiation (irradiance) and (2) for both spectra the most reliable localization of the 'continuum'-level yet produced.

The internal accuracy of the data is defined by the 'scatter' in the FTS spectra, which is on the order of 0.1%. Local systematic deviations exceeding 0.5% are not to be expected. The absence of a significant neutral or wavelength-dependent systematic error in the absolute data has already been well established; now it is confirmed again.

A detailed paper has been published (Neckel and Labs 1984), as has a summary (Neckel 1984).

REFERENCES

Labs, D. and Neckel, H. 1962, Z. für Astrophys., 55, 269.
_____ 1963, ibid., 57, 283.
_____ 1967, ibid., 65, 133.

D. S. Hayes et al. (eds.), Calibration of Fundamental Stellar Quantities, 473–474.
© 1985 by the IAU.

Neckel, H. 1984, Space Sci. Rev., 38, 187.
Neckel, H. and Labs, D. 1984, Solar Phys., 90, 205.
Pierce, A. K. and Slaughter, C. D. 1977a, Solar Phys., 51, 25.
                              1977b, ibid., 52, 179.

DISCUSSION

JASCHEK: Thank you for this very nice paper. For once we have a calibration which has errors smaller than 1%. That is marvelous.

HAYES: Do you plan any further work on the absolute calibration of the Sun?

NECKEL: We have new solar irradiance measurements which extend from 2000 Å to 3 microns which were made in December of last year in collaboration with some French and Belgian colleagues on the Spacelab I flight. It will take some time before we finally receive all the data. I should say that there is also another good set of solar irradiance data which agree very well with ours. Those are the data published exactly 60 years ago by Minnaert, who took his data from the famous observations by Abbot made in 1922 and 1923. There is really amazing agreement between our measurements and these old ones, which were, for many decades, the standard for solar radiation.

BESSELL: When will the KPNO FTS Atlas of Brault be available for use?

NECKEL: You can get it (on tape) from James Brault at Kitt Peak. We have published the polynomials which you can use to calibrate the data.

A NEW OBSERVING STATION AT THE PIC DU MIDI OBSERVATORY FOR THE ABSOLUTE CALIBRATION OF STELLAR RADIATION

Roger H. Peyturaux

Institute for Astrophysics, Paris

ABSTRACT. A new observing station designed for the absolute calibration of stellar and solar radiation has been built in the last few years at the Pic du Midi Observatory, France, at an elevation of 2860 meters. The stellar observations may begin next summer. The main improvements with respect to the calibration experiments carried out during the last decades are the use of a new type of blackbody source and an optical arrangement which is free of systematic errors in the measurements of the star/blackbody ratio. The aim of the experiment is to establish a set of homogeneous standards (about 10 B and A stars) covering the whole northern sky.

1. INTRODUCTION

In the last decade a number of fundamental stellar calibrations have been made by Oke and Schild (1970), Hayes (1970), Hayes, et al. (1975), Hayes and Latham (1975), Tüg, et al. (1977), Tüg (1979), Terez and Terez (1979), and Kharitonov, et al. (1980). Despite the number and quality of these studies, important discrepancies still exist. For example, in the case of Vega, discrepancies as high as 7-8 percent are found. These discrepant values appear too large when compared with the accuracy reached in metrological measurements and with the accuracy of stellar energy distributions that are desirable for astrophysical applications.

For these reasons and to take into account the long tradition of radiation measurement of the Institut d'Astrophysique, we decided to undertake a new experiment in stellar calibration. Our attempt was not to reproduce the preceding experiments, but to produce an instrumentation free from systematic errors. Moreover, our aim involved establishing a set of homogeneous standards covering the whole northern sky. Therefore, a permanent high altitude location was needed which would be available for at least several years. Such a location was found at the Pic du Midi Observatory (2860 m).

D. S. Hayes et al. (eds.), Calibration of Fundamental Stellar Quantities, 475–478.
© *1985 by the IAU.*

In the following sections of this article, we shall describe the method and instrument used and estimate the precision that can be obtained in the measurement of absolute fluxes.

2. THE RADIATION SOURCE

2.1. Choice of the source.

In recent work blackbody sources operating at the melting point of a metal like copper (1358K) or platinum (2045K) have often been used. However, it seems that a better solution would be the use of a blackbody source with adjustable temperature. Of course, this requires the use of an accurate pyrometer. With such a source we can adjust the temperature according to the star brightness and the spectral range in order to obtain a star/blackbody ratio not too far from unity (e.g., between .1 and 10). For that reason we decided to build a new type of blackbody specifying that its temperature stability would not be worse than that of a blackbody operating at the melting point of a metal.

2.2 Description of the blackbody source.

The essential part is a graphite tube containing two main cavities, as seen in Fig. 1. The central cavity (a) produces the radiation to be observed through a quartz window. The cavity (b) emits radiation through another quartz window to a photodiode used for the temperature servo control. The tube is surrounded by heat insulating graphite felt and is heated by the Joule effect to 2500K in a vacuum (10^{-2}mm Hg) achieved by a pump operating continuously. To fulfill the blackbody conditions the temperature of the cavity (a) must be uniform and the hole must be small enough with respect to the internal area of the cavity. To achieve the first condition the profiles of the tube and of the surrounding felt are not cylindrical. They must be adjusted empirically until the central hole is no longer visible when the tube is hot. Then the gradient along

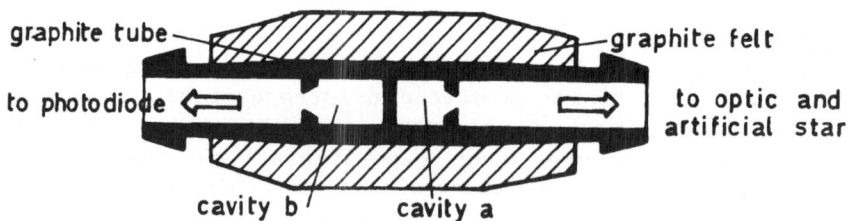

Fig.1. The blackbody source – The radiation from the cavity (a) is used for the absolute calibration. The one from the cavity (b) is used for the servo control of the temperature.

the cavity is less than 0.2K at 1800K.

The hole cannot be too small and a theoretical calculation is necessary to evaluate the error introduced. Several methods have been proposed for this calculation but none is perfect. Averaging the results we can say that ε is better than .9995 for the visible and the near UV.

2.3. Measurement of the temperature.

The temperature is measured with a polychromatic photoelectric pyrometer that we have designed and built. It is calibrated against a blackbody at the melting point of gold. It can measure the brightness temperature at several wavelengths through the optical system associated with the blackbody. Some corrections are necessary to take into account the spectral transmission of the optical system.

3. OPTICAL ARRANGEMENT OF THE OBSERVING STATION

The photometric equipment includes a 55 cm telescope built for that purpose and a spectrometer at the f/3 prime focus. This spectrometer includes a 1200 1/mm grating for the wavelength range 3000 – 6800Å in the first order. The grating can be replaced easily by another one for another wavelength range. The dispersion is 60 Å/mm and the true resolution is about 1Å. The detector is a photomultiplier with a S-20 response curve. Two modes are possible: scanning with 2Å or longer steps and measurements at discrete wavelengths with a typical band width of 30Å.

The artificial star is a hole illuminated by the blackbody through an optical system. The hole is located at the focus of a collimating mirror which has the same diameter as the telescope and a focal length of 5.5 m. The blackbody and its associated optical system are located in the dome. The collimating mirror is outside moving on a railway in a 10 m horizontal tunnel. With this arrangement the mirror can be moved from the position for calibration to the bottom of the tunnel. Then the center of curvature can be reached for the measurement of the reflectivity of the mirror. This measurement is achieved with the aid of a specially designed photometer. The hole has a diameter of 50µ in order to obtain a size of image at the focus of the telescope comparable to the size of a stellar image. This hole is calibrated by a photometric method by comparison with several diaphragms of much larger diameters which can be measured on a measuring machine. The overall accuracy of this calibration is about 2×10^{-3}.

4. CONCLUSION

In this experiment, the beam sizes for the star and blackbody source are exactly the same and the need for a horizontal extinction determination is removed. Of course, the elimination of this problem incurs the introduction of the transmission of the 55 cm collimating mirror. However, the essential difference is that the horizontal extinction is not a measurable quantity, while the reflectivity of a concave mirror can be determined with a high degree of precision. Taking into account the several causes of errors, especially in the measurements of the temperature of the blackbody (.5K at 2000K) and of the vertical extinction by the Bouguer's law, we feel that a global accuracy of about 1% in stellar absolute fluxes is not beyond the scope of the experiment.

REFERENCES

Hayes, D. S. 1970, Astrophys. J., 159, 165.

Hayes, D. S., Latham, D. W. and Hayes, S. H. 1975, Astrophys. J., 197, 587.

Hayes, D. S. and Latham, D. W. 1975, Astrophys. J., 197, 593.

Kharitonov, A. V., Tereshchenko, V. M., Knyazeva, L. N. and Boiko, P. N. 1980, Astron. Zh., 57, 287.

Oke, J. B. and Schild, R. E. 1970, Astrophys. J., 161, 1015.

Terez, C. A. and Terez, E. I. 1979, Astron. Zh., 56, 800.

Tüg, H., White, N. M. and Lockwood, G. N. 1977, Astron. Astrophys., 61, 679.

Tüg, H. 1979, Astron. Astrophys., 82, 195.

DISCUSSION

NECKEL: Why do you use a blackbody instead of a lamp calibrated against a blackbody? It is much easier to handle.

PEYTERAUX: The black body with adjustable temperature is an alternative to other methods (striplamp or a blackbody at the melting point of a metal) and I have a long experience of this kind of method. On the other hand I don't think the lamp is easier to handle in a dome.

NECKEL: Why do you run the black body in a vacuum and not e.g., in an argon atmosphere of normal pressure? This would enable much higher temperatures and operation without any window, which is in any case a source for systematic errors (contamination, etc.).

PEYTERAUX: The use of argon is not excluded for the future. The window is included in the optical system and I measure the temperature for several wavelengths through the optical system.

NECKEL: What is the lifetime of the blackbody cavity?

PEYTERAUX: Typically 100 hours at 2000 K. The tube can be replaced very easily by another one.

THE STATUS OF THE ABSOLUTE CALIBRATION OF STELLAR FLUXES BETWEEN
912 AND 1200 Å

R. S. Polidan and J. B. Holberg

Lunar and Planetary Laboratory. University of Arizona

ABSTRACT.   Recent results have shed new light on the status of the
calibration of absolute stellar fluxes between 912 and 1200 Å.
Observations of hot white dwarfs, subdwarfs and planetary nebula
nuclei with the Voyager ultraviolet spectrometers provide evidence
that the current calibration agrees very well with extrapolations of
IUE energy distributions shortwards of 1200 Å. Voyager observations
of main sequence B-stars used as flux calibration sources have
revealed that many are variable in brightness in the 912 - 1200 Å
region. We conclude there is no current observational motivation for
any revision of the 912 to 1200 Å calibration described by Holberg et
al. (1982).

1. INTRODUCTION

     The calibration used to determine absolute fluxes for the Voyager
1 and 2 ultraviolet spectrometers (UVS) have been discussed by Holberg
et al. (1982). At that time the status of absolute stellar fluxes in
the spectral region between 912 and 1200 Å was in some doubt. The
discrepancies were clear. The Voyager results were in overall
disagreement with those of Brune, Mount, and Feldman (1979) and in
partial disagreement with those of Carruthers, Heckathorn, and Opal
(1981) (CHO). The latter two experiments, although in mutual
disagreement, both indicated that the fluxes predicted by model
atmospheres were substantially in excess of observation. A major
conclusion of Holberg et al. was that model predictions appeared to
adequately represent observations and were not in need of substantial
revision. Recently, two new developments have occurred which clarify
the situation regarding absolute fluxes in the 912 and 1050 Å
region. Two new calibration flight results have been reported and an
analysis of a much wider range of Voyager stellar observations has
revealed wide spread photometric variability in B-stars in the far UV.

D. S. Hayes et al. (eds.), Calibration of Fundamental Stellar Quantities, 479–483.
© 1985 by the IAU.

2. NEW CALIBRATION RESULTS

Carruthers and Heckathorn (1982) have reported preliminary results from a reflight of the CHO instrumentation, employing osmium coated optics. They find the observed flux distributions of early type stars to be in closer agreement with the Kurucz (1979) model atmospheres than the results from the initial flight which employed LiF coated optics. Until actual calibrated fluxes are available for the new Carruthers and Heckathorn observations, the precise relation of these results to the Voyager observations remains unclear. But it would appear that the major discrepancy between Voyager and CHO in the 912 - 1050 Å region may have abated. In an independent experiment Opal and Weller (1984) report agreement between Voyager 2 and their extreme-ultraviolet photometer on board the STP 72-1 satellite. This single channel photometer measures stellar flux in an effective band pass covering the 912 - 1050 Å region. These authors report that absolute Voyager 2 fluxes from three isolated bright stars (α Vir, η UMa and ε Per) when convolved with the preflight (absolute) response curve of the photometer yield excellent agreement with the observed count rate signal from these stars.

3. B-STAR VARIABILITY

Since the publication of Holberg et al. (1982) the Voyager data base has been greatly expanded. One aspect of this program has been to make repeated high signal-to-noise observations of standard B-stars in order to refine the far-UV flux calibration. Analysis of these data has indicated that, in general, the observed 912 - 1200 Å flux distributions are in good agreement with predictions from model atmospheres (see Peters and Polidan 1985). However, the majority of these "standard" stars were found to be variable in their FUV flux level and distribution.

The largest variations have been observed in the B0.5 III star ε Per. Seven observations of ε Per obtained over three years show a 40% change in flux. Analysis of the FUV flux distribution also indicates a change in the shape of the distribution during this time. Comparison with Kurucz (1979) model atmospheres implies a change in T_{eff} of 1500 ± 500 K from maximum to minimum flux. Thus, the observed variability may be associated with changes in the effective temperature of the star. The majority of B-stars observed with Voyager exhibit similar variability, though with lower amplitude (typically ~10 - 20%). A few stars, notably α Vir, are observed to be effectively constant (Δflux $\lesssim$ 5%).

The source of this variability in B-stars appears to be 53 Persei type non-radial pulsations. In a joint program with Dr. M. Smith of the National Solar Observatory we are monitoring a sample of B-stars for FUV photometric variability and visual photospheric line profile changes. Preliminary results indicate that all the stars for which we find UV-flux variability do show non-radial pulsations. These stars are, in general, not known to be light variables in the visual region. This can be reconciled with the observed FUV variability through the realization that small changes in effective temperature will produce larger changes in the FUV, where most of the stellar flux is emitted, than in the visual region. A better understanding of the nature of B-star variability in the FUV waits further data. We can clearly state, however, the B-stars are unsuitable as photometric calibration sources in the FUV.

4. SUB-LUMINOUS STARS

The expanded Voyager stellar program has also included observations of sub-luminous stars: white dwarfs, subdwarfs, and planetary nebula nuclei. These objects offer numerous advantages over B-stars as FUV calibration sources. First, from our limited set of observations, they are constant in brightness. Second, their flux distributions, in general, have less line blocking and are frequently simpler (e.g. power law) than those of B-stars. They also offer an important third advantage. Some are known to be EUV sources (e.g. HZ 43) and, hence, will allow accurate calibration of stellar sources at wavelengths shorter than 912 Å.

Examples of these objects can be found in Figures 1 through 3. The hot (T_{eff} ~80,000 K) sdO star BD+28°4211 is presented in Figure 1. It exhibits a power law spectrum from 950 to 8000 Å. The nucleus of the planetary nebula NGC 246 (T_{eff} ~100,000 K) (Figure 2) also exhibits a power law spectrum. The nearby DA white dwarf CoD-38° 10980 (T_{eff}= 24,750 ± 250 K) is compared with pure hydrogen models of Wesemael et al. (1980) in Figure 3. In this case the flux distribution is not a power law but can be adequately fit with the pure hydrogen model atmosphere.

It is apparent from Figures 1 and 2 that Voyager fluxes shortward of 1200 Å fall along the same power laws which characterize the energy distributions longward of 1200 Å. To the extent that it is reasonable to extrapolate such power laws down to the Lyman limit, Voyager results can be considered to be in satisfactory agreement with accepted absolute calibrations. It is important to note that this conclusion is independent of any reliance on model atmosphere predictions. It is also stressed that various independent data sets in Figures 1 and 2 have not been forced into agreement or corrected in any non-standard manner. For example, no reddening corrections are applied to either BD+28°4211 or NGC 246, since estimates of the extinction for both objects are generally very low.

5. CONCLUSIONS

Our principal conclusion is that the results of Holberg et al. (1982) adequately represent the existing data and no change in the FUV

calibration is warranted at this time. We propose that future
experiments abandon B-stars as their principal calibration sources
because of their observed FUV flux variability and adopt sub-luminous
stars (white dwarfs, subdwarfs, and planetary nebula nuclei). These
latter stars offer numerous advantages over B-stars, particularly in
their constancy of flux, simple energy distributions, and chance for
observable EUV emission.

REFERENCES

Breger, M.,1976, Astrophys. J. Suppl., 32, 7.
Brune, W. H., Mount, G. H., and Feldman, P. D. 1979, Astrophys. J.,
    227, 884.
Carruthers, G. R., Heckathorn, H. M., and Opal, C. B. 1981, Astrophys.
    J., 243, 855.
Carruthers, G. R., and Heckathron, H. M. 1982, Bull. Amer. Astron.
    Soc., 14, 651.
Holberg, J. B., Forrester, W. T., Shemansky, D. E. and Barry, D. C.
    1982, Astrophys. J., 257, 656.
Kurucz, R. L. 1981, Astrophys. J. Suppl., 40, 1.
Opal C. B. and Weller, C. S. 1984, Astrophys. J., 282, 445.
Peters, G. J. and Polidan, R. S. 1985, in IAU Symposium No. 111:
    Calibration of Fundamental Stellar Quantities, ed, D. S. Hayes, L.
    E. Pasinetti and A. G. Davis Philip. (Reidel: Dordrecht), p.417.
Wesemael, F., Auer, L. H., Van Horn, H. M., and Savedoff, M. P. 1980,
    Astrophys. J. Suppl., 43, 159.

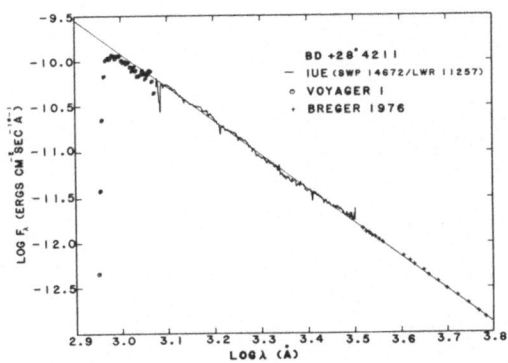

Fig. 1. The unadjusted absolute fluxes observed for the hot O sub-dwarf BD +28°4211. The straight line represents a least squares fit to the IUE and ground based data (excluding Voyager).

Fig. 2. The unadjusted absolute fluxes observed for the central star of the planetary nebula NGC 246. The error bars associated with the Voyager data are representative of counting statistics at various wavelengths. The straight line was computed in Fig. 1.

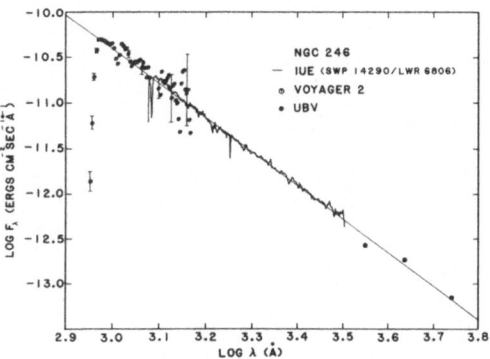

Fig. 3. The observed Voyager 2 and IUE fluxes for the near-by white dwarf CoD -38°10980. The model shown here employs an effective temperature of 24,500 K and a surface gravity of log g = 8.0.

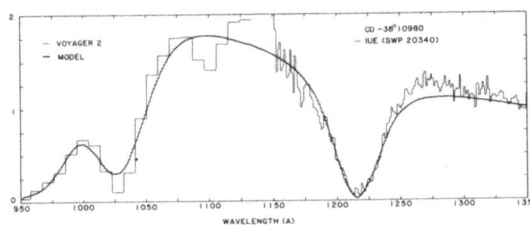

DISCUSSION

JASCHEK: Is a 5% variation in UV flux not a little bit large for non-radial pulsation?

POLIDAN: Remember that we are observing near the maximum of the flux distribution of these stars. Small changes in visual flux translate to large changes in the far UV. As an example, consider the radial pulsator BW Vul. It has an observed change in V mag. of 0.15 mag. while at 1000 Å the amplitude of the variation is 1.2 mag. (Barry et al. 1984 Astrophys. J., in press). The limit of <5% variation for Alpha Vir implies a temperature variation of less than 500 K.

SECONDARY STANDARD STARS IN THE FOUR-COLOR SYSTEM

A. G. Davis Philip

Van Vleck Observatory and Union College

Now that larger telescopes are being used to make photometric measures of very faint stars, fainter standard stars are needed. Most of the stars listed in the standard star list of Crawford and Barnes (1970) are too bright, even for a 1.5 m telescope. Over the past two decades observations have been made of B, A and F-type stars in the magnitude range V = 6 to 13 in the northern and southern hemispheres in a program of four-color photometry of field and blue horizontal-branch stars. An earlier list of secondary standards was published in Philip and Philip (1973). The present paper reports on over twice as many standard stars and the secondary standards are increased by about 20%. The number of observations per star is substantially increased.

Over eighty standard stars from Crawford and Barnes have been used as standards in these observing runs. The zero point differences for the group of stars relative to the standards are 0.000, +0.002 and −0.001 in (b−y), c_1 and m_1 respectively (in the sense P − CB). The rms errors of a single observation are ±0.003, 0.005 and 0.006 in (b−y), c_1 and m_1 respectively.

The fifty secondary standards include Population I stars and field horizontal branch stars. The four prototype FHB stars (HD 2857, 86986, 109995 and 161817) have been measured an average of over 100 times each and are among the best measured stars in the new standard star list. In Fig. 1 a plot is shown of the delta (b−y) magnitude for HD 161817 over a period of 17 years. The rms error is ±0.009 mag. and no trend can be seen in the residuals. This behavior is typical for the FHB stars in the secondary standard star list; these stars exhibit a very stable nature.

Four plots (Figs. 2 - 5) show the distribution of the primary and secondary standards in the (b−y), (c_1) and (b−y), (m_1) planes. In the standard list, the great majority of the stars are main sequence Pop I stars. In the secondary standard star list there is a

D. S. Hayes et al. (eds.), Calibration of Fundamental Stellar Quantities, 485–490.
© *1985 by the IAU.*

prominent group of FHB stars whose indices fall away from the main sequence. In Fig 6a, b and c the differences in (b-y), c_1 and m_1 between Philip and Crawford-Barnes are shown. There are no differences greater than ±0.02 and most of the differences are smaller than ±0.01. In Fig. 7a, b and c the rms errors of the secondary standard stars are plotted. The typical average rms error is near ±0.015 with the largest rms error falling under ±0.03 mag.

The details concerning these stars are being prepared for publication in a journal article. The four-color photometric indices will be presented in two tables; the first will list the indices for the Crawford-Barnes standards and the second will list the indices for the secondary standards. Those who would be interested in obtaining the list of secondary standards prior to publication can obtain it on request.

REFERENCES

Crawford, D. L. and Barnes, J. V.   1970 Astron. J. <u>75</u>, 978.
Philip, A. G. D. and Philip, K. D.   1973 Astrophys. J. <u>179</u>, 855.

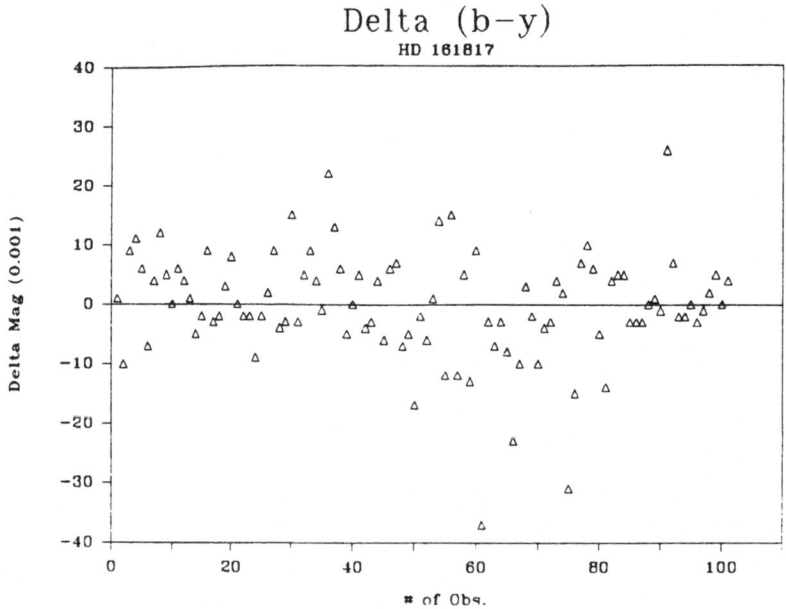

Fig. 1.   Delta (b-y) for observations made over a 17 year period of HD 161817.

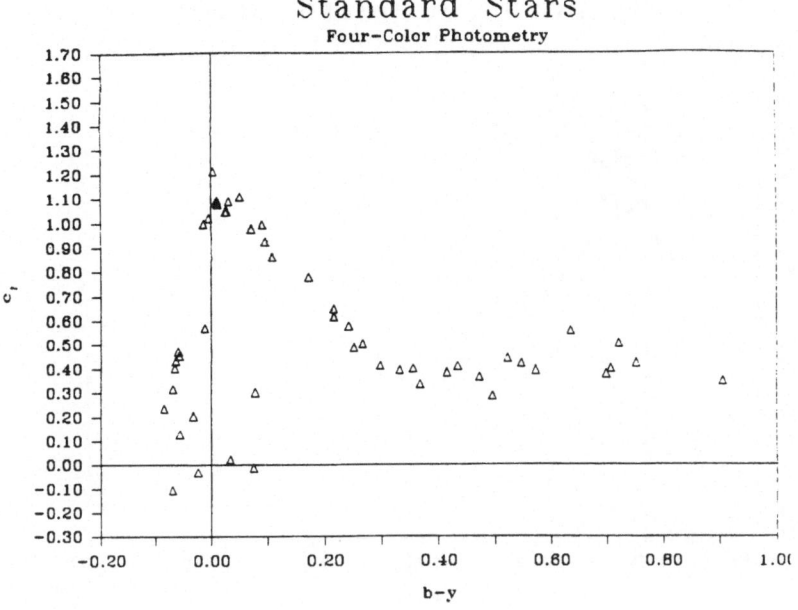

Fig. 2. The distribution of the standard stars in the c_1, (b-y) plane.

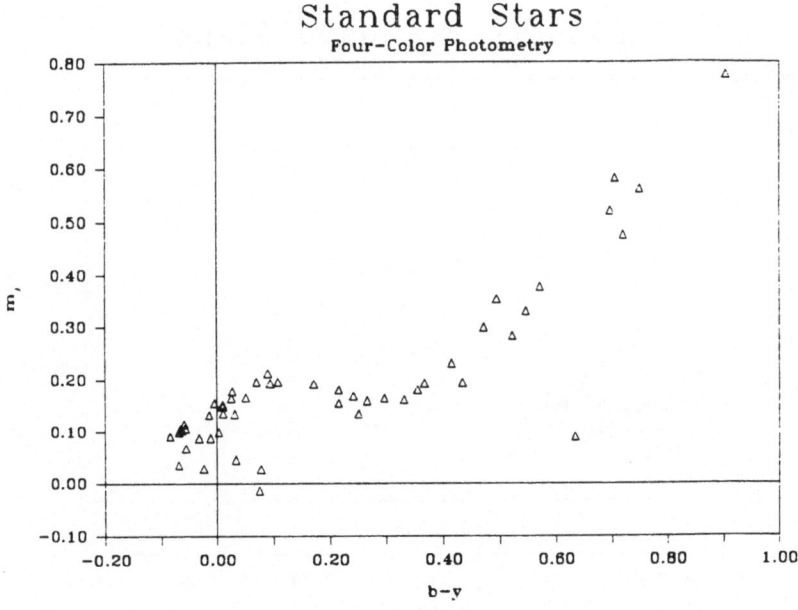

Fig. 3. The distribution of the standard stars in the m_1, (b-y) plane.

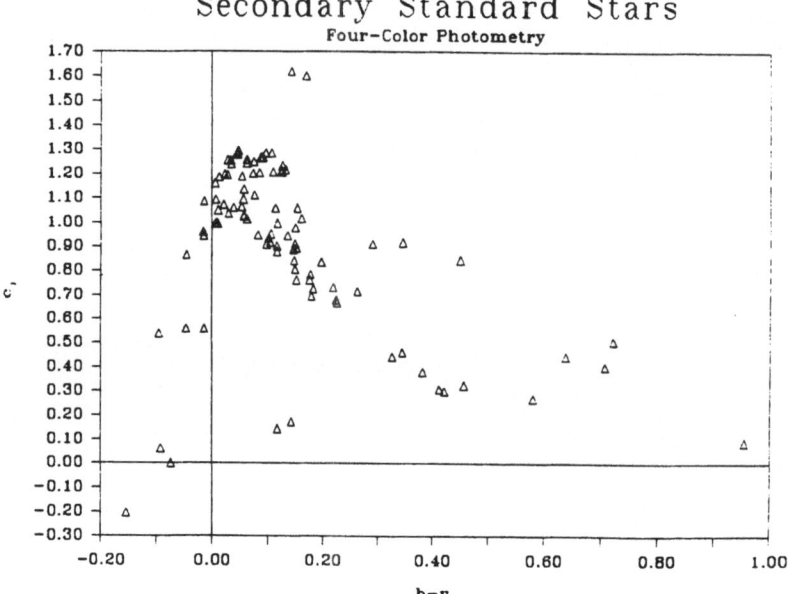

Fig. 4. The distribution of the secondary standard stars in the c_1,
         (b-y) plane.

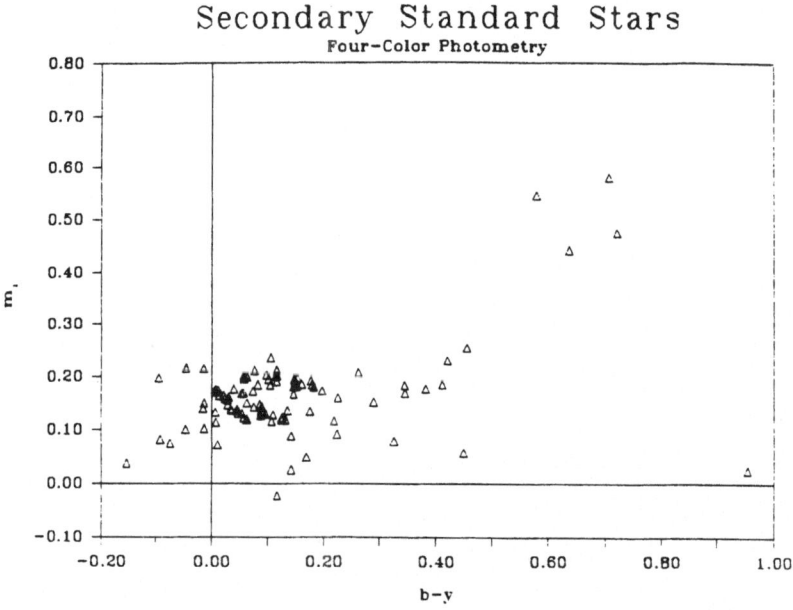

Fig. 5. The distribution of the secondary standard stars in the m_1,
         (b-y) plane.

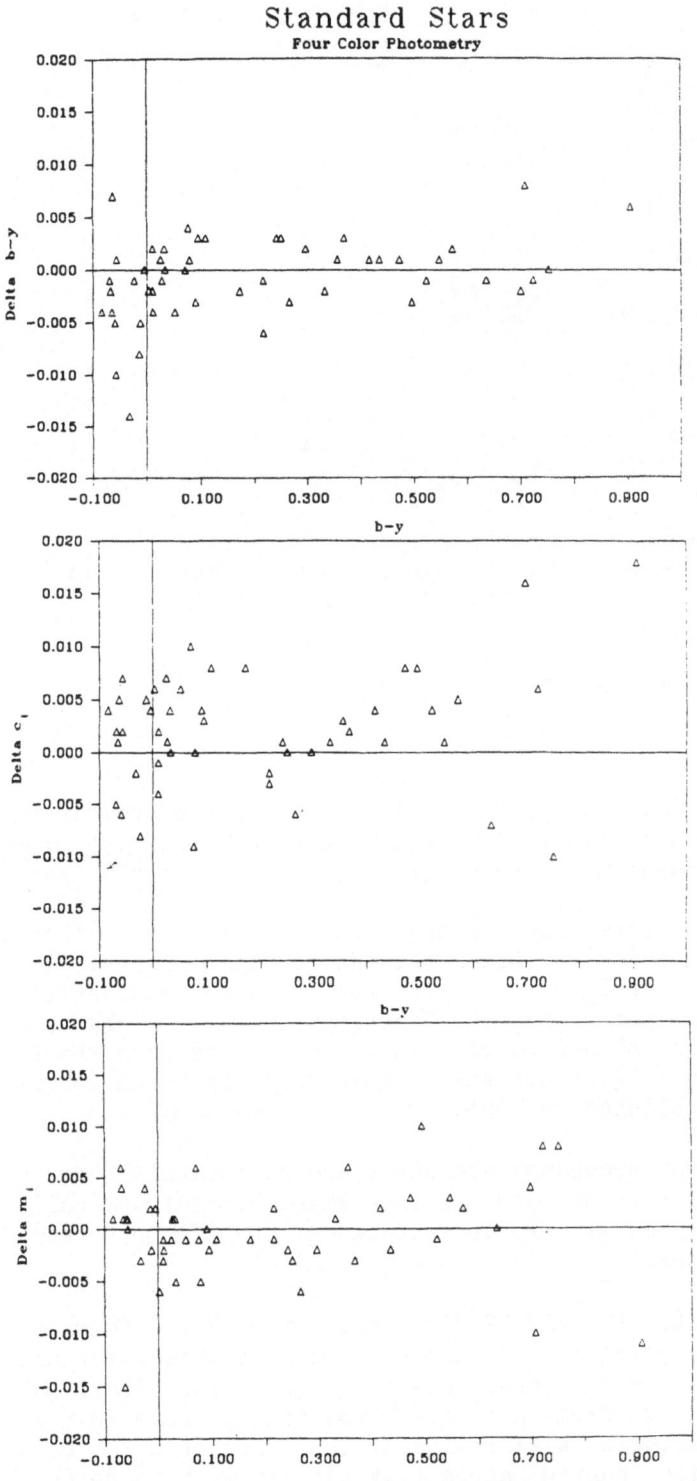

Fig. 6   Delta (b-y), c_1 and m_1 are plotted for the standard stars.

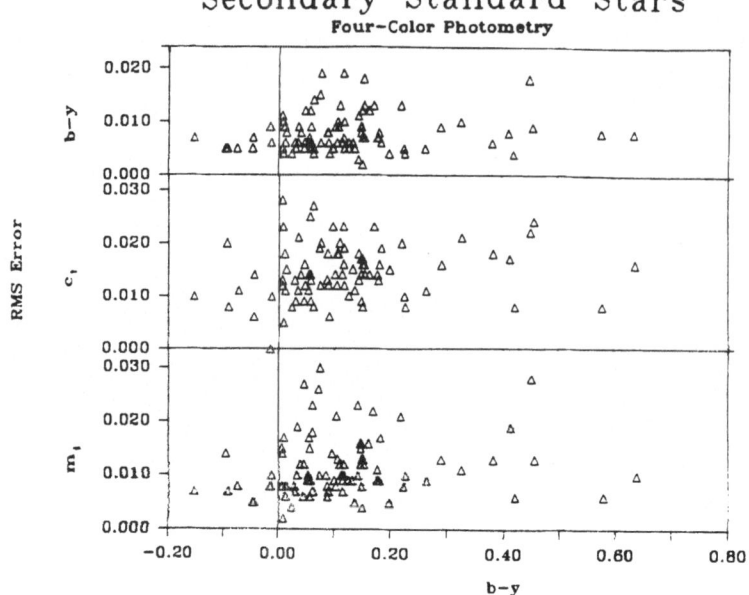

Fig. 7. The rms errors of the secondary standard stars.

DISCUSSION

JASCHEK: I would like to insist on the importance of providing detailed charts for all stars not contained in the BD or CoD, with proper scale, orientation and magnitude (for at least one star) clearly indicated.

PHILIP: When I published the finding lists of field horizontal-branch stars I made charts dotted from Schmidt spectral plates. These charts are fine for telescopes of 1 m but for large telescopes better charts are needed. As part of VVO Contribution No. 2 charts are being presented for the 60 FHB stars with reproductions from the Palomar Sky Atlas. The nearest BD or CoD star is marked on each chart. These charts will be published in 1985.

TOBIN: Many of your secondary standards are horizontal-branch stars. Is there any possibility that your indices might be systematically in error as a result of bandpass differences between your filters and those of Crawford and Barnes?

PHILIP: No. Up until the mid 1970's I used the No. 1 four-color filter set at Kitt Peak, which is the same set used by Strömgren and Crawford in setting up the system. When this filter set was "retired" I used a filter set obtained from Schott. These filters have been measured on the spectrophotometer at Kitt Peak and the shapes and the maxima of the filter transmission curves agree very closely with those of the No. 1 set.

A PRELIMINARY LUMINOSITY CALIBRATION OF THE uvby,β SYSTEMS FOR SUPERGIANT F-TYPE STARS

E. Antonello

Astronomical Observatory of Brera, Milano-Merate

1. INTRODUCTION

Calibrations of the uvby,β photometric systems have been made by Crawford (1975, 1978, 1979) for F, B and late A-type stars of luminosity class V-III; the calibration for early A-type stars has been made by Hilditch et al. (1983). One of the remaining areas of the HR diagram to be calibrated is that of the bright giant to supergiant region (luminosity class II-I). We have attempted to obtain a calibration in terms of luminosity for supergiant F-type stars. To this end we have used the statistical method of multiple stepwise regression analysis. Before doing the calibration for supergiants, we show the usefulness of this method by a comparison of the results obtained for dwarf F-type stars with the previous calibration given by Crawford (1975).

2. DWARF F-TYPE STARS

The multiple stepwise regression analysis is a method of the multivariate statistical analysis which allows us to correlate a dependent variable with a set of significant independent parameters selected from a larger set of variables (see e.g., Antonello 1983). We have used a program by Buzzi-Ferraris (1975) which is more general than the original one by Efroymson (1964). This program contains an F-test for the introduction and removal of variables in the statistical relation.

We have considered the forty-three F-type stars used by Crawford (1975) for deriving the calibration in luminosity (absolute magnitude M_v). The initial set of parameters known for each star contained the indices β, m_1, c_1, $b-y$, δm_1, δc_1 and their combinations of second and third order (that is powers and cross products). Several tests have given some good relations with only the indices β, c_1 and their combinations; we have selected the following simple one:

D. S. Hayes et al. (eds.), Calibration of Fundamental Stellar Quantities, 491–494.
© *1985 by the IAU.*

$$M_V = - 87.8 \ c_1 + 10.93 \ \beta^2 \cdot c_1 - 13.89 \ \beta + 44.93 \tag{1}$$
$$\pm 0.31 \quad \pm 17.6 \qquad \pm 2.42 \qquad \pm 5.58$$

The standard error in M_V is slightly higher than the error given by Crawford for his calibration. However, we think that the two calibrations are statistically equivalent because an analysis of the (O-C) data has given similar results for both. Indeed, the absolute values of (O-C) have a mean value of 0.23 in our case and 0.24 in Crawford's case.

Relation (1) should be considered as another way of writing the formula

$$M_V = M_V \ (ZAMS, \beta) - f \cdot \delta c_1, \tag{2}$$

given by Crawford, because $M_V \ (ZAMS, \beta)$ depends on β, f is related to β and δc_1 depends on c_1 and β.

Using $[c_1] = c_1 - 0.20 \ (b-y)$ rather than c_1, we obtain a good calibration in terms of luminosity by means of indices which are independent of interstellar extinction. The following formula,

$$M_V = - 73.8 \ [c_1] + 8.85 \ \beta^2 [c_1] + 7.71, \tag{3}$$
$$\pm 0.32 \quad \pm 13.8 \qquad \pm 1.81$$

allows us to obtain M_V without a previous determination of the interstellar reddening E(b-y) and its effect on c_1. The error in M_V is close to the respective error for relation (1).

Finally, we have tried to obtain a relation using only the parameters $[m_1]$ and $[c_1]$ (independent of interstellar extinction), because for several distant stars the β index has not been measured.

We have obtained the following equation

$$M_V = -45.63 [m_1][c_1] + 18.19 [m_1]^2 + 5.00 [c_1]^2 + 5.76, \tag{4}$$
$$\pm 0.35 \quad \pm 7.87 \qquad \pm 4.24 \qquad \pm 2.10$$

where $[m_1] = m_1 + 0.32 \ (b-y)$. The error in M_V is slightly higher than the error given for relations (1) and (3).

We point out that equations (1) and (3) can be used instead of the calibration given by Crawford (1975), taking into account the same limitations in the indices.

3. SUPERGIANT F-TYPE STARS

3.1. The observational data.

We have searched for supergiant stars in binary stars and multiple systems with spectral type from about A9 to G2 and with known uvby, β indices (Hauck and Mermilliod 1980). We have found only nine stars in open clusters; three of them are Cepheids (Fernie and McGonegal 1983). We have considered also thirteen bright Cepheids whose absolute magnitude has been obtained by means of the Period-Luminosity relation given by Fernie and McGonegal (1983). The sample contains peculiar objects such as Y Oph and the second overtoue pulsator HR 7308 and Cepheids probably in binary systems; moreover, the photometric indices used in the analysis are generally the mean of a very small number of measures; hence they could give poor estimates of the true mean values. In spite of these limitations, however, the analysis has yielded some interesting results.

3.2. The calibration.

In order to avoid problems due to interstellar extinction, we have considered only the indices β, $[m_1]$, $[c_1]$ and their combinations (second order), that is, a set of nine variables.

A good relation we have obtained is the following one,

$$M_V = 121.01[m_1] - 168.6[m_1]^2 - 20.68[m_1][c_1] - 19.46, \quad (5)$$
$$\pm0.54 \quad \pm9.52 \qquad \pm14.8 \qquad \pm3.60$$

which contains only $[m_1]$, $[c_1]$ and their combinations. A better relation contains also β as follows:

$$M_V = 108.98[m_1] - 146.8[m_1]^2 - 27.22[m_1][c_1] \qquad (6)$$
$$\pm0.42 \quad \pm8.32 \qquad \pm13.4 \qquad \pm3.45$$
$$+20.51\beta \quad - 70.61.$$
$$\pm6.11$$

We have made several tests in order to check the reliability of equations (5) and (6). The effect of β is weak, and, for our purposes, equation (5), taking into account our previous discussion on the observational data, should be sufficient.

We can conclude that, in spite of the fact that the data are not of excellent quality, it is possible to evaluate the absolute magnitude of a supergiant star, and equations (5) or (6) can be considered as preliminary calibrations in terms of luminosity for stars with spectral type from F0 to G2 and luminosity class II-I. Our sample contains a small number of stars; in a future paper we will extend the sample, taking into account the photometric data for other Cepheids.

4. REFERENCES

Antonello, E. 1983, in Proc. Coll. Statistical Methods in Astronomy, (ESA: Strasbourg SP-201), p.145.
Buzzi-Ferraris, G. 1975, Analisi ed identificazione di modelli, (Milano: CLUP), p.37.

Crawford, D. L. 1975, Astron. J., 80, 955.
Crawford, D. L. 1979, Astron. J., 83, 48.
Crawford, D. L. 1979, Astron. J., 84, 1858.
Efroymson, M. A. 1964, in Mathematical Methods for Digital Computers,
     eds. A. Ralston and H. S. Wilf, (New York: J. Wiley), p.191.
Fernie, J. D. and McGonegal, R. 1983, Astrophys. J., 275, 732.
Hauck, B. and Mermilliod, M. 1980, Astron. Astrophys. Suppl., 40, 1.
Hilditch, R. W., Hill, G. and Barnes, J. V. 1983, Monthly Notices Roy.
     Astron. Soc., 204, 241.

DISCUSSION

PHILIP: What is the error in your absolute magnitudes?

ANTONELLO: The standard deviation is 0.4 in M_v.

JASCHEK: What standards did you use for the calibration of absolute magnitude?

ANTONELLO: I used nine stars in open clusters and 13 bright Cepheids.

JASCHEK: Could you deredden all your stars with the same procedure?

ANTONELLO: I've used indices entirely independent of reddening.

JASCHEK: Are there no local differences in reddening?

ANTONELLO: No.

ON PHOTOMETRIC STANDARDS AND COLOR TRANSFORMATION

J. Manfroid

Department of Astrophysics, University of Liège

ABSTRACT. We have carried out a preliminary investigation of a method of observation and reduction which leads to a large improvement in the accuracy of the color transformation of photometric measurements. This method involves the use of extra filters for some of the observations. It is argued that the better precision obtained is well worth the additional telescope work.

1. INTRODUCTION

Exact matching of photometric instrumental systems is very difficult to achieve. Hence color transformation procedures are needed in order to get results which can be compared with those obtained in the standard system. Color transformations are valid for a family of spectral distributions. For example it is easy to find the transformation, over a limited range of wavelengths, adapted to a family of spectra such as:

$$f(\alpha,T,\lambda) = \lambda^{\alpha} B(T,\lambda) \qquad (1)$$

where B is the Planck function and α a parameter. Actual spectra however are irregular and satisfactory color transformations do not exist over more than a small range of spectral types. This is amply demonstrated by theory (see e.g. Young 1974) as well as experience: reduction of many observing runs in the *uvby* system with various equipment shows that errors as high as .05 magnitude, and more, are not uncommon. Similar results have been obtained by other authors (e.g. Graham and Slettebak, 1973).

In order to solve those difficulties, we have developed an idea proposed by Young. The subsequent analysis is devoted to the Strömgren photometry, but would apply equally well to any other system.

2. METHOD

In order to transform an instrumental system into another one, the local behavior of the spectra around the filter wavelengths is needed.

D. S. Hayes et al. (eds.), Calibration of Fundamental Stellar Quantities, 495–498.
© *1985 by the IAU.*

To obtain this information we add a second set of filters with slightly different wavelengths. An interpolation algorithm restores the standard values from the two sets of data. Both data sets are first reduced to the standard system by use of a conventional method (see Manfroid and Heck, 1983, for one which performs well: it allows one to incorporate the eight filters and to account for additional constraints between the filters of each pair with only slight modifications). The interpolation is then straightforward.

3. RESULTS

We have tested the method in extensive numerical simulations, the complete results of which will be published elsewhere (Manfroid 1984). The instrumental systems were made as close as possible to real equipment. In order to check the atmospheric effects we simulated the extinction of the spectra through various airmasses. The data created in that way had the form of observational data obtained with systems differing by the filter shapes and/or the cathode response. Since no source of random errors was included, the variance of the final data, after reductions, reflected the atmospheric color effects and reached .01 mag. in the u filter for blue objects.

The first investigations were made with spectra represented by (1). Objects with spectral anomalies (lines, depressions) were added. The results were very satisfying even with the most critical objects. This was confirmed by simulations using the data from the spectrophotometric catalog by Gunn and Stryker (1983). See Fig. 1. When other sets of filters with different slopes are used or when other photomultipliers are taken, similar results are obtained. Adding any amount of interstellar reddening does not modify the results. Evidently, some filter combinations are to be avoided. A look at the tracings of the filter transmissions allows us to make good choices.

4. CONCLUSION

A considerable improvement in the absolute photometry of any star can be obtained by using a second filter set. The observing time is certainly not doubled. In most cases it could be sufficient to spend a small fraction of the observing runs on the double measurements. Only if different stars are observed every night (which is also the worst case for conventional reductions, Manfroid and Heck, 1983) or if the transmission curve of the filters varies, is it necessary to double every observation. That few double measurements are necessary comes from the differential nature of the corrections. Many effects cancel in the process (uncertainties on the atmospheric extinction, on the zero points).

To work properly the method requires a knowledge of the standard values in each filter. It will be necessary to check this problem in the $uvby$ system where no standard exists for y. An arbitrary decision about such a standard could lead to an impossible standard system, i.e. one which could not be reproduced by any instrumental system.

A byproduct of the method concerns the stellar classification. A great accuracy is already advantageous, but the four additional measurements give very useful information on the spectral distribution. This will be discussed in a forthcoming paper.

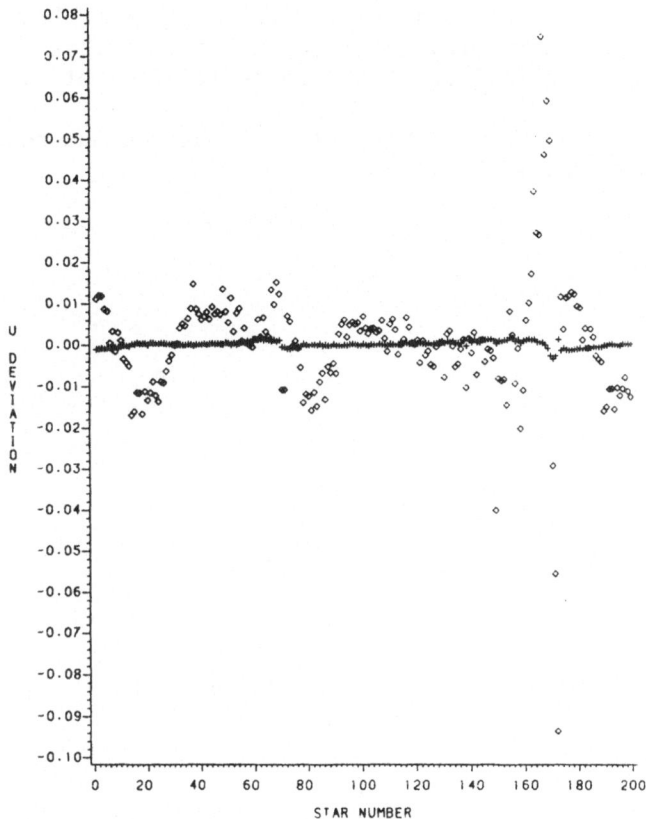

Fig.   1. Comparison of the deviations to the standard values for stars of all types and classes. ◊: conventional reductions when the filter u is shifted by 20 Å; +: reductions with our method.

The author wishes to thank Dr. W.H. Warren Jr. and the Astronomical Data Center at the NASA Goddard Space Flight Center for providing Gunn and Stryker's spectrophotometric atlas.

5. RFERENCES

Graham, J.A., Slettebak, A. 1973, Astron. J. 78, p. 295
Gunn, J.E., Stryker, L.L. 1983, Astrophys. J. Suppl. 52, p. 121
Manfroid J. 1984, in preparation
Manfroid J., Heck, A. 1983, Astron. Astrophys. 120, p. 302
Young, A.T. 1974, in Methods of Experimental Physics, Vol. 12, Part
    A, N. Carleton, ed. (Academic Press, New York), p.123.

DISCUSSION

WALKER: This is a general comment. It is now possible to achieve, routinely 1% precision photometry with CCD's. Their high sensitivity is leading photometrists to use narrower bands. To what extent do the single star, photomultiplier-based calibrations we have been hearing about provide a good basis for more extensive CCD photometry given that there are problems with blocking filters, bandpass definition, etc.?

CAYREL: I agree: the era is now coming when we must think about how to calibrate future photometry done with CCD techniques.

AN EMPIRICAL CALIBRATION OF THE STRÖMGREN SYSTEM FOR LATE-TYPE STARS

B. Nelles, Th. Richtler and W. Seggewiss

Observatorium Hoher List, Bonn University

1. INTRODUCTION

The applicability of the Strömgren system for abundance determination of late-type stars has been shown by several authors (e.g. Eggen
1978b, Bond 1980, Ardeberg and Lindgren 1981). The aim of our contribution is not to present a completely new calibration. Rather we want to
discuss the $(b-y)-m_1$ diagram with an increased sample of spectroscopically analysed stars, hoping to eliminate some unsatisfying properties of
previous approaches and to make improvements in terms of generality and
reliability.
We want to show:
  (1) the insensitivity of the $(b-y)-m_1$ diagram to log g effects,
  (2) the existence of a uniform relation between metallicity
      and a suitable parameter for stars in the color range
      $0.3 < b-y < 1.0$.

2. THE DATA

Our calibration sample consists mainly of stars taken from the
[Fe/H] catalog of Cayrel de Strobel et al (1980), for which vby colors are
available (the u band shall be ignored). Most of these colors come from
the lists of Eggen (1978a) and Olsen (1983). Although Eggen uses a narrower v filter than that of the standard system, one can see from 19
stars in common with Olsen that the mean differences in b-y and m_1 are
just 0^m006 and 0^m03, respectively. We therefore treat the colors as
homogeneous.
A few additional, mostly metal-poor stars were measured by Nelles
and Richtler (1983). Also included are some further spectroscopically
analysed stars, listed by Bond (1980). We end up with 136 stars.
Reddening has been taken into account for 43 stars. The values are
from Bond (1980) and Neckel et al. (1980). For most of the other stars
we expect that the reddening can be neglected due to their small
distances.

D. S. Hayes et al. (eds.), Calibration of Fundamental Stellar Quantities, 499–501.
© *1985 by the IAU.*

3. THE CALIBRATION

According to the late-type star models of Bell and Gustafsson (1978) the $(b-y)-m_1$ diagram should be insensitive to gravity effects. In their empirical study, Ardeberg and Lindgren (1981) arrive at a different conclusion. However, they do not make sure that they compare stars of equal metallicity.

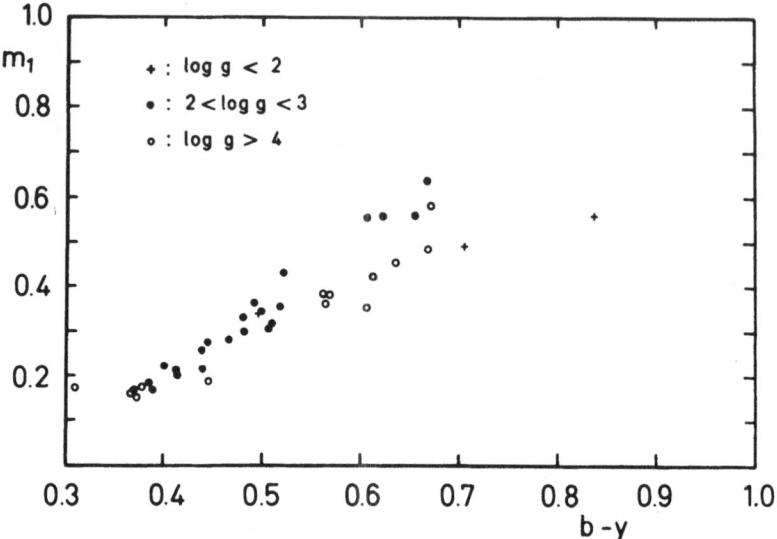

Fig. 1. Stars of different *log* g in the metallicity range $-0.2 < [Fe/H] < 0.2$

Fig. 1 shows all stars in our sample which fall in the metallicity range $-0.2 < [Fe/H] < 0.2$. A gravity effect in the sense of Ardeberg and Lindgren is not seen (one may even suspect the opposite behavior). Furthermore, Eggen's (1983) list of high luminosity stars supports also the theoretical prediction that surface gravity does not affect the locus of a late-type star in the $(b-y)-m_1$ diagram. (Without that assumption, no simple approach to abundance calibration would be possible.)

We assume that the photometric loci of equal metallicity can be represented by straight lines ($m_1 = a_1 + a_2(b-y)$), whose slopes and intersections depend linearly on metallicity.(The slopes decrease with decreasing abundance, leading to a lower metallicity resolution for blue stars.)

Thus the dependence of $[Fe/H]$ from m_1 and $b-y$ takes the form

$$[Fe/H] = A + B \frac{m_1 + C}{(b-y) + D} \qquad \text{with} \qquad \mu \equiv \frac{m_1}{(b-y) + D} \; .$$

The four constants A, B, C, and D are determined by least-squares' fit using the photometric data and the catalog values for [Fe/H]. Weights were given to the individual values of μ according to the expected photometric errors of m_1 and b-y. In this way allowance is made for the lower metallicity resolution at blue colors.

One yields

$$A = -3.34 \pm 0.11, \quad B = 2.93 \pm 0.13, \quad D = -0.22 \pm 0.01.$$

C can assumed to be zero (C = 0.016 $\pm$ 0.017).

Fig. 2 shows the relation between $[Fe/H]$ and μ for our sample. The scatter of the blue stars (crosses) seems not to be larger than that of the red stars (open circles). We take this as a hint that the overall scatter is essentially a consequence of the inhomogeneities in the $[Fe/H]$ data than in the photometric data. Although a proper discussion of individual reddenings, $[Fe/H]$ values etc. would reduce the scatter, we do not expect that the mean relation will change significantly.

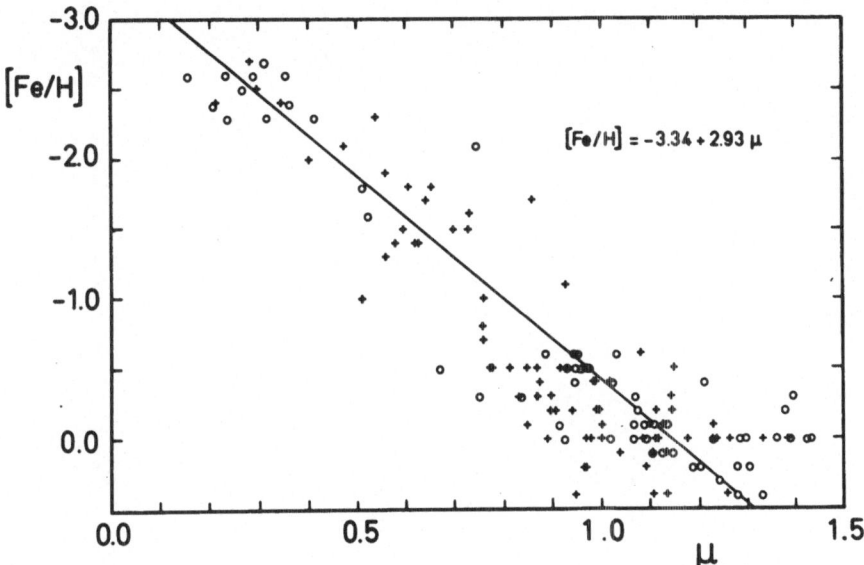

Fig. 2. Calibration of μ against metallicity for our sample of 136 stars. Open circles represent stars in the color range 0.5 < b-y < 1.0, crosses in the range 0.3 < b-y < 0.5.

REFERENCES

Ardeberg, A., Lindgren, H. 1981, Rev. Mexicana Astron.Astrophys. 6, 173
Bell, R.A., Gustafsson, B. 1978, Astron. Astrophys. Suppl. 34, 229
Bond, H. 1980, Astrophys. J. Suppl. 44, 517
Cayrel de Strobel. G., Bentolila, C., Hauck, B., Curchod, A.: 1980,
    Astron. Astrophys. Suppl. 41, 405
Eggen, O.J. 1978a, Astrophys. J. Suppl. 37, 251
Eggen, O.J. 1978b, Astrophys. J. 221, 881
Eggen, O.J. 1983, Astron. J. 88, 1187
Neckel, Th., Klare, G., Sarcander, M. 1980, Bull. Inform. CDS 19, 61
Nelles, B., Richtler, Th. 1983, unpublished
Olsen, E. H. 1983, Astron. Astrophys. Suppl. 54, 55

ANALYSIS OF uvby PHOTOMETRY AND LOW-RESOLUTION SPECTROPHOTOMETRY OF B STARS

William Tobin

Laboratory for Space Astronomy, CNRS, Marseille

A method of analysis of Strömgren photometry and low-resolution spectrophotometry of B stars is being developed in collaboration with J.P. Kaufmann (Technische Universität, Berlin). A full account will be published elsewhere.

The method is similar in spirit to that employed by Tobin & Kaufmann (1984) but aims at a more careful consideration of systematic effects. Stellar effective temperature and gravity are determined by the inter-section in a 'Kiel diagram' of the loci of gravities and temperatures permitted by the observed colors and Hγ and Hδ equivalent widths. Kurucz (1979) models are used in this part of the analysis, and a major difficulty is that systematic effects in Balmer equivalent widths measured from low-resolution spectra can affect $\log g$ by $\gtrsim 0.5$!   We are investigating using empirical multiplicative factors for our Balmer equivalent widths in order to tie our mean results onto values determined from Sinnerstad's (1980) analysis of the photoelectric β index.

The helium content can next be determined (at least for $T_{eff} \gtrsim$ 20 000 K) using the method of Kaufmann & Hunger (1972) on Hγ, Hδ, HeI 402.6 and HeI 447.1 nm. Pure hydrogen-helium 'Berlin models' are used in this part of the analysis, and differences in the blanketing and thus effective temperature scale must be allowed for. Projected rotational velocity can be determined from measured widths of HeI 402.6, 438.8 and 447.1 nm absorptions and calculations of solid-body rotation of the Berlin atmospheres. Allowing for both temperature *and* gravity dependence of the HeI lines, values of rotational velocity can be obtained to ±30 km/s, or about half the error of an earlier calibration which only allowed for the temperature dependence of the HeI lines (Balona 1975).

REFERENCES

Balona, L. 1975, *Memoirs Roy. Astron. Soc.* 78, p. 51.
Kaufmann, J.P., Hunger, K. 1972, *Mitt. Astron. Gesellschaft* 31, p. 185.
Kurucz, R.L. 1979, *Astrophys. J. Suppl. Ser.* 40, p. 1.
Sinnerstad, U. 1980, *Astron. Astrophys. Suppl. Ser.* 40, p. 395.
Tobin, W., Kaufmann, J.P. 1984, *Monthly Notices Roy. Astron. Soc.* 207 p.269.

D. S. Hayes et al. (eds.), Calibration of Fundamental Stellar Quantities, 503–504.
© *1985 by the IAU.*

DISCUSSION

PHILIP: I have a question concerning the program that you mention in your paper, the runaway B stars. You may know that I have been working on A stars at high galactic latitudes and I have found a similar problem. There are too many so-called normal A-type stars far off the galactic plane. Rodgers worked on a list that Sanduleak and I published of A-Type stars at the South Galactic Pole and he claimed that there were 19 out of 62 A-stars in the list that seemed to be normal, Population I stars. John Drilling and I have just completed a paper concerning four-color measures of all 62 stars in the list and we have reclassified some of the Rodgers classifications. But we still end up with something like 12 out of 62 stars seem to be normal Population I A-type stars. Do you know the statistics for the B-type stars?

TOBIN: No. The whole problem is that what has been done with these stars has been very unsystematic. I have no idea about completeness of the sample, etc.

PHILIP: At the South Galactic Pole we measured all the early-type stars in the area down to a limiting magnitude of V = 15.

TOBIN: Yes, but the problem is that we depend on published material.

PHILIP: How many B-type stars of this type are there?

TOBIN: Many scores, certainly, if you were to have a sample complete to something like 9th magnitude. I would guess that if you were complete down to the 10th magnitude you might have perhaps 1000 stars. For instance, I have just submitted a paper on the total galactic reddening using B-type stars more than 10 from the galactic plane as tracers. I was able to compile a catalogue of 1100 such B-type stars with uvby photometry without any problem. And then, under the assumption that they had normal luminosity, and – because of the reddening interest – after eliminating many stars with possible peculiar intrinsic colors (because of emission, etc.) I was still left with 72 stars more than 250 parsecs from the plane.

JASCHEK: Could you use the H-gamma line for your absolute magnitude calibration?

TOBIN: I am not attempting an M_v calibration. For the plate material I am using the H-gamma line that has already been calibrated in terms of M_v by Balona and Crampton (1974 Mon. Not. R. Astr. Soc. <u>166</u>, 203).

THE STABILITY AND ACCURACY OF THE uvby SYSTEM

J. Manfroid

Department of Astrophysics, University of Liège

ABSTRACT.   The recent analysis by Olsen (1983) of uvby data
demonstrates the difficulty of comparing observations obtained with
different equipment.   He detected systematic errors in the standard
stars of Crawford and Barnes (1970), who merged several sources. The
introduction of new photometers with quasi-rectangular passbands
(Nielsen 1983), which give theoretically nontransformable output,
(Young 1974), could make the problem more acute.

In order to evaluate the transformation errors for stars of
various types and luminosities, we have simulated observations made
with   standard   and   non-standard   equipment.   Large   uncorrectable
deviations are observed mainly for objects outside the range
0.0 <b-y< 0.4 and in the u filter. The rectangular passbands have
many advantages, however. They can be very stable and accurately
reproducible.   Hence the problem of transformability may not exist
between such systems.   Since many uvby observations are to be
collected with such equipment, it could be advantageous to define a
new standard "rectangular" uvby system along with the conventional
one.

REFERENCES

Crawford, D. L. and Barnes, J. V. 1970, Astron. J., 75, 978.
Nielsen, R. F. 1983, Inst. Theo. Astrophys., Oslo, Report No. 59, 141.
Olsen, E. H. 1983, Astron. Astrophys. Suppl., 54, 55.
Young, A. T. 1974, in Methods of Experimental Physics, Vol. 12, Part
     A, N. Carleton, ed. (Academic Press, New York), p.123.

D. S. Hayes et al. (eds.), Calibration of Fundamental Stellar Quantities, 505–507.
© 1985 by the IAU.

DISCUSSION

LYNGÅ: The effects of interstellar extinction are different in the two systems (interference and rectangular) but this can be handled as long as the amount of extinction is known.

ADELMAN: It would be easier to synthesize uvby or other photometric colors from spectrophotometry if rectangular rather than conventional filters were used. One must be very careful in placing the bandpasses so that either errors in wavelength centering or large radial velocities will not bring strong lines into or out of the bandpass.

PHILIP: When Relyea and I did our work on matching the four-color indices to the predictions of atmospheric models, we had to take the shape of the passband into account when we calculated the relationship between the theoretical and observed colors. At this point in the work it would have been easier to have used a slot.

JASCHEK: How precisely can you define the windows of the rectangular filter? Can you repeat the positioning of the filter, say within 10 Å?

ARDEBERG: Square filters are, of course, in themselves very appropriate – at least in principle. However, for spectra that are line rich, the definition of square filters are normally quite critical. therefore, grating plus pre-selecting slot plus filter is often the best solution.

PHILIP: Erik Olsen has done four-color photometry with a grating and a slot for thousands of early-type stars and his rms errors are very small.

CAYREL: How can the Strömgren temperature index (b-y) still be a good index at spectral classes as late as M?

PHILIP: The idea of extending the four-color system to later type stars is not a new one. Bell has used the predictions of atmospheric models to construct synthetic spectra that indicate that the m_1 index should be a good indicator of metal abundance for late type stars. I used the 16 inch telescope at Kitt Peak National Observatory and observed some red giants in some of the brighter globular clusters. I calculated the delta m_1 index and found a good relation between it and [Fe/H].

ARDEBERG: For main sequence, as well as for giant stars, our comparisons between (b-y) and MK spectral type give very tight relations. This is true down to stars of middle M type.

PHILIP: When one works with Population II early-type stars it is obvious that there are great differences in the c_1 and m_1 indices for these stars and Population I stars of the same (b-y) index. If I observed one night with Crawford's filters and the next night with rectangular slots and I was observing some peculiar stars I would not know how much of the effect was due to the star and how much was instrumental. The danger, when you go to a new system, is that you may misinterpret results for peculiar stars. A new system must be calibrated very carefully and the differences from the standard four-color system must be well determined.

PARTHASARATHY: (b-y) may not be a good T_{eff} indicator for stars with anomalous CN and CH strength.

ARDEBERG: Our tight relations in the (b-y), c-(1) diagram is valid only for "normal" stars. Stars with CNO anomalies are not included. They certainly should be treated separately.

TOBIN: The question of bandpass shape is also important for single-channel photometers because manufacturers may now deliver interference filters with rectangular and not bell-shaped transmissions.

ON THE CALIBRATION OF THE uvby PHOTOMETRIC SYSTEM FOR LATE-TYPE STARS

A. Ardeberg and H. Lindgren

Lund Observatory, Sweden, and European Southern Observatory

ABSTRACT.   An attempt has been made to calibrate the indices of the uvby photometric system in terms of MK classes equal to and later than that of the Sun. Results are presented for stars of luminosity classes V and III; b-y and c_1 data are given. For stars on the main sequence, the relation between m_1 and b-y is discussed for stars of solar type.

1. INTRODUCTION

It has become increasingly evident that the uvby system provides an excellent basis for determination of effective temperature and luminosity as well as the abundance of heavy elements not only for F-type stars. Recent work has shown it to be of highest value also for cooler stars (Ardeberg and Lindgren 1981, 1982). We have attempted uvby calibration of stars ranging from solar type to M stars. Here, some results are given for a sample of stars with high-class MK and metallicity data.

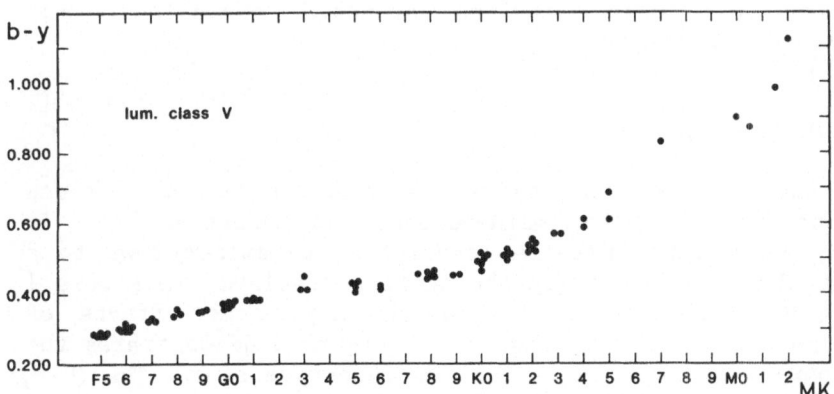

Fig. 1. b-y versus spectral type for main-sequence stars.

D. S. Hayes et al. (eds.), Calibration of Fundamental Stellar Quantities, 509–512.

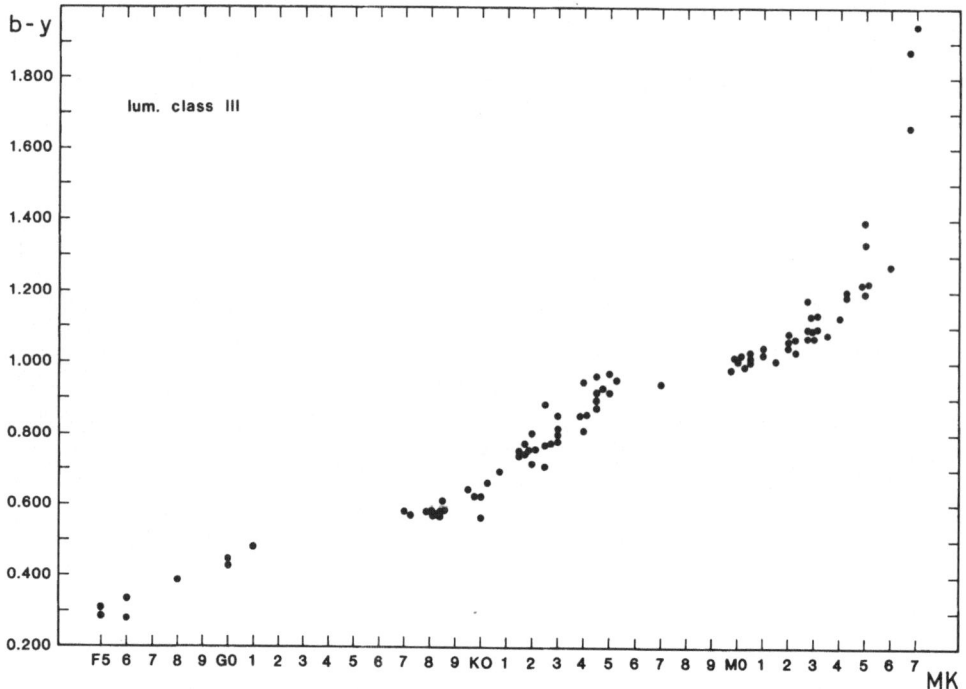

Fig. 2. b—y versus spectral type for giant stars.

2. SELECTION AND OBSERVATIONS OF PROGRAM STARS

We have used only stars with MK classifications by groups including Keenan and/or Morgan. For abundance data we selected investigations from compilations by Cayrel et al. (1980). Most of the stars have been observed in uvby several times. In all cases we used the Strömgren grating photometer on the Copenhagen 50cm telescope on La Silla. For the color indices standard deviations only exceptionally exceed 0.010 mag.

3. RESULTS AND DISCUSSION

Figures 1 and 2 show that there are tight relations between spectral type and b—y index for main-sequence and giant stars. It appears that b—y is a very suitable temperature parameter down to a spectral type of about M5. Whereas the scatter displayed in Figure 1 should be close to the intrinsic spread in b—y, some effects of reddening are probably present in Figure 2. Figure 3 demonstrates the high resolving power of the uvby system for late-type stars. The (b—y, c_1) diagram provides solid data for effective temperature and luminosity. It is shown in Figure 4 that the metallicity also can be determined with high accuracy from uvby observations. The relation for solar-type stars is well defined. The same is true for giant stars. For G/K-type stars $d[Fe/H]/dm_1$ is typically around 8.

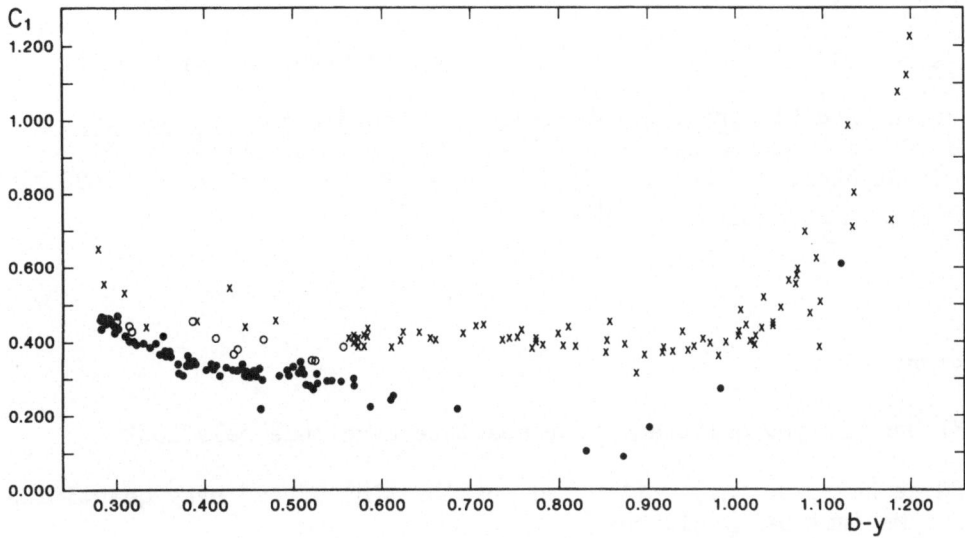

Fig. 3. b-y, c_1 diagram. Crosses denote stars of luminosity class
III, open circles IV and filled circles V.

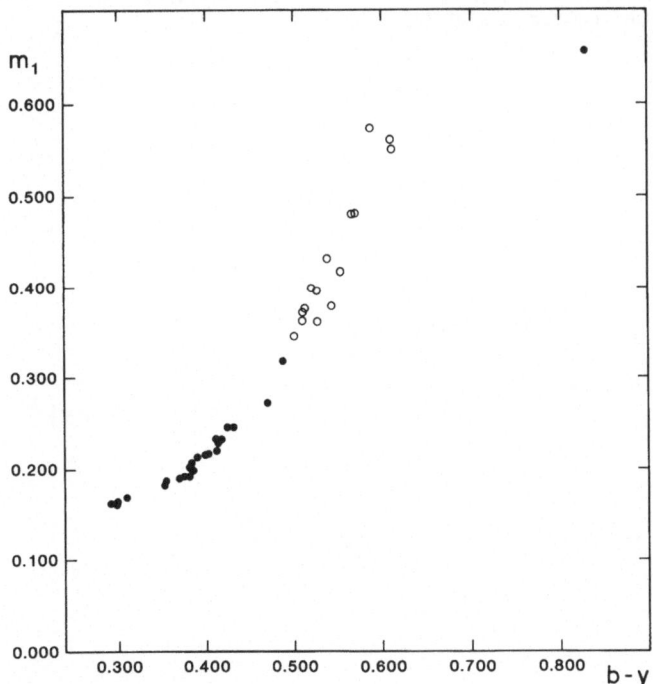

Fig. 4. b-y, m_1 diagram for solar-type stars of luminosity class V.
Filled circles denote stars with [Fe/H]>-0.10, open circles
stars with high quality MK data but with weak or no [Fe/H]
data.

REFERENCES

Ardeberg, A. and Lindgren, H. 1981, Rev. Mex. Astron. Astrophys., 6,
      173.
Ardeberg, A. and Lindgren, H. 1982, in Proc. Nordic Astron. Meeting,
      ed. G. Larsson-Leander (Lund. Obs. Rept. No. 18, Lund), p. 81.
Cayrel de Strobel, G., Bentolila, C., Hauck, B. and Curchod, A. 1980,
      Astron. Astrophys., 41, 405.

DISCUSSION

JASCHEK: How did you calibrate your absolute magnitude relation?

ARDEBERG We have, so far, attempted calibration of our uvby quantities only with MK luminosity classes.

JASCHEK: What explanation do you propose for the $(b-y)$, c_1 relation found?

ARDEBERG: The Balmer discontinuity explains the $(b-y)$, c_1 relation only down to spectral type around late K.

STEPS TOWARD A PHYSICAL CALIBRATION OF UBV PHOTOMETRY

Roland Buser

Astronomical Institute, University of Basel

Robert L. Kurucz

Center for Astrophysics, Cambridge

1. INTRODUCTION

The large data base of photoelectric measurements on the Johnson UBV system has been a primary source of information in many fields of astrophysical interest. The availability of UBV data for virtually all types of stars known to make up the stellar populations in galaxies requires a continued effort toward establishing a fully physical calibration of these data in order to propagate effectively our improving knowledge on stellar evolution and stellar atmospheres (i.e., the HR diagram) through the observations relevant to the structure and evolution of the galaxies. One of the major links in this long chain of scientific progress is provided by the synthesis of stellar photometric properties from theoretical model atmospheres. This paper will briefly address some of the basic problems involved in the synthetic color calculations and discuss the theoretical calibration of UBV photometry as obtained from various grids of model atmospheres covering a large range of stellar types.

2. IMPORTANCE OF PHOTOMETRIC PASSBANDS

An accurate knowledge of the photometric passbands is vital to synthetic photometry in the same way as the knowledge of the standard stars and the proper reduction procedure is vital to observational photometry. A unique set of passbands and normalization equations should be established which apply to all (grids of) model atmospheres used to compute the quantities observed on a given standard photometric system. Unless this basic requirement is met, a full interpretaion of the data in terms of physical parameters as well as an assessment of the relative merits and defects of the theoretical models is impossible and may be achieved only with recourse to independent data.

D. S. Hayes et al. (eds.), Calibration of Fundamental Stellar Quantities, 513–518.
© *1985 by the IAU.*

The passbands of the UBV system have been studied by Buser (1978) who reevaluated the effective wavelength and width of the U-filter using a large sample of spectrophotometric data. It was shown that the mean observed U-B and B-V colors can be computed to within the observational uncertainties from the adopted set of response functions and normalization equations simultaneously for the whole variety of stellar types of normal chemical composition. This normalization was fully confirmed by Buser and Kurucz (1978) in a study of theoretical UBV colors calculated from Kurucz' (1979a) grid of model atmospheres. The revised response function for the U passband was demonstrated to adequately monitor the nonlinear color effects produced by the Balmer lines in early-type spectra, thus eliminating hitherto existing large systematic discrepancies between theory and observation.

Since the ultraviolet colors provide most sensitive measures of different atmospheric parameters for different types of star, a more systematic investigation of various grids of model atmospheres appears justified.

3. THE U-B COLORS OF VARIOUS GRIDS OF MODEL ATMOSPHERES

We have computed theoretical colors from three grids of model atmospheres for DA white dwarfs, F-G stars, and G-K giants, respectively.

3.1. DA White Dwarfs

For the hotter DA stars ($T_{eff} > 12500$ K), the U-V color has been known as a sensitive temperature parameter. While Terashita and Matsushima (TM, 1969) failed to calibrate these colors from their models and the UBV response functions of Matthews and Sandage (MS, 1963), because the large systematic discrepancies between theory and observation were left unexplained, Schulz (1978) successfully reconciled the discrepant results by adjusting the transformation coefficients. This adjustment, however, is unnecessry for synthetic colors calculated using the improved U response function, which accurately accounts for the confluence of the Balmer lines in the DA-spectra. The temperature calibration of the U-V index thus obtained from the TM models ($\theta_{eff} < 0.4$) is almost identical to that given by Weidemann (1982).

3.2. F-G Stars

The normalized ultraviolet excess, $\delta(U-B)_{0.6}$, has been widely used as a metallicity parameter for F- and G-type dwarf stars. But in this temperature range incomplete coverage of atomic and/or molecular line opacity and problems with the treatment of convection conspire to produce large systematic errors (< 0.1 mag) in the synthetic colors of model atmospheres (Relyea and Kurucz 1978, Buser and Kurucz 1978).

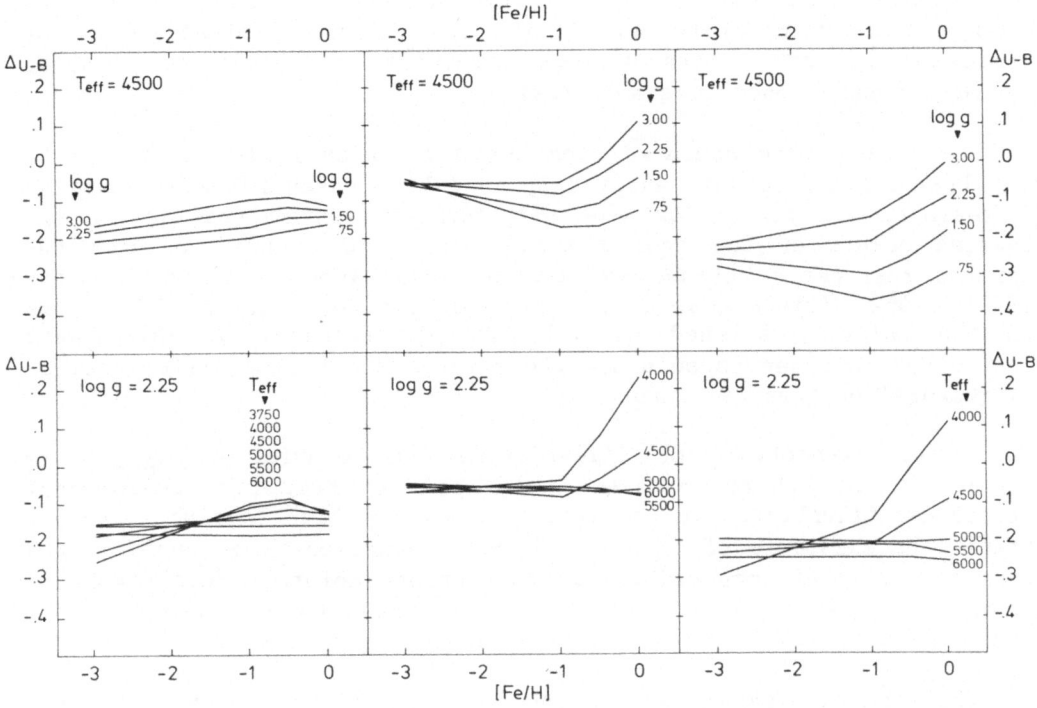

Fig. 1. Effects of instrumental response and normalization (left), molecular or atomic line opacity input (center), and combined effects (right) on the U-B colors of G-K giant model atmospheres.

We have investigated the photometric properties of an extended grid of models (Kurucz 1979b) which incorporates improved convection treatment (Lester et al. 1982). We found that for the log g = 4.5 models the theoretical uv-excess as a function of metal abundance is in very good agreement with the most recent empirical relations (Carney 1979, Cameron 1984) as determined for the dwarf stars listed in the catalog of spectroscopic [Fe/H]-determinations by Cayrel de Strobel et al. (1980). For the zero-metallicity models $\delta(U-B)_{0.6}$ = 0.32, which exactly matches the observational limit. The new models thus appear to provide sufficiently accurate fluxes in the ultraviolet and blue spectral ranges to be useful in differential abundance analyses of F-G dwarf colors.

3.3. G-K Giants

The "ultraviolet discrepancy" shown by Gustafsson and Bell (GB, 1979) to exist between observed and computed colors of G-K-type giants has motivated us to study the nature of this phenomenon by considering the effects on the U-B colors produced by the improved U response function and by different opacity source input. The grid of model atmospheres by Gustafsson et al. (1975) was used along with the updated Kurucz and Peytremann (1975) atomic lines list to compute new model fluxes. While GB's own synthetic spectrum calculations include a large list of molecular opacities but a rather limited number of atomic lines, our spectra neglect molecules completely but instead involve a massive body of atomic data.

U-B colors were computed from our new fluxes in two ways: first, we followed the procedure adopted by GB (which essentially employs the MS response functions for one air mass and normalization to the observed colors of the cool star ϕ^2 Ori), and second, we used the improved response functions and the normalizations as given by Buser (1978). The differences, Δ_{U-B}, between our two sets of U-B colors and the colors published by Bell and Gustafsson (BG, 1978) were constructed in order to separate the effects due to the differences in instrumental or physical input.

A representative plot illustrating the behavior of Δ_{U-B} as a function of atmospheric parameters is shown in Fig. 1. Instrumental effects are illustrated in the left-hand panel, which (again) demonstrates the existence of (nonlinear) color equations that apply in the transformations of computed colors to standard colors. Furthermore, a zero-point error of about -0.2 mag is indicated to be present in the BG colors (which is what GB considered likely to be the case).

The central panel shows the effects of different physical input. For the cooler models Δ_{U-B} strongly depends on [Fe/H] in the sense that our model colors are in general bluer than BG's for the lower metallicities but tend to become redder than BG's at solar abundances.

The right-hand panel of our Fig. 1 combines the instrumental and physical effects and should be compared to GB's Figure 16, where Δ_{U-B} = (observed)-(computed) is plotted as a function of [Fe/H] for a sample of about fifty giants. The striking similarity between these two figures suggests that while the neglect of molecular opacities in model atmospheres of cool stars cannot possibly be correct, atomic line blocking due to the known opacity sources compiled in the Kurucz-Peytremann list may actually account for much of the systematic uv-discrepancy, which is further diminished by using proper instrumental response functions in the calculation of theoretical U-B colors.

ACKNOWLEDGMENTS

This work has been partially supported by the Swiss National Science Foundation.

REFERENCES

Bell, R. A., Eriksson, K., Gustafsson, B. and Nordlund, Å. 1976,
    Astron. Astrophys. Suppl. 23, 37.

Bell, R. A. and Gustafsson, B. 1978, Astron. Astrophys. Suppl. 34, 229.

Buser, R. 1978, Astron. Astrophys. 62, 411.

Buser, R. and Kurucz, R. L. 1978, Astron. Astrophys. 70, 555.

Cameron, L. M. 1984, in preparation.

Carney, B. W. 1979, Astrophys. J. 233, 211.

Cayrel de Strobel, G., Bentolila, C., Hauck, B. and Curchod, A. 1980, Astron. Astrophys. Suppl. 41, 405.

Gustafsson, B., Bell, R. A., Eriksson, K. and Nordlund, Å. 1975, Astron. Astrophys. 42, 407.

Gustafsson, B. and Bell, R. A. 1979, Astron. Astrophys. 74, 313.

Kurucz, R. L. 1979a, Astrophys. J. Suppl. 40, 1.

Kurucz, R. L. 1979b, in Problems of Calibration of Multicolor Photometric Systems, A. G. Davis Philip, ed., Dudley Obs. Rep. No. 14, p.363.

Kurucz, R. L. and Peytremann, E. 1975, Smithsonian Astrophys. Obs. Spec. Rep. No. 362.

Lester, J. B., Lane, M. C. and Kurucz, R. L. 1982, Astrophys. J. 260, 272.

Matthews, T. A. and Sandage, A. R. 1963, Astrophys. J. 138, 30.

Relyea, L. J. and Kurucz, R. L. 1978, Astrophys. J. Suppl. 37, 45.

Schulz, H. 1978, Astron. Astrophys. 68, 75.

Terashita, Y. and Matsushima, S. 1969, Astrophys. J. 156, 203.

Weidemann, V. 1982, in Landolt-Börnstein, Neue Serie, Gruppe VI, Bd. 2b, K. Schaifers, H. H. Voigt, eds., Berlin-Heidelberg-New York, p.373.

DISCUSSION

GUSTAFSSON: The basic philosophy behind the line list of Kurucz and Peytremann is different from that of Bell, on which our models and colors are based. The Kurucz-Peytremann list contains more than 10^6 calculated atomic lines, while the Bell list was based on laboratory data for atomic lines ($<10^5$); on the other hand it contains a very great number of molecular lines.

When we analyzed the "UV discrepancy" for the red giants, we found it reasonable to relate it to a "veil" of very weak metal lines, not included in our line list. It is therefore very interesting to see the indications, presented by Buser and Kurucz, that a significant fraction, if not all, of this discrepancy may possibly vanish when the extensive Kurucz-Peytremann list is merged with adequate list of molecular lines.

BUSER: I should add that the (B–V) colors of our new model flux distributions are slightly bluer (0.02 mag.) than those calculated from models including many molecules, but there is nothing like a systematic dependence on [Fe/H] which is so pronounced in (U–B). For (B–V), we probably have only a small zero-point correction.

PHILIP: Does one of your next steps towards a calibration include a calibration for Population II Stars?

BUSER: Well, yes, the present theoretical calculations actually do include Population II characteristics, i.e. F-G dwarf and G-K giant model atmospheres with solar to 10^{-3} times solar abundances. Work on Population II stars is, however, being continued using scanner observations and UBVRI data of globular cluster and high-velocity field stars, which should help us in progressing towards a more complete understanding of their atmospheres and in establishing a consistent calibration for a large range of stellar types.

CALIBRATION OF PHOTOMETRIC SYSTEMS FROM HOMOGENEOUS SPECTROPHOTOMETRIC DATA

Lukas Labhardt and Roland Buser

Astronomical Institute, University of Basel

1. INTRODUCTION

The atlas of stellar spectrophotometric data published by Gunn and Stryker (GS, 1983) constitutes an extremely valuable tool for the evaluation and calibration of photometric systems. The resolution of the scanner fluxes (10 or 20 Å), given for a broad wavelength range ($\lambda\lambda$ 3160-10620 Å) and the systematic coverage of stellar types are (almost) ideal for calculating synthetic colors for Becker's (1946) photographic RGU system which has been extensively applied in galactic structure work (cf. Buser 1981). Since RGU photometry is tightly linked to UBV data (Steinlin 1968, Buser 1978b), the Gunn-Stryker atlas has been used in the present paper to evaluate these two systems and subsequently investigate the resulting calibration of the RGU colors in terms of MK spectral classification.

2. UBV COLORS

Our evaluation of UBV colors differ from that by Gunn and Stryker in three basic respects.

2.1 UV Fluxes

Unlike GS, who absorbed incomplete wavelength coverage in their UV fluxes into the transformation of synthetic to standard colors (cf. section 2.2 below), we have been interested in the most adequate representation of the U filter passband (cf. also Buser and Kurucz 1985). We have therefore used available OAO (Code and Meade 1979, Meade and Code 1980) and IUE (Heck et al. 1984) data of about ten early-type stars of the GS sample to extend their fluxes to the atmospheric cutoff of the U filter(s) at 3000 Å. For most stars the UV data could be joined smoothly with the GS data, and the U magnitudes computed from the Matthews and Sandage (MS, 1963) response functions (for one air mass) were found to be brighter by about 0.05 mag (for O-stars) than the corresponding magnitudes obtained from the

D. S. Hayes et al. (eds.), Calibration of Fundamental Stellar Quantities, 519–521.
© *1985 by the IAU.*

incomplete scans, declining to 0.03 mag for an A star. Extrapolation of these corrections to later spectral types and different luminosity classes was checked using the library of mean stellar energy distributions given in the catalog by Straižys and Sviderskienė (1972), and the synthetic (U-B) colors for the whole GS sample were corrected accordingly. For the U passband as given by Buser (Paper I, 1978a), these corrections never exceeded 0.01 mag due to its slightly redder short-wavelength cutoff.

2.2. Transformations

In contrast to GS, who first corrected the synthetic scan colors and the observed colors for interstellar reddening (by adopting an interstellar extinction model of the Galaxy) in order to then establish the transformations, our interest focussed on the direct comparison of the scan colors (uncorrected for interstellar reddening) and the observations, because systematic errors introduced by inadequate response functions in the synthetic calculations are augmented by interstellar reddening effects (Paper I).

More than 110 GS stars for which homogeneous UBV colors exist in the catalog of Nicolet (1978) have been used in our study of the transformations (GS excluded stars with low-weight photometry as well as those marked for duplicity and/or variability to end up with about 70 objects used in their corresponding calculations). Linear regressions between synthetic and observed colors confirm that the UBV response functions adopted in Paper I are clearly superior to the MS functions. The former give tighter fits in the U-B and transformation coefficients closer to unity for both B-V and U-B. Most importantly, the systematic nonlinearities present in the residuals of the $(U-B)_{MS}$ colors are completely eliminated on the improved U-B system of Paper I.

2.3. Dereddening

While GS adopted an interstellar extinction model to deredden the fluxes and colors, we have decided to study extinction using the color excesses derived by comparing the program star colors with the mean intrinsic colors as given by FitzGerald (1970). Homogenization of the spectral type and luminosity class (using essentially the Bright Star Catalog) assigned to each star was important for estimating the intrinsic scatter of the resulting E_{B-V} vs. spectral type/luminosity class plot. A uniform scatter independent of spectral type and luminosity class of ±0.05 mag was found, which is comparable to the standard deviations of the mean intrinsic colors associated with a given spectral type (FitzGerald 1970). We therefore chose to deredden only those stars with $E_{B-V} > 0.05$ mag and to exclude the few objects with $E_{B-V} < -0.05$ mag. Dereddening of the fluxes was then accomplished employing Whitford's (1958) law as adopted in Paper I.

3. RGU COLORS

The intrinsic colors on the photographic RGU system were computed applying the response functions given in Paper I. Plots of color vs. spectral class were then used to establish smooth relations from weighted averages for each stellar type. For (U-G), the resulting relations are in good accord with those obtained in Paper I from averaged spectral scans. As yet unexplained systematic discrepancies exist however for the (G-R) colors of intermediate-type dwarfs and late-type giants. Further work including the comparison of synthetic colors with photoelectrically determined RGU data will be necessary for an improved evaluation of the photographic system.

ACKNOWLEDGEMENTS

This work was supported by the Swiss National Science Foundation.

REFERENCES

Becker, W. 1946, Veröff. Univ. Sternwarte Göttingen, **79.**
Buser, R. 1978a, Astron. Astrophys., **62,** 411.
Buser, R. 1978b, Astron. Astrophys., **62,** 425.
Buser, R. (ed.) 1981, Galaktische Struktur und Entwicklung, preprint
    Astron. Inst. Univ. of Basel, No. 2, 271 pp.
Buser, R. and Kurucz, R. L. 1985, in IAU Symposium No. 111,
    Calibration of Fundamental Stellar Quantities, D. S. Hayes, L. E.
    Pasinetti, and A. G. Davis Philip eds. (Reidel, Dordrecht), p.513.
Code, A. D. and Meade, M. R. 1979, Astrophys. J. Suppl., **39,** 195.
FitzGerald, M. P. 1970, Astron. Astrophys. 4, 234.
Gunn, J. E. and Stryker, L. L. 1983, Astrophys. J. Suppl., 52, 121.
Heck, A., Egret, D., Jaschek, M. and Jaschek, C. 1984, IUE Low-
    Dispersion Spectra Reference Atlas - Part I. Normal Stars. ESA
    SP-1052. Paris.
Matthews, T. A. and Sandage, A. R. 1963, Astrophys. J., **138,** 30.
Meade, M. R. and Code, A. D. 1980, Astrophys. J. Suppl., 42, 283.
Nicolet, B. 1978, Astron. Astrophys. Suppl., **34,** 1.
Steinlin, U. W. 1968, Z. Astrophys., **69,** 276.
Straižys, V. and Sviderskienė, Z. 1972, Bull. Vilnius Astron. Obs. 35.
Whitford, A. E. 1958, Astron. J., **63,** 201.

AN IMPROVEMENT IN THE VISUAL SURFACE BRIGHTNESS SCALE FOR B5-F5 MAIN SEQUENCE STARS

L. Pastori and G. Malaspina

Astronomical Observatory of Milano-Merate

Angular diameters of 593 B5-F5 main sequence stars listed in the "Catalogue of apparent diameters and absolute radii of stars" (CADARS; Fracassini et al. 1981) have been analysed in order to improve the precision of the visual surface brightness S_V. The new relations between this quantity and the color index $(B-V)_0$ turn out to be in good agreement with those found with the interferometric method (Barnes et al. 1978). Moreover, the results suggest that surface gravity effects may bias the S_V-$(B-V)_0$ relations.

REFERENCES

Barnes, T. G., Evans, D. S. and Moffett, T. J. 1978, Mon. Not. Roy. Astron. Soc., 183, 285.
Fracassini, M., Pasinetti, L. E. and Manzolini, F. 1981, Astron. Astrophys. Suppl., 45, 145.

D. S. Hayes et al. (eds.), Calibration of Fundamental Stellar Quantities, 523–524.

DISCUSSION

EVANS: What is the source of your angular diameters?

PASTORI: The catalogue by Fracassini et al.

EVANS: But these are not direct determinations.

PASTORI. No.

EVANS: I think your argument is circular.

PASTORI: The method of Fracassini et al. is an <u>indirect</u> method derived from the original method by Chalonge and Divan applied to about thirty stars by Cayrel (1980) In the method of Fracassini et al. the parameters are obtained by means of Geneva photometry. In our work we compare our determinations of visual surface brightnesses with those of Barnes et al. (1978) which come from a <u>direct</u> method. So there is no circularity.

THE RESPONSE FUNCTION S(λ) OF THE UBV COLOR SYSTEM - A PROBLEM AGAIN AND AGAIN?

Erich E. Lamla

Observatory, University of Bonn

ABSTRACT. By comparing the numerical data for the response functions of the UBV color system originally given by Matthews and Sandage (1963) with the improved functions published by Ažusienis and Straižys (1969), taking into account of the extinction values given by Melbourne (1960), the response function of U and B is found corrected for a printing error and other mistakes.

As is well known, the UBV color system was defined originally by Johnson and Morgan (1951). However, they did not publish the complete numerical data, S(λ), of their photometer, used at their telescope under different observing conditions. Therefore, one may find in the literature several data lists of S(λ) without knowing all the details of the reconstruction of the color system used, which are more or less similar to the original one.

The numerical data, given by Matthews and Sandage (1963) have been used very often, although Hayes (1975) found some 'numerical noise' in the figures. In preparing the new Landolt-Börnstein compilation Lamla (1982) detected as a reason for that 'noise' an apparent printing error and errors in reading off the numerical values of the response functions of U and B by a bad interpolation in the overlapping wavelength range around λ = 3900 Å.

Ažusienis and Straižys (1966) improved the original S(λ) of the UBV system published by Johnson and Morgan (1951) by using the extinction of the earth's atmosphere given by Melbourne (1960). One can follow that reconstruction step-by-step in their English version of it, published in 1969.

By comparing the differences, Δ, between the S(λ) given by Matthews and Sandage (1963) and those reconstructed by Ažusienis and Straižys (1966) I found the differences violate the Gaussian distribution at some wavelengths (see Table I and II, column 5). By making the biggest differences smaller, I could correct the values of

D. S. Hayes et al. (eds.), Calibration of Fundamental Stellar Quantities, 525–528.

the response functions U and B which are then in accord with the even curves of $S(\lambda)$. The corrected values are given in column 6 of the Tables I and II.

TABLE I

The response function $S(\lambda)$ of the U magnitude. Comparison of the figures given by Matthews and Sandage (1963) and by Ažusienis and Straižys (1966), and the corrected values.

λ Å	U_1	U_{cal}	U_{tab}	Δ (10^{-3})	$U_{1,corr}$
3100	0.250	0.078	0.060	$-$ 18	
3200	0.680	0.237	0.170	$-$ 47	
3300	1.137	0.442	0.375	$-$ 67	
3400	1.650	0.720	0.675	$-$ 45	
3500	2.006	0.975	1.000	$+$ 25	
3600	2.250	1.220	1.250	$+$ 30	
3700	2.337	1.390	1.390	0	
3800	1.925	1.234	1.125	$-$ 109	1.800
3900	0.650	0.440	0.600	$+$ 160	0.750
4000	0.197	0.139	0.140	$+$ 1	
4100	0.070	0.051	0.030	$-$ 21	

U_1 = U for sec Z = 1; Matthews and Sandage (1963).

U_{cal} = $U_1 \cdot P_1$ = U_2 = U for sec z = 2;

P_1 = extinction coefficient; Ažusienis and Straižys (1966).

U_{tab} = U_2; Ažusienis and Straižys (1966).

REFERENCES

Ažusienis, A. and Straižys, V. 1966 Bulletin Vilnius Astron. Obs. No.16.

Ažusienis, A. and Straižys, V. 1969, Soviet Astron. A. J., 13, 316.

Hayes, D. S. 1975, Multicolor Photometry and the Theoretical HR Diagram. A. G. Davis Philip and D. S. Hayes (eds.) (Dudley Obs. Reports No. 9, Albany), p.309.

Johnson, H. L. and Morgan, W. W. 1951, Astrophys. J., 114, 522.

Lamla, E. 1982, "Magnitudes and Colors". Landolt-Börnstein, New Series, Group VI, Vol. 2b, K. Schaifers and H. H. Voigt (eds.) (Springer Verlag Berlin).

Matthews, Th. A. and Sandage, A. R. 1963, Astrophys. J., 138, 30.

Melbourne, W. G. 1960, Astrophys. J., 132, 101.

TABLE II

The response function S(λ) of the B magnitude. Comparison between the values given by Matthews and Sandage (1963) and by Ažusienis and Straižys (1966), and the corrected values.

λ Å	B_o	B_{cal}	B_{tab}	Δ (10^{-3})	$B_{tab,cor}$
			S(λ)		
3571	0	0	0	0	
3600	0.015	0.008	0.006	− 2	
3700	0.100	0.058	0.080	+ 22	
3800	0.500	0.312	0.337	+ 25	
3900	1.800	1.185	1.425	+ 240	1.270
4000	3.620	2.484	2.253	− 231	2.523
4100	3.910	2.789	2.806	+ 17	
4200	4.000	2.950	2.950	0	
4300	3.980	3.012	3.000	− 12	
4400	3.780	2.919	2.937	+ 18	
4500	3.500	2.745	2.780	+ 35	
4600	3.150	2.507	2.520	+ 13	
4700	2.700	2.175	2.230	+ 55	
4800	2.320	1.892	1.881	− 11	
4900	1.890	1.558	1.550	− 8	
5000	1.530	1.273	1.275	+ 2	
5100	1.140	0.956	0.975	+ 19	
5200	0.750	0.633	0.695	+ 62	
5300	0.500	0.425	0.430	+ 5	
5400	0.250	0.214	0.210	− 4	
5500	0.070	0.060	0.055	− 5	
5560	0	0	0	0	

B_o = B(sec z = 0) ; Ažusienis and Straižys (1966)

B_{cal} = B_1 = B(sec Z = 1) = $B_o \cdot P_1$

B_{tab} = B_1 ; Matthews and Sandage (1963) .

DISCUSSION

JASCHEK: Does the analysis of Matthews and Sandage, and Ažusiensis and Straižys refer to the same sample of observations of the same stars?

LAMLA: There is no need for that! One has to measure the sensitivity function of the filter, the multiplier, the optics (reflectivity of the mirrors as a function of wavelength) and one has to put in the transmission of the Earth's atmosphere. This was done earlier by Matthews and Sandage, and Ažusiensis and Straižys in the same way. A comparison of both values for the sensitivity function then shows differences (in one case it can be a printing error!). And I made these differences smaller; the new values for V and B are in agreement with those given by Arp (1961 Astrophys. J. 133, 874). What can then be done is to calculate, with an energy distribution of a star, color differences with the new sensitivity function and see how good the values are compared to the observed color indices.

CODE: The Matthews-Sandage UBV sensitivity function came from actual laboratory measurements of 1P21 response and filter transmission carried out by Harold Johnson and published by him (1951 Astrophys. J. 114, 511). To these measurements I added the reflectivity of two aluminum surfaces, which was adopted by Matthews and Sandage. Thus the UBV sensitivity functions are the measured response of the original UBV photometer of H. Johnson and W. W. Morgan and not a deduced function.

LAMLA: I agree.

HAUCK: Did you compare your response functions with those of Buser?

LAMLA: Yes, I did. There are no big differences.

BOHLIN: Art Code says that the original UBV transmissions (Matthews and Sandage) were measured on a monochromator for the filters and 1P21 and then multiplied by typical aluminum reflectances for two mirrors. Were the new curves measured using monochromatic light or were they inferred from the need to make the photometry consistent internally?

LAMLA: I used the old UBV sensitivity functions corrected as Code said. These new curves were again corrected only by calculation: to make the differences between the values given by Ažusiensis and Straižys and those given by Matthews and Sandage smaller. I did not try to calculate synthetic color indices to see if one found the same or other values. The sensitivity functions, with the corrected values, are exactly the same as those given by Arp. There are differences between Arp and Matthews and Sandage.

BESSELL: I have also measured the ultraviolet response of several 1P21 and several ultraviolet filters, 1 mm UG2 and standard thickness 9863. The red cut off measured was identical to that of Buser, but the blue cut off was about 100 Å bluer, although not so blue as that of Matthews and Sandage. However, to fit synthesized (U-B) colors from spectrophotometry one must use the Buser blue cut off. Clearly, the original Johnson system U appears to require more UV absorption such as would be supplied by several mm of soda glass.

ON THE OBSERVATIONAL DETERMINATION OF OB STANDARDS

J. Krełowski and A. Strobel

Institute of Astronomy, N. Copernicus University, Torun

ABSTRACT. Observational evaluation of intrinsic energy distributions in OB spectra is difficult as these stars are usually reddened and any dereddening procedure may change from place to place in the Galaxy. We compare two-color diagrams (UV vs. visual) from two stellar complexes of the same Sp/L class. The observed differences may be caused by additional far-UV reddening of distant objects as well as by intrinsic differences between stars belonging to different complexes. We argue in favor of the first possibility which would allow one to use the same standards in all of the Galaxy.

1. INTRODUCTION

The observational determination of the intrinsic parameters of OB stars is rather difficult. In the absence of existing unreddened stars we have to apply a "dereddening procedure" in order to find their intrinsic color indices. Then, the basic questions are: is the extinction law identical in every part of the Galaxy and may it be applied after a single extinction parameter (e.g. E_{B-V}) is determined.

Recent attempts at observational determination of the extinction curve in the far-UV (FUV) (Krełowski and Strobel 1983) present a negative answer to the above questions. The extinction seems to be caused by at least two agents (grain populations of different sizes). Different proportions of these agents along possible lines of sight make the extinction law variable. E_{B-V} must not be used any longer as a single extinction parameter especially in the FUV, where reddening is caused almost exclusively by small (bare) grains which do not affect the visual light.

As was concluded by Krełowski and Strobel (1983), bare particles are related to the interstellar space by being absent inside of OB associations. Moreover, based on the most recent investigations (Kiszkurno et al. 1984), we incline to the conclusion that the

529

D. S. Hayes et al. (eds.), Calibration of Fundamental Stellar Quantities, 529–532.
© 1985 by the IAU.

extinction law inside a given spiral structure of the Galaxy is relatively uniform.

The present paper attempts to examine this conclusion because of the basic importance of the observational determination of the intrinsic colors of early type stars. The colors based on nearby stars (e.g. Wesselius et al. 1980) or on all observed objects of a Sp/L class (Gałecki et al. 1983) may not be applicable to all stars of our Galaxy.

2. DATA REDUCTION AND ANALYSIS

In the presence of local variations of the extinction law (Krełowski and Strobel 1983), the intrinsic colors of early-type stars may depend on the space positions of the objects considered. To check if this is true we decided to compare the relations of UV color indices (based on the ANS bands) vs. (B-V) taken for stars of chosen Sp/L class from different regions of the Galaxy: one containing local and Sgr arms and the second the Per arm$_0$ (see Table I), being independent "units" in the galactic disc (Lynga 1982).

For this purpose we have selected a sample of B0-1, V-III stars. Hotter stars are inconvenient for their total sample in the ANS Catalogue (Wesselius et al. 1982) is too small. Supergiants are found very seldom in our vicinity which makes the planned comparison impossible. Dwarfs of spectral type later than B2 become too faint to be observed at large distances in the Per arm. We decided to mix dwarfs and giants since their intrinsic UV colors seem to be almost identical (e.g. Bless and Savage 1972).

We decided to compare the relations of the color indices (33-22) and (15-18) to (B-V) for the chosen regions. We have omitted the (22-18) color index since the relations of the excesses E_{22-18} to E_{B-V} seem to be identical in all OB associations (Kiszkurno et al. 1984). If the relations (33-22) and (15-18) vs. (B-V) differ from complex to complex then observationally evaluated intrinsic colors may be affected by the contents of the samples considered and proper de-reddening procedures should differ from region to region in the Galaxy.

3. RESULTS AND CONCLUSIONS

The stars listed in Table I are displayed as open circles (Sgr+ local) and dots (Per) in two figures below. Figure 1 presents the relation between the UV color index (33-22) and (B-V). The dots representing Per arm stars are clearly situated above the relation for the second complex. They seem to form a parallel sequence. Then the intrinsic colors calculated separately for both complexes using the method of Gałecki et al. (1983) would differ by approximately 0^m25.

The (15-18) color index vs. (B-V) is plotted in Figure 2. We observe here no correlation in any single complex. The complexes are rather characterized by different average values of (15-18). The poor correlation observed for all points in Figure 2 is then meaningless.

The results presented above seem to suggest the following two interpretations: the Per arm stars are additionally reddened in the FUV or they are intrinsically fainter in the FUV than those in our vicinity. We prefer the first interpretation, as an intrinsic FUV faintness of distant stars should affect their Sp/L. We summarize our considerations as follows:

i) the spectral energy distributions of OB stars should be based on stars of the Sgr or local arm which are practically free from FUV obscuration.

ii) different stellar complexes require different dereddening procedures.

iii) OB supergiants, very rare in our vicinity, may be dereddened only using extinction curves obtained for neighboring dwarfs.

TABLE I

Photometric Data for B0-1, V-III Stars

HD/BD	B-V	15-18	22-33	HD/BD	B-V	15-18	22-33
L 34748	-0.11	-0.19	-0.73	13745	+0.17	+0.04	+0.73
34989	-0.13	-0.26	-0.87	14014	+0.14	+0.05	+0.86
35149	-0.15	-0.16	-0.90	14250	+0.32	-0.07	+1.60
35299	-0.21	-0.22	-1.21	14331	+0.17	+0.13	+0.76
35439	-0.20	-0.18	-1.15	14605	+0.27	-0.01	+0.87
36351	-0.18	-0.25	-1.11	14707	+0.55	+0.25	+2.57
36591	-0.20	-0.24	-1.15	15642	+0.08	+0.04	+0.27
36695	-0.17	-0.22	-1.18	56°473	+0.25	+0.01	+1.00
36960	-0.25	-0.24	-1.35	59°357	+0.47	-0.03	+2.24
37303	-0.22	-0.25	-1.27	59°456	+0.55	+0.11	+2.55
37481	-0.23	-0.24	-1.27	56°482	+0.30	+0.07	+1.47
37744	-0.20	-0.24	-1.16	56°484	+0.32	+0.08	+1.29
113791	-0.19	-0.22	-1.17	56°589	+0.41	+0.13	+1.95
141637	-0.05	-0.21	-0.58	236633	+0.40	+0.17	+1.67
144217	-0.07	-0.23	-0.55	236815	+0.22	-0.01	+0.95
144470	-0.04	-0.27	-0.38	237007	+0.33	-0.09	+1.54
S 164018	+0.65	+0.02	+2.61	59°510	+0.56	+0.11	+2.66
164637	-0.05	-0.16	-0.45	63° 87	+0.52	+0.02	+2.17
165285	+0.31	-0.16	+1.03	53°2820	+0.10	+0.02	+0.21
166539	+0.25	-0.23	+0.86	53°2833	+0.23	+0.09	+0.91
152217	+0.16	-0.07	+0.45	54°2790	+0.20	-0.03	+0.66
152245	+0.12	-0.14	+0.22	55°2795	+0.31	+0.10	+1.42
P 1544	+0.16	-0.04	+0.54	56°2903	+0.22	-0.12	+1.16
2451	+0.19	-0.05	+0.68	57°2581	+0.61	+0.11	+2.53
13036	+0.53	+0.24	+2.09	60°2525	+0.29	0.00	+1.20

REFERENCES

Bless, R. C. and Savage, B. D. 1972, Astrophys. J., **171**, 293.
Gałecki, Z., Braczyk, M., Janaszak, E., Kołos, R., Krełowski, J. and
    Strobel, A. 1983, Astron. Astrophys., **122**, 207.
Kiszkurno, E., Kołos, R., Krełowski, J. and Strobel, A. 1984, Astron.
    Astrophys., 135, 337.
Krełowski, J. and Strobel, A. 1983, Astron. Astrophys., **127**, 271.
Lynga, G. 1982, Astron. Astrophys.,**109**, 213.
Wesselius, P. R., van Duinen, R. J., Aalders, J.W.G. and Kester, D.
    1980, Astron. Astrophys., 85, 221.
Wesselius, P. R., van Duinen, R. J., de Jonge, A.R.W., Aalders,
    J.W.G., Luinge, W. and Wildeman, K. J. 1982, Astron. Astrophys.
    Suppl., **49**, 427.

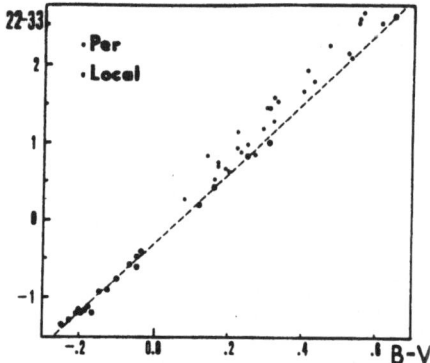 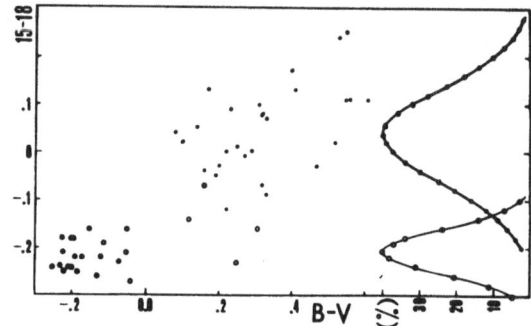

Fig. 1. Two-color diagram for   Fig. 2. Gaussian curves charac-
         stars from the com-   terize the distributions of
         plexes considered.   FUV colors for both
                                            complexes.

DISCUSSION

TOBIN: Problems resulting from interstellar reddening can be minimized by studying the unreddened or little-reddened apparently normal B stars found at high galactic latitudes.

KREŁOWSKI: These studies can not solve problems resulting from local discrepancies caused by the dust associated with young stars.

AN EFFECTIVE TEMPERATURE CALIBRATION OF A UV-VISUAL PHOTOMETRIC INDEX FOR "NORMAL" NON-SUPERGIANT STARS

C. Morossi and M. L. Malagnini

Astronomical Observatory, Trieste

ABSTRACT.   The determination of the effective temperature of each individual star requires observations of the flux distribution in the largest possible wavelength range.   Unfortunately, owing to observational constraints, this requirement severely limits the number of stars for which the direct determination of T_{eff} can be achieved. However, an estimate of T_{eff} for a very large number of stars can be obtained through a calibration of a photometric index versus the stellar effective temperature.   The UV-Visual dereddened index, R=log(F 1965/F 5445), has been shown to be suitable for "normal" non-supergiant stars in the spectral type range B5 - F7 (Malagnini et al. 1984).   The analysis, with the method described in Malagnini et al. (1983), of stars in the spectral type range B5 - B0, enables us to extend and to refine the already proposed calibration of R versus T_{eff}.   By using the complete calibration, the determination of T_{eff} for all the "normal" non-supergiant stars sampled from the 31215 objects in the Catalogue of Stellar Ultraviolet Fluxes (Thompson et al. 1978) is being carried out.

REFERENCES

Malagnini, M. L., Faraggiana, R. and Morossi, C. 1983, Astron. Astrophys., 128, 375.
Malagnini, M. L., Morossi, C. and Faraggiana, R. 1984, in The MK Process and Stellar Classification, ed. R. F. Garrison (David Dunlap Observatory, Toronto), p.321.
Thompson, G. I., Nandy, K., Jamar, G., Monfils, A., Houziaux, L., Carnochan, D. J. and Wilson, R. 1978, Catalogue of Stellar Ultraviolet Fluxes (The Science Research Council).

D. S. Hayes et al. (eds.), Calibration of Fundamental Stellar Quantities, 533–534.

DISCUSSION

JASCHEK: What do you mean by "normal" non-supergiant stars?

MOROSSI: The "normal" non-supergiant stars are those quoted as not chemically peculiar according to the catalogue of M. Jaschek.

JASCHEK: Does your calibration have any advantage over the other ones based on the usual photometric system?

MOROSSI: I think that we can obtain greater accuracy in determining T_{eff} using our photometric index since it takes into account the behavior of the stellar flux in the ultraviolet region.

DETERMINATION OF T_{eff} FOR CANDIDATE STANDARD STARS BASED ON COMPARISON WITH MODEL ATMOSHPERES

M. L. Malagnini, C. Morossi and M. Ramella

Astronomical Observatory, Trieste

ABSTRACT. A systematic analysis, as complete as possible, of early type standard stars has been undertaken at the Astronomical Observatory of Trieste, in order to obtain a sequence of carefully determined spectroscopic data to be used in the comparison of "normal" and "chemically peculiar" stars. As an intermediate and necessary step, effective temperatures are derived. The determination of T_{eff} is performed either by comparing observed and computed flux distributions, or by using a calibration of UV-visual photometric index versus T_{eff}. The results for a sample of stars in the spectral type range B2-F8 are presented, and an analysis of the influence of the adopted value of log g on the derived T_{eff} values is reported. As a check on the validity of the results, the T_{eff} and log g values are used to construct a synthetic spectrum which is compared with the IUE high resolution observations of the stars in our list.

1. INTRODUCTION

The sample of candidate standard stars, proposed by Vladilo et al. (1983), has been selected in order to arrive at a reference sequence to be analyzed homogeneously. All these stars are bright stars, with quite low Vsin i values, already known to have solar chemical composition. To arrive at a quantitative analysis of the spectra, an accurate determination of the most important atmospheric parameter, T_{eff}, is required. We report here on the T_{eff} determination for the stars listed in Table I, and discuss the influence of the gravity values on its accuracy. The atmospheric parameters are used as input for the computation of synthetic spectra in the IUE UV region, in order to derive information about the chemical composition.

D. S. Hayes et al. (eds.), Calibration of Fundamental Stellar Quantities, 535–538.
© 1985 by the IAU.

TABLE I.

Parameters for Non-Supergiant Stars

HR	HD	Sp.	Type	E(B-V)	T1	T2	log g1	Log g2
39	886	B2	IV	0.01	21.77	21.78	3.64	4.0
153	3360	B2	IV	0.04	21.79*			
269	5448	A5	V	0.00	8.01	8.14	3.52	4.0
811	17081	B7	V	0.00	13.16	13.17	3.83	4.0
1034	21278	B5	V	0.07	15.05	15.32	3.27	4.0
1292	26462	F4	V	0.00	6.72	6.56	3.30	4.5
1380	27819	A7	V	0.00	7.99	8.22	3.56	4.5
1637	32537	F0	V	0.01	7.18*			
1810	35708	B2.5	IV	0.07	18.62	19.72	2.97	4.0
2010	38899	B9	IV	0.00	10.71	10.73	4.50	4.0
2085	40136	F1	III	0.00	7.05*			
2421	47105	A0	IV	0.02	9.43	9.28	3.39	4.0
2818	58142	A1	V	0.00	9.37*			
2943	61421	F5	IV-V	0.00	6.60	6.44	3.59	4.5
4049	90277	F0	V	0.00	7.25*			
4141	91480	F1	V	0.00	7.24*			
4359	97633	A2	V	0.00	9.55	9.38	3.20	4.0
4399	99028	F2	IV	0.04	6.86	6.94	3.58	4.5
4540	102870	F8	V	0.02	6.14	5.95	3.24	4.5
4564	103578	A3	V	0.03	-	8.12	-	4.0
5404	126660	F7	V	0.00	6.20	6.09	3.54	4.5
5447	128167	F3	V	0.00	7.03	7.03	4.33	4.5
6092	147394	B5	IV	0.01	14.84	15.00	3.72	4.0
6396	155763	B6	III	0.02	-	13.11	-	4.0
6588	160762	B3	V	0.02	17.50	17.46	3.89	4.0
7001	172167	A0	V	0.01	9.67	9.53	3.33	4.0
7371	182564	A2	III	0.00	8.65*			
7773	193432	B9.5	V	0.00	9.93	9.97	4.09	4.0
8641	214994	A1	IV	0.00	9.59	9.56	3.92	4.0
8805	218470	F5	V	0.00	6.95*			

2. MATERIALS AND METHODS

The list of 30 non-supergiant stars, with spectral types between B2 and F8, is given in Table I, together with the adopted E(B-V) corrections. For the determination of the atmospheric parameters we apply the method described in Malagnini et al. (1984), based on the comparison between observed and computed flux distributions. The observational data refer to the visual region in the wavelength range 3187-10000 Å, and are derived from the Breger Catalogue (Breger 1976). The computed data are from the grid of Kurucz' models (Kurucz 1979), with solar chemical composition. We performed the analysis keeping the gravity value fixed and leaving the three parameters, T_{eff}, log g, and angular diameter free. The Breger Catalogue lists 22 of the 30 program stars; for the 8 remaining stars, the T_{eff} values are derived by applying the calibration of the dereddened UV-Visual index, R=log (F[1965]/F[5445]), versus temperature, as proposed by Malagnini et al. (1984).

3. RESULTS

The results labelled "1" in Table I refer to the fit performed when the parameters T_{eff}, log g, and angular diameter are left free. Those labelled "2" refer to the fit performed by assuming a fixed value for log g. Temperatures are given in thousands of degrees K. The 8 results marked by * are only provisional, in the sense that a complete calibration of R versus T_{eff} is in progress. For the remaining 22 stars, the fit performed by leaving the gravity free produces results that are, in general, different from those achieved by keeping the gravity fixed. In particular, log g1 is lower than log g2, except in two cases. The differences between T1 and T2 are generally on the order of the uncertainties in the solutions, but there are some stars for which the difference is significant. Since the flux distribution for non-supergiant stars is largely independent of gravity, the log g1 values may not be very significant. Therefore, the dependence of the T_{eff}'s on log g has been analyzed by comparing the solutions obtained at log g = 3.0, 3.5, 4.0, and 4.5, respectively. For 20 out of 22 stars, there is a monotonic trend of T_{eff} with log g. This trend is positive for $T_{eff} > 13000$ K and between 7000 and 9000 K; it is negative otherwise. The percentage of the range in T_{eff}, for different log g's, with respect to the T2 values, increases with T_{eff}, and reaches the maximum value of 11% at T2 = 21780 K.

4. REFERENCES

Breger, M. 1976, Astrophys. J. Suppl., 32, 7.

Kurucz, R. L. 1979, Astrophys. J. Suppl., 40, 1.

Malagnini, M. L., Faraggiana, R. and Morossi, C. 1983, Astron. Astrophys., 128, 375.

Malagnini, M. L., Morossi, C. and Faraggiana, R. 1984, in the MK Process and Stellar Classification, ed. R. F. Garrison (David Dunlap Obs., Toronto), p.321.

Vladilo, G., Morossi, C., Ramella, M., Rusconi, L. and Sedmak, G. 1983, Inform. Bull. CDS. No. 24.

DISCUSSION

ADELMAN: In your list of candidate stars, you include Omicron Peg. It is a hot Am star which is slightly metal-rich and as such does not belong with the normal stars. The increase in the metallicity has a slight effect on the derived effective temperature. Model atmospheres for such stars are not as certain as for solar composition stars.

MOROSSI: We will start with the assumption of solar abundance, but if we find a problem we will not use that star.

THE EFFECT OF METALLICITY AND PULSATION ON THE INFRARED COLORS OF LUMINOUS M GIANTS

M. S. Bessell and J. Brett

Mt. Stromlo and Siding Spring Observatory

M. Scholz

Institute for Theoretical Astrophysics, Heidelberg

P. R. Wood

Mt. Stromlo and Siding Spring Observatory

The temperature calibration for cool stars and in particular the Miras continues to be contentious. Lunar occultations have provided radii for many K and M stars and a good temperature calibration has been derived for the hotter non-variable M stars (Ridgway et al. 1980). The situation for the Miras and carbon stars and the metal-rich and metal-poor M stars is, however, not so clear cut. Observations are generally made in some broad-band color   such as (R-I), (V-K) or (J-K) and a temperature derived using either the Ridgway et al. (1980) empirical scale or a black-body scale; differences can amount to several hundred degrees. We decided to theoretically explore the effects that extension, metallicity and pulsation could have on colors.

Using the code of Scholz and Tsuji (1984) we have computed spherically extended model atmospheres for two sets of luminosity and mass (L = 4 x $10^4 L_\odot$, M = 4 $M_\odot$; 10^4 $L_\odot$, 1 $M_\odot$) at three different stellar radii (R = 273 $R_\odot$ [T_{eff} = 3500°K], 371 $R_\odot$ [3000°K], 535 $R_\odot$ [2500°K] and three different metal abundances [Z/H] = +0.5, 0, -1. These models should be relevant to the oxygen-rich asymptotic giant branch stars in the galactic center, the solar neighborhood, the Magellanic Clouds and the dwarf spheroidals. In addition, for 4 $M_\odot$, 6 models with modified CN abundances were computed to investigate how the dredge-up of nitrogen rich material in upper AGB stars, as suggested by Wood, Bessell and Fox (1983), could affect the spectrum. Finally, we have commenced the investigaton of Mira-type atmospheres by computing the structure and flux of several atmospheres with L = 10^4 $L_\odot$, M = 1 $M_\odot$, R = 371 $R_\odot$ but imposing density-radius profiles with two discontinuities, to simulate the presence of shock fronts in the atmosphere. Such a profile was produced with   pulsation

539

D. S. Hayes et al. (eds.), Calibration of Fundamental Stellar Quantities, 539–542.

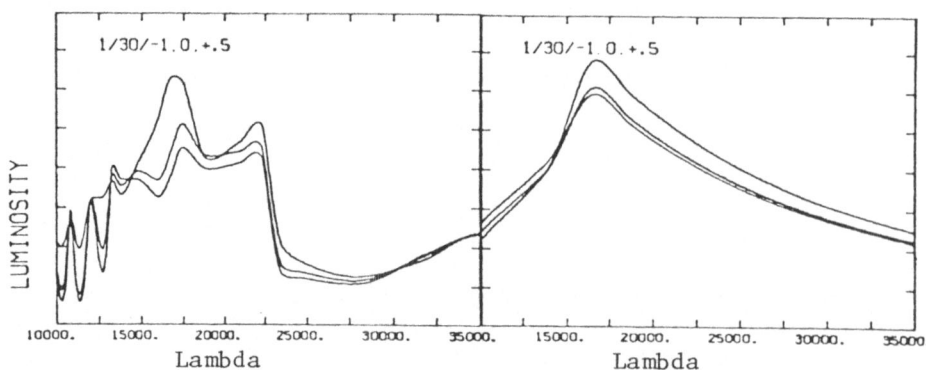

<u>Fig. 1.</u> Luminosity vs. wavelength for extended models.

modelling by Wood (1979), and possible values of density and velocity
at the discontinuities were determined from observatons by Hinkle,
Scharlach and Hall (1984) - see also Fox, Wood and Dopita (1984).

    Details of the model computations and opacities are given in
Scholz and Tsuji (1984). More complete results and details of the
modified code for handling the Mira-type models will be published
elsewhere.

    In Fig. 1 are shown the infrared fluxes for the extended models
of mass $1M_\odot$, temperature 3000°K and abundances [Z] = -1, 0, +.5. The
blanketed fluxes are shown to the left and the continuum fluxes are
shown to the right, computed with all band opacities switched off. At
this temperature the spectra are dominated by bands of TiO and CO,
which weaken with decreasing metallicity. But unexpectedly the
continuum colors also change, the more metal-deficient models being
significantly redder. In order to assess quantitatively these color
changes we have computed broad-band and narrow-band colors from these
fluxes. Unfortunately, it appears that the band opacities used for
the models, while adequate for the construction of the atmospheres,
are not calculated at high enough wavelength resolution nor with
sufficient precision to produce realistic spectra or colors; however,
examination of the continuum colors does provide insight into the
scale of the temperature-color differences that could occur.

    In normal models constructed under hydrostatic equilibrium the
density profile of the atmosphere exhibits a monotonic decrease of
density with height. In a Mira variable, however, dynamical effects
can be considerable and grossly non-monotonic density profiles can
occur. In the "mira-models" we have attempted to represent this
effect by imposing two discontinuous density steps, corresponding to
shocks, onto the normal density profile. The location and separation
of these steps with respect to the photosphere was different in the
two atmospheres, in an attempt to model a Mira at two different
phases, near ϕ = 0.9 and ϕ = 0.15.

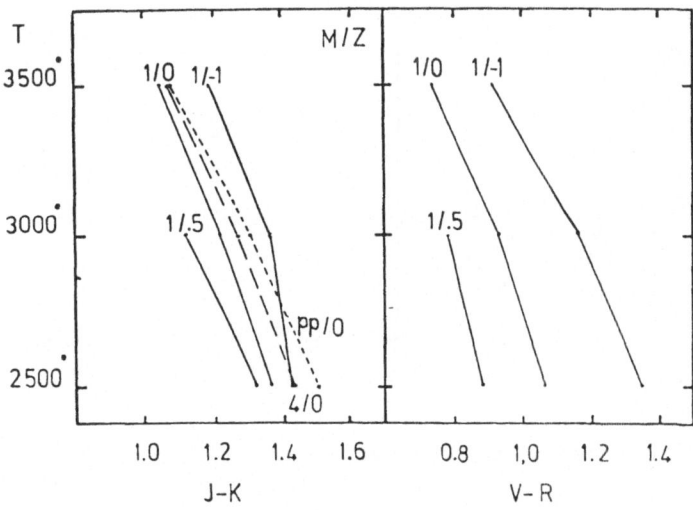

Fig. 2. Continuum colors for the models.

Hot H_2O absorption is a feature of the spectra of the cooler normal stars, but in a Mira it seems that regions favorable to the formation of polyatomic molecules exist in a higher temperature atmosphere. Carbon-Miras also show this phenomenon, their spectra differing from those of non-mira carbon stars in the occurrence of strong HCN or C_2H_2 bands. We hope to extend this preliminary work to different temperatures, phases and possibly different C/O abundances.

In Fig. 2 are plotted the (J-K) and (V-R) continuum colors for the various models. It can be clearly seen that color differences due to extension and metallicity are quite large and would result in significantly different temperatures being derived from an inappropriate color-temperature relation. For example, the extended model for $1M_\odot$ and 3000 K has the same (J-K) color as the extended model for $4M_\odot$ and 3100 K, and the plane-parallel model for 3220 K. Metallicity effects are even larger, models with (T,[Z]) = (3500,-1), (3100,0) and (2850,+.5) having the same (J-K) color. Similar color differences occur in other spectral regions such as (V-R) and (V-K), although the plane-parallel and extended model colors are much closer in VRI. Clearly, to obtain accurate temperatures from fitting observed and theoretical colors it is necessary to compute more realistically blanketed colors or observe in band-free windows, and to have some knowledge as to the mass, luminosity and abundance of the stars. A further complication in such cool stars is whether or not the star is a Mira (or Long-Period) variable as will be discussed in the next section.

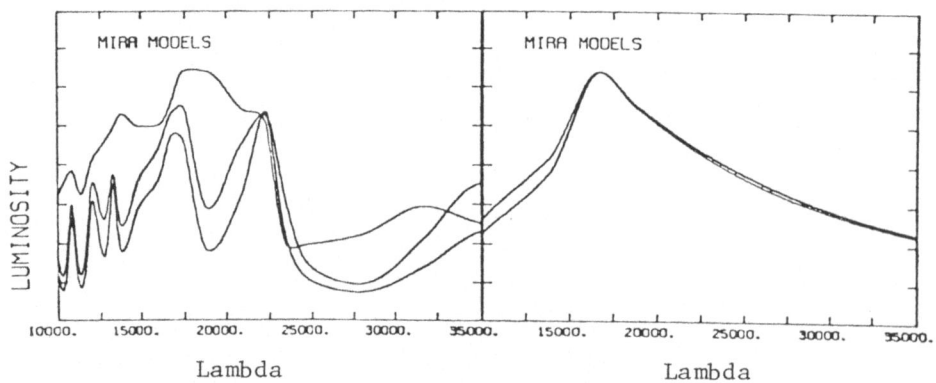

Fig. 3.   Luminosity vs. wavelength for mira-models.

Fig. 3 shows the resultant fluxes for the two mira-models and for comparison, the flux for the normal model of the same temperature and mass.   The very obvious difference is the occurrence of strong H_2O bands in the mira-models (a well noted feature of all M-type Mira variables), which change with phase, although there was little change in continuum color.   The H_2O absorption occurs most strongly in the K-band, causing the (J-K) colors to be redder.

REFERENCES

Fox, M. W., Wood, P. R. and Dopita, M. A. 1984, Astrophys. J., submitted.

Hinkle, K. H., Scharlach, W. W. G. and Hall, D. N. B. 1984, preprint.

Ridgway, S. T., Joyce, R. R., White, N. M. and Wing, R. F. 1980, Astrophys. J., 235, 126.

Scholz, M. and Tsuji, T. 1984, Astron. and Astrophys., 130, 11.

Wood, P. R. 1979, Astrophys. J., 227, 220.

Wood, P. R., Bessell, M. S. and Fox, M. W. 1983, Astrophys. J., 272, 99.

THE TEMPERATURE SCALE OF G AND K SUBGIANTS

U. O. Frisk

Space Science Department, ESA

R. A. Bell

National Science Foundation

ABSTRACT: This investigation of nearby Population I stars was
motivated by our need for accurate effective temperatures in order to
be able to obtain reliable N abundances from lines belonging to the CN
red system.   These determinations are based on narrowband photometry
calibrated by means of synthetic spectra.   It is believed that the
temperatures are accurate to within 100 K.   The results show that
recent calculations of evolutionary tracks of G and K subgiants
predict the observations quite well.   However, a value for the mixing
length parameter l/H_P close to 1.6 is required.

1.   THE PHOTOMETRIC SYSTEM

The selection of bandpasses is based on the work on the effective
temperature of Arcturus by Frisk et al. (1982).   Three 0.01 micron
wide bandpasses were selected; their central wavelengths are 0.59,
0.78 and 1.06 micron.   The bandpasses fit well into the atmospheric
windows and avoid uncertain or large stellar blocking.   The
observations were normalized against Vega and have maximum total
errors of less than 5%.

Detailed synthetic spectra were calculated using the Synthetic
Spectrum Generating (SSG) program as described by Bell and Gustafsson
(1978).   These spectra were calculated for a grid of models (Bell et
al. 1975, plus additional models calculated using the same methods)
covering effective temperatures from 4000 K to 6000 K in steps of 500
K, surface gravity from log g equal 3.00 to 5.25 in steps of 0.75 and
for metallicities [M/H] from −2.0 to solar.   Spectra were also
calculated for a model of Vega (Dreiling and Bell 1980) and for a grid
model with solar parameters.   The theoretical colors were produced by
evaluating the integral of the product of the spectra and the filter
passbands.

D. S. Hayes et al. (eds.), Calibration of Fundamental Stellar Quantities, 543–545.
© *1985 by the IAU.*

The solar flux was also computed at the wavelength of the passbands using a grid model from Bell et al. (1975, BEGN) and for two semi-empirical models, one from Holweger and Müller (1975, HM) and one due to Vernazza et al. (1976, VAL). These are compared to the observed flux as by Arvesen et al. (1967, AGP) and Labs and Neckel (1968, LN) in Table I.

TABLE I

CALCULATED AND OBSERVED SOLAR FLUXES

Flux ratio	model flux			obs. flux	
	BEGN	HM	VAL	AGP	LN
$2.5 \text{ LOG}_{10} F_{590}/F_{780nm}$	0.42	0.44	0.43	0.44	0.43
$2.5 \text{ LOG}_{10} F_{590}/F_{1060}$	1.10	1.13	1.11	1.12	1.10

2. THE EFFECTIVE TEMPERATURES

The determination of the effective temperatures assumes that the metal content is proportional to the iron abundance which is based on the calibration of weak ($\log W/\lambda < -4.8$) Fe I lines. The surface gravity is based on the assumption of solar mass and the trigonometric parallax. The temperature calibration is not very sensitive to small changes of the metallicity or surface gravity.

The temperatures determined from the theoretical calibration were used to calibrate the Johnson R-I colors. According to our scale a star like the Sun would have an R-I index close to 0^m35. This high value is in agreement with direct measurements of the solar B-V color. For the stars in this sample the excitation temperatures based on weak Fe I lines do not disagree much with the color-based temperatures. The mean difference is 30 ± 30 K. The excitation temperatures are hotter.

The dominating uncertainty in the temperature scale comes from the calibration source, Vega. Changing the effective temperature of the Vega model by 300 K corresponds to a shift of 60 K for a star with an effective temperature of 5000 K.

3. COMPARISON WITH EVOLUTIONARY TRACKS

Since we know the effective temperature of each star and we can infer the luminosity from the parallax together with the V-magnitude and a bolometric correction it is of interest to compare these values to evolutionary tracks of Population I stars. These were taken from VandenBerg (1983). Bolometric corrections were adopted from Bell and Gustafsson (1978). The results have been plotted in Fig. 1. The set of tracks has not been shifted to agree with the position of the Sun. One should be aware that the selection of stars was not unbiased. From Fig. 1 it is clear that the stars cluster around a line which we believe to be the base of the red giant branch.

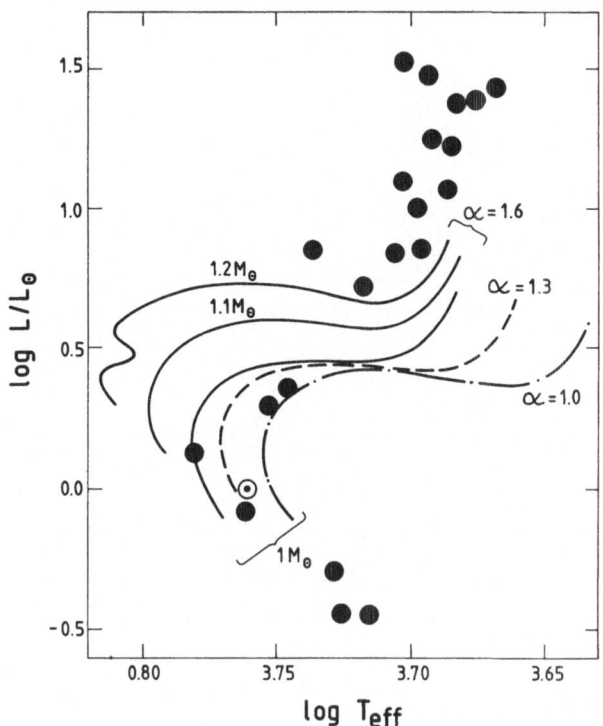

Fig. 1. The position of the program stars in a temperature-luminosity diagram. For comparison calculated evolutionary tracks for stars of solar composition (VandenBerg 1983) have been included.

However, the stars were not selected to cluster, and on the contrary an attempt was made to sample a range both in temperature and luminosity. If we believe that the mass of a red giant is close to solar then we get a reasonable agreement with our determination for a value of the mixing length parameter $1/H_p$ close to 1.6.

REFERENCES

Arvesen, J. C., Griffin, R. N. and Pearson, B. D., Jr. 1969, Appl. Optics 8, 2215.

Bell, R. A., Eriksson, K., Gustafsson, B. and Nordlund, A. 1975, Astron. Astrophys. Suppl. 23, 37.

Bell, R. A. and Gustafsson, B. 1978, Astron. Astrophys. Suppl. 34, 229.

Dreiling, L. A. and Bell, R. A. 1980, Astrophys. J. 241, 736.

Frisk, U. O., Bell, R. A., Gustafsson, B., Nordh, H. L. and Olofsson, S. G. 1982, Mon. Not. R. Astr. Soc. 199, 471.

Holweger, H. and Müller, E. A. 1974, Solar Phys. 39, 19.

Labs, D. and Neckel, H. 1968 Zeits. f. Astrophys. 69, 1.

VandenBerg, D. A. 1983, Astrophys. J. Suppl. 51, 29.

Vernazza, J. E., Avrett, E. H. and Loeser, R. 1976, Astrophys. J. Suppl. 30, 1.

ON THE ACCURACY OF INFRARED PHOTOMETRY

R. Papoular and B. Pégourié

Department of Astrophysics, CEN, Saclay

ABSTRACT. Photometric errors due to scintillation are considered in detail. Given noise characteristics, the standard deviations, as deduced from observations, are computed for the quantities measured at successive steps of the photometric procedure. This allows us to understand better the errors computed on-line in observatories, and to understand better the overall error. The latter can be minimized by suitable changes in the time sequence of the measurement.

1. INTRODUCTION

Many reports concerning IR photometry are not explicit as to the accuracy of their data, the definition of error bars and/or the possible systematic errors. This probably reflects a real problem in IR photometry, which results from a combination of various sources of noise, whose statistical properties differ from each other and may vary between observations. Thus, although it is highly desirable to obtain average atmospheric characteristics from extended monitoring at each and every observatory, this must be complemented by efforts to characterize fluctuations at each observing session. Resident astronomers have realized this and provided on-line data pretreatment that may be very efficient, if adequately used. We propose here to analyze the meaning of on-line results, to show how to deduce from them an estimate of the accuracy and to discuss how to improve this accuracy. Only extinction fluctuations are considered here because they supersede sky emission fluctuations when the observed object is not too faint. The case where emission noise is dominant was treated elsewhere (Papoular 1983). It is assumed that extinction is independent of direction and that image wandering is negligible.

2. STATISTICAL BACKGROUND

First recall the basic statistical definitions and relations that will be needed. Let X be the random variable to be measured and $W_X(f)$, its noise power spectrum. Then, its variance is $\sigma_X^2 = \int_0 W_X(f)\, df$ and an estimate of the error on X is the standard deviation, σ_X. If $W_X(f)$ is not known, σ_X has to be evaluated by repeating the measurement a large number of times. Usually, however, a small number, N, of measurements are

547

D. S. Hayes et al. (eds.), Calibration of Fundamental Stellar Quantities, 547–551.

performed, from which the sample variance S_X is computed using $S_X^2 = (1/(N-1))\Sigma^N(X-\overline{X})^2$, where $\overline{X} = \Sigma^N X/N$, and S_X is considered as an estimate of σ_X. But we showed (Papoular and Pegourie 1983) that this is often overly optimistic because the low frequency fluctuations are not properly taken into account. Using the results of Barnes et al. (1971), it can be shown that the most probable value of S_X^2 is not σ_X^2 but

$$\delta_X^2 = \frac{N}{N-1} \int_0^\infty df. \; W_x(f) . \left[1 - \frac{\sin^2(\pi fNT)}{N^2 \sin^2(\pi fT)} \right] \quad , \qquad (1)$$

where T is the interval between successive measurements. The difference between δ_X^2 and σ_X^2 is not negligible unless all the frequencies in $W_X(f)$ are such that $f > 1/\pi NT$. The monitoring of atmospheric emission and extinction fluctuations in a number of observatories (Allen and Barton 1981, Papoular 1983) yielded a conspicuous "1/f$^\alpha$-noise" spectrum with $W(f)$ still increasing for $f < 0.001$ Hz. As a result, the measuring time NT should, in principle, be very long in photometry ($\sim 1/2$ hour).

The best estimate of X is taken to be its average, $\overline{X}$. The error in X is often taken to be $\sigma_X/\sqrt{N}$. Again, this is shown in Papoular and Pegourie (1983) to be too optimistic because it neglects correlations between measurements. In fact, the averaging procedure does not uniformly reduce the noise spectrum $W_X(f)$, but preferentially its higher components (Barnes, et al. 1971):

$$W_{\overline{X}}(f) = W_x(f) . \frac{\sin^2(\pi fNT)}{N^2 \sin^2(\pi fT)} \qquad (2)$$

On the other had, integration over a time NT alters the noise according to

$$W_{\overline{X}}(f) = W_x(f) . \frac{\sin^2(\pi fNT)}{(\pi fNT)^2} \qquad (3)$$

Thus, integration acts as a real low-pass filter, with a bandwidth $\sim 1/NT$, while averaging acts like a multiple-band pass filter.

By definition, $\sigma_{\overline{X}}$ is the integral, over frequency space, of expression (2) or (3), the computation of which requires a knowledge of the noise spectrum $W_x(f)$. It is usually implicitly assumed that this spectrum is white, in which case $S_x/\sqrt{N}$ is a good estimate of $\sigma_{\overline{X}}$. As stated above, this is not acceptable in astronomical photometry. Moreover, the exponent α (in the 1/f$^\alpha$ spectrum) changes with time and photometric band.

Photometric procedures also include back-ground subtraction and comparison with standard stars. In these cases, sums and differences of random variables arise, with a constant time difference, T, between the measurements of two variables. If both variables (X,Y) can be assumed to have the same noise spectrum W_x, then the following relations hold (Papoular 1983, Appendix C):

$$W_{x+y}(f) = 4 . W_x(f) . \cos^2(\pi fT) \qquad (4a)$$

$$W_{x-y}(f) = 4 . W_x(f) . \sin^2(\pi fT) \qquad (4b)$$

It is possible to compute the variance at any stage of the process, using equations of the form (2) to (4). For purposes of comparison, we may normalize variances by dividing them with the total noise energy, $\sigma_o^2 = \int W_x(f) \, df$ over the available spectrum.

3. COMPUTATION OF VARIANCES IN PHOTOMETRY

The sequence of measurements, as well as the symbols used, are summarized in Figure 1. A, B, a, b represent signals; s^2, estimated sample variances; indices $*$ and Σ are for program star and standard star respectively; τ, θ and T are time intervals. Each line of the figure represents an element of the sequence with its name and outcome. Let I be a quantity proportional to the spectral brightness of the point object observed, and $\chi(t)$ be the atmospheric extinction at the same wavelength and direction, and at the time of the observation. Then,

$$a(t) = I \cdot exp\left[-\chi(t)\right] \quad or \quad b(t) = -I \cdot exp\left[-\chi(t)\right] \tag{5}$$

Here we overlook the sky emission, for reasons stated in the introduction. χ is a random variable with a non-zero average: $\chi(t) = \chi_o + \tilde{\chi}(t)$. It usually is so small that $exp[-\chi(t)] \approx 1 - \chi(t)$ and the i^* average of a over a time τ is

$$a_i = I \cdot \left[1 - \chi_o - \tilde{\chi}_i\right] \quad , \tag{6}$$

where the average of $\tilde{\chi}_i$ (over a long time) is zero. Let $W_x(f)$ be the (unknown!) noise power density of $\tilde{\chi}(t)$. Because of the linearity of this relation, the noise spectra of the photometric signals will be proportional to W_x, with the coefficient I^2. Table I represents, in a self-explanatory way, the signals at the various levels of the sequence of Figure 1, together with their respective averages, noise spectra and estimated variances, according to the rules of section 2. Descending this sequence, we find a number of quantities that usually are computed on-line during observations: A,B (A-B)/2 (or A-B), I and $\ddot{I}$. Also computed are sample variances, such as

$$S_a^2 \equiv \sum_1^n (a_j - \bar{a})^2 / (n-1)$$

$$S_{(A-B)/2}^2 \equiv \sum_1^N \left[\frac{(A-B)_i}{2} - I\right]^2 / (N-1) \tag{7}$$

$$S_I^2 \equiv \sum_1^M (I - \bar{I})^2 / (M-1)$$

By way of an example, at ESO (La Silla), the on-line computer gives E1, E2 and STD, where $E1 \simeq S_a \sqrt{2/n}$; $E2 = 2 S_{(A-B)/2}$; $STD = 2 S_I$.

Note that their IT and n correspond to our $n\tau$ and N. Also $\%E1 = 100 \cdot \frac{E1}{I} \cdot \frac{1}{(N-1)^{1/2}}$,

$$\%E2 = 100 \cdot \frac{E2}{I} \frac{1}{(N-1)^{1/2}} \quad , \quad \%STD = 100 \cdot \frac{STD}{I} \cdot \frac{1}{(M-1)^{1/2}} \quad .$$

The quantity of interest for one star (at a given wavelength) is $\overline{I}$. Since there is usually no time available to determine its variance experimentally, the latter is often estimated by $S_{\overline{I}}^{2} = S_{I}^{\nu}/M$ or $S_{I}^{\nu}/(M-1)$. As stated in section 2, this usually is not valid.

The final step usually involves a comparison with a standard star of known spectral brightness $I_{\Sigma,\lambda}$. At a given wavelength, we have (skipping subscript λ):

$$I_{*\Sigma} = \left(\overline{I}_{*}/\overline{I}_{\Sigma}\right) \cdot I_{\Sigma} = I_{*} \, exp\left(\tilde{\chi}_{\Sigma} - \tilde{\chi}_{*}\right) \tag{8}$$

where $\overline{I}_{*}$ and $\overline{I}_{\Sigma}$ are the results of the above sequence of measurements for star $*$ and standard Σ, respectively. I_{*} is the true spectral brightness of the star and $\tilde{\chi}_{\Sigma}$, $\tilde{\chi}_{*}$ the average atmospheric extinctions during the respective sequences. Then, the relative error in $I_{*\Sigma}$ is

$$\epsilon \equiv \left(I_{*\Sigma}/I_{*}\right) - 1 \simeq \tilde{\chi}_{*} - \tilde{\chi}_{\Sigma} \tag{9}$$

which is also to be considered as a random variable, with zero mean value. Assuming the directions of $*$ and Σ are close to one another, the standard deviation (or overall error) σ_{ϵ} can be determined by noting that $\tilde{\chi}_{*}$ and $\tilde{\chi}_{\Sigma}$ are the same functions of time, taken at intervals $T = LMN\theta$. Then eq. (4b) is applicable and gives the last line in Table I; σ_{ϵ} can then be determined by integrating $W_{I_{*\Sigma}}$ over all frequencies.

4 DISCUSSION

Let us assume a simple model for the noise spectrum:

$$\begin{aligned} W_{\chi}(f) &= W_{0}\,(f_{0}/f)^{\alpha} &\quad for &\quad f_{0} \leqslant f \leqslant f_{M} \\ W_{\chi}(f) &= W_{0} &\quad for &\quad f \leqslant f_{0} \end{aligned} \tag{10}$$

Then it is easy to compute (numerically) all standard deviations in units of σ_{0} as a function of α, for different values of f_{0} and of the parameters, τ, n, N.... It is found, e.g., that $<E1/E2> \simeq 1$ only got $\alpha \simeq 0$, but decreases notably as α increases. Experimentally, it is found that $<E1/E2) \sim 1$ in the middle of photometric nights and falls to ~ 0.1 in the presence of cirrus clouds. Similarly, Figure 2 shows the same trend for $<(\% E2/(\%STD)>$. Observations yield values of the order of 0.5 in normal weather at La Silla, which corresponds to $\alpha \sim 1$ (i.e. 1/f noise) around $f \sim 0.01$ Hz. Photometry of bright stars at Mauna Kea (Barnes et al. 1971) indicated dominant noise components down to at least 0.002 Hz. This, together with Milone and Robb (1983) and Allen and Barton (1981) compels us to investigate thoroughly the behavior of errors in the presence of 1/f noise. As an example, Figure 3 shows clearly the advantage of performing the comparison with a standard as early as possible in the sequence of Figure 1, for instance for each wave band instead of once, after completing the whole sequence for the program star. The higher α, the larger the advantage, either in accuracy or in total measuring time, or both (for $\alpha = 0$, the accuracy depends only on the total time spent for each wavelength, whatever the order of operations in the sequence). The total time in each case is equal to $nNM\tau$ as indicated in the Table under Figure 3. This was clearly demonstrated with RADS (Milone and Robb 1983), which performs the comparison immediately after step 1 in Figure 1. Quantitatively, this effect can be traced to the factor $\sin^2(\pi ft)$ in eq. (4b).

REFERENCES

Allen, D., and Barton, J. 1981, Publ. Astron. Soc. Pacific, 93, 381.
Barnes, J. A., et al. 1971, IEEE Trans. IM-20, No. 2, 105.
Maillard, J. P. 1983, private communication.
Milone, E. F., and Robb, R. M. 1983, Publ. Astron. Soc. Pacific, 95, 666.
Papoular, R. 1983, Astron. Astrophys., 117, 46.
Papoular, R., and Pégourié, B. 1983, in Statistical Methods in Astronomy, ESA Internat. Colloq. SP-201, ed. E. J. Rolfe (ESA, Noordwijk), p. 161.

TABLE I

X	$\bar{X}$	$W_X(f)$	$(\sigma_X^2 = \int W_X \, df)$	$\delta_X^2 = (N/N-1)\int df \cdot W_X \left[1 - \frac{\sin^2(\pi f NT)}{N^2 \sin^2(\pi f T)}\right]$
a	$\bar{a} = \int_0^\tau dt \cdot a(t)/\tau$	$W_a = I^2 W_x \frac{\sin^2(\pi f \tau)}{(\pi f \tau)^2}$		$\frac{n}{n-1} \int df \cdot W_a \left[1 - \frac{\sin^2(\pi f n \tau)}{n^2 \sin^2(\pi f \tau)}\right]$
	$\bar{a} \equiv A_i$ $\sum_1^n a_j/n$	$W_{\bar{a}} = W_a \frac{\sin^2(\pi f n \tau)}{n^2 \sin^2(\pi f \tau)}$		$\frac{N}{N-1} \int df \cdot W_{\bar{a}} \left[1 - \frac{\sin^2(\pi f N\theta)}{N^2 \sin^2(\pi f \theta)}\right]$
$\left(\frac{A-B}{2}\right)_i$	$I = \sum_1^N \frac{(A-B)_i}{2N}$	$W = W_{\bar{a}} \cos^2(\pi f \theta/2)$		$\frac{N}{N-1} \int df \cdot W_{\bar{a}} \cos^2\left(\frac{\pi f \theta}{2}\right) \left[\quad\right]$
I	$\bar{I} = \sum_1^M I_k/M$	$W_I = W_{\bar{a}} \cos^2\left(\frac{\pi f \theta}{2}\right) \frac{\sin^2(\pi f N\theta)}{N^2 \sin^2(\pi f \theta)}$		$\frac{M}{M-1} \int df \cdot \left[1 - \frac{\sin^2(\pi f MLN\theta)}{M^2 \sin^2(\pi f LN\theta)}\right]$
$\bar{I}$		$W_{\bar{I}} = W_I \frac{\sin^2(\pi f MLN\theta)}{M^2 \sin^2(\pi f LN\theta)}$		
$\epsilon \equiv \bar{I}_a/\bar{I}_\Sigma$ -1		$W_\epsilon = 4 W_{\bar{I}} \sin^2(\pi f MLN\theta)$	$\sigma_\epsilon^2 = \int W_\epsilon \cdot df$	

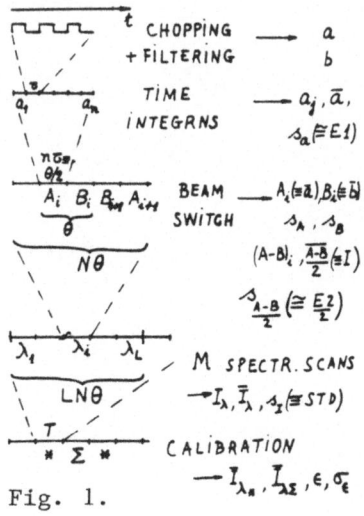

Fig. 1.

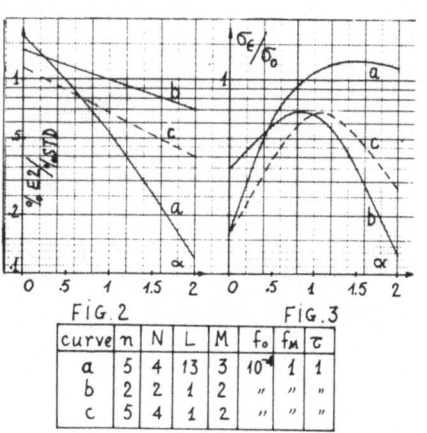

curve	n	N	L	M	f_0	f_M	τ
a	5	4	13	3	10^4	1	1
b	2	2	1	2	"	"	"
c	5	4	1	2	"	"	"

M_V AND T_{eff} OF B-TYPE STARS AS DERIVED FROM 13-COLOR PHOTOMETRY

P. Conconi and L. Mantegazza

Astronomical Observatory of Brera, Milano-Merate

1. INTRODUCTION

The 13 color medium-narrow band photometric system (Johnson and Mitchell 1975) potentially contains a great deal of information, but so far only a little work has been done in order to recover some physical parmeters from its color indices. Our aim is to obtain relations to be used in calibrating this system in terms of physical quantities, making use of the information given by this system alone.

We begin with B type stars as there is an excellent paper of Underhill et al. (1979) which evaluates many physical quantities for 160 O-B type stars which are among the stars observed in the 13 color system. The present paper describes the procedure we adopted to get preliminary relations for deriving T_{eff}'s and M_V's.

Our first attempt has been to calibrate all classes of B type stars, but as this fit was unsatisfactory, we have restricted ourselves to the stars of luminosity classes III-IV-V. Independent relations are probably needed for supergiants. Moreover, some Be stars, which were in extremely active stages when they were observed, fail to fit our relations; therefore they have been excluded from our sample. Finally a group of 108 stars has been adopted.

2. THE METHOD

The interstellar extinction is not generally negligible for these stars. As there is some uncertainty connected with the estimation of this coefficient, especially for the Be stars which may have circumstellar absorption, and in order to avoid the use of information not supplied by the 13 color photometry, we have defined the 'reddening free indices':

$$[\lambda_i - \lambda_{i+1}] = (\lambda_i - \lambda_{i+1}) - \alpha_i \cdot (45-52) \quad i=1,12 \tag{1}$$

D. S. Hayes et al. (eds.), Calibration of Fundamental Stellar Quantities, 553–556.
© *1985 by the IAU.*

TABLE I

Coefficients for the calculation of the reddening-free indices	
color index	α_i
(33-35)	0.47
(35-37)	0.32
(37-40)	0.53
(40-45)	0.84
(45-52)	1.00
(52-58)	0.89
(58-63)	0.37
(63-72)	0.79
(72-80)	0.37
(80-86)	0.37
(86-99)	0.58
(99-110)	0.42

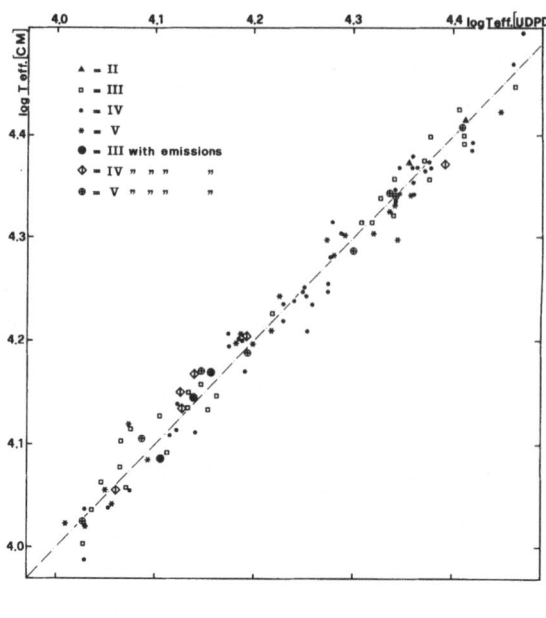

Fig. 1

where the α_i's have been calculated by means of the interstellar reddening law given by Scheffer (1982) and they are reported in Table I.

Then we have reduced the reddening — free indices into 'standardized variables' (zero mean and unit variance) and we have constructed 11 orthogonal indices by means of factor analysis (Whitney, 1983a,1983b). This procedure yields linear combinations of the indices that are themselves linearly independent. This is an important step, because the indices are strongly correlated, being in almost all cases strong functions of temperature, and this can result in an ill-conditioned coefficient matrix and meaningless coefficients when one performs a least-squares fit.

Multiple linear regressions have been performed between the orthogonal indices and their quadratic and bi-linear functions as independent variables.

The independent variables contained in each relation and their number have been selected with a FORWARD procedure; the number of independent variables has been increased starting from one, selecting at each stage that variable which gives the most important contribution to the residual variance of the dependent variable. The procedure ends when any further contribution is no longer statistically significant (Buzzi-Ferraris 1975). Finally a STEPWISE procedure has been applied to the results of the FORWARD procedure in order to get a further improvement (i.e., if some independent variables are no longer significant because of the introduction of successive terms in the relations, they are discarded, and, if in consequence of this, some excluded variables become significant, they are included in the relations). Figs. 1 and 2 show the correlations between the values of Underhill et al. (1979; UDPD) and ours (CM) for T_{eff} and M_V respectively. The rms residual is 0.017 for log T_{eff} and 0.287 for M_V.

3. THE CRITERION FOR SELECTING THE STARS

We have searched for a criterion which permits one to choose, by means of the 13 color photometry alone, the stars to which our relations apply. If a star falls inside the cloud constituted by the 108 stars of our sample in the 11-dimensional space of orthogonal indices, then our relations may be applied with confidence to it. As this cloud has not a simple geometrical structure, an empirical criterion must be adopted to verify if a star belongs to it. We have tried a few. One which works quite satisfactorily consists in verifying if the examined star falls inside a hyper-sphere with a radius of the order of the mean distance between adjacent stars of the sample and centered on the nearest star of the sample.

4. CONCLUSIONS

Our attempt at calibrating the 13-color photometric system seems promising. Now we are working to improve our relations, making a more accurate selection of the stars in our sample and defining better the selection criterion. Moreover, we intend to extend these relations to other types of stars and to study the correlations of the indices of 13 color photometry with other quantities such as log g, intrinsic brightness and rotational velocity.

REFERENCES

Buzzi-Ferraris, G. 1975, _Analisi ed Identificazione di Modelli_, (CLUP, Milano).
Johnson, H. L. and Mitchell, R. I. 1975, _Rev. Mex. Astron. Astrophys._ 1, 299.
Scheffer, H. 1982, in 'Numerical Data and Fundamental Relationships in Sciences and Technology,' Group VI, Vol. 2c, 45, Landolt-Börnstein.
Underhill, A. B., Divan, L., Prevot-Burnichon, M. L. and Doazan, V. 1979, _Mon. Not. R. Astron. Soc._ 189, 601.
Whitney, C. A. 1983a, _Astron. Astrophys. Suppl._ 51, 443.
Whitney, C. A. 1983b, _Astron. Astrophys. Suppl._ 51, 463.

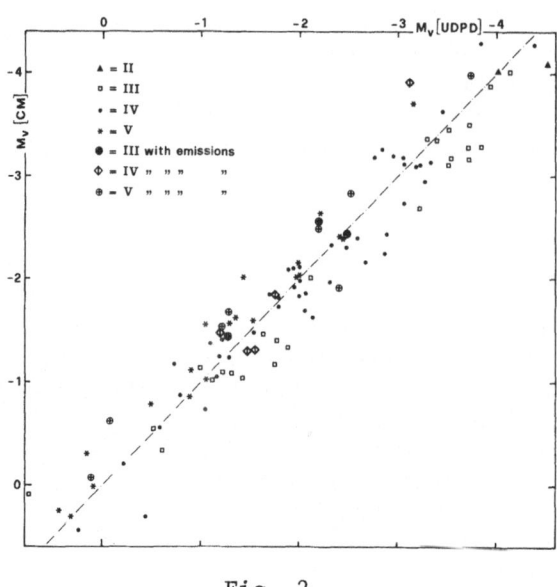

Fig. 2

DISCUSSION

HAUCK: We already have good calibrations to determine T_{eff} and M_v for B stars with the four-color system (Philip and Relyea, e.g.) and the Geneva system (Maeder and Cramer). What is the advantage of using the 13-color system for this kind of star? Do you obtain more precise values of T_{eff} and M_v?

MANTEGAZZA: The more color indices you have the more independent physical parameters you can get from them. Moreover, you may expect to be able to evaluate these parameters for peculiar objects too (i.e. Be, Ap, Am, Of, etc.) without referring to corrections for metal blanketing and so on.

ROUNTREE: In calibrating the 13-color photometry with the stars on the list of Underhill et al., did you make any use of the spectral types quoted in that paper? Many of these spectral types are incorrect, because the original MK type was changed to match the colors.

MANTEGAZZA: We did not make any use of the spectral types or luminosity classes quoted in that paper because our aim was to derive all we needed from the information contained in the photometric indices only.

TOBIN: Remie and Lamers (1982 Astron. Astrophys. 105, 85) have reanalyzed the Underhill et al. stars using almost the same data and method and they find temperatures 1000 - 1500 K cooler. People using the Underhill et al. results should be aware that they may suffer important systematic effects. (See also Tobin, 1983 Astron. Astrophys. 125, 168.)

THE ESTIMATION OF THE ELECTRON DENSITY n_e WITH A NARROW-BAND PHOTOMETRIC SYSTEM CALIBRATED BY MODEL ATMOSPHERES

M. Gerbaldi[1,2] and N. Morguleff[1]

[1]Institute for Astrophysics, Paris

[2]University of Paris

1. INTRODUCTION

A spectrophotometer was originally designed by Barbier (1960) for the study of early-type stars. The band passes were chosen to measure the features of the hydrogen spectrum. To the first definition of the system (Barbier and Morguleff 1964), two more bands were added, specifically devoted to the peculiar stars of type Ap and Am (Gerbaldi 1972). The interpretation of these observations has already been given separately. We will concentrate here on the calibration of this spectrophotometric system in terms of T_{eff}, log g and electron density n_e.

The properties of this system, including technical details of the equipment, can be found in Gerbaldi (1977). Table I gives, for each band pass, the mean wavelength and the equivalent width. Seven color indices are defined as being the difference between two magnitudes: $C(J) = m(J) - m(2)$, $J \in [1,8]$.

2. CALIBRATION OF THE COLOR INDICES

The principle is the following: knowing the response functions of the system, we calculate color indices for stars for which energy distributions are known, and then we determine the transformation coefficients between those computed color indices and the observed ones.

Table I.

Band passes n°	1	2	3	4	5	6	7	8
Mean wavelength (Å)	5937	4951	4859	4350	4043	3946	3753	3618
Equivalent width (Å)	93	99	89	70	60	40	60	73

D. S. Hayes et al. (eds.), Calibration of Fundamental Stellar Quantities, 557–560.
© 1985 by the IAU.

We determined the response function of this system for each band pass by defining : - the transmission of the atmosphere - the reflectivity of the mirrors of the telescope - the transmission of the optics of the spectrograph - the sensitivity of the photomultiplier (Gerbaldi 1977).

This spectrophotometric system being a narrow band system, we must avoid introducing numerical inaccuracy in the computation of the theoretical color indices by using low resolution energy distributions for the stars. Unfortunately published scanner observations are not continuous and even avoid regions of strong lines so they cannot be used to calibrate the bands centered on hydrogen or calcium lines. For the calibration of thoses bands, we computed the detailed flux of Vega from a model atmosphere (Kurucz 1979).

The relation between the observed and the computed indices is then $C(J) = aC(J)_S + b$. For the parameter a we have the value 1. The values of b are in Gerbaldi (1977).

3.   CALIBRATION IN T_{eff} AND LOG G

We have already shown (Gerbaldi and Morguleff 1978) that from these color indices we could define parameters which are related to the MK spectral classification. These parameters are : measurement of the intensity of the H_γ line (HGAMMA) and the Balmer Jump (DB).

HGAMMA = C(4) - 0.618 C(5)
DB   = C(8) - 1.644 C(5)

The theoretical indices and the parameters were calculated using the extensive grid of fluxes of Kurucz (1979) to yield a calibration of these parameters in terms of effective temperature (T_{eff}) and surface gravity (log g). Zero-point corrections were applied. Fig. 1 presents only these calibrations for the early-type stars of types B and A0. We investigated to what extent our theoretical calibration in T_{eff} and log g resembles the observations.

Rapid rotation affects the color of a star. This effect has not been taken into account here, but according to a previous analysis (Gerbaldi 1977) we are confident that the parameters are altered only in case of extremely rapid rotational velocity.

As we do not have enough stars in common with the results of Code (1975), which are as free as possible from theoretical models, we compared the values of T_{eff} and log g, determined from Fig. 1, with the values collected by Cayrel, et al. (1981).

The agreement, based on 6 stars in common : HD 24368, 97633, 147394, 179761, 214994, 216735, is sufficiently good to derive with this grid first-order values of T_{eff} and log g. The star's effective temperature can be determined to within $\sim$ 800 K and its surface gravity to within $\sim$ 0.3 in log g. Such a dispersion between values can also be seen in the fine analyses by different authors.

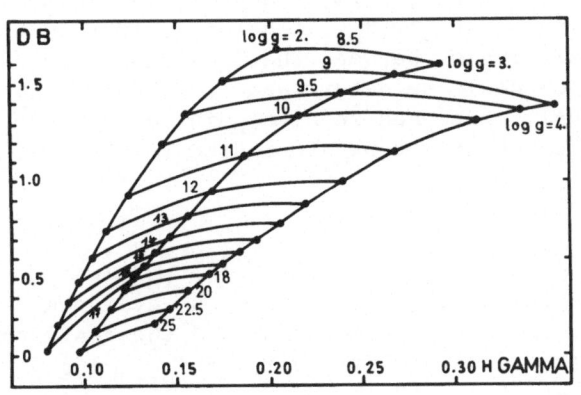

Fig. 1. The calibrated DB vs HGAMMA diagram, for models with [m/H] = 0.0. The grids are identified according to the value of log g and Teff (in 10^3 K).

4. CALIBRATION IN ELECTRON DENSITY n_e

In large-scale programs, it is customary to apply the Inglis-Teller formula (1939) or the Unsöld method (1955) to derive a mean electron density. In our spectrophotometric system, we can, with the band pass (7) measure the intensity of the confluence of the Balmer series. The corresponding reddening-free parameter is : DELTA = C(7) - 1.421 C(5).

We correlate this parameter with the electron density n_e given by the model atmosphere calculations of Kurucz (1979). As the optical depth $\tau \sim 0.1$ corresponds, for B type stars, to the layers where weak and moderate strength lines are formed, we took the value of n_e at τ_{5000} = 0.1.

Figure 2 presents the theoretical relationship between log n_e and DELTA, for the early type stars B and AO.

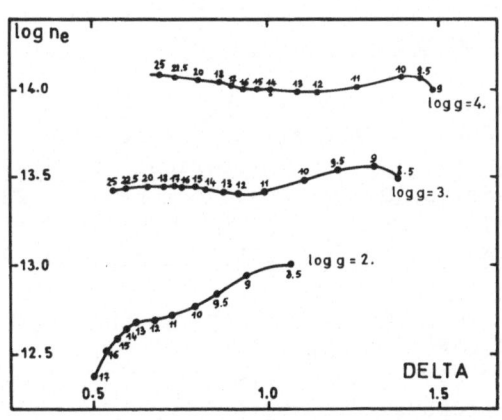

Fig. 2.- The calibration log n_e vs DELTA diagram for models with [m/H] = 0.0 and values of n_e at τ_{5000} = 0.1. The points are identified according to the value of log g and Teff (in 10^3 K).

We cannot compare directly the values of n_e obtained with this grid to other ones because of the sensitivity of n_e to the value of τ. Nevertheless we shall mention the extensive work performed by Kopylov (1961, 1966) in this field.

By entering Fig. 1 with observed values of DB and HGAMMA, an estimate of T_{eff} and log g can be made.

Fig. 2, with observed values of DELTA, permits a determination of n_e, corresponding to the preceding values of T_{eff} and log g. But with Fig. 2 we can also detect inconsistencies between the T_{eff} and log g values from Fig. 1 and those which could be determined from DELTA. From the sample of stars previously mentioned we have pointed out such an inconsistency for HD 214994, which cannot be attributed to observational errors.

5. CONCLUSIONS

Spectrophotometric measurements of the features of the hydrogen spectrum, such as H_γ equivalent width, the Balmer jump and the confluence of the last Balmer lines have been calibrated in terms of T_{eff}, log g and log n_e for the early-type stars B and A0.

Intercomparison between these three parameters have shown that some stars do not have a value of n_e compatible with their values of T_{eff} and log g. Such a situation has been detected for the first time by Fischel and Klinglesmith (1973) for some helium-weak stars. We have observed six HeW stars : HD 21699, 49606, 51688, 182568, 183339, 212454. Three of them (HD 21699, 49606, 183339) have values of Teff and log g obtained from spectroscopic observations and collected by Underhill and Doazan (1982), which are compatible with our determination from Fig. 1. But the value of n_e from Fig. 2 is inconsistent with values of T_{eff} and log g from Fig. 1 for HD 21699, 49606 and 212454.

This means that the behavior of these three hydrogen features can be properly reproduced by model atmospheres in some cases, but that these models can fail in other situations.

REFERENCES

Barbier, D. 1960, Ann. Astrophys., 23, 421.
Barbier, D. and Morguleff, N. 1964, Comptes Rendus Acad. Sc. Paris, 258, 4925.
Cayrel, G., Bentolila, C., Hauck, B., Lovy, D. 1981, A Catalogue of (Fe/H) Determination. (Centre de Données Stellaires, Strasbourg).
Code, A.D. 1975, in Multicolor Photometry and the Theoretical H R diagram, ed. A.G.D. Philip & D.S. Hayes, (Dudley Obs. Rept.n°9), p. 221.
Fischel, D. and Klinglesmith, D. 1973 Astrophys. J., 181, 841.
Gerbaldi, M. 1972, Comptes Rendus Acad. Sc. Paris, 275, Série B, 295.
Gerbaldi, M. 1977, Thesis. Paris.
Gerbaldi, M. and Morguleff, N. 1978, in Spectral Classification for the Future IAU Colloquium 47, ed. Mc Carthy M.F., Philip A.G.D., and Coyne, G. V. (Vatican Observatory), p. 508.
Inglis, D.R. and Teller, E. 1939, Astrophys. J., 90, 439.
Kopylov, I. 1961, Izv. Krym. Astrofiz. Obs., 26, 232.
Kopylov, I. 1966, Izv. Krym. Astrofiz. Obs., 35, 11.
Kurucz, R. 1979, Astrophys. J. Suppl., 40, 1.
Underhill, A. and Doazan, V. 1982, Monograph series on nonthermal phenomena in stellar atmospheres : B Stars With and Without Emission Lines, NASA Sp-456, CNRS-NASA.
Unsöld, A. 1955, Physik der Sternatmosphären, 2nd ed., (Springer, Berlin).

A REDDENING AND METALLICITY–FREE TEMPERATURE ESTIMATOR FOR LATE M GIANTS

M. O. Mennessier

Montpellier University

M. Grenon

Geneva Observatory

ABSTRACT.   Lockwood's   measurements   in   five-color   narrow-band
photometry of spectroscopic standard M stars is used to define an
estimator of temperature.   The TiO and VO indices deduced from this
photometry depend in the same way on the metallicity but vary in a
different way with the temperature: their ratio is only influenced by
the temperature.   A relation between T/V and the temperature is
calibrated for M giants between 2000 and 3000° K.   Both the T and V
indices are indicators of metallicity and gravity which allow us to
check that the relation (T/V)/temperature is independent of the
metallicity.   This relation leads to a precise determination of the
temperature and can be applied to a wide variety of M giants,
independently of their reddening and chemical composition.

1. INTRODUCTION

M giants are generally distant stars, most are variable, and a
great number of them have an extended dust envelope; thus many are
affected by interstellar or circumstellar reddening.   Their kinematics
show that they are mainly old stars; a large dispersion in their
metallicity and surface gravity may then be expected.

The   search for a reddening and metallicity – free temperature
estimator is fully justified for these stars.   Such an estimator can
be derived from the near–infrared molecular band intensities.   The
method is based on the fact that the TiO and VO bands vary in a
distinct way with the temperature but depend in the same way on the
metallicity and gravity.   An application of this method to Lockwood's
narrow-band photometry is presented here.

D. S. Hayes et al. (eds.), Calibration of Fundamental Stellar Quantities, 561–564.
© *1985 by the IAU.*

2. NEAR INFRARED NARROW-BAND PHOTOMETRY

Wing(1967) established a 27-color photometric system which measured continuum magnitudes and the strengths of molecular absorptions in late-type stars in the spectral region 0.75 - 1.11μm. For M stars, only TiO and VO band strengths have to be measured, and, on the basis of Wing's scanner study, Lockwood (1972) reduced the number of band passes to five: two continuum regions (filters 87 and 104), one VO band (filter 105) and two TiO bands (filters 78 and 88). Then color-free indices of TiO and VO strengths are defined:

$$T_1 = (78-87) - 0.6(87-104) \text{ and } V_1 = (105-104) + 0.1(87-104).$$

The effect of reddening on these indices is very small (0.03 mag. for T_1 and 0.003 mag. for V_1 per magnitude of visual absorption): they may be considered as free from reddening effects. Filter 104 is blanketing-free, but filter 87 is affected by TiO blanketing after M5. T_1 is a function of metallicity and gravity (Mould 1976). T_1 and V_1 depend in the same way on the metallicity and gravity but vary in a different way with the temperature. T_1/V_1 is influenced only by the temperature (Grenon 1981) and can be used to estimate it in the range where both the indices are well defined (M5-M9).

3. BLANKETING-FREE INDICES

For most of the standard stars measured by Lockwood (1972) we know a V magnitude, which is affected by a TiO absorption δ_v estimated by Smak and Wing (1979). A black-body fit between $V + \delta_v$ and 104 leads to an estimate of $87 + \delta_{87}$. The normalization of $V - 104 = 0$ for α Lyr was used. The comparison of the calculated $87 + \delta_{87}$ and the measured 87 magnitudes gives an estimate of the TiO blanketing δ_{87} in the filter 87 as a function of the spectral type. This estimate of δ_{87} is quite consistent with the measurement of TiO absorption at 7540 Å by Piccirillo et al. (1981).

Then we can define blanketing free indices:
- two in the continuum region: $(87)_c = 87 + \delta_{87}$ and 104
- one of TiO strength $(T_1)_c = T_1 + 1.6 \delta_{87}$
- one of VO strength $(V_1)_c = V_1 - 0.1 \delta_{87}$

4. REDDENING AND METALLICITY-FREE TEMPERATURE ESTIMATOR FOR M GIANTS.

The spectral-type standard stars selected by Lockwood are small amplitude variable late-type stars from Keenan's (1963) list supplemented by few semi-regular small-range variables later than M6. To define spectral class M10, the mira, R Cas, at its minimum, is

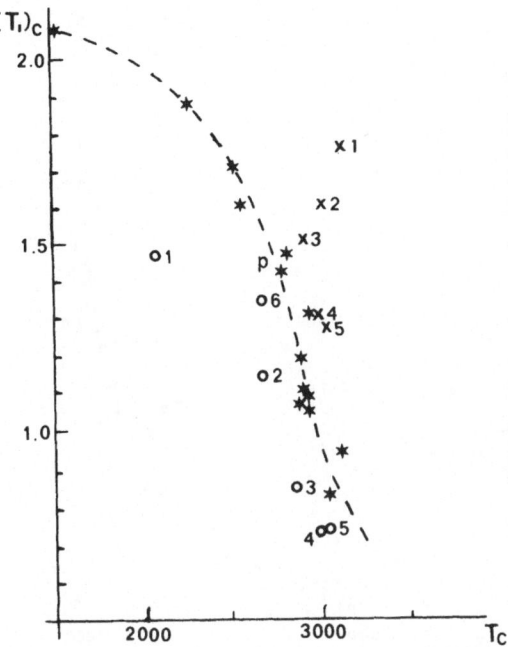

Fig. 1. Relation between the corrected TiO strength and the color temperature. The probable metal rich or lower gravity stars are noted x and the deficient or larger gravity stars by o.

used as adopted by Lockwood (1970), so the results concerning this spectral type has to be taken with caution.

From $(87)_c$ and 104 we can deduce a color temperature T_c. The indices $(T_1)_c$ and $(V_1)_c$ are mainly functions of metallicity and temperature. Fig. 1 shows the relation $(T_1)_c/T_c$. At a given T_c, the stars above and below the mean relation are probably metal rich or lower gravity stars and metal deficient or greater gravity stars respectively. The stars occupy the same relative location in the $(V_1)_c/T_c$, diagram and in the $(T_1)_c/T_c$ diagram. This confirms our interpretation of the specific nature of these stars. Only one star, denoted "p" on the figures, is not located similarly in both diagrams: it is likely a vanadium-poor star (TY Dra).

Fig. 2 gives the relation of the ratio $(T_1)_c/(V_1)_c$ versus the temperature T_c. This relation is somewhat noisy but the noise is lower than when no blanketing correction is made. The noticeable property is that no systematic effect is seen with abundance or gravity.

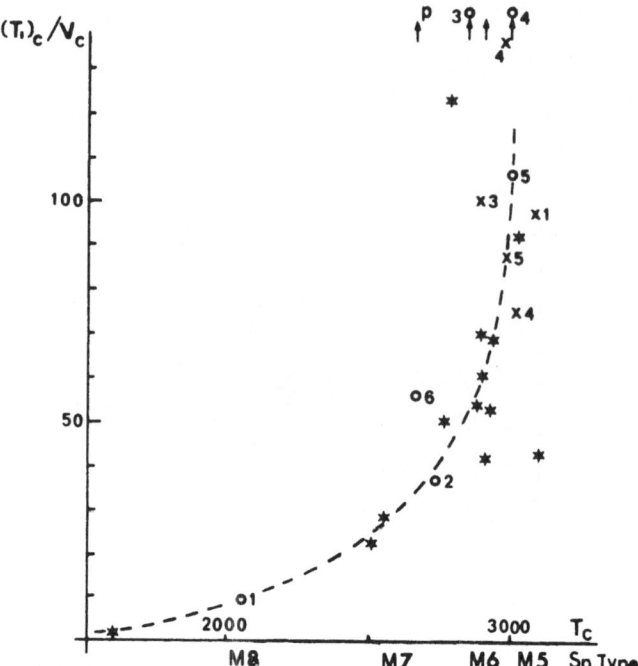

Fig. 2.   Relation between the temperature T_c and the ratio of molecular TiO and VO strengths for spectral type standard stars measured by Lockwood (1972).   The symbols are the same as in Fig. 1.

5. CONCLUSION

The ratio of the indices T_1 and V_1 of TiO and VO strengths leads to a precise determination of color temperature in the range 1900 – 2900°K.   This T_c may also be transformed into T_{eff} using stars with known apparent diameters.   The deduced relation can be applied to a wide variety of late M giants independently of their reddening, chemical composition or mass.

6. REFERENCES

Grenon, M. 1981, Ann. Phys. Fr., 6, 127.
Keenan, P. C. 1963, Basic Astronomical Data, ed. K. Aa. Strand, U. of
     Chicago, p.78.
Lockwood, G. W. 1970, Astrophys. J., 160, L47.
Lockwood, G. W. 1972, Astrophys. J. Suppl. 24, 375.
Mould, J. R. 1976, Astrophys. J., 207, 535.
Piccirillo, J., Bernat, A. P. and Johnson, H. R. 1981, Astrophys.
     J., 246, 246.
Smak, J. and Wing, R. F. 1979, Acta Astron., 29, 187.
Wing, R. F. 1967, in Colloquium on Late-Type Stars, ed. M. Hack
     (Astron. Obs., Trieste), p.231.

ON THE HOMOGENIZATION OF PHOTOMETRIC DATA

J. Manfroid

Department of Astrophysics, University of Liège

A. Heck

Astronomical Observatory, Strasbourg

ABSTRACT. Catalogs of averaged photometric data have been published for several photometric systems. The homogenizing procedures used to produce them are not without pitfalls. We question the accuracy of these methods with the available data.

1. INTRODUCTION

    The quantity of photometric data available in many systems is growing steadily. The General Catalogue of Photometric Data (Hauck 1982; Mermilliod 1984) already lists more than 100,000 stars and 75 photometric systems. While many systems have been used by only one author and for a few hundred stars, several others (UBV, Geneva, uvby) have from 20,000 to 70,000 stars. The measurements are scattered among hundreds of publications and involve different equipment, telescopes and observers. The Geneva system shows a very good homogeneity, but this is not true for most of the others. The variations between equipment used for the same system gives a wide dispersion in the resulting data. Poor reduction techniques and observing procedures also occasionally account for part of this scatter.

    This situation has lead to the compilation of homogenized catalogues of averaged values (see for instance in uvby: Lindemann and Hauck, 1973; Hauck and Mermilliod 1975, 1980; in UBV: Nicolet 1978). However, these catalogues cannot be completely satisfactory because of the data on which they are based.

2. HOMOGENIZING IMMISCIBLE DATA

    The largest causes of deviation between different observations are instrumental: filter profiles, temperature variations affecting

D. S. Hayes et al. (eds.), Calibration of Fundamental Stellar Quantities, 565–567.
© 1985 by the IAU.

the filters and producing summer-winter variations, cathode response,
humidity, altitude, etc. This leaves some doubts concerning the
difficulties of transforming all of the data to a common standard
system (see for instance Bessell 1983; Straižys 1983; Manfroid 1985a
and b).

Published observations have generally been reduced to a standard
system by a regression method using standard stars. The first step in
the homogenizing method (Lindemann and Hauck 1973) is to compare the
published values with a reference list and to use color-by-color
regressions to improve the original transformations. Except for rare
cases of bad reductions we do not see how this method could give
better results than the original one used by the astronomer since (1)
it uses the same kind of regression analysis, or even a cruder one (no
intercolor terms) and (2) the original measurements and especially
those of the standard stars are not often available to the
homogenizers.

The quality of the data is then estimated (Lindemann and Hauck
1973) (1) by the standard error obtained in the color transformation,
which is mainly a measure of the departure between the reference and
the instrumental system of the observer and (2) in the case of the
uvby system, by the slope and the y intercept, which is more a measure
of the size of the data set: only very small data sets could deviate
from the 1 and 0 values.

The second step of the homogenization consists in averaging the
observations weighted according to their estimated quality. The role
of a quality index is to select preferentially (1) instrumental
systems which are close to the standard one or (2) observational
material which overlaps with the reference lists only over easily
transformable groups of stars. In the latter case data can be
included which deviate significantly from the standard values even
though their quality index is high.

Those errors due to the peculiarities of the instrumental systems
can be very large. They will remain undetected. They can propagate
throughout the catalogues if those data are used as secondary
references, a very likely hypothesis since they concern
underrepresented stellar groups.

The overall divergence would be smaller if the original
reductions by the observers were preferred, since they involved
standard stars covering the whole range of program stars. More
generally all data introduced with small weighting factor are not
transformable and their contribution is to degrade the average value
because the number of independent data for any star is usually very
small.

3. CONCLUSION

The homogenizing procedure generally does not improve individual measurements, unless the latter are badly reduced by their authors, or were kept in the instrumental system. The resulting homogenized catalogues can be considered to be lists of values averaged between the larger catalogues already firmly tied to the standard system. Hence their usefulness can be questioned.

All other observational data, even if very accurate, spanning a wide range of stellar types and classes and, hence, showing a large scatter after color transformation, are either eliminated or (worse) included with small weighting coefficients. Some categories of stars can show large systematic deviations from the standard system.

It would be advantageous to compile separate lists for each instrumental system and any set of observational material showing good internal accuracy. All standard stars should be listed and the best available information concerning the system (filters, photomultiplier, etc.) should be given. This would leave open the possibility of retrieving a great deal of high quality data.

REFERENCES

Bessell, M. S. 1983, Publ. Astron. Soc. Pac., 95, 480.
Hauck, B. 1982, Inform. Bull. CDS, 22, 67.
Hauck, B. and Mermilliod, M. 1975, Astron. Astrophys. Suppl. 22, 235.
Hauck, B. and Mermilliod, M. 1980, Astron. Astrophys. Suppl. 40, 1.
Lindemann, E. and Hauck, B. 1973, Astron. Astrophys. Suppl. 11, 119.
Manfroid, J. 1985a, I.A.U. Symposium No. 111, Calibration of
    Fundamental Stellar Quantities, D. S. Hayes, L. E. Pasinetti and
    A. G. Davis Philip, eds., (Reidel, Dordrecht), p.493.
Manfroid, J. 1985b, I.A.U. Symposium No. 111, Calibration of
    Fundamental Stellar Quantities, D. S. Hayes, L. E. Pasinetti and
    A. G. Davis Philip, eds., (Reidel, Dordrecht) p.505.
Mermilliod, J. C. 1984, Inform. Bull. CDS, 26, 3.
Nicolet, B. 1978, Astron. Astrophys. Suppl. 34, 1.
Straižys, V. 1983, Inform. Bull. CDS, 25, 41.

SPECTROSCOPIC GRAVITY ESTIMATES FOR LATE-TYPE GIANTS: ARCTURUS AS AN EXAMPLE

R. A. Bell

Astronomy Program, University of Maryland

B. Edvardsson

Uppsala University

B. Gustafsson

Stockholm University

The surface gravity of Arcturus is estimated from the strength of the MgH features (the Mg abundance being derived from MgI lines), from strong metal lines and from the FeI/FeII ionization equilibrium. The MgH lines give log g = 1.7 (cgs units) and 4375 K for the effective temperature. This value of log g is consistent with the gravity derived from the sample of strong pressure-broadened lines from FeI, CaI and NaI which gives log g = 1.6, and what we obtain from the ionization equilibrium of Fe, log g = 1.4. The corresponding estimates of the maximum error are 0.3, 0.2 and 0.5 dex, respectively. The mass of Arcturus is found to be in the interval 0.6 - 1.0 solar masses. It is concluded that the MgH features offer good possibilities for determining gravities of late-type stars, when good estimates of effective temperatures are available.

D. S. Hayes et al. (eds.), Calibration of Fundamental Stellar Quantities, 569.

THE TEMPERATURES OF G AND K STARS

Robert F. Wing

Astronomy Department, Ohio State University

Bengt Gustafsson* and Kjell Eriksson

Astronomical Observatory, Uppsala

ABSTRACT. Effective temperatures have been determined for G and K stars by comparing synthetic colors computed from model atmospheres to observed colors measured at near-infrared continuum points. Results are presented for giant stars in the range K0 III - K4 III.

Effective temperatures of cool stars can be derived from interferometric or occultation measurements of angular diameter, or from calibrations based on model atmospheres. Although the "direct" methods based on angular diameters avoid the many assumptions that must be made in constructing a model atmosphere, the available measurements of angular diameters are in such short supply for certain classes of stars that the model-atmosphere results are more reliable. This is, at present, the case for the G and K stars.

The atmospheres of dwarf and giant stars of types G and K, with temperatures in the range $4000 - 6000$ K, can be well represented by atmospheric models (Gustafsson et al. 1975; Bell et al. 1976). The spectrum emitted by a model can be computed in detail and combined with filter response functions to obtain synthetic color indices (Bell and Gustafsson 1978; Gustafsson and Bell 1979). Color indices computed from a sequence of models of specified T_{eff} then allow the effective temperature to be found for stars with observed color indices. Any well-calibrated photometric system which provides a temperature-sensitive color index can be used in this manner, but the accuracy of the temperatures derived depends upon the ability of the color index to discriminate against the effects of all parameters other than temperature (e.g. composition, gravity, and microturbulence) as well as upon the ability of the model to represent the actual atmosphere of the star.

The eight-color system of narrow-band photometry described by Wing (1971) is particularly well suited for temperature determinations in G

---

* Present address: Stockholm Observatory.

D. S. Hayes et al. (eds.), Calibration of Fundamental Stellar Quantities, 571–574.
© 1985 by the IAU.

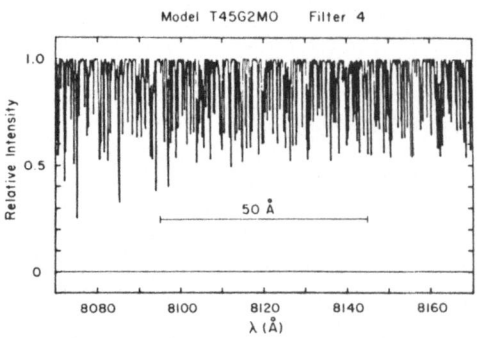
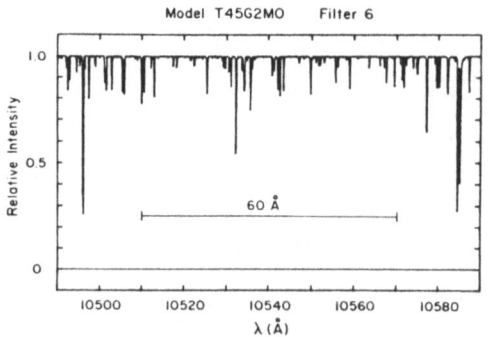

Fig. 1. Synthetic spectra in the regions of two filters of the eight-color system (half-power points indicated) for a solar-composition, 4500 K giant. Filter 4 (*left*) is strongly affected by CN, while filter 6 (*right*) is an excellent continuum point.

and K stars. Its filters lie in the 7000 - 11000 Å region where atomic lines are very weak, so that effects caused by atomic line absorption are minimal and the incompleteness of atomic line lists is not an important problem. In G and early K stars, the only significant absorber within the eight-color filter bandpasses is the CN molecule, the spectrum of which is well known. The system includes continuum points centered at 7810 Å (FWHM = 40 Å) and 10540 Å (FWHM = 60 Å), which are only slightly affected by CN, as well as filters located within strong CN bands. The system has been calibrated absolutely by fitting the photometry of Vega (A0 V) to a 9500 K model. Although there remains some uncertainty in the temperature of Vega, an error of 100 K in this choice introduces an error of only 25 K in the derived temperatures of K stars.

In Figure 1, two examples of synthetic spectrum calculations are shown. On the left is the region of filter 4 (8120 Å, FWHM = 50 Å), which is strongly affected by CN lines from the (3,1) and (2,0) bands. On the right is the region of filter 6, one of our continuum points.

We have computed synthetic eight-color photometry for most of the solar-composition models published by Bell *et al.* (1976) and for a number of additional models computed by the same procedures subsequently. The model temperatures range from 3500 to 5780 K and the gravities from log g = 0.75 to 4.5 (cgs units). A two-dimensional grid is needed to determine whether the same calibration of color index into T_{eff} can be used at all luminosities. We also considered several metal-deficient models to verify that our continuum points are immune to metallicity effects. To explore the effects of CN we constructed new models with altered nitrogen abundance, and we also computed spectra with the oscillator strength of the CN red system arbitrarily raised or lowered.

Synthetic eight-color photometry for a 5000 K giant and a 4500 K supergiant are shown in Figure 2 with their model continua. Most of the absorption in all filters in due to CN. The model continua are flatter than blackbody curves because H⁻ continuous opacity depresses the 8000 Å region more than the one-micron region. At the bottom of Figure 2 is

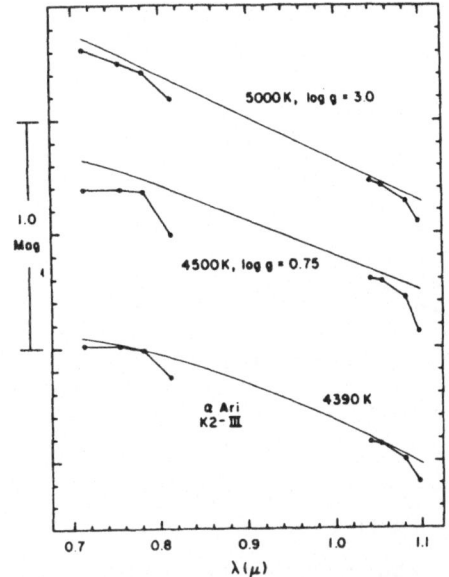

Fig. 2. Eight-color spectra of two models and the star α Ari.

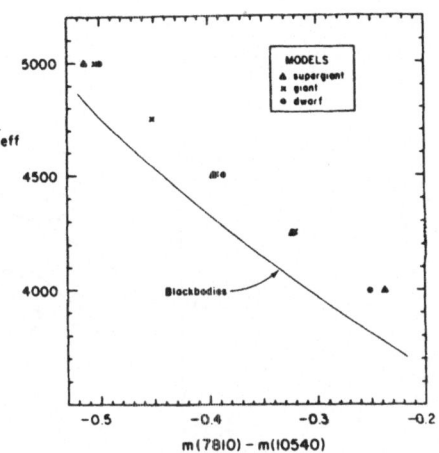

Fig. 3. Synthetic continuum colors *vs*. effective temperature. Opacity from H⁻ makes all models redder than blackbodies.

plotted the photometry of the standard star α Ari (K2⁻ III), fitted with a blackbody curve for 4390 K which passes through the continuum points at 7810 and 10540 Å. Thus 4390 K is the *color* temperature of α Ari on the eight-color system; its *effective* temperature — i.e. the effective temperature of the model which has the same value of the color index $m(7810) - m(10540)$, as well as similar values of gravity, microturbulence and CN strength — is about 200 K higher.

In Figure 3, the computed $m(7810) - m(10540)$ colors are plotted against the effective temperatures of the models from which they were derived. All models are redder, in this index, than blackbodies of the same temperature, as a result of H⁻ opacity. Differential effects due to gravity and CN strength are seen to be small.

Eight-color observations of MK standard stars were used to establish the mean $m(7810) - m(10540)$ color index for normal, unreddened giant stars of each spectral type from K0 III to K4 III, and the calibration indicated in Figure 3 for giants (x's) was used to obtain the corresponding effective temperatures. The results are given in Table I. Further work is underway to evaluate formally the uncertainties in these results and to extend them to other luminosities and temperatures.

Table I. Temperatures for K Giants

Spectral Type	T_{eff}
K0 III	4970 K
K1 III	4780
K2 III	4590
K3 III	4400
K4 III	4210

Our new temperatures for K giants are higher than most previous values (Figure 4) but are consistent with the spectroscopic results of Lambert and Ries (1981). The scales of Kuiper (1938) and Johnson (1966)

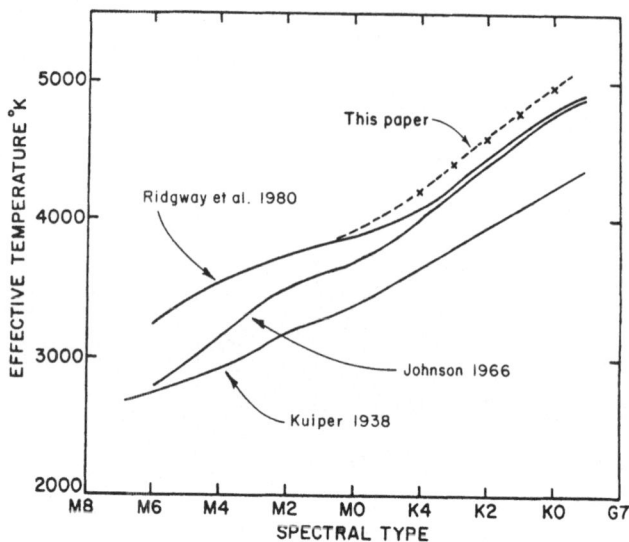

Fig. 4. Effective temperature scales from three
previous studies and from the present work.

are both based on the same half-dozen angular-diameter stars (including
only one K giant), while the lunar-occultation scale of Ridgway et al.
(1980), which is well determined for M giants, is also weak for the G
and K stars. The results of Frisk and Bell (1985) also indicate that an
upward revision of the temperature scale of Johnson (1966) is necessary
for G-K giants, although the revision suggested by Frisk and Bell is
somewhat smaller than that suggested here. A more definitive recommendation as regards the proper temperature scale for G-K giants will be discussed in a later paper.

    We would like to thank the Nordic Institute for Theoretical Nuclear
Physics (NORDITA) for a travel grant to Wing, and we are pleased to
acknowledge the program for synthetic spectra written by Dr. R. A. Bell.

REFERENCES

Bell, R. A., Eriksson, K., Gustafsson, B., and Nordlund, Å. 1976,
    Astron. Astrophys. Suppl. 23, 37.
Bell, R. A., and Gustafsson, B. 1978, *Astron. Astrophys. Suppl.* 34, 229.
Frisk, U. and Bell, R. 1985, in IAU Symp. No. 111: Calibration of Fund-
    amental Stellar Quantities, ed. D. S. Hayes, L. Pasinetti and A. G.
    D. Philip (Reidel, Dordrecht), p. 543.
Gustafsson, B., and Bell, R. A. 1979, *Astron. Astrophys.* 74, 313.
Gustafsson, B., Bell, R. A., Eriksson, K., and Nordlund, Å. 1975,
    Astron. Astrophys. 42, 407.
Johnson, H. L. 1966, *Ann. Rev. Astron. Astrophys.* 4, 193.
Kuiper, G. P. 1938, *Astrophys. J.* 88, 429.
Lambert, D. L., and Ries, L. M. 1981, *Astrophys. J.* 248, 228.
Ridgway, S., Joyce, R., White, N. and Wing, R. 1980, Ap. J., 235, 126.
Wing, R. F. 1971, in *Proc. Conf. Late-Type Stars*, ed. G. W. Lockwood
    and H. M. Dyck (Kitt Peak Nat'l Obs. Contrib. No. 554), p. 145.

THE CHEMICAL EVOLUTION OF THE GALACTIC DISC, INVESTIGATED BY ABUNDANCE ANALYSIS OF F STARS: A PROGRESS REPORT[*]

B. Edvardsson

Uppsala Astronomical Observatory

P. E. Nissen

Astronomical Institute, Aarhus

B. Gustafsson

Stockholm Observatory

J. Andersen

Astronomical Observatory, Brorfelde

ABSTRACT. An extensive project for the investigation of abundances, especially of light elements, in F-type stars is described and some preliminary results are given.

1. PURPOSE OF THE PROJECT

Several models of the chemical evolution of the Galaxy exist at present, but only limited and often specialized observational tests have been performed.

We have undertaken to make a large statistically significant investigation of chemical abundances, especially of the lighter elements, for stars of different ages and different over-all metal abundances ([M/H]). For this purpose we have selected some 200 bright F stars, close to the main sequence, distributed over the whole sky. The stars were selected from uvby photometry, giving estimates of Teff, log g, [M/H] and age for each star. The stars were divided into nine groups according to [M/H] and a significant number of low-metal abundance stars was included in the investigation.

---

[*] Based on observations carried out at the European Southern Observatory, La Silla, Chile.

D. S. Hayes et al. (eds.), Calibration of Fundamental Stellar Quantities, 575–579.
© *1985 by the IAU.*

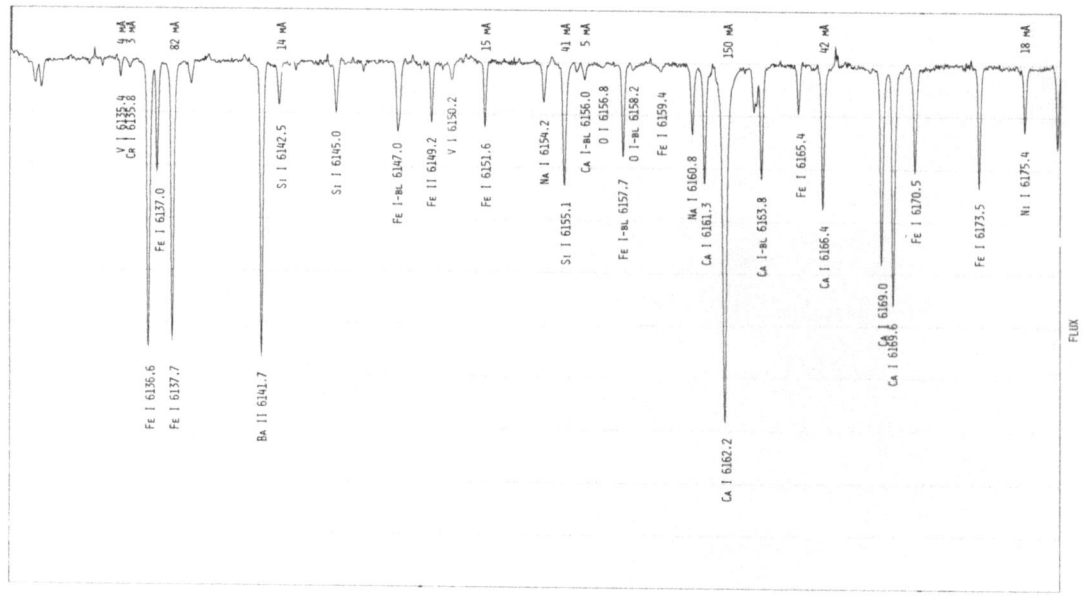

Fig. 1. This is a typical example of a spectrum used in the analysis. It is a 45 minute exposure of the star BS 3018, V=5.4, in the region 6130 - 6177 Å. The resolving power is 100,000. A signal of $350 \cdot 10^3$ photons is typical for one of the about 1830 channels shown.

2. OBSERVATIONS AND REDUCTIONS

The stars in the southern sky are observed at ESO with the 1.4-m Coudé Auxillary Telescope and the Coudé Echelle Spectrograph using a 1870 diode Reticon array as the detector. The resolving power is about 80,000 and a signal-to-noise ratio of about 100 for each spectral resolution element is typical.

In all five different wavelength regions are observed, each about 50 Å wide, containing weak spectral lines from O, Na, Mg, Al, Si, Ca, Ti, V, Cr, Fe, Co, Ni, Y and Ba.

The reductions of the spectra have been carried out with the ESO IHAP system, and the equivalent widths were measured with the very convenient PHYS program written in Uppsala. The equivalent widths measured refer to a continuum defined by a number of narrow spectral regions selected to be free of lines in the solar and the Procyon spectra.

A typical spectrum, of BS 3018, V=5.4, is shown in Fig. 1. This star has been observed independently on two different occasions with different Reticon arrays, and the two spectra have been reduced and measured independently by the two observers. The resulting sets of equivalent widths,ranging from 3 to 87 mÅ, are compared in Fig. 2. The mean difference in the widths obtained is 0.4 mÅ, with a standard deviation of 3.3 mÅ for a line. We conclude that it seems possible to obtain very accurate equivalent widths with the present instrumentation.

The full project is carried out in collaboration with D. Lambert and J. Tomkin, who perform the corresponding observations of the stars in the northern sky at the McDonald Observatory

3. ANALYSIS

The stars are compared with blanketed and convective model atmospheres, calculated with an updated version of the program presented in Gustafsson, et al. (1975; cf. also Nissen and Gustafsson 1978). The abundance analysis is differential relative to the Sun (the reflected light of which was also observed with the same instrumentation).

The analysis is based on the assumption of LTE; however, the effects of departures from LTE on the abundances derived are being studied theoretically (cf. Saxner 1984) and observationally. Another important source of uncertainty is the mixing-length theory adopted for convection. More detailed simulations, such as those made for the solar convection by Nordlund (cf. Dravins et al. 1981), will be attempted.

The temperature scale is also uncertain as a result of the uncertainty of the solar colors and the possibility of errors due to incomplete spectral line blocking in the calculated b band or due to the convection theory (cf. Gehren 1981). Although a recent study of the effective temperature scale of F stars using the integrated-flux method of Blackwell and Shallis (1977) by Saxner (1984) indicates that our scale should not be seriously in error, this source of error in the abundances must be studied further.

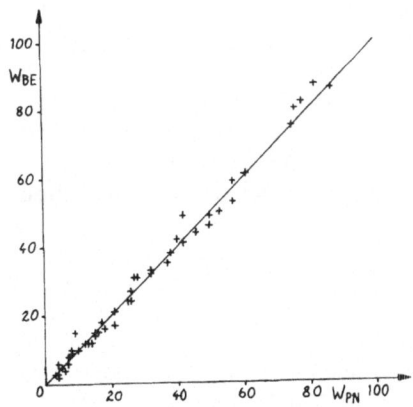

Fig. 2. Comparison of equivalent widths (in mÅ) obtained from different spectra, taken by different observers with different Reticon arrays and separately reduced and measured by the two observers. The mean difference is 0.4 mÅ with a standard deviation of 3.3 mÅ for a line. One of the four spectra used is that of Fig. 1.

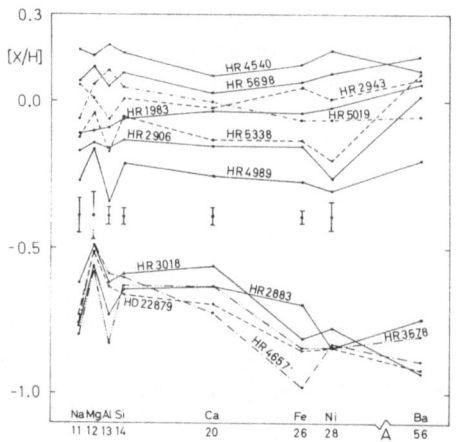

Fig. 3. Logarithmic abundances relative to the Sun, $[X/H]=lg(N_X/N_H)_*-$ $-lg(N_X/N_H)_\odot$, for the 13 stars analysed so far. Only 30 of the about 100 lines were used in this preliminary test. We will also determine abundances of O, Ti, V, Cr, Co and Y. Typical errors (standard deviation of the mean) are also shown in the plot.

4. RESULTS AND CONCLUSIONS

In Fig. 3 logarithmic abundances relative to the Sun, $[X/H]=$ $lg(N_X/N_H)_*-lg(N_X/N_H)_\odot$, are plotted for various elements. Only a minority of the spectral lines were used in this preliminary analysis, and we shall also later derive abundances of O, Ti, V, Cr, Co and Y.

From Fig. 3 it is seen that the underabundances are less pronounced for the lighter elements than for the iron-peak elements. If this tendancy prevails when more stars have been analysed this will support the idea that the lighter elements relative to the heavier ones were more abundantly produced in stars when the galactic disc was less metal-rich. Also note the tendency for an odd-even effect, varying with overall metal abundance, for Na, Mg, Al and Si.

Further results of this study will be published in Astronomy and Astrophysics.

REFERENCES

Blackwell, D.E. and Shallis, M.J. 1977, Mon. Not. R. astr. Soc. 180, 177
Dravins, D., Lindegren, L. and Nordlund, Å, 1981, Astron. Astrophys. 96, 345
Gehren, T. 1981, Astron. Astrophys. 100, 97
Gustafsson, B., Bell, R.A., Eriksson, K. and Nordlund, Å. 1975, Astron. Astrophys. 42, 407
Nissen, P.E. and Gustafsson, B. 1978, Astronomical Papers dedicated to Bengt Strömgren, Eds. A. Reiz, T. Andersen,(Copenhagen University Observatory), p. 43

Saxner, M. 1984, Thesis, Uppsala University; to be submitted to Astron.
  Astrophys.

DISCUSSION

HEINTZE: Please tell us how you disentangle abundance and NLTE effects.

GUSTAFSSON: This is not easy or even possible observationally if only
lines from neutral atoms are measured, and "over-ionization", due to hot
UV radiation from deeper atmospheric layers, is of importance. If so,
as indicated by the study of Saxner (1984) for Fe in metal-poor F
dwarfs, one has to use Fe II lines or try to calibrate the Fe I lines by
detailed statistical-equilibrium calculations. These are, however,
quite uncertain due to uncertain UV fluxes, collision cross-sections and
necessarily primitive model atoms.

LTE-BLANKETED AND UNBLANKETED MODELS FOR SDO STARS

L. Rossi

Institute for Space Astrophysics, CNR, Frascati

SdO stars have particular importance in astrophysics due to their position in the HR diagram between the hot horizontal-branch stars and the precursors of the white dwarfs. A careful determination of their parameters is therefore highly desirable.

Unfortunately, their effective temperatures ($\geq$30000K) put them in the region where NLTE effects are no longer negligible. Line blanketing, on the other hand, largely dominates the UV spectral region of these stars, and its effect on the temperature distribution in the atmosphere is opposite to that of NLTE. So, since NLTE blanketed models are not yet available, it seems that presently the LTE unblanketed models may better describe the actual situation.

We present in this paper the preliminary estimates of sdO stellar parameters by means of fitting of the low-dispersion IUE spectra to the energy distributions of blanketed and unblanketed LTE model atmospheres. A comparison of the results from the different models is discussed.

D. S. Hayes et al. (eds.), Calibration of Fundamental Stellar Quantities, 581.
© *1985 by the IAU.*

THE ZERO-POINT OF THE IAU STANDARD VELOCITY SYSTEM

C. D. Scarfe

Department of Physics, University of Victoria

ABSTRACT. Radial velocities of bright IAU standards have been obtained
photographically over the past decade using the long camera of the DAO
1.2 meter telescope's coudé spectrograph. Most of the stars observed
have been found to be constant in velocity to better than 0.15 km/s
over that interval. The mean velocities agree with the IAU velocities,
on the average, within 0.10 km/s, although mean velocities of some
individual stars differ considerably more than this from the IAU value.
A preliminary determination of the zero point of the long camera
system, and hence of the IAU system, has been made from observations of
the asteroid Vesta, whose actual radial velocity has been calculated
from its orbital elements.

One of the optical configurations of the coudé spectrograph of the
Dominion Astrophysical Observatory (DAO) 1.2 meter telescope includes a
mosaic of four 150 mm by 175 mm gratings with 830.77 grooves per mm,
used with a camera mirror of focal length 2433 mm (96 inches). This
"long camera" provides a reciprocal dispersion of 2.4 Å/mm in the
second order blue region; it is described in detail by Richardson
(1968). Light is admitted to the spectrograph via an image slicer
designed specifically for this camera, and the resolution is about $\lambda/\Delta\lambda$
$\simeq$ 100,000. The spectrograph has been used chiefly photographically
since its first use in the 1960's, but in recent years the radial-
velocity spectrometer described by Fletcher et al. (1982) has
superseded photographic plates as the most frequently-used detector.

Experience has shown that this spectrograph provides highly
consistent radial velocities over several years of operation (Scarfe
1983). Both Scarfe et al. (1983) and Batten et al. (1984) have
published studies of visual binary stars in which the scatter about the
velocity curve defined by the adopted elements is of order 0.25 km/s
for photographic observations obtained over at least thirteen years.
It thus seems that the spectrograph is sufficiently stable for a study

This work is based on observations obtained at the Dominion
Astrophysical Observatory.

D. S. Hayes et al. (eds.), Calibration of Fundamental Stellar Quantities, 583–586.
© *1985 by the IAU.*

of the reliability of the system of standard velocity stars in current use.

The IAU system of standard velocity stars is basically that of Pearce (1957), which was adopted by the Union at the General Assembly in 1955. Additional fainter stars chosen by Heard and Fehrenbach (1973) were adopted by the IAU in that year. Most of the observations that were used in selecting the standard stars were obtained using Cassegrain spectrographs whose dispersion and resolution were considerably inferior to those available in modern coudé instruments, and which were subject to temperature fluctuations and instrument flexure, from which a coudé spectrograph is largely free. Several papers criticizing the IAU system have already been published (Griffin 1975, Batten et al. 1983). But these too were based on observations with equipment less stable and precise than the long camera, although much superior to that used to obtain the original data for the standards.

As a further check on the long camera's reliability, IAU standard stars have been observed by the author on most of the nights over the past ten years in which the instrument has been used. Because the stability of the spectrograph has been established, the data may also be used to provide a test of the standard star system. The stars observed are largely from Pearce's Table I (bright standards) with only two of the brighter objects from his Table II (faint standards) being included, in order to keep exposure times under an hour, and so to leave time for program stars. Although none of them have been followed for as long as ten years, as recommended by Griffin (1975), it seems worthwhile to provide this progress report.

All the spectra are on IIaO photographic plates and have been measured using the DAO's ARCTURUS oscilloscopic measuring machine, using a set of lines chosen by J.M. Fletcher in the region 4325-4525 Å, and listed by Scarfe et al (1983). The results to date, from 62 spectra of 12 stars, are summarized in Table I, where the standard error in the fifth column is that of a single observation about the mean for the star. The r.m.s. value of this standard error is 0.15 km/s, and one observation of γ Aquilae, which differed from the mean of the other three by more than six times this r.m.s. value, has been omitted. The mean value of the difference $\bar{V} - V_{IAU}$, with each star weighted by either the number of observations or its square root, is

$$\bar{V} - V_{IAU} = -0.07 \pm 0.10 \text{ km/s}$$

where the uncertainty is the standard error of the mean. Individual stars, in particular 5 Serpentis and β Virginis, give mean velocities substantially and consistently different from the IAU value, but show no sign of variability.

Although all the stars observed, with the possible exception of γ Aquilae, show constancy of velocity, their true radial velocities remain unknown. Thus although they may be used to check the consistency of a spectrograph, they cannot be used to determine the absolute size of any systematic error which may be present, and which may or may not vary with the velocity of the object being observed. This may be done only using objects in the solar system, whose radial velocity may be calculated from their orbital elements and those of

Table I. Mean Velocities

Star	HD	No. Obs.	$\bar{V}$ (km/s)	SE (km/s)	$\bar{V}-V_{IAU}$ (km/s)	Interval (years)
α Cas	3712	4	-4.21	0.13	-0.31	1980-1984
α Ari	12929	5	-14.51	0.14	-0.21	1980-1984
10 Tau	22484	2	27.76	0.16	-0.14	1979
α Tau	29139	1	54.18		0.08	1982
β Gem	62509	17	3.19	0.15	-0.11	1976-1983
β Vir	102870	7	4.38	0.16	-0.62	1976-1984
α Boo	124897	7	-5.32	0.14	-0.02	1977-1984
5 Ser	136202	2	54.48	0.24	0.98	1983
β Oph	161096	3	-12.28	0.06	-0.28	1982-1984
γ Aql	186791	3	-1.97	0.05	0.13	1976-1983
β Aqr	204867	1	6.52		-0.18	1982
ι Psc	222368	10	5.58	0.11	0.28	1975-1983

the earth. The brighter asteroids are perhaps the most suitable of these objects since their images are usually smaller than the seeing disk, and thus they illuminate the spectrograph collimator much as stars do. (This advantage is less marked for a spectrograph using an image slicer since the collimator illumination is determined mainly by the slicer rather than by the image falling upon it, and is thus less subject to the effects of seeing than is that of a slit spectrograph). Objects in the solar system behave as moving mirrors and the radial velocity measured for them is to a very high approximation the scalar sum of their heliocentric and geocentric velocities plus a correction for the earth's rotation (Sher 1968).

As a first attempt to use asteroids to detect a systematic error in the radial velocities obtained with the coudé spectrograph, two plates of Vesta were obtained close to the time of its opposition in 1983 December. The exposures were close to eight hours, but the resulting broadening of the lines due to the change of the relative velocity of the asteroid and the telescope is negligibly small. They were compared with velocities kindly provided by B.G. Marsden, calculated from the osculating elements from the 1983 Minor Planet Ephemerides, and accurate to ±0.01 km/s. The results are as follows

U.T. Date	V_{Obs} (km/s)	V_{Calc} (km/s)	O-C (km/s)
1983 Dec 24.31	5.43	5.14	0.29
1984 Jan 15.25	14.55	14.30	0.25

The first observation is better exposed than the second, and the weighted mean residual (O-C) is 0.28 km/s, with an uncertainty close to 0.15 km/s, the r.m.s. value for the stellar observations.

The spectrograph thus appears, on the basis of these preliminary data, to give velocities too positive by 0.28 ± 0.15 km/s. It follows that the IAU system as defined by the sample of twelve stars so far observed, is too positive by 0.35 ± 0.18 km/s. This agrees with the result obtained by Batten et al. (1983), using the spectrometer, for the difference between the system of the bright IAU standards and the mean residual for asteroids. It is clear, of course, that the present result is based on very few observations, and more are needed to confirm or revise it. This in turn requires continued effort to maintain the stability of the spectrograph and if possible to improve it. This becomes increasingly important since observations of potentially greater accuracy than those described here are now being made (Campbell 1983).

I am much indebted to B.G. Marsden for computing the velocities of Vesta from its orbital elements, to J.B. Tatum both for arranging to do this by direct computer link to the Smithsonian Astrophysical Observatory as well as for a thoughtful discussion of Sher's paper, and to A.H. Batten for helpful discussion and encouragement throughout this work.

REFERENCES

Batten, A.H., Harris, H.C., McClure, R.D. and Scarfe, C.D. 1983. Publ. Dom. Astrophys. Obs. 16, 143.

Batten, A.H., Fletcher, J.M., and Campbell, B. 1984. Publ. Astron. Soc. Pacific 96, 903.

Campbell, B. 1983. Publ. Astron. Soc. Pacific 95, 577.

Fletcher, J.M., Harris, H.C., McClure, R.D. and Scarfe, C.D. 1982. Publ. Astron. Soc. Pacific 94, 1017.

Griffin, R.F. 1975. Mon. Not. Roy. Astron. Soc. 171, 407.

Heard, J.F. and Fehrenbach, C. 1973. Trans. IAU. 15a, 409.

Pearce, J.A. 1957. Trans. IAU. 9, 441.

Richardson, E.H. 1968. J. Roy. Astron. Soc. Canada 62, 313.

Scarfe, C.D. 1983. in IAU Coll. No. 62, Current Techniques in Double and Multiple Star Research, ed. R.S. Harrington and O.G. Franz. Lowell Obs. Bull. 9, 93.

Scarfe, C.D., Funakawa, H., Delaney, P.A., and Barlow D.J. 1983. J. Roy. Astron. Soc. Canada 77, 126.

Sher, D. 1968. J. Roy. Astron. Soc. Canada 62, 105.

PRECISE RADIAL VELOCITIES AND RADIAL VELOCITY STANDARDS

G. A. H. Walker, J. Amor and S. Yang

University of British Columbia, Vancouver

B. Campbell

Dominion Astrophysical Observatory

ABSTRACT. By imposing absorption lines of HF in stellar spectra we can measure changes in r.v. with a precision of ~10m/s from a single spectrum, provided stellar line profiles are not distorted by atmospheric motions. The precision of absolute radial velocities is currently limited to ~100m/s by knowledge of rest wavelengths. Representative results are presented from our three, active PRV programs: velocity variations of δ Scuti stars; a search for unseen companions to late-type stars; and routine observations of certain IAU velocity 'standards'.

1. INTRODUCTION

Radial velocity and its variation are fundamental elements for any star. The recent detection of Solar oscillations has provided a new probe of its structure, and the larger scale oscillations of various classes of variable stars have been known for a long time. The solution of binary systems, particularly those with a large mass ratio, requires precise radial velocities. For a sufficiently high precision it should even be possible to measure dynamical parallaxes from the accelerations of nearby, 'non-variable' stars.

Compared to classical methods, our technique, which imposes the (3,0) band system of HF at 870nm in the stellar spectra, improves the precision with which radial velocity variations can be measured by two orders of magnitude. It has been described by Campbell and Walker (1979) and by Campbell, Walker, Johnson, Lester, Yang, and Auman (1981). The results discussed here were all obtained with the coudé spectrograph of the Canada France Hawaii 3.6m telescope. The detector is a refrigerated 1872 Reticon.

Line displacements are measured with a resolution of 10^{-3} of a diode spacing (= 0.015 microns = 2.4m/s). Our results to date suggest that we can achieve an external precision of ~10m/s for spectra with a S/N >1000/diode based on some 16 stellar lines.

D. S. Hayes et al. (eds.), Calibration of Fundamental Stellar Quantities, 587–589.

2. RESULTS

(a) δ Scuti variables: results for β Cas have already been published by Yang, Walker, Fahlman and Campbell (1982). In Figure 1 velocity curves derived from a mean of the Fe I lines and from the Ca II 866.2nm line are shown with only the data points for the Fe I lines plotted. The velocity residual between the two curves is also shown but at 8 times the scale. Apart from the obvious difference in velocity amplitude there appears to be a difference in phase between the two curves. Precision in this case is limited by the line profile variations accompanying the stellar pulsations and a S/N <1000 arising from the need to adequately sample the period. The data are for ρ Puppis

(b) Unseen companion search and IAU velocity standards: the program stars are listed in the table below. The I magnitude is the most appropriate for the 870nm region. The stars have been observed for three and a half years and the number of observations per star is listed in the last column. Figure 2 shows, combined, the results for one star from each program, β Virginis (triangles), and α Hydrae (crosses). Residual velocities from the mean are plotted for each star on each night where the mean velocities are: β Vir +4542m/s, α Hya -4601m/s.

Publication of the first results from this program is expected early in 1985.

Planetary Companion Program Stars

STAR	HR #	Sp Type	I	# OBSERVED
Tau Cet	0509	G8Vp	2.41	18
Iota Per	0937	G0V	3.25	11
Kappa Cet	0996	G5V	3.95	5
Epsil. Eri	1084	K2V	2.54	17
Omic.2 Eri	1325	K1V	3.29	11
Chi Ori	2047	G0V	3.61	8
Alpha CMi	2943	F5IV	-0.27	37
36 UMa	4112	F8V	4.10	12
Beta Vir	4540	F8V	2.86	19
Beta Com	4983	G0V	3.46	15
61 Vir	5019	G6V	3.82	12
Xi Boo	5544	G8V	3.75	6
36B Oph	6401	K1V	2.41	3
36A Oph	6402	K0V	3.99	4
70A Oph	6752	K0V		11
Sigma Dra	7462	K0V	3.66	12
Beta Aql	7602	G8IV	2.59	12
Gamma2 Del	7948	K1IV	2.84	9
Eta Cep	7957	K0IV	2.27	11
61A Cyg	8085	K5V	3.54	9
61B Cyg	8086	K7V	4.03	1
	8832	K3V	4.23	2
Gamma Cep	8974	K0IV	1.93	13

The IAU Radial Velocity standards

Alpha Ari	0617	K2III	0.54	16
Alpha Tau	1457	K5III	-1.31	21
Beta Gem	2990	K0III	-0.11	13
Alpha Hya	3748	K4III	0.16	22
Alpha Boo	5340	K2III	-1.67	28
Delta Sag	6859	K2III	0.82	5
Epsil. Peg	8308	K2Ib	0.56	14

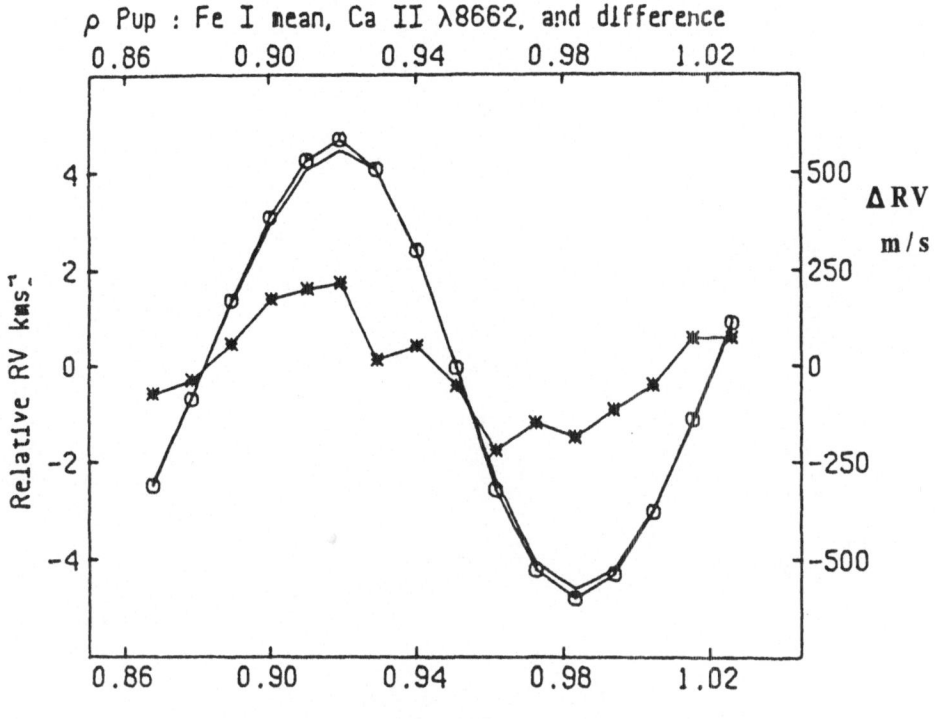

Fig. 1.   Relative Velocity vs. Jusian Day for Rho Pup.

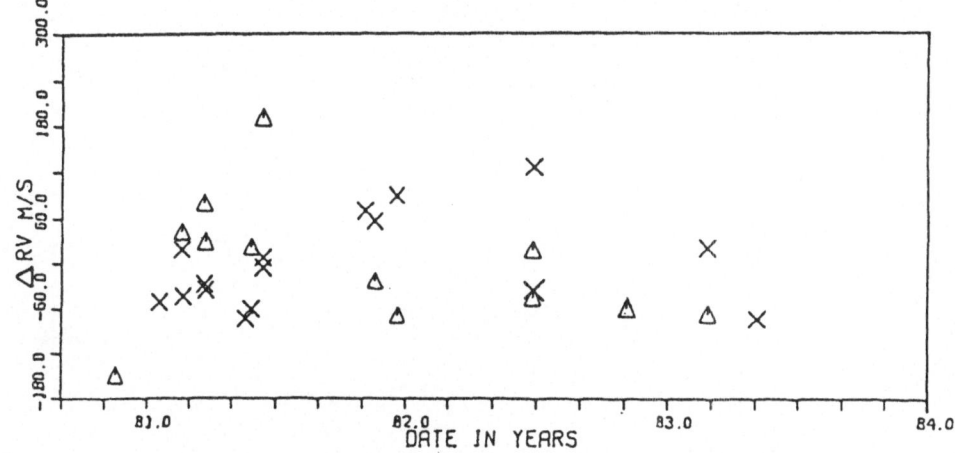

Fig. 2.   Velocity Residuals.

REFERENCES

Campbell, B. and Walker, G.A.H. 1979, Publ. Astron. Soc. Pacific, 91,
    540.
Campbell, B. Walker, G.A.H., Johnson, R., Lester, T., Yang, S. and
    Auman, J. 1981, Proc. S. P. I. E., 290, 215.
Yang, S., Walker, G.A.H., Fahlman, G.G., and Campbell, B. 1982, Publ.
    Astron. Soc. Pacific, 94, 317.

STANDARD PULSARS: PROBES FOR THE CALIBRATION OF THE GALACTIC CONTINUUM
BACKGROUND TEMPERATURE AND THE ELECTRON DENSITY OF THE INTERSTELLAR
MEDIUM

M. Fracassini and L. E. Pasinetti

Department of Physics, University of Milan

ABSTRACT. From the list of observed and derived parameters of 330 pulsars published by Manchester and Taylor (1981), we have selected sixteen
"standard" pulsars whose position and radio luminosity fit well a statistical relation found by Antonello and Fracassini (1984). This relation
shows also that these pulsars may be considered as probes of the galactic
places where the observed 408 MHz continuum background temperature, T_{408},
and the computed electron densities, n_e, of the interstellar medium, ISM,
are best correlated.

1. INTRODUCTION

A Symposium devoted to the calibration of fundamental stellar quantities should consider also the ISM where the stars are born and die.
The knowledge of the physical properties of the ISM is undoubtedly necessary for the understanding of these important stages of the stellar evolution.
A large contribution to this knowledge (space distribution, chemical
composition, temperature, density, magnetic field, polarization, etc.)
is due to the radioastronomy and, in particular to the discovery of pulsars. In the present paper we consider the "standard" pulsars, probes of
the galactic places where some physical properties of the ISM (temperature and electron density) may be considered as standards: namely, the galactic places where the observed T_{408} are best correlated to the n_e derived from the adopted models (Lyne, 1982).

2. STATISTICAL ANALYSIS AND RESULTS

Two papers were of primary importance for the present study: a) the
observed and derived parameters for 330 pulsars published by Manchester
and Taylor (1981), at present the most complete and updated compilation
of the principal observational parameters of pulsars; b) the 408 MHz All-
Sky Continuum Survey, Atlas of Contour Maps, published by Haslam et Al.
(1982), which provides an accurate and complete data base for large-scale

D. S. Hayes et al. (eds.), Calibration of Fundamental Stellar Quantities, 591–594.
© *1985 by the IAU.*

comparisons with other observations (see f.i. Phillipps et Al. 1981).

Our study began with a preliminary multivariate analysis of the parameters reported in a) (Fracassini et Al., 1983). A linear correlation (R = 0.60) between the T_{408} observed in the direction of the pulsars and their dispersion measures, DM, was found. A subsequent multiple stepwise regression analysis (Antonello and Fracassini, 1984) gave the best relation (R = 0.90) between T_{408} and DM, the position (Z, R) and radio luminosity (L) of the pulsars.

On the ground of the scattering around this relation (O-C = log T_{obs}- log T_{calc}) we have selected the following groups:
1) <u>peculiar</u> pulsars (|O-C| > 2.5σ), which are set in the galactic places where the observed T_{408} and derived n_e are higher or lower than the averaged ones;
2) <u>normal</u> pulsars (|O-C| < 2.5σ), which are set in the galactic places where the observed T_{408} and derived n_e are in agreement on the average;
3) <u>standard</u> pulsars (|O-C| < 0.01), which are set in the galactic places where the observed T_{408} and derived n_e are in good agreement.

The list of sixteen "standard" pulsars and the corresponding parameters are reported in Table I.

Fig. 1.   The relation log T_{408} vs. log n_e. Note that the pulsars numbered 1 and 6, which should have good distance estimates (Manchester and Taylor 1981), deviate from the average relation.

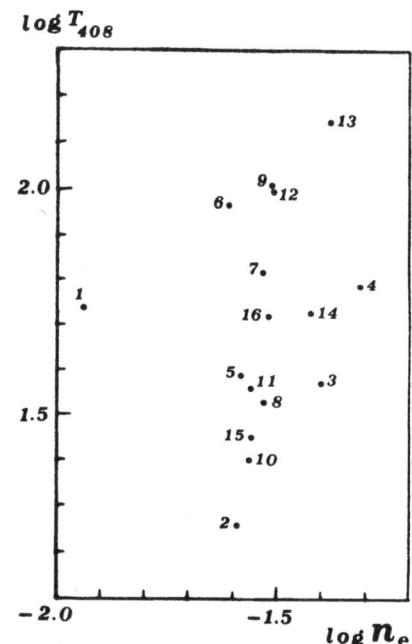

3. DISCUSSION AND CONCLUSION

Several papers concerning the distribution (homogeneity and cloudiness) of the ISM have been published recently. One of them in particular, regarding the variation of a scattering index parameter related to the existence of a three-phase structure of the ISM (Akujor and Okeke, 1982), is related to our results.

As a matter of fact this index is proportional to the mean n_e and

TABLE I. "Standard" Pulsars

N	PSR	l	b	T_{408}	DM	Z	R	log L	n_e	d
1	0138+59	129.1	− 2.1	55	34.80	−0.11	12.12	28.26	0.0116	3.00
2	0447−12	211.1	−32.6	18	39	−0.82	11.12	27.23	0.0260	1.50
3	0808−47	263.3	− 8.0	37	228.3	−0.78	12.02	29.18	0.0401	5.70
4	0840−48	267.2	− 4.1	62	197.0	−0.28	10.93	27.42	0.0493	4.00
5	1119−54	290.1	+ 5.9	39	204.7	+0.78	10.27	28.77	0.0269	7.60
6	1240−64	302.1	− 1.5	94	297.4	−0.32	10.80	29.72	0.0248	12.00
7	1256−67	303.7	− 4.8	66	95	−0.27	8.65	27.08	0.0297	3.20
8	1604−00	10.7	+35.5	34	10.72	+0.21	9.71	26.27	0.0298	0.36
9	1647−52	335.0	− 5.2	114	179.1	−0.52	5.34	28.44	0.0309	5.80
10	1811+40	67.4	+24.0	25	44	+0.67	9.53	26.69	0.0275	1.60
11	1839+56	86.1	+23.9	36	27	+0.40	9.98	27.35	0.0276	0.98
12	1900−06	28.5	− 5.7	104	190	−0.62	5.43	28.71	0.0306	6.20
13	1924+16	51.9	+ 0.1	140	178	0.00	8.10	27.79	0.0414	4.30
14	1944+17	55.3	− 3.5	54	16.3	−0.03	9.76	26.80	0.0379	0.43
15	2043−04	42.7	−27.4	28	36	−0.61	9.17	26.68	0.0277	1.30
16	2224+65	108.6	+ 6.8	53	36.5	+0.14	10.43	27.12	0.0304	1.20

PSR = Pulsar, l = Galactic Longitude, b = Galactic Latitude, T_{408} = 408 MHz Galactic Background Continuum Temperature, DM = Dispersion Measure, Z = Distance from the Galactic Plane (kpc), R = Distance from the Galactic Axis (kpc), L = Absolute Radio Luminosity, n_e = Electron Density, d = Heliocentric Distance (kpc) seem to show that the galactic places with mean values of T_{408} and n_e, located by the "standard" pulsars, should belong to the phase II (warm phase) of the ISM. This phase includes regions of different properties: the ionized hydrogen HII regions, which mainly contribute to the average electron density of the ISM, and the intercloud medium.

Therefore, the above-mentioned large-scale physical properties could be considered as characteristic ones of what we have found and meant as the standard ISM.

REFERENCES

Akujor, C.E., Okeke, P.N. 1982, Astrophys. Space Sci. 88, 487.

Antonello, E., Fracassini, M. 1984, in preparation.

Fracassini, M., Manzotti, G., Pasinetti, L.E., Raffaelli, G., Antonello, E., Pastori, L. 1983, Coll. on Statistical Methods in Astronomy,(ESA SP – 201), 21.

Haslam, C.G.T., Salter, C.J., Stoffel, H., Wilson, W.E. 1982, Astron. Astrophys. Suppl. Ser. 47, 1.

Lyne, A.G. 1982, Supernovae: A Survey of Current Researches, M.J. Rees and R.J. Stoneham eds.,(Reidel Publ. Co., Dordrecht), p. 405.

Manchester, R.N., Taylor, J.H. 1981, Astron. J. 86, 1953.

Phillipps, S., Kearsey, S., Osborne, J.L., Haslam, C.G.T., Stoffel, H. 1981, Astron. Astrophys. 103, 405.

THE CALIBRATION OF STELLAR SPECTRA USED IN MEASUREMENTS OF THE GENERALIZED COMPTON SHIFT

M. Missana

Astronomical Observatory of Brera, Milano-Merate

ABSTRACT. In the study of the changes induced by the Compton effect and Thomson scattering on the shape of spectral lines in light traversing a chromosphere or a planetary nebula it is necessary to have accurate wavelength measurements, central intensities and half widths (FWHM) of the lines. In the comparison of FWHM measures belonging to different spectra it is useful to have the intensities of the continuum at the wavelengths of the lines. The studied spectra should have a dispersion higher than 10 Å/mm and a spectral range larger than 700 Å. A short recommendation is also given about the comparison spectrum and the calibration plate.

1. THEORETICAL RESULTS

The numerical values of the generalized Compton effect in stellar spectra are determined from the equation:

$$\lambda_s - \lambda_l = \Delta(U,\lambda_l) + \beta\lambda_l; \qquad (1)$$

where λ_s and λ_l are the stellar and laboratory wavelengths of the spectral line, $\beta = V/c$ is the mean radial velocity of the stellar atmosphere observed from the Earth, divided by the light velocity c and $\Delta(U,\lambda)$ is the generalized Compton shift. Δ depends on the profile of the spectral line (here defined by means of the width U), on the wavelength λ and on the temperature and thickness of the scattering layer. The function $\Delta(U,\lambda)$ can be obtained from the solution of the radiative-transfer equation (Chandrasekhar 1960, Missana 1977) and depends on the scattering cross-section of the photons with the bound and free electrons and with the atomic nucleus; it depends also on the Delbrück scattering cross-section (Schweber 1966). The rigorous computation of the $\Delta(U,\lambda)$ function is still an unresolved problem. However an approximate solution has been found for wavelengths in the visible spectral range far from a resonance of the scattering atoms for free or bound electrons with a cross-section

595

D. S. Hayes et al. (eds.), Calibration of Fundamental Stellar Quantities, 595–597.
© *1985 by the IAU.*

independent of wavelength, as in the Thomson cross-section.

In that simpler case, for lines having a Gaussian shaped profile before the scattering, one obtains $\Delta(U,\lambda) = \Delta(U)$, where U is the full half width of the line (FWHM) or the normalized width of the line, which is defined as the equivalent width W divided by central intensity I of the line, (U=W/I). $\Delta(U)$ is a crescent function of U which becomes nearly constant for $U>U_o$; the value U_o depends on the optical thickness of the scattering layer. This theoretical result is in fair accord with the measurements of the solar spectrum (Missana 1982). Hence in the study of the stellar spectra by means of equation (1) it is necessary to know the U values of the lines in addition to the usual data.

2. THE MEASUREMENTS

The practical measurement of U is quite sensitive to the adopted definition of the continuum; so in the comparison of values of U measured in different spectra it is useful to know the intensity of the continuum at the wavelength of the line.

In the computation of the generalized Compton effect by means of equation (1) the spectral lines, free of apparent blends, are ordered in an increasing sequence with respect to U; then they are divided in n classes with at least 9 spectral lines in each class, if the wavelengths are distributed in a spectral interval of at least 700 Å; all the lines with the same U have to be in the same class, of course. Then we assume that the Δ-function is constant in each class and the n+1 unknowns Δ and β are computed by means of the method of least squares, together with the standard deviation S of the residuals of the equation (1). It is useful to notice that a few measures, with large errors, change the results of equation (1) in a quite remarkable way, as underlined by Trumpler et al. (1953); hence a careful examination of S and an accurate sampling of the spectral lines is recommended.

From the results of the article by Missana et al. (1975) it follows that Δ is smaller than 0.05 Å except in the case of the stars surrounded by a thick layer of interstellar matter or embedded in a planetary nebula etc. Then the division of the spectral lines into 2 or 3 classes of the width U is sufficient. Please note that some formulae of the quoted article contain typographical errors. From the study of some spectra from the Haute-Provence Observatory the writer has deduced that dispersion of 9 Å/mm for the stellar spectra and a spread of the wavelengths in an interval of 800 Å is sufficient for a reliable computation of the Δ and β values. These spectra have been measured with a Gaertner comparator and the wavelengths of the comparison spectra have been taken from Crosswhite (1958). The normalized widths have been obtained at the Nice Observatory with a quite large error mainly depending on the definition of the continuum.

3. CONCLUSIONS

About sixty years ago the first study of the Compton effect was published; now at last it is possible to measure that effect also in stellar spectra. For a reliable computation of the generalized Compton shift it is necessary to have at least 25 spectral lines in a spectral interval of at least 700 Å; for each line one must know the wavelength with an error smaller than about 0.02 Å, the central intensity and the equivalent width or the half width FWHM. It is also useful to know the intensity of the continuum at the wavelength of each line.

When the photographic spectrophotometry is used one should have 2 calibration plates at different exposure times and, in my opinion, a lamp with a known intensity distribution should be used. The use of a hollow cathode lamp for the comparison spectrum and of an impersonal oscilloscope comparator for the measurements are recommended. Less accurate measurements are needed for spectra of light passing through a thick layer of interstellar matter or that are observed in a wider interval of wavelength.

REFERENCES

Chandrasekhar, S. 1960, Radiative Transfer, (Dover Publications Inc.,
    New York), p.328.
Crosswhite, H. M. 1958, The Spectrum of Iron I, (The Johns Hopkins
    University, Baltimore).
Missana M. and Piana, A. 1975, Astrophys. Space Sci., 37, 263.
Missana, M. 1977, Astrophys. Space Sci., 50, 409.
Missana, M. 1982, Astrophys. Space Sci., 85, 137.
Schweber, S. 1966, Relativistic Quantum Field Theory, (Harper & Row,
    New York), p.705.
Trumpler, R. J. and Weaver, H. F. 1953, Statistical Astronomy, (Dover
    Publications Inc., New York), p.183.

RECENT TRIGONOMETRIC PARALLAX MEASUREMENTS OF MK LATE-TYPE SPECTRAL STANDARD GIANT STARS

Philip A. Ianna

Dept. of Astronomy, University of Virginia

Roger B. Culver

Dept. of Physics, Colorado State University

ABSTRACT. Trigonometric parallaxes are reported for ten bright, southern late-type MK giant stars. The plate material was obtained with the 66 cm Yale-Columbia refractor at the Mt. Stromlo Observatory and measured with the PDS 1010A at Mt. Stromlo. The ten stars include HR 794 (KO III), HR 1247 (M2 IIIab), HR 2245 (M2.5 III), HR 2773 (K3 Ib), HR 3518 (K3 III), HR 3803 (K5 III), HR 5287 (K2 III-IIIb), HR 5603 (M3 IIIa), HR 6832 (M3.5 III), and HR 6913 (K1 IIIb). The modern parallaxes are compared with earlier results and the luminosity calibration for these stars is discussed.

1. INTRODUCTION

The problem of the calibration of absolute magnitudes of stars on the red-giant branch has generally been approached through statistical parallaxes, and trigonometric parallaxes have contributed only to a limited extent. This is a consequence of the relatively small parallaxes of most of the astrometrically observed giants, and that the size of these parallaxes are comparable to the average error of older parallaxes (0.010 - 0.020 arcsec). Most recently maximum-likelihood statistical methods have been applied to the giants by Mikami and Heck (1982), and Egret, Keenan, and Heck (1982), the latter including the most up-to-date MK classifications.

Keenan (1971) has advocated determining more trigonometric parallaxes for bright G, K and M class III stars since the spectral classes for these stars have been improved to more closely define their luminosities (Keenan and Pitts 1980). However these objects remain neglected in modern parallax programs in part owing to their observational difficulty. There are additional reasons for adding the

D. S. Hayes et al. (eds.), Calibration of Fundamental Stellar Quantities, 599–602.
© *1985 by the IAU.*

apparently bright stars to parallax programs. Since a number of these
bright stars have never had parallax determinations, it would be useful
to complete the sample, and many of the existing parallaxes, especially
in the southern hemisphere, have been measured at one observatory only.
Single observatory parallaxes are known to be less reliable. The
prospect of substantially improved trigonometric parallaxes suggests
this most fundamental method may make a greater contribution to the
calibration of spectroscopic luminosity classes for the giants over
the next few years.

2. RESULTS AND DISCUSSION

The BSC data and final absolute magnitudes for our ten stars are
displayed in Table I and Figure 1. The statistical corrections to
absolute parallax were taken from the tables of Binnendijk (1943).
In several cases these corrections are of the same size as the relative
parallaxes. The errors in the absolute magnitudes are estimated as
$\sigma_M = 2.7\sigma_\pi/\pi$ using the errors in the relative parallaxes. The early
relative parallaxes listed for comparison in Table I are all Yale results
with the exception of the first entry for the last star which is an early
McCormick parallax. Given the large errors in these parallaxes, they
are mostly in agreement with our new results.

Of the stars in this list of southern hemisphere giants, four have
trigonometric parallax absolute magnitudes which are discordant with
published spectral classes: HR 794 (Iot Eri), HR 3518 (Gam Pyx),
HR 5287 (Pi Hya), and HR 5603 (Sig Lib). The parallax values obtained
herein for these four objects lead to the absolute magnitudes presented
in Table I. The absolute magnitudes for Iot Eri and Gam Pyx are substantially in agreement with the absolute magnitudes expected for their
respective spectral classes. The absolute magnitude of the star Pi
Hya is in good agreement with previous determinations, but is somewhat
fainter at + 1.8 than one would expect for a luminosity class III/IIIb
star. The results for Sig Lib (M3 IIIa) indicate a luminosity of − 0.9
which seems about right for this class. The earlier parallax measurement
for this star gives an absolute magnitude of + 2.3 which is much too faint
for this class.

HR 3803 (N Vel) and HR 6832 (Eta Sgr) have measured parallaxes fairly
consistent with their spectral classes. The parallax measured for the
star HR 6913 (Lam Sgr) yields an absolute value of +2.2 compared with
a value of + 1.4 from the BSC data where the parallax is a mean of two
earlier determinations. The mean of all three parallaxes would still
give an absolute magnitude rather fainter than expected for the spectral
classification.

The stars HR 1247 (Del Ret) and HR 2245 (Eta^2 Dor) have parallaxes
which are essentially zero giving large uncertainties in the absolute
magnitudes, but nevertheless the absolute magnitudes are not inconsistent with their spectral classes. The last star in our sample HR 2773
(Pi Pup) is classified as a K3 Ib star (Garrison 1984). Its derived
absolute magnitude of − 1.9 is, again, not inconsistent with its spectral
class.

Table I. Early Parallaxes and Final Results

HR	Sp	GCTSP	Parallaxes		Abs. Par.	M_v
794	KOIIIb*	547.	+0.''0.32	+0.015	0.''317	+1.62 +0.25
1247	M2IIIab	889.	$-$ 7	16	56	$-$1.70 1.82
2245	M2.5III	1460.	+ 4	15	61	$-$1.06 1.42
2773	K3Ib*	1716.	26	16	124	$-$1.83 0.60
3518	K3III	2114.	28	15	151	$-$0.10 0.46
3803	K5III	2276.	17	15	382	+1.04 0.23
5287	K2III-IIIb	3205.	42	18	506	+1.79 0.13
5603	M3IIIa	3400.	59	19	144	$-$0.92 0.63
6832	M3.5III	4191.	41	16	299	+0.49 0.28
6913	K1IIIb	4239.	30	15	752	+2.19 0.13
			67	16		

* Spectral classification from Garrison (1984).

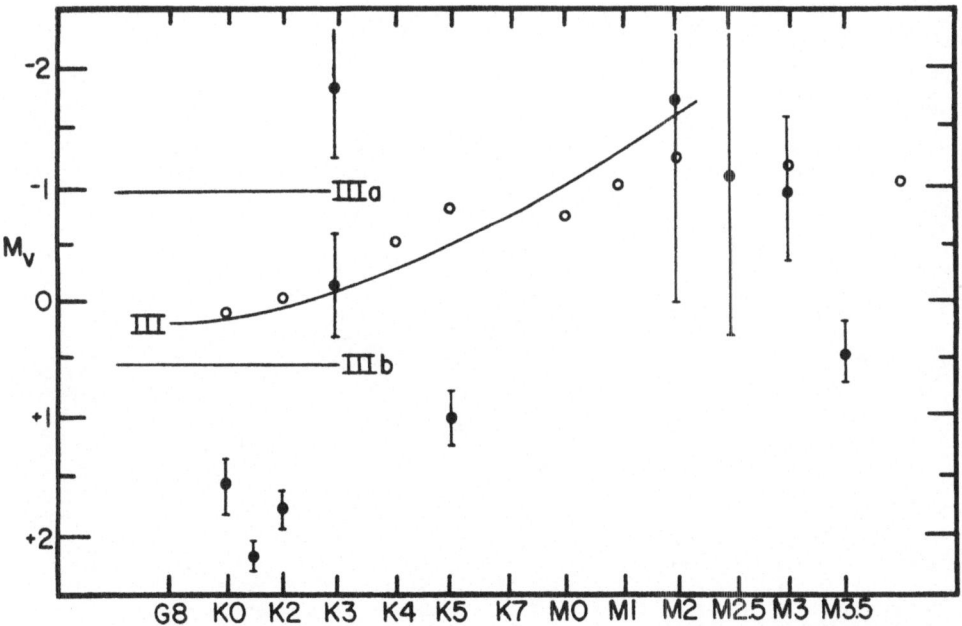

Fig. 1 Absolute magnitudes for ten late-type giants; the error
bars are the standard errors. Calibrations of Mikami and Heck
(1982) are shown as open circles; solid lines are from Egret,
Keenan and Heck (1982).

3. ACKNOWLEDGEMENTS

This work is supported by the National Science Foundation through grant
AST-82-00232 and by the Estate of Leander J. McCormick. We are grate-
ful to Prof. Don Mathewson, Director of the Mt. Stromlo and Siding
Spring Observatories for his continuing support of this program.
Roger Culver was a Visiting Fellow at Mt. Stromlo while part of this
work was carried out.

4. REFERENCES

Binnendijk, L. 1943, *Bull. Astr. Inst. Netherlands*, 10, 9.
Egret, D., Keenan, P.C., and Heck, A. 1982, *Astron. Astrophys.* 106, 105.
Garrison, R.F. 1984, private communication.
Hoffleit, D. and Jaschek, C. 1982. *The Bright Star Catalogue*. Yale
      University, New Haven.
Keenan, P.C. 1971. *The Fourth Astrometric Conference*, Publ. McCormick
      Obs. XVI. ed. P.A. Ianna, p. 249.
Keenan, P.C. and Pitts, R. E. 1980, *Astrophys. J. Suppl.* 42, 541.
Mikami, T. and Heck, A. 1982, *Publ. Astron. Soc. Japan* 34 529.

THE DETERMINATION OF EXTINCTION AND TEMPERATURE FOR THE CENTRAL STAR
OF THE PLANETARY NEBULA NGC 40

L. Bianchi

Astronomical Observatory of Turin

M. Grewing

Astronomical Institute, University of Tubingen

ABSTRACT. NGC 40 is an extended, very inhomogeneous planetary nebula,
with a WC8 central star. Very discrepant determinations exist in the
literature for the extinction towards this object and the temperature
of its nucleus. We review here the various methods which can be used
to derive these quantities, discussing the assumptions underlying each
method and their inherent limitatons, and the uncertainties that arise
when they are applied to this particular object. The results are
compared to the values which we derive from far UV
spectrophotometry: $E(B-V) = 0.50$ and $T_* \approx 90000K$.

1. INTRODUCTION

     NGC 40 is characterized by an extremely irregular and
inhomogeneous structure. It is classified as a low excitation
nebula. As noticed by Aller and Czyzak (1979), however, the
excitation level changes dramatically from point to point.

     The problem of determining a correct value for the reddening is
particularly severe in the analysis of far ultraviolet data: for
example, a value of $A_V = 1$ mag would correspond to an extinction of
about 2 to 14 mag. from 300 to 100 nm. Therefore, a small error in
$E(B-V)$ can introduce large errors in the UV-line flux corrections and
can greatly change the slope of the continuum spectrum, resulting in a
wrong color temperature determination.

     In the literature, values of $E(B-V)$ ranging from 0.2 to 0.82 can
be found for NGC 40. Given these discrepancies and the importance of
using a correct value for $E(B-V)$, we shall briefly discuss in Section
2 the uncertainties that can have affected the individual
determinations. The resultant $E(B-V)$ has a direct bearing on the
determination of the temperature of the central star of NGC 40, which

D. S. Hayes et al. (eds.), Calibration of Fundamental Stellar Quantities, 603–609.

TABLE I

Determinations of the extinction towards NGC 40

E(B-V)	Ref.	Method
0.50	this work	UV (IUE) (220nm bump)
0.38	Pottasch et al. 1978	UV (ANS)
0.40	Benvenuti et al. 1982	UV (IUE)
0.34	Clegg et al. 1983	fit of UV nebular continuum
0.49, 0.51	Clegg et al. 1983	avg. Balmer line ratios
0.20	Cahn 1976	avg. Balmer line ratios
0.50	Cahn et al. 1977	radio/H-beta flux
0.45	Pottasch et al. 1977	radio/H-beta flux
≤ 0.65	Cahn and Kaler 1971	radio/H-beta flux
0.82	Lang 1974	?

is classified as a WC8 star by Hiltner and Schild (1966). We discuss its temperature in Section 3.

2. DETERMINATION OF THE EXTINCTION

     In Table I we summarize determinations of the extinction towards NGC 40 that we have found in the literature. We also include our own determination. The individual methods are briefly discussed below.

2.1. Strength of the 220nm extinction bump

     The high value of $A_\lambda/E(B-V)$ in the far UV makes the extinction determination very critical, as we stressed in the introduction. Conversely, the non-monotonic shape of the extinction curve in this region, and in particular the strong "bump" at around 220nm due to graphite grains, allow a very precise determination of E(B-V), on the assumptions that (a) the continuum spectrum is smooth across the 220nm band, and (b) that the standard UV extinction law applies.

     From the strength of the 220nm dip we derive for NGC 40 E(B-V) = 0.50. In this case, the assumption that the observed spectrum must be smooth in the far UV is supported by the fact that the contribution of the nebular continuum is negligible, as can be seen from the IUE image, where the spectrum is clearly not extended. Also the nebular continuum observed offset from the central star is fainter than the stellar spectrum by about two orders of magnitude.

     The possibility should also be investigated that part of the extinction is of circumstellar origin. Circumstellar dust shells are found to surround some WR stars, especially of WC8-9 type. They are revealed by IR excesses measured around 10μ, showing cool (800 to 1800K) blackbody emission, on top of the free-free radiation distribution (see e.g., van der Hucht et al. 1984, van der Hucht et al. 1981 and references therein). If there is a circumstellar component, the standard interstellar extinction law might not apply, especially in the region of the 220nm bump, since this depends on the grain composition. For some WC stars it is argued that the dust could be of iron rather than graphite grains (e.g., Hackwell et al. 1979, van der Hucht et al. 1982). However, in the SWP high resolution spectrum of NGC 40 we do not detect any evidence for the Fe III (UV 34) absorption features which denote the "iron curtain" found by van der Hucht et al. (1982) in two WC9 stars.

Moreover, our spectrum appears very smooth when dereddened using the average galactic extinction law, and can be fitted very well with a blackbody distribution (see next section) so that it seems justified to use the standard extinction law in the present case. For a more detailed discussion see Grewing and Bianchi (1984) and Bianchi and Grewing (1984).

2.2 Balmer line ratios

E(B-V) can be derived from the Balmer line intensities, using the known relation

$$E(B-V) = 2.5 \log (\hat{I}_i/\hat{I}_{i,o}) / (X_i - X_\beta),$$

where $\hat{I}_i$ and $\hat{I}_{i,o}$ are the theoretical and observed intensities normalized to $H\beta$, and $X_i = A(\lambda_i)/E(B-V)$. In this method, it is essential to use line intensities that apply to the nebula proper, avoiding any contribution from the central star, if the pure recombination theory is applied. Even then, however, the observations of NGC 40 do not give a unique answer. This can be seen e.g., by comparing the intensities (from Minkowski and Aller 1956) for a weak and a strong knot in the nebula. In Table II we list the relative intensities and the E(B-V) values derived using theoretical predictions for the line ratios for $T = 10^4$K.

The numbers clearly illustrate the uncertainties involved. If we ignore the values derived from the H-alpha fluxes because these might be affected by [NII] emission, the averages become E(B-V) = 0.48 and 0.12 in front of the weak and the strong knot, respectively. The first result is consistent with our previous result and the ones derived from the radio/H-beta ratio discussed next; the second result clearly is not.

This difference in extinction is too large to be simply explained as an extinction variation across the object. Using the standard relation between E(B-V) and the hydrogen column density N(H), the latter would have to vary by as much as $\Delta N(H) = 2 \times 10^{21} cm^{-2}$, which seems improbable.

TABLE II

Relative Intensities and Colors

	Relative intensities					E(B-V)			
	F(Hα)	F(Hβ)	F(Hγ)	F(Hδ)	F(Hε)	Hα	Hγ	Hδ	Hε
weak knot	581	100	37.2	19.8	12.0	0.64	0.55	0.46	0.42
strong knot	904	100	44.0	24.1	15.1	1.04	0.15	0.12	0.08

2.3. Flux ratio F(radio)/F(H-beta)

By comparing the ratio of the observed 10 cm and H-beta fluxes with theoretical predictions, Cahn and Kaler (1971) derived E(B-V) $\lessgtr$ 0.65 assuming an electron temperature of T_e = 5000K and taking the value of the fractional ionization of helium from Kaler (1970). Using the same method but a different T_e Cahn (1976) derived E(B-V) = 0.50, and Pottasch et al. (1977) found E(B-V) = 0.45, using a larger value for H-beta.

In this method, the possibility exists that the radio flux is contaminated by unresolved nearby sources. Also, the H-beta flux can be contaminated by emission from the central star atmosphere; this is particularly true in the present case given the relative strength of the stellar and the nebular component (see e.g., Aller 1968 and Bianchi and Grewing 1984). Indeed, three values for the H-beta line absolute intensity are quoted in the literature (e.g., Higgs 1971, Carrasco et al. 1983). Further sources of uncertainties are the electron temperature and the abundance and fractional ionization of Helium. The ratio of the free-free radio continuum observed at 6 cm wavelength to the H-beta flux for an optically-thin nebula can be expressed as (cf. e.g., Milne and Aller 1975),

$$S(6cm)/F(H\text{-}beta) = 1.15 \cdot 10^7 \cdot T^{0.4} \cdot (\ln T - 3.08) \cdot (1 + He^+/H^+ + 3.7He^{++}/H^+)$$

if there is no extinction. In this equation S is measured in Jy, F(H-beta) in erg/cm^{2}s, and T in K. As S and F have the same density dependence, the ratio is independent of density fluctuations, a fact which is particularly important for such an inhomogeneous object. The two quantities do, however, differ slightly in their temperature dependence. Moreover, the fractional ionization of He enters explicitly into the calculation of S.

Assuming T_e = 10^4 and using S(6 cm) = 0.57 Jy and F(H-beta) = 2.3x 10^{-11} erg/cm^2 s, we find:

$A_\lambda(H\beta)$ = 2.02 for He^+ = He = 0.1 H^+(case I)

  = 2.26 for He^{++} = He = 0.1 H^+(case II)

Using a ratio R = $A_V/E(B-V)$ = 3.2, the two values correspond to E(B-V) = 0.46 (case I) and E(B-V) = 0.53 (case II). As helium is likely to be partially in the form of He^+ and He^{++}, the true value should be bracketed by these numbers, and it is then consistent with the value we derived from the UV spectrum.

3. DETERMINATION OF THE CENTRAL STAR TEMPERATURE

A detailed discussion of the resulting temperature of the central star of NGC 40 cannot be given here in view of the limited space available (see, however, Grewing and Bianchi, 1984). Here we confine ourselves to summarizing some of the results.

3.1. Comparison with blackbody distribution

The IUE spectrum dereddened for E(B-V) = 0.50 is very well represented by a blackbody distribution with temperature between 80000 and 100000K. The spectrum and the fit is shown in Bianchi and Grewing (1984), Fig. 1. This temperature is in agreement with theoretical predictions based on evolutionary calculations.

3.2. Comparison with model atmospheres

In principle, this method should give more precise information than the blackbody approximation. However, model atmospheres for WR stars should take into account the effects of the dense extended atmosphere, the high Helium abundance, and the high mass-loss rate (not hydrostatic, but expanding atmosphere). Unfortunately, very few such models exist in the literature, and they are all flatter than our observed spectrum (see Grewing and Bianchi 1984, for further discussion). Moreover, the existing calculations show that models with very different T_{eff} can have similar emerging flux distributions, depending on the extension of the atmosphere. Therefore, in the case of WR stars this method does not give unique results, unless the physical parameters of the atmosphere can be determined independently.

3.3. Energy-balance methods

For low excitation nebulae, the temperature of the central star can be inferred from the [O III] 5007/[O II] 3727 line ratio (see e.g., Köppen and Tarafdar 1978), or with the Zanstra method. However, this can be applied only to nebulae which are optically thick in all directions, which is not the case here. For NGC 40, in fact, this method would give a temperature T_* around 35000K, which is discrepant with the previous determination.

3.4. Excitation-class

Webster (1976) finds that the excitation class is primarily determined by the temperature of the central star and by the optical depth of the nebula, and gives a relation between T and the excitation class for optically thick nebulae. In the case of NGC 40, which is not optically thick, this method must be applied with caution, also in view of the very large inhomogeneities within the nebulae. The temperature of the central star one would find from the Webster relation is 40000 - 50000K, which must be considered as a lower limit.

REFERENCES

Aller, L. H. 1968, in IAU Symposium No. 34, Planetary Nebulae, ed. D. E. Osterbroch and C. R. O'Dell (Reidel, Dordrecht), p.339.

Aller, L. H. and Czyzak, S. J. 1979, Astrophys. Space Sci., 62, 397.

Aller, L. H., Czyzak, S. J., Buerger, E. G. and Lee, P. 1972, Astrophys. J., 172, 361.

Benvenuti, P., Perinotto, M. and Willis, A. 1982, in IAU Symposium No. 99, Wolf-Rayet Stars: Observations, Physics, Evolution, ed. C. W. H. de Loore and A. J. Willis (Reidel, Dordrecht), p.453.

Bianchi, L. and Grewing, M. 1984, in Future of Ultraviolet Astronomy, Based on Six Years of IUE Research, ed. J. M. Mead, R. D. Chapman and Y. Kondo (NASA, Washington, D. C.), p.262.

Cahn, J. H. 1976, Astron. J., 81, 407.

Cahn, J. H. and Kaler, J. B. 1971, Astrophys. J. Suppl. 22, 319.

Clegg, R., Seaton, M., Peimbert, M. and Torres-Peimbert, S. 1983, Mon.Not.R.Astron. Soc., 205, 417.

Carrasco, L., Senano, A. and Costero, R. 1983, Rev. Mexicana Astron. Astrof., 8, 187.

Hackwell, J., Gehrz, R. and Grasdalen, G. 1979, Astrophys. J., 234, 133.

Higgs, L. A. 1971, Mon.Not.R.Astron.Soc., 153, 315.

Hiltner, W. and Schild, R. 1966, Astrophys. J., 143, 770.

van der Hucht, K., Conti, P. and Willis, A. 1982, in IAU Symposium No. 99, Wolf-Rayet Stars: Observations, Physics, Evolution, ed. C. W. H. de Loore and A. J. Willis (Reidel, Dordrecht), p.277.

van der Hucht, K., Williams, P. and The, P. 1984, preprint.

Kaler, J. B. 1970, Astrophys. J., 160, 887.

Köppen, J. and Tarafdar, S. P. 1978, Astron. Astrophys., 69, 363.

Lang, K. R. 1974, in: Astrophysical Formulae, p.108.

Milne, D. K. and Aller, L. H. 1975, Astron. Astrophys., 38, 183.

Minkowski, R. and Aller, L. H. 1956, Astrophys. J., 124, 93.

Pottasch, S., Wesselius, P., Wu, C. and van Duinen, R. 1977, Astron. Astrophys., 54, 435.

Pottasch, S., Wesselius, P., Wu, C., Feiten, H. and van Duinen, R. 1978, Astron. Astrophys., 62, 95.

Webster, B. L. 1979, Mon.Not.R.Astron.Soc., 174, 157.

DISCUSSION

TOBIN: How much does the secular change in IUE LWR sensitivity affect estimates of E(B-V) obtained by "ironing out" the 2200 Å bump, and how can someone using IUE fluxes correct for the LWR sensitivity changes?

BIANCHI: In the particular case of the data I have shown, these were taken soon after the data on which the revised absolute calibration is based (Bohlin and Holm, 1980). In general, to know the amount of change in sensitivity at the time of given observation, one can see the reports on IUE sensitivity monitoring which appear in the IUE newsletter. The overall sensitivity degradation from the time of the acquisition of the calibration data to now is at maximum about 20% in the LWR and less than 5% in the overall range of the SWP camera.

BOHLIN: The use of mean extinction curves to determine the amount of reddening by "ironing out" the 2200 Å feature can lead to rather large errors in corrected flux distributions because of large variations in the UV extinction curves, even for stars without any evidence for dust that could be associated with the star itself. For example, in the case of E(B-V) = 0.5, I estimate that different observed extinction curves could give UV fluxes that are different by more than one magnitude, while equally well ironing out the 2200 Å bump.

BIANCHI: The possibility that some of the extinction is of circumstellar origin is discussed by Grewing and Bianchi (1984). If circumstellar dust shells are present surrounding the central star, they can be detected by IR observations or UV high-resolution observations. Moreover, in the case that the interstellar extinction law deviates from the average galactic one, usually the largest difference is in the shape of the 2200 Å bump. Therefore, one can realize that the extinction curve is wrong because the dereddened spectrum would never look smooth.

THE MICROFICHE OF STANDARD STARS

In a pocket inside the back cover of this volume is a microfiche, prepared by A. G. Davis Philip of Van Vleck Observatory and Union College and Daniel Egret of the Stellar Data Center, Strasbourg, containing standard stars for a variety of types of measurement. A description of the contents of the microfiche can be found in the paper by Philip and Egret on p. 353.

RESOLUTION

The following resolution was proposed and accepted by the participants in IAU Symposium No. 111:

"IAU Symposium No. 111 participants (The Calibration of Fundamental Stellar Quantities, Como, Italy, May 24-29, 1984) call the attention of the directors of observing facilities and members of the program selection committees to the basic importance of observing standard stars together with program objects, and urge them to allocate sufficient time for this purpose.

It is also necessary to allocate time for the establishment of new standards in some cases: for example, faint standards for large telescopes, or ultraviolet standards for space-borne telescopes. Failure to establish and re-observe frequently standard objects ultimately will compromise the quality of the work done at any observatory."

DISCUSSION OF SUGGESTION BY CODE

During the symposium, Dr. A. D. Code suggested that the assembled participants discuss the question: "For the non-specialist user, how does he know which catalogue to use, and how to use it? Should we get specialists to write critiques of catalogues, which perhaps, could be included with the material sent out by the Centre de Donées Stellaires?" This question was suggested as the basis of a possible resolution. It was the sense of the meeting that this was a very important question, and that any critiques of catalogues would have to be very carefully done. The discussion was concluded with acceptance of the suggestion by Hauck that the question be taken up by the Working Group for Astronomical Data, and be discussed by the General Assembly.

CLOSING REMARKS BY R. HANBURY BROWN, PRESIDENT OF THE IAU

I'll keep you only for a moment away from the fresh air. You have
to come to Italy to know what "al fresco" really means! Last night we
thanked the Local Organizing Committee, and Dr. Laura Pasinetti, its
Chairman. My job now is to thank the Scientific Organizing Committee.
I've enjoyed attending a meeting at which the "nuts and bolts" of astronomy were discussed. As a radio astronomer, a meeting for me is not a
meeting unless someone gets up and tells us the purpose and origin of
the Universe. Nobody, so far, has done that, thank goodness. In point
of fact, Dr. Gustafsson has completely misinterpreted the picture on the
ceiling above us (see p.328), because, of course, he is an optical
astronomer. It actually represents a group of radio astronomers - you
can see them holding their antennas - explaining to the granting agency,
which is in the purple coat, the purpose and origin of the Universe,
which is pictured above. As far as the Scientific Organizing Committee
is concerned, their best thanks will not be anything that I may say;
their best thanks can only be the success of the meeting. And its not,
of course, the success of the meeting that we have attended here, but it
is the success of the influence of the meeting, in the future, which will
be extended through the medium of its publication. And, certainly, on
behalf of the Executive of the IAU, I can't overemphasize the importance
of that. We help fund these meetings partly because we are interested in
their being published so that their influence will extend beyond the
rather small group that actually attended the meeting. And I think that
part of your thanks will come from the subsequent success and the influence of this meeting through its publication. Nevertheless, for us as a
group here, I would like to extend our thanks to the Scientific Organizing Committee for organizing this Symposium.

D. S. Hayes et al. (eds.), Calibration of Fundamental Stellar Quantities, 615.
© *1985 by the IAU.*

PICTURES OF PARTICIPANTS

Kaj Strand

Massimo Fracassini

Laura Pasinetti

R. Hanbury Brown

David Evans

F. Rufener

Arthur Code and William Tobin

Arthur Upgren

Donald and Julie Hayes

Jacek Krełowski

Philip Keenan and Karl Rakos

Daniel Popper

Mary Bongiovanni

The Editors, Donald Hayes,
Davis Philip, Laura Pasinetti

Laura Pasinetti and Alan Batten

The Iris - Sunday Cruise

Daniel Popper and

R. Hanbury Brown and Mrs. Batten

Giusa Cayrel and Ken Janes

Kristina Philip and Hans Heintze

Robert Garrison and
I. N. Glushneva

Alan Batten and Daniel Popper

Wilhelm Gliese

Michael Bessell and Ken Janes

Saul Adelman

Richard Boyle

James Rose

Roger Culver

Gerrie Peters

Hartmut Jahreiss

Brigitta Nordström   Claude Lacy

RANDOM QUOTES

It works! I am astonished!

A spectral type is a spectral type is a spectral type.

I don't favor one way or another.

It's not very easy to change hardware when it works.

I think I have forgotten something -- my talk.

You don't know your new husband! So, you don't know all the
spikes he has! It's so exciting to have a new husband ......

I have not said anything important, yet.

The use of large telescopes makes this problem more accute.

To find a single star is more difficult than you might think.

I don't like to use a lamp.

Never meddle in the affairs of specialists.

One half of a tenth arc millisecond.

Fekel does speckle.

I can see emission spikes! I can see them from here!

What we do, in two words, is to use ....(three minutes of
    additional explanation).

The interstellar astronomers.....

Around the Sun there are other suns....

Perhaps we should go back to the Babylonian method - clay tablets!

Probably their pinhole is not perfect.

INDICES

If a page number is underlined it indicates that the reference is
to an author of a paper (Name Index) or to a subject listed in the title
of a paper (Subject Index).

OBJECT INDEX